# Lecture Notes in Computer Science 6190

*Commenced Publication in 1973*
Founding and Former Series Editors:
Gerhard Goos, Juris Hartmanis, and Jan van Leeuwen

James F. Peters   Andrzej Skowron
Roman Słowiński   Pawan Lingras
Duoqian Miao   Shusaku Tsumoto (Eds.)

# Transactions on Rough Sets XII

Springer

Editors-in-Chief

James F. Peters
University of Manitoba, Winnipeg, Manitoba, Canada
E-mail: jfpeters@ee.umanitoba.ca

Andrzej Skowron
Warsaw University, Warsaw, Poland
E-mail: skowron@mimuw.edu.pl

Guest Editors

Roman Słowiński
Poznań University of Technology, 60-965 Poznań, Poland
E-mail: roman.slowinski@cs.put.poznan.pl

Pawan Lingras
Saint Mary's University, Halifax, NS, B3H 3C3, Canada
E-mail: pawan@cs.smu.ca

Duoqian Miao
Tongji University, 201804 Shanghai, P.R. China
E-mail: miaoduoqian@163.com

Shusaku Tsumoto
Shimane University, Izumo 693-8501, Japan
E-mail: tsumoto@computer.org

Library of Congress Control Number: 2010930335

CR Subject Classification (1998): F.4.1, F.1.1, H.2.8, I.5, I.4, I.2

| | |
|---|---|
| ISSN | 0302-9743 (Lecture Notes in Computer Science) |
| ISSN | 1861-2059 (Transaction on Rough Sets) |
| ISBN-10 | 3-642-14466-7 Springer Berlin Heidelberg New York |
| ISBN-13 | 978-3-642-14466-0 Springer Berlin Heidelberg New York |



springer.com


Printed in Germany

Typesetting: Camera-ready by author, data conversion by Scientific Publishing Services, Chennai, India
Printed on acid-free paper 06/3180

# Preface

Volume XII of the *Transactions on Rough Sets* (TRS) includes articles that are part of a special issue on *Rough Set Structuring of Knowledge.* These articles are extended versions of papers accepted for presentation at the Rough Set and Knowledge Technology Conference (RSKT 2008) organized in Chengdu, China, in May 2008. In fact, this conference did not take place because of the earthquake that dramatically hit the Chengdu province just before the event. The editors selected some papers accepted for RSKT 2008 and invited their authors to submit extended versions to this issue. The 11 submissions received after this invitation were sent to reviewers and, finally, 8 papers were accepted for publication in the special issue.

The editors of the special issue are particularly grateful to all the authors of submitted papers, and to the editors of *Transactions on Rough Sets*, James F. Peters and Andrzej Skowron, who provided friendly and valuable assistance in the reviewing process. Special thanks are due to the following reviewers: Mohua Banerjee, Anna Gomolińska, Ryszard Janicki, Julia Johnson, Maneesh Joshi, Jusheng Mi, Sonajharia Minz, Sushmita Mitra, Tuan Trung Nguyen, Georg Peters, Andrzej Skowron, Genlou Sun, Marcin Szczuka, Dominik Ślęzak, Guoyin Wang, Hai Wang, Marcin Wolski, Haiyi Zhang, Yan Zhao, William Zhu. Their laudable efforts made possible a careful selection and revision of submitted manuscripts.

The articles of the special issue on *Rough Set Structuring of Knowledge* introduce a number of new advances in both the foundations and the applications of rough rets. The advances in the foundations concern rough mereological logics and reasoning, topological properties of a dominance-based rough set approach to ordinal classification, and rough approximations based on information coming from multiple sources. Methodological advances influence rough set-based methodology with respect to knowledge reduction, rule induction, and classification. The novel applications include gene selection and cancer classification, face recognition, and path planning for autonomous robots.

In addition, this volume contains articles introducing advances in rough set theory and its applications. These include: perceptually near Pawlak partitions, hypertext classification, topological space vs. rough set theory in terms of lattice theory, feature extraction in interval-valued information systems, jumping emerging patterns (JEP) and rough set theory. Our special thanks to our contributors and to reviewers of regular articles: Aboul Ella Hassanien, Jouni Järvinen, Sheela Ramanna, Piotr Wasilewski and Marcin Wolski.

The editors and authors of this volume extend their gratitude to Alfred Hofmann, Anna Kramer, Ursula Barth, Christine Reiss, and the LNCS staff at Springer for their support in making this volume of the TRS possible.

The Editors-in-Chief were supported by research grants N N516 368334, N N516 077837 from the Ministry of Science and Higher Education of the Republic of Poland, and the Natural Sciences and Engineering Research Council of Canada (NSERC) research grant 185986, Canadian Network of Excellence (NCE), and a Canadian Arthritis Network (CAN) grant SRI -BIO-05.

April 2010

Roman Słowiński
Pawan Lingras
Duoqian Miao
Shusaku Tsumoto
James F. Peters
Andrzej Skowron

# LNCS Transactions on Rough Sets

The *Transactions on Rough Sets* series has as its principal aim the fostering of professional exchanges between scientists and practitioners who are interested in the foundations and applications of rough sets. Topics include foundations and applications of rough sets as well as foundations and applications of hybrid methods combining rough sets with other approaches important for the development of intelligent systems. The journal includes high-quality research articles accepted for publication on the basis of thorough peer reviews. Dissertations and monographs up to 250 pages that include new research results can also be considered as regular papers. Extended and revised versions of selected papers from conferences can also be included in regular or special issues of the journal.

# Table of Contents

# Granular Rough Mereological Logics with Applications to Dependencies in Information and Decision Systems*

Lech Polkowski[1] and Maria Semeniuk–Polkowska[2]

[1] Polish–Japanese Institute of Information Technology
Koszykowa 86, 02008 Warsaw, Poland
Department of Mathematics and Computer Science
University of Warmia and Mazury
Zolnierska 14, 10560 Olsztyn, Poland
[2] Chair of Formal Linguistics, Warsaw University
Browarna 8/12, 00650 Warsaw,Poland
polkow@pjwstk.edu.pl, m.polkowska@uw.edu.pl

**Abstract.** We are concerned with logical formulas induced from data sets, in particular, with decision rules. Contrary to the standard practice of many–valued logics in which formulas are semantically interpreted as their states /values of truth and logical calculi consist essentially in finding functional interpretations of logical functors, in the considered by us case, the semantic interpretation takes place in the universe of entities/objects and formulas are interpreted as their meanings, i.e., subsets of the object universe. Yet, the final evaluation of a formula should be its state of truth. In search of an adequate formal apparatus for this task, we turn to rough mereology and to the idea of intensionality vs. extensionality. Rough mereology allows for similarity measures (called rough inclusions) which in turn form a basis for the mechanism of granulation of knowledge. Granules of knowledge, defined as classes of satisfactorily similar objects, can be regarded as worlds in which properties of entities are evaluated as extensions of logical formulas. Obtained in this way granular rough mereological intensional logics reveal essential properties of rough set based reasoning.

**Keywords:** rough sets, granulation of knowledge, rough mereology, logics for reasoning about knowledge.

## 1 Introductory Notions

We assume that the reader is familiar with basics of rough sets, see, e.g., [19], [20]; the context in which our considerations are set is an information system $(U, A)$ or a decision system $(U, A, d)$. We recall that an information system $(U, A)$ consists of a set of *entities/objects* $U$ along with a set $A$ of *attributes* each of which

* This article extends the contribution "Reasoning about Concepts by Rough Mereological Logics" by the authors at RSKT 2008, May 2008 at Chengdu, Sichuan, China.

J.F. Peters et al. (Eds.): Transactions on Rough Sets XII, LNCS 6190, pp. 1–20, 2010.

is constructed as a mapping on $U$ with values in a set $V$ of values (we will use also the symbol $V_a$ for the more specific set of values of the attribute $a$); in any decision system $(U, A, d)$, a singled–out attribute $d$, the *decision*, represents the external knowledge by an oracle/expert.

*Concepts* with respect to a given data are defined formally as subsets of the universe $U$. Concepts can be written down in the language of *descriptors.*

For an attribute $a$ and its value $v$, the *descriptor* defined by the pair $a, v$ is the atomic formula $(a = v)$. Descriptors can be made into formulas by means of sentential connectives: $\vee, \wedge, \neg, \Rightarrow$: formulas of descriptor logics are elements of the smallest set which contains all atomic descriptors and is closed under the mentioned above sentential connectives. Introducing for each object $u \in U$ its *information set* $Inf(u) = \{(a = a(u)) : a \in A\}$, we can define the basic *indiscernibility relation* $IND(A) = \{(u, v) : Inf(u) = Inf(v)\}$. Replacing $A$ with a subset $B$ of attribute set, we define the *B–indiscernibility relation* $IND(B)$.

A descriptor $(a = v)$ is interpreted semantically in the universe $U$; the meaning $[a = v]$ of this descriptor is the concept $\{u \in U; a(u) = v\}$. Meanings of atomic descriptors are extended to meanings of formulas of descriptor logic by recursive conditions,

$[p \vee q]=[p] \cup [q]$;
$[p \wedge q]=[p] \cap [q]$;
$[\neg p]=U \setminus [p]$;
$[p \Rightarrow q]=(U \setminus [p]) \cup [q]$.

In the language of descriptor formulas, one can express relations among concepts. As proposed by Pawlak, [19], [20], the buildup of those relations should start with entities: any relation IND(B) partitions the universe $U$ into blocks – equivalence classes $[u]_B$ of IND(B), regarded as *elementary B–exact concepts.* Unions of families of elementary $B$–exact concepts constitute *B–exact concepts.*

In terms of exact concepts, one can express dependencies among attributes [20]: in the simplest case, a set $D$ of attributes *depends functionally* on a set $C$ of attributes if and only if $IND(C) \subseteq IND(D)$; the meaning is that any class $[\bigwedge_{a \in C}(a = v_a)]$ is contained in a unique class $[\bigwedge_{a \in D}(a = w_a)]$ so there is a mapping $U/IND(C) \rightarrow U/IND(D)$. We write down this dependency as $C \mapsto D$.

Dependency need not be functional; in such case, the relation $IND(C) \subseteq IND(D)$ can be replaced [20] with a weaker notion of a *(C,D)–positive set* which is defined as the union $Pos_C(D) = \bigcup\{[u]_C : [u]_C \subseteq [u]_D\}$; clearly then, $IND(C)|Pos_C(D) \subseteq IND(D)$. In [20] a factor $\gamma(B, C) = \frac{|Pos_B(C)|}{|U|}$ was proposed as the measure of degree of dependency of $D$ on $C$, where $|X|$ is the number of elements in $X$. This form of dependency is denoted symbolically as $C \hookrightarrow_{\gamma} D$.

Dependencies have a logical form in logic of descriptors as sets of implications of the form

$$\bigwedge_{a \in C}(a = v_a) \Rightarrow \bigwedge_{a \in D}(a = w_a); \tag{1}$$

in a particular case of a decision system $(U, A, d)$, dependencies of the form $C \hookrightarrow_{\gamma} \{d\}$ are called decision algorithms and individual relations of the form $\bigwedge_{a \in C}(a = v_a) \Rightarrow (d = w_d)$ are said to be *decision rules.* There have been proposed various measures of the truth degree of a decision rule, under the name of a *rule quality*, see, e.g., [20].

In descriptor logic setting, a decision rule $r : \alpha_C \Rightarrow \beta_d$ is said to be *true* if and only if the meaning $[r] = U$ which is equivalent to the condition that $[\alpha_C] \subseteq [\beta_d]$.

The above introduced constituents: entities, indiscernibility relations, concepts, dependencies, form building blocks from which knowledge is discovered as a set of statements about those constituents.

It is our purpose to construct a formal logical system in which one would be able to define values of truth states of formulas of knowledge, decision rules in particular, in a formal manner, preserving the notion of truth as recalled above, but in a localized version, with respect to a particular exact concept of entities.

Rough set theory discerns between *exact concepts* which are unions of indiscernibility classes and *rough concepts* which are not any union of indiscernibility classes. Passing from rough to exact concepts is achieved by means of approximations: the *lower approximation* to a concept $W \subseteq U$ is defined as $\underline{W} = \{u \in U : [u]_A \subseteq W\}$, and, the *upper approximation* $\overline{W} = \{u \in U : [u]_A \cap W \neq \emptyset\}$.

## 2 Many–Valued Approach to Decision Rules

Any attempt at assigning various degrees of truth to logical statements places one in the realm of many–valued logic. These logics describe formally logical functors as mappings on the set of truth values/states into itself hence they operate a fortiori with values of statements typically as fractions or reals in the unit interval $[0, 1]$, see in this respect, e.g., [12], [13], [14], [15], and, as a survey, see [8].

### 2.1 t–norms

In many of those logics, the functor of implication is interpreted as the *residual implication* induced by a continuous t–norm. We recall, cf., e.g., [8] or [21] that a *t–norm* is a mapping $t : [0, 1]^2 \rightarrow [0, 1]$ which satisfies the conditions,

(TN1) $t(x, y) = t(y, x)$ (symmetry);

(TN2) $t(x, t(y, z)) = t(t(x, y), z)$ (associativity);

(TN3) $t(x, 1) = x; t(x, 0) = 0$ (boundary conditions);

(TN4) $x > x'$ implies $t(x, y) \geq t(x'), y)$ (monotonicity coordinate–wise);

and additionally,

(TN5) $t$ can be continuous.

A continuous t–norm is *Archimedean* in case $t(x,x) = x$ for $x = 0,1$ only, see [8] or [21], or, [18] for original results by Mostert and Shields; for such t–norms, it was shown, see [11], that a formula holds,

$$t(x,y) = g(f(x) + f(y)), \tag{2}$$

with a continuous decreasing function $f : [0,1] \to [0,1]$ and $g$ – the pseudo–inverse to $f$, proofs are given also in [21].

Examples of Archimedean t–norms are,

**The Łukasiewicz t–norm** $L(x,y) = max\{0, x + y - 1\}$;

**The product (Menger) t–norm** $P(x,y) = x \cdot y$.

The two are up to an automorphism on [0,1] the only Archimedean t–norms [18].

An example of a t–norm which is not any Archimedean is

**Minimum t–norm** $Min(x,y) = min\{x,y\}$. It is known, see [2], that for $Min$ the representation (2) with $f$ continuous does not exist.

### 2.2 Residual Implications

*Residual implication* $x \Rightarrow_t y$ induced by a continuous t–norm $t$ is defined as,

$$x \Rightarrow_t y = max\{z : t(x,z) \leq y\}. \tag{3}$$

As $t(x,1) = x$ for each t–norm, it follows that $x \Rightarrow_t y = 1$ when $x \leq y$ for each t–norm $t$.

In case $x > y$, one obtains various semantic interpretations of implication depending on the choice of $t$, see, e.g., [8] for a review. Exemplary cases are,

**The Łukasiewicz implication** $x \Rightarrow_L y = min\{1, 1 - x + y\}$, see [13];

**The Goguen implication** $x \Rightarrow_P y = \frac{y}{x}$;

**The Gödel implication** $x \Rightarrow_{Min} y = y$.

### 2.3 Logics of Residual Implications vs. Logical Containment in Decision Rules

In logics based on implication given by residua of t–norms, negation is defined usually as $\neg x = x \Rightarrow_t 0$. Thus, the Łukasiewicz negation is $\neg_L x = 1 - x$ whereas Goguen as well as Gödel negation is $\neg_G x = 1$ for x=0 and is 0 for $x > 0$. Other connectives are defined with usage of the t–norm itself as semantics for the strong conjunction and ordinary conjunction and disjunction are interpreted semantically as, respectively, *min, max.* Resulting logics have been a subject of an intensive research, cf., a monograph [8].

In this approach a rule $\alpha \Rightarrow \beta$ is evaluated by evaluating the truth state $[\alpha]$ as well as the truth state $[\beta]$ and then computing the values of $[\alpha] \Rightarrow_t [\beta]$ for a chosen t–norm $t$. Similarly other connectives are evaluated.

In the rough set context, this approach would pose the problem of evaluating the truth state of a conjunct $\alpha$ of descriptors; to this end, one can invoke the idea of Łukasiewicz [12] and assign to $\alpha$ a value $[\alpha]_L = \frac{||\alpha||}{|U|}$, where $[\alpha]$ is the meaning already defined, i.e., the set $\{u \in U : u \models \alpha\}$. Clearly, this approach does not take into account the logical containment or its lack between $\alpha$ and $\beta$, and this fact makes the many–valued approach of a small use when data mining tasks are involved.

For this reason, we propose an approach to logic of decision rules which is based on the idea of measuring the state of truth of a formula against a concept constructed as a granule of knowledge; concepts can be regarded as "worlds" and our logic becomes intensional, cf., e.g., [3], [17]: logical evaluations at a given world are extensions of the intension which is the mapping on worlds valued in the set of logical values of truth.

To implement this program, we need to develop the following tools:

1 a tool to build worlds, i.e, a granulation methodology based on a formal mechanism of granule formation and analysis;

2 a methodology for evaluating states of truth of formulas against worlds.

In both cases 1, 2, our approach exploits tools provided by *rough mereology*, see, e.g., [24], [25] or [26] for the latest surveys, a theory of reasoning induced from considerations of similarity among objects and granules of objects. Similarity measures – rough inclusions – provide means for all necessary definitions of relevant notions.

## 3 On Mereology and Rough Mereology: The Basic Facts

Rough mereology is an extension of the theory of mereology. We have to begin with a nutshell account of the latter.

### 3.1 Mereology

In mereological theory of concepts proposed by Leśniewski [9], the basic notion of a *part* is a relation $part(x, y)$ (read *x is a part of y*) which is transitive and non–reflexive, i.e.,

P1. $part(x, y), part(y, z)$ imply $part(x, z)$;

P2. $part(x, x)$ holds for no $x$.

Associated with the relation of *part*, is a relation of an ingredient *ingr* which is defined as

$$ingr(x, y) \text{ if and only if } part(x, y) \text{ or } x = y. \quad (4)$$

The important property of the ingredient relation is the inference rule established by Leśniewski [9],

If for each z [if $ingr(z, x)$ there is $w$ s.t. $ingr(w, z)$ and $ingr(w, y)$]then $ingr(x, y)$. (5)

This means that if for each $z$ such that $z$ is an ingredient of $x$ there exists an object $w$ such that $w$ is an ingredient of both $z$ and $y$ then $x$ is an ingredient of $y$.

In turn, the ingredient relation is essential in definition of the class operator $Cls$ whose role is to make collections (properties) of objects into single objects representing those collections. The class operator is defined in the following way,

$$\begin{array}{c} C1.\text{ If } x \in F \text{ then } ingr(x, ClsF); \\ C2.\text{ If } ingr(x, ClsF) \text{ then there are } w, z \text{ such that} \\ ingr(w, x), ingr(w, z), z \in F. \end{array} \tag{6}$$

This defines the class $ClsF$ for each non–empty collection $F$.

To visualize the working of the operator $Cls$, we remind that the strict containment relation $\subset$ is a part relation, the corresponding ingredient relation is the containment relation $\subseteq$, and the class $ClsF$ for a non–empty family of sets is then according to (6) the union $\bigcup F$ of $F$.

### 3.2 Rough Mereology

The basic primitive notion of Rough Mereology is that of a *rough inclusion*, see [30] for the original source of the notion.

We recall that a rough inclusion $\mu$ is a relation $\mu \subseteq U \times U \times [0, 1]$ which satisfies the conditions,

RI1. $\mu(u, v, 1)$ if and only if $ingr(u, v)$.

RI2. $\mu(u, v, 1), \mu(w, u, r)$ imply $\mu(w, v, r)$.

RI3. $\mu(u, v, r), s < r$ imply $\mu(u, v, s)$.

The relation $ingr$ is an ingredient (element) relation of Mereology [9], see also [24], cf., sect. 3.1.

Rough inclusions can be regarded as similarity measures on sets of objects, $\mu(u, v, r)$ meaning that the object $u$ is *similar (is part of) to the object $v$ to a degree of* $r$. The formal definition (RI1) – (RI3) of a rough inclusion along with technical features of mereology, see [9], cf., sect. 3.1 allow for a formalization of rough set based reasoning concerning similarity.

We now recall some means for inducing rough inclusions along with some new results.

### 3.3 Rough Inclusions

The part relation of mereology was given an extension in the form of a relation $\mu$ of part to a degree, see [24]. The relation $\mu$ was required to satisfy conditions (RI1–RI3) of sect. 1.

Rough inclusions in information systems have been described in many places, one of most recent, e.g., in [24]. For the present purpose, we begin with rough inclusions on the interval $[0, 1]$.

**Rough inclusions on [0,1].**

**Proposition 1.** *For each continuous t–norm t, the residual implication $\Rightarrow_t$ induced by t does satisfy conditions (RI1) – (RI3) defining rough inclusions in sect. 3.2 with $\mu_t(x, y, r) \Leftrightarrow x \Rightarrow_t y \geq r$.*

Indeed, consider a residual implication $\Rightarrow_t$ along with the generating continuous t–norm $t$. The fact that $\mu_t(x, x, 1)$ follows by (TN1) of sect.2.1.

Assume $\mu(x, y, 1)$, i.e., by (TN3), cf. also, sect. 2.2, $x \leq y$. In case $\mu(z, x, r)$, we have $t(z, r) \leq x$ by (3), hence $t(z, r) \leq y$ and thus $z \Rightarrow_t y \geq r$, i.e., $\mu(z, y, r)$ proving (RI2). (RI3) is satisfied obviously.

This proposition opens up a possibility of defining many rough inclusions, by a choice of a continuous t–norm with the resulting implication.

There exist rough inclusions not definable in this way, e.g., the *drastic rough inclusion*,

$$\mu_0(x, y, r) \text{ if and only if either } x = y \text{ and } r = 1 \text{ or } r = 0. \tag{7}$$

Clearly, $\mu_0$ is associated with the ingredient relation $=$ and, a fortiori, the part relation is empty, whereas any rough inclusion $\mu$ induced by a residual implication in the sense of Prop. 1, is associated to the ingredient relation $\leq$ with the part relation $<$.

The upshot of this little discussion is that the notion of a rough inclusion does generalize the notion of a residual implication on the universe $[0, 1]$ of reals. One may attempt to, conversely, define a *rough norm* $t_\mu$ on the interval $[0, 1]$ by letting, in analogy to (3), where the rough inclusion $\mu$ is associated with the ingredient relation $\leq$,

$$t_\mu(x, z) \leq y \text{ if and only if } \mu(x, y, z). \tag{8}$$

One can establish the following properties of the functor $t_\mu$.

**Proposition 2.** *(TM1) $t_\mu(x, 1) = x$;*

*(TM2) $t_\mu$ is monotonically increasing in the second coordinate.*

Clearly, $t_\mu(x, 1) \leq y$ if and only if $\mu(x, y, 1)$, i.e., when $ingr(x, y)$, i.e., when $x \leq y$; from the equivalence $t_\mu(x, 1) \leq y \Leftrightarrow x \leq y$ it follows that $t_\mu(x, 1) = x$ proving (TM1).

For (TM2), consider $t_\mu(x,z)$ and let $z > z'$. Assuming that $t_\mu(x,z) \leq y$, we have, $\mu(x,y,z)$ hence $\mu(x,y,z'$ and finally $t_\mu(x,z') \leq y$. Letting $y = t_\mu(x,z)$, we get $t_\mu(x,z') \leq t_\mu(x,z)$. This proves (TM2).

A more general result is also true: we can replace the relation $\leq$ with a general *ingr* relation, and then arguing on the same lines, we obtain a proposition.

**Proposition 3.** *(TM1ingr)* $ingr(t_\mu(x,1),y)$ *if and only if* $ingr(x,y)$*;*

*(TM2ingr)* $ingr(y,y')$ *implies* $ingr(t_\mu(x,y),t_\mu(x,y'))$.

In case of Archimedean t–norms, it is well–known that a representation formula holds for them, see (2), which implies the residual implication in the form,

$$x \Rightarrow y = g(f(x) - f(y)). \tag{9}$$

This formula will be useful in case of information systems which is going to be discussed in the next section.

### 3.4 Rough Inclusions in Information Systems

Assuming an information system $(U,A)$ given, it is our purpose to extend the results of the preceding section to the case of $(U,A)$.

In this case, the most important rough inclusion is defined by adopting (9) to this case,

$$\mu(u,v,r) \text{ if and only if } g(\frac{|\{a \in A : a(u) \neq a(v)\}|}{|A|} \geq r, \tag{10}$$

where $g, \mu$ correspond to a chosen Archimedean t–norm $t$, $u,v \in U$.

In the case of the Łukasiewicz t–norm $t_L(x,y) = max\{0, x+y-1\}$, for which $g(y) = 1 - y$, one specifies (10) to

$$\mu_L(u,v,r) \text{ if and only if } \frac{|IND(u,v)|}{|A|} \geq r, \tag{11}$$

where $IND(u,v) = \{a \in A : a(u) = a(v)\}$.

An important property of this rough inclusion is its transitivity.

**Proposition 4.** *The rough inclusion* $\mu_L$ *induced from the Łukasiewicz t–norm* $t_L$ *by means of (11) satisfies the transitivity property,*

$$\mu_L(u,v,r), \mu_L(v,w,s) \textit{ imply } \mu_L(u,w,t_L(r,s)). \tag{12}$$

Proposition 4 follows by (11) and the containment $DIS(u,w) \subseteq DIS(u,v) \cup DIS(v,w)$, where

$$DIS(u,v) = A \setminus IND(u,v) = \{a \in A : a(u) \neq a(v). \tag{13}$$

Thus, $g(|DIS(u,w)|/|A|) = 1 - \frac{|DIS(u,w)|}{|A|} \geq (1 - \frac{|DIS(u,v)|}{|A|}) + (1 - \frac{|DIS(v,w)|}{|A|}) - 1 = t_L(r,s)$ hence $\mu_L(u,w,t_L(r,s))$ follows.

In the case of the interval [0,1], formula (10) boils down to the formula for the residual implication $\Rightarrow_L$ induced by the Łukasiewicz t–norm. Information systems, however, allow for a more diverse approach.

Let us observe that introducing the notation $ind(u,v) = \frac{|IND(u,v)|}{|A|}$, we can write down the formula (11) in the form,

$$\mu_L(u,v,r) \Leftrightarrow 1 \Rightarrow_t ind(u,v) \geq r. \tag{14}$$

This formula can be generalized to the formula,

$$\mu^*(u,v,r) \Leftrightarrow dis(u,v) \Rightarrow_t ind(u,v), \tag{15}$$

where $dis(u,v) = \frac{|DIS(u,v)|}{|A|}$, cf., (13).

The predicate $\mu^*$ satisfies (RI1), (RI3) however it may fail (RI2). Its usefulness has been tested in [27] in classifier synthesis problems.

Yet, we can repeat our earlier observation that taking into account only values of state truth misses important aspect of logical containment between the premises and the conclusion of an implication, and we should consider inclusions on sets which are meanings of formulas.

### 3.5 Rough Inclusions on Sets

For our purpose it is essential to extend rough inclusions to sets; we use the t–norm $t_L$ of Łukasiewicz, along with the representation $t_L(r,s) = g(f(r) + f(s))$ already mentioned in (2), which in this case is $g(y) = 1 - y$, $f(x) = 1 - x$. We denote these kind of inclusions with the generic symbol $\nu$.

For sets $X, Y \subseteq U$, we let,

$$\nu_L(X,Y,r) \text{ if and only if } g(\frac{|X \setminus Y|}{|X|}) \geq r; \tag{16}$$

as $g(y) = 1 - y$, we have that $\nu_L(X,Y,r)$ holds if and only if $\frac{|X \cap Y|}{|X|} \geq r$. Let us observe that $\nu_L$ is *regular*, i.e., $\nu_L(X,Y,1)$ if and only if $X \subseteq Y$ and $\nu_L(X,Y,r)$ only with $r = 0$ if and only if $X \cap Y = \emptyset$.

Thus, the ingredient relation associated with a regular rough inclusion is the improper containment $\subseteq$ whereas the underlying part relation is the strict containment $\subset$.

Other rough inclusion on sets which we exploit is the 3–valued rough inclusion $\nu_3$ defined via the formula, see [22],

$$\nu_3(X,Y,r) \text{if and only if} \begin{cases} X \subseteq Y \text{ and } r = 1 \\ X \cap Y = \emptyset \text{ and } r = 0 \\ r = \frac{1}{2} \text{ otherwise,} \end{cases} \tag{17}$$

The rough inclusion $\nu_3$ is also regular.

Finally, we consider the drastic rough inclusion on sets, $\nu_1$,

$$\nu_1(X, Y, r) \text{ if and only if} \begin{cases} X = Y \text{ and } r = 1 \\ X \neq Y \text{and } r = 0. \end{cases} \tag{18}$$

Clearly, $\nu_1$ is not regular.

## 4 Granulation of Knowledge

The paradigm of granulation was proposed by Zadeh [33]; within rough set framework, the attention to it was brought for in Lin [10]. Many authors have devoted attention to various aspects of granulation, see surveys [25], [26]. Among approaches to granulation, one can single out the generalization of granules defined by indiscernibility relations to granules defined by general binary relations (Lin, Yao) and the mereological approach which we are going to present below. In [10], a topological aspect of granulation was hinted at, which can be formally exhibited in the approach by us.

Our approach to granulation differs from those by other authors because we exploit rough inclusions as a basic tool, not using directly indiscernibility relations or inverse images of binary relations. We begin with an inventory of technical means for forming granules and establishing their basic properties.

### 4.1 Granule Formation

Granules of knowledge are defined by means of a rough inclusion $\mu$ in the universe $U$ of an information/decision system $(U, A)/(U, A, d)$ as classes (see C1, C2, sect. 3.1) of appropriate similarity property, see [23],

$$g(u, r, \mu) \; is \; ClsP_\mu(u, r), \tag{19}$$

where $g(u, r, \mu)$ is the granule about an object $u$ of the radius $r$, and $P_\mu(u, r)$ is the property (collection) $\{v \in U : \mu(v, u, r)\}$.

Of particular interest in this work are granules induced by the rough inclusion $\mu_L$. Their important property relies in the lack of synergic effects which could be caused by general definition of the class, see [23],

$$ingr(v, g(u, r, \mu_L)) \text{ if and only if } P_\mu(u, r)(v). \tag{20}$$

On the strength of this property, one can express the granule $g(u, r, \mu_L)$ as a list of those objects $v$ which satisfy $\mu_L(v, u, r)$. In order to show the reader the flavor of reasoning by mereological notions, we offer a short proof of (20).

Assume that an object $v$ is an ingredient of the granule $g(u, r, \mu_L)$; by definition (19) of a granule, and by definition (6) of the class operator, there exist objects $w, z$ such that:

(a) $ingr(w, v)$;
(b) $ingr(w, z)$;
(c) $z \in P_{\mu_L}(u, r)$,
i.e., $\mu_L(z, u, r)$, where $ingr$ is the ingredient relation associated with $\mu_L$.
By condition RI1 in Sect. 1, it follows from (a) and (b) respectively that
(a') $\mu_L(w, v, 1)$;
(b') $\mu_L(w, z, 1)$.
By symmetry of $\mu_L$, one obtains from (a') that $\mu_L(v, w, 1)$ which together with (b') yields by the transitivity property (12) that $\mu_L(v, z, L(1, 1))$, i.e., $\mu_L(v, z, 1)$. The last fact combined with (c) by means of the transitivity property, gives that $\mu_L(v, u, L(r, 1))$, i.e., $\mu_L(v, u, r)$, as required. The converse implication holds obviously by condition C1 in (6).

This pattern of reasoning can be applied as well in establishing a number of properties of granules.

### 4.2 Basic Properties of Granules

We know by (20) that any granule $g(u, r)$ can be represented as the set $\{v : \mu_L(v, u, r)\}$. Whence we have,

G1. If $\mu(w, v, s)$ and $ingr(v, g(u, r))$ then $ingr(w, g(u, t_L(r, s)))$. In particular, if $ingr(w, v)$ and $ingr(v, g(u, r))$ then $ingr(w, g(u, r))$.

This follows from transitivity property (12) of $\mu_L$.

G2. If $s > r$ then $ingr(g(u, s), g(u, r))$.

In establishing this property, we avail ourselves with the inference rule (5), so we consider $ingr(v, g(u, s))$. By class definition C1-2, sect.3.1, there exist $w, t$ such that $ingr(w, v)$, $ingr(w, t)$, $\mu_L(t, u, s)$, hence $\mu_L(t, u, r)$ by (RI3), i.e., $ingr(t, g(u, r)$. By (5) $ingr(g(u, s), g(u, r))$ follows.

G3. If $ingr(v, g(u, r))$ and $\mu(v, u, s)$ with $s > r$ then for some $\delta > 0$ we have $ingr(g(v, \delta), g(u, r))$.

Indeed, mimicking the proof of G2, and making use of the inference rule (5), we infer from assumptions $ingr(v, g(u, r))$ and $\mu(v, u, s)$ that it is sufficient to take $\delta$ such that $t_L(\delta, s) \geq r$, i.e., $\delta \geq 1 + (r - s)$. It is possible to choose such value of $\delta$ because of $s > r$.

Properties G2, G3 do suggest that the system of granules $g(u, r)$ for $u \in U$ and $r \in [0, 1]$ is a weak topology on the universe $U$.

## 5 Granular Rough Mereological Logics

The idea of a granular rough mereological logic, see [28], [23], consists in measuring the meaning of a predicate (we restrict ourselves here to the case of unary predicates) in the model which is a universe of an information system against a granule defined to a certain degree by means of a rough inclusion. The result can be regarded as the degree of truth (the logical value) of the predicate with respect to the given granule.The obtained logics are intensional as they can be regarded as mappings from the set of granules (possible worlds) to the set of logical values in the interval $[0, 1]$, the value at a given granule regarded as the extension at that granule of the generally defined intension, see [3] for a general introduction to intensional logics and [17] for an excellent their application.

We assume that an information system $(U, A)$ is given, along with a rough inclusion $\nu$ on the subsets of the universe $U$; for a collection of predicates (unary) $Pr$, interpreted in the universe $U$ (meaning that for each predicate $\phi \in Pr$ the meaning $[\phi]$ is a subset of $U$), we define the intensional logic $grm_\nu$ on $Pr$ by assigning to each predicate $\phi$ in $Pr$ its intension $I_\nu(\phi)$ defined by the family of extensions $I_\nu^\vee(g)$ at particular granules $g$, as,

$$I_\nu^\vee(g)(\phi) \geq r \text{ if and only if } \nu(g, [\phi], r). \tag{21}$$

With respect to the rough inclusion $\nu_L$, the formula (21) becomes,

$$I_{\nu_L}^\vee(g)(\phi) \geq r \text{ iff } \frac{|g \cap [\phi]|}{|g|} \geq r. \tag{22}$$

The counterpart for $\nu_3$ is specified by definition (17).

We say that a formula $\phi$ interpreted in the universe $U$ of an information system $(U, A)$ is *true* at a granule $g$ with respect to a rough inclusion $\nu$ if and only if $I_\nu^\vee(g)(\phi) = 1$.

**Proposition 5.** *For every regular rough inclusion $\nu$, a formula $\phi$ interpreted in the universe $U$, with meaning $[\phi]$, is true at a granule $g$ with respect to $\nu$ if and only if $g \subseteq [\phi]$. In particular, for a decision rule $r : p \Rightarrow q$ in the descriptor logic, the rule $r$ is true at a granule $g$ with respect to a regular rough inclusion $\nu$ if and only if $g \cap [p] \subseteq [q]$.*

Indeed, truth of $\phi$ at $g$ means that $\nu(g, [\phi], 1)$ which in turn, by regularity of $\nu$ is equivalent to the inclusion $g \subseteq [\phi]$.

We will say that a formula $\phi$ is a *theorem* of our intensional logic if and only if $\phi$ is true at every world $g$.

The preceding proposition implies that

**Proposition 6.** *For every regular rough inclusion $\nu$, a formula $\phi$ is a theorem if and only if $Cls(GranCon) \subseteq [\phi]$, where $GranCon$ are all granules under consideration; in the case when granules considered cover the universe $U$ this condition simplifies to $[\phi] = U$. This means for a decision rule $p \Rightarrow q$ that it is a theorem if and only if $[p] \subseteq [q]$.*

### 5.1 Relations to Many–Valued Logics

Here we examine some axiomatic schemes for many–valued logics with respect to their meanings under the stated in introductory section proviso that $[p \Rightarrow q] = (U \setminus [p]) \cup [q]$, $[\neg p] = U \setminus [p]$.

We examine first axiom schemes for 3–valued Łukasiewicz logic investigated in [32] (Wajsberg schemes).
(W1) $q \Rightarrow (p \Rightarrow q)$;
(W2) $(p \Rightarrow q) \Rightarrow ((q \Rightarrow r) \Rightarrow (p \Rightarrow r))$;
(W3) $((p \Rightarrow \neg p) \Rightarrow p) \Rightarrow p$;
(W4) $(\neg q \Rightarrow \neg p) \Rightarrow (p \Rightarrow q)$.

We have as meanings of those formulas,

$[W1] = (U \setminus [q]) \cup (U \setminus [p]) \cup [q] = U$;
$[W2] = ([p] \setminus [q]) \cup ([q] \setminus [r]) \cup (U \setminus [p]) \cup [r] = U$;
$[W3] = (U \setminus [p]) \cup [p] = U$;
$[W4] = ([p] \setminus [q]) \cup [q] = U$.

Thus, *all instances of Wajsberg axiom schemes for 3–valued Łukasiewicz logic are theorems of our intensional logic in case of regular rough inclusions on sets.*

The deduction rule in 3–valued Łukasiewicz logic is Modus Ponens: $\frac{p, p \Rightarrow q}{q}$.

In our setting this is a valid deduction rule: if $p, p \Rightarrow q$ are theorems than $q$ is a theorem. Indeed, if $[p] = U = [p \Rightarrow q]$ then $[q] = U$.

We have obtained

**Proposition 7.** *Each theorem of 3–valued Łukasiewicz logic is a theorem of rough mereological granular logic in case of a regular rough inclusion on sets.*

In an analogous manner, we examine axiom schemes for infinite valued Łukasiewicz logic, proposed by Łukasiewicz [15], with some refinements showing redundancy of a scheme due to Meredith [16] and Chang [5], cf., in this respect [21] for an account of the reasoning.

(L1) $q \Rightarrow (p \Rightarrow q)$;
(L2) $(p \Rightarrow q) \Rightarrow ((q \Rightarrow r) \Rightarrow (p \Rightarrow r))$;
(L3) $((q \Rightarrow p) \Rightarrow p) \Rightarrow ((p \Rightarrow q) \Rightarrow q)$;
(L4) $(\neg q \Rightarrow \neg p) \Rightarrow (p \Rightarrow q)$.

As (L1) is (W1), (L2) is (W2) and (L4) is (W4), it remains to examine (L3). In this case, we have $[(q \Rightarrow p) \Rightarrow p]$=$(U \setminus [q \Rightarrow p]) \cup [p])$=$(U \setminus ((U \setminus [q]) \cup [p])) \cup [p]$=$([q] \setminus [p]) \cup [p]$= $[q] \cup [p]$. Similarly, $[(p \Rightarrow q) \Rightarrow q]$ is $[p] \cup [q]$ by symmetry, and finally the meaning $[L3]$ is $(U \setminus ([q] \cup [p])) \cup [p] \cup [q] = U$.

It follows that,

*all instances of axiom schemes for infinite–valued Łukasiewicz logic are theorems of rough mereological granular logic.*

As Modus Ponens remains a valid deduction rule in infinite–valued case, we obtain, analogous to Prop. 7,

**Proposition 8.** *Each theorem of infinite–valued Łukasiewicz logic is a theorem of rough mereological granular logic in case of a regular rough inclusion on sets.*

It follows from Prop.8 that all theorems of *Basic logic*, see [8], i.e. logic which is intersection of all many–valued logics with implications evaluated semantically by residual implications of continuous t–norms are theorems of rough mereological granular logic.

The assumption of regularity of a rough inclusion $\nu$ is essential: considering the drastic rough inclusion $\nu_1$, we find that an implication $p \Rightarrow q$ is true only at the world $(U \setminus [p]) \cup [q]$, so it is not any theorem – this concerns all schemes (W) and (L) above as they are true only at the global world $U$.

### 5.2 Graded Notion of Truth

As already stated, the usual interpretation of functors $\vee, \wedge$ in many–valued logics is $[p \vee q] = max\{[p], [q]\}$ and $[p \wedge q] = min\{[p], [q]\}$, where $[p]$ is the state of truth of $p$. In case of concept–valued meanings, we admit the interpretation which conforms to accepted in many valued logics (especially in the context of fuzzy set theory) interpretation of $min$ as $\cap$ and $max$ as $\cup$.

The formula $\nu(g, [\phi], 1)$ stating the truth of $\phi$ at $g, \nu$ with $\nu$ regular can be regarded as a condition of orthogonality type, with the usual consequences.

1. If $\phi$ is true at granules $g, h$ then it is true at $g \cup h$.
2. If $\phi$ is true at granules $g, h$ then it is true at $g \cap h$.
3. If $\phi, \psi$ are true at a granule $g$ then $\phi \vee \psi$ is true at $g$.
4. If $\phi, \psi$ are true at a granule $g$ then $\phi \wedge \psi$ is true at $g$.
5. If $\psi$ is true at a granule $g$ then $\phi \Rightarrow \psi$ is true at $g$ for every formula $\phi$.
6. If $\phi$ is true at a granule $g$ then $\phi \Rightarrow \psi$ is true at $g$ if and only if $\psi$ is true at $g$.

The graded relaxation of truth is given obviously by the condition, a formula $\phi$ is *true to a degree at least* $r$ *at* $g, \nu$ if and only if $I_\nu^\vee(g)(\phi) \geq r$, i.e., $\nu(g, [\phi], r)$ holds. In particular, $\phi$ is *false* at $g, \nu$ if and only if $I_\nu^\vee(g)(\phi) \geq r$ implies $r = 0$, i.e. $\nu(g, [\phi], r)$ implies $r = 0$.

The following properties hold.

1. For each regular $\nu$, a formula $\alpha$ is true at $g, \nu$ if and only if $\neg\alpha$ is false at $g, \nu$.
2. For $\nu = \nu_L, \nu_3$, $I_\nu^\vee(g)(\neg\alpha) \geq r$ if and only if $I_\nu^\vee(g)(\alpha) \geq s$ implies $s \leq 1 - r$.
3. For $\nu = \nu_L, \nu_3$, the implication $\alpha \Rightarrow \beta$ is true at $g$ if and only if $g \cap [\alpha] \subseteq [\beta]$ and $\alpha \Rightarrow \beta$ is false at $g$ if and only if $g \subseteq [\alpha] \setminus [\beta]$.
4. For $\nu = \nu_L$, if $I_\nu^\vee(g)(\alpha \Rightarrow \beta) \geq r$ then $\Rightarrow_L (t, s) \geq r$ where $I_\nu^\vee(g)(\alpha) \geq t$ and $I_\nu^\vee(g)(\beta) \geq s$.

The functor $\Rightarrow$ in 4. is the Łukasiewicz implication of many–valued logic: $\Rightarrow_{t_L} (t, s) = min\{1, 1 - t + s\}$.

Further analysis should be split into the case of $\nu_L$ and the case of $\nu_3$ as the two differ essentially with respect to the form of reasoning they imply.

# 6 Reasoning with $\nu_L$

The last property 4. shows in principle that the value of $I_\nu^\vee(g)(\alpha \Rightarrow \beta)$ is bounded from above by the value of $I_\nu^\vee(g)(\alpha) \Rightarrow_{t_L} I_\nu^\vee(g)(\beta))$.

This suggests that the idea of collapse attributed to Leśniewski can be applied to formulas of rough mereological logic in the following form: for a formula $q(x)$ we denote by the symbol $q^*$ the formula $q$ regarded as a sentential formula (i.e., with variable symbols removed) subject to relations:

$(\neg q(x))^*$ is $\neg(q(x)^*)$ and $(p(x) \Rightarrow q(x))^*$ is $p(x)^* \Rightarrow q(x)^*$. As the value $[q^*]_g$ of the formula $q(x)^*$ at a granule $g$, we take the value of $\frac{|g \cap [q(x)]|}{|g|}$, i.e, $argmax_r\{\nu_L(g, [q^*]_g, r)\}$. Thus, item 4 above can be rewritten in the form.

$$I_\nu^\vee(g)(\alpha \Rightarrow \beta) \leq [\alpha^*]_g \Rightarrow_{t_L} [\beta^*]_g. \tag{23}$$

The following statement is then obvious:

*if $\alpha \Rightarrow \beta$ is true at g then the collapsed formula has the value 1 of truth at the granule g in the Łukasiewicz logic.*

This gives a necessity condition for verification of implications of rough mereological logics:

*if $\Rightarrow_L ([\alpha^*]_g, [\beta^*]_g) < 1$ then the implication $\alpha \Rightarrow \beta$ is not true at $g$.*

This concerns in particular decision rules:

*for a decision rule $p(v) \Rightarrow q(v)$, the decision rule is true on a granule g if and only if $[p^*]_g \leq [q^*]_g$.*

## 6.1 Rough Set Reasoning: Possibility and Necessity

Possibility and necessity are introduced in rough set theory by means of approximations: the upper and the lower, respectively. A logical rendering of these modalities in rough mereological logics exploits the approximations. We define two modal operators: M (possibility) and L (necessity) by means of their semantics.

To this end, we let

$$\begin{aligned} I_\nu^\vee(g)(M\alpha) \geq r &\text{ if and only if } \nu_L(g, \overline{[\alpha]}, r) \\ I_\nu^\vee(g)(L\alpha) \geq r &\text{ if and only if } \nu_L(g, \underline{[\alpha]}, r). \end{aligned} \tag{24}$$

Then we have the following criteria for necessarily or possibly true formulas.

A formula $\alpha$ is *necessarily true at a granule* $g$ if and only if $g \subseteq \underline{[\alpha]}$; $\alpha$ is *possibly true at* $g$ if and only if $g \subseteq \overline{[\alpha]}$.

This semantics of modal operators $M, L$ can be applied to show that rough set structures carry the semantics of S5 modal logics, i.e., the following relations hold at each granule $g$.

1. $L(\alpha \Rightarrow \beta) \Rightarrow [(L\alpha) \Rightarrow L(\beta)]$.
2. $L\alpha \Rightarrow \alpha$.
3. $L\alpha \Rightarrow LL\alpha$.
4. $M\alpha \Rightarrow LM\alpha$.

We recall proofs given in [28]. We need to show that the meaning of each of formulas 1-4 is $U$. Concerning the formula 1 (modal formula (K)), we have $[L(\alpha \Rightarrow \beta) \Rightarrow [(L\alpha) \Rightarrow L(\beta)]=(U \setminus \underline{(U \setminus [\alpha]) \cup [\beta]}) \cup (U \setminus [\alpha]) \cup [\beta]$. Assume that $u \notin (U \setminus [\alpha]) \cup [\beta]$; thus, (i) $[u]_A \subseteq [\alpha]$ and (ii) $[u]_A \cap [\beta] = \emptyset$. If it were $u \in \underline{(U \setminus [\alpha]) \cup [\beta]}$ then we would have $[u]_A \subseteq (U \setminus [\alpha]) \cup [\beta]$, a contradiction with (i), (ii). Thus, the meaning of (K) is $U$.

For formula 2, modal formula (T), we have $[L\alpha \Rightarrow \alpha]=((U \setminus \underline{[\alpha]}) \cup [\alpha])$; as $\underline{[\alpha]} \subseteq [\alpha]$, it follows that the meaning of (T) is $U$.

In case of formula 3, modal formula (S4), the meaning is $(U \setminus \underline{[\alpha]}) \cup \underline{\underline{[\alpha]}}=(U \setminus \underline{[\alpha]}) \cup \underline{[\alpha]}= U$.

The meaning of formula 4, modal formula (S5), is $(U \setminus \overline{[\alpha]}) \cup \underline{\overline{[\alpha]}}=(U \setminus \overline{[\alpha]}) \cup \overline{[\alpha]}=U$.

It follows that the logic S5 is satisfied within logic induced by $\nu_L$ and more generally in logic induced by any regular rough inclusion $\nu$.

## 7 Reasoning with $\nu_3$

In case of $\nu_3$, one can check on the basis of definitions that $I_\nu^\vee(g)(\neg\alpha) \geq r$ if and only if $I_\nu^\vee(g)(\alpha) \leq 1 - r$; thus the negation functor in rough mereological logic based on $\nu_3$ is the same as the negation functor in the 3–valued Łukasiewicz logic. For implication, the relations between granular rough mereological logic and 3–valued logic of Łukasiewicz follow from truth tables for respective functors of negation and implication.

Table 1 shows truth values for implication in 3–valued logic of Łukasiewicz. We recall that these values obey the implication $x \Rightarrow_L y = min\{1, 1 - x + y\}$. Values of $x$ correspond to rows and values of $y$ correspond to columns in Table 1.

**Table 1.** Truth values for implication in $L_3$

| $\Rightarrow$ | 0 | 1 | $\frac{1}{2}$ |
|---|---|---|---|
| 0 | 1 | 1 | 1 |
| 1 | 0 | 1 | $\frac{1}{2}$ |
| $\frac{1}{2}$ | $\frac{1}{2}$ | 1 | 1 |

Table 2 shows values of implication for rough mereological logic based on $\nu_3$. Values are shown for the extension $I_\nu^\vee(g)(p \Rightarrow q)$ of the implication $p \Rightarrow q$. Rows correspond to $p$, columns correspond to $q$.

We verify values shown in Table 2. First, we consider the case when $I_{\nu_3}^\vee(g)(p) = 0$, i.e., the case when $g \cap [p] = \emptyset$. As $g \subseteq (U \setminus [p]) \cup [q]$ for every value of $[q]$, we have only values of 1 in the first row of Table 2.

Assume now that $I_{\nu_3}^\vee(g)(p) = 1$, i.e., $g \subseteq [p]$. As $g \cap (U \setminus [p]) = \emptyset$, the value of $I_\nu^\vee(g)(p \Rightarrow q)$ depends only on a relation between $g$ and $[q]$. In case $g \cap [q] = \emptyset$, the value in Table 2 is 0, in case $g \subseteq [q]$ the value in Table 2 is 1, and in case $I_{\nu_3}^\vee(g)(q) = \frac{1}{2}$, the value in Table 2 is $\frac{1}{2}$.

**Table 2.** Truth values for implication $p \Rightarrow q$ in logic based on $\nu_3$

| $\Rightarrow$ | $I^{\vee}_{\nu_3}(g)(q) = 0$ | $I^{\vee}_{\nu_3}(g)(q) = 1$ | $I^{\vee}_{\nu_3}(g)(q) = \frac{1}{2}$ |
|---|---|---|---|
| $I^{\vee}_{\nu_3}(g)(p) = 0$ | 1 | 1 | 1 |
| $I^{\vee}_{\nu_3}(g)(p) = 1$ | 0 | 1 | $\frac{1}{2}$ |
| $I^{\vee}_{\nu_3}(g)(p) = \frac{1}{2}$ | $\frac{1}{2}$ | 1 | 1 when $g \cap [\alpha] \subseteq [\beta]$; $\frac{1}{2}$ otherwise |

Finally, we consider the case when $I^{\vee}_{\nu_3}(g)(p) = \frac{1}{2}$, i.e., $g \cap [p] \neq \emptyset \neq g \setminus [p]$. In case $g \cap [q] = \emptyset$, we have $g \cap ((U \setminus [p]) \cup [q]) \neq \emptyset$ and it is not true that $g \subseteq ((U \setminus [p]) \cup [q])$ so the value in table is $\frac{1}{2}$. In case $g \subseteq [q]$, the value in Table is clearly 1. The case when $I^{\vee}_{\nu_3}(g)(q) = \frac{1}{2}$ remains. Clearly, when $g \cap [p] \subseteq [q]$, we have $g \subseteq (U \setminus [p]) \cup [q]$ so the value in Table is 1; otherwise, the value is $\frac{1}{2}$.

Thus, negation in both logic is semantically treated in the same way, whereas treatment of implication differs only in case of implication $p \Rightarrow q$ from the value $\frac{1}{2}$ to $\frac{1}{2}$, when $g \cap [p]$ is not any subset of $[q]$.

It follows from these facts that given a formula $\alpha$ and its collapse $\alpha^*$, we have,

$$I^{\vee}_{\nu_3}(g)(\neg\alpha) = [(\neg\alpha)^*]_{L_3}, I^{\vee}_{\nu_3}(g)(\alpha \Rightarrow \beta) \leq [(\alpha \Rightarrow \beta)^*]_{L_3}. \tag{25}$$

A more exact description of implication in both logics is as follows.

**Proposition 9.** *1. If $I^{\vee}_{nu_3}(g)(\alpha \Rightarrow \beta) = 1$ then $[(\alpha \Rightarrow \beta)^*]_{L_3} = 1$;*

*2. If $I^{\vee}_{nu_3}(g)(\alpha \Rightarrow \beta) = 0$ then $[(\alpha \Rightarrow \beta)^*]_{L_3} = 0$;*

*3. If $I^{\vee}_{nu_3}(g)(\alpha \Rightarrow \beta) = \frac{1}{2}$ then $[(\alpha \Rightarrow \beta)^*]_{L_3} \geq \frac{1}{2}$ and this last value may be 1.*

We offer a simple check–up on Proposition 9. In case 1, we have $g \subseteq ((U \setminus [\alpha]) \cup [\beta])$. For the value of $[(\alpha \Rightarrow \beta)^*]$, consider some subcases. Subcase 1.1: $g \subseteq U \setminus [\alpha]$. Then $[\alpha^*] = 0$ and $[(\alpha \Rightarrow \beta)^*] =[\alpha^*] \Rightarrow [\beta^*]$ is always 1 regardless of a value of $[\beta^*]$. Subcase 1.2: $g \cap [\alpha] \neq \emptyset \neq g \setminus [\alpha]$ so $[\alpha^*] = \frac{1}{2}$. Then $g \cap [\beta] = \emptyset$ is impossible, i.e., $[\beta^*]$is at least $\frac{1}{2}$ and $[(\alpha \Rightarrow \beta)^*]$=1. Subcase 1.3: $g \subseteq [\alpha]$ so $[\alpha^*] = 1$; then $g \subseteq [\beta]$ must hold, i.e., $[\beta^*] = 1$ which means that $[(\alpha \Rightarrow \beta)^*]$=1.

For case 2, we have $g \cap ((U \setminus [\alpha]) \cup [\beta]) = \emptyset$ hence $g \cap [\beta] = \emptyset$ and $g \subseteq [\alpha]$, i.e., $[\alpha^*] = 1, [\beta^*] = 0$ so $[\alpha^*] \Rightarrow [\beta^*]$=0.

In case 3, we have $g \cap ((U \setminus [\alpha]) \cup [\beta]) \neq \emptyset$ and $g \cap [\alpha] \setminus [\beta] \neq \emptyset$. Can $[\alpha^*] \Rightarrow [\beta^*]$ be necessarily 0? This would mean that $[\alpha^*] = 1$ and $[\beta^*] = 0$, i.e., $g \subseteq [\alpha]$ and $g \cap [\beta] = \emptyset$ but then $g \cap ((U \setminus [\alpha]) \cup [\beta]) = \emptyset$, a contradiction. Thus the value $[\alpha^*] \Rightarrow [\beta^*]$ is at least $\frac{1}{2}$. In the subcase: $g \subseteq [\alpha]$, $g \cap [\beta] \neq \emptyset \neq g \setminus [\beta]$, the value of $[\alpha^*] \Rightarrow [\beta^*]$ is $0 \Rightarrow_L \frac{1}{2} = 1$, and the subcase is consistent with case 3.

### 7.1 Dependencies and Decision Rules

It is an important feature of rough set theory that it allows for an elegant formulation of the problem of dependency between two sets of attributes, cf., [19], [20], in terms of indiscernibility relations.

We recall, see sect.1 that for two sets $C, D \subseteq A$ of attributes, one says that $D$ *depends functionally on* $C$ when $IND(C) \subseteq IND(D)$, symbolically denoted $C \mapsto D$. Functional dependence can be represented locally by means of functional dependency rules of the form

$$\phi_C(\{v_a : a \in C\}) \Rightarrow \phi_D(\{w_a : a \in D\}), \tag{26}$$

where $\phi_C(\{v_a : a \in C\})$ is the formula $\bigwedge_{a \in C}(a = v_a)$, and $[\phi_C] \subseteq [\phi_D]$.

Clearly, *if* $\alpha : \phi_C \Rightarrow \phi_D$ *is a functional dependency rule as in* (26), *then* $\alpha$ *is a theorem of logic induced by* $\nu_3$.

Indeed, for each granule $g$, we have $g \cap [\phi_C] \subseteq [\phi_D]$. Let us observe that the converse statement is also true: if a formula $\alpha : \phi_C \Rightarrow \phi_D$ is a theorem of logic induced by $\nu_3$ then this formula is a functional dependency rule in the sense of (26). Indeed, assume that $\alpha$ is not any functional dependency rule, i.e., $[\phi_C] \setminus [\phi_D] \neq \emptyset$. Taking $[\phi_C]$ as the witness granule $g$, we have that $g$ is not any subset of $[\alpha]$, i.e, $I^{\vee}_{\nu_3}(g)(\alpha) \leq \frac{1}{2}$, so $\alpha$ is not true at $g$, a fortiori it is no theorem.

Let us observe that these characterizations are valid for each regular rough inclusion on sets $\nu$.

A more general and also important notion is that of a local proper dependency: a formula $\phi_C \Rightarrow \phi_D$ where $\phi_C(\{v_a : a \in C\})$ is the formula $\bigwedge_{a \in C}(a = v_a)$, similarly for $\phi_D$, is a local proper dependency when $[\phi_C] \cap [\phi_D] \neq \emptyset$.

We will say that a formula $\alpha$ is *acceptable with respect to a collection* $M$ *of worlds* when $I^{\vee}_{\nu_3}(g)(\alpha) \geq \frac{1}{2}$ for each world $g \in M$, i.e, when $\alpha$ is false at no world $g \in M$. Then,

*if a formula* $\alpha : \phi_C \Rightarrow \phi_D$ *is a local proper dependency rule, then it is acceptable with respect to all C-exact worlds.*

Indeed, for a C–exact granule $g$, the case that $I^{\vee}_{\nu_3}(g)(\alpha) = 0$ means that $g \subseteq [\phi_C]$ and $g \cap [\phi_D] = \emptyset$; as $g$ is C–exact and $[\phi_C]$ is a C–indiscernibility class, either $[\phi_C] \subseteq g$ or $[\phi_C] \cap g = \emptyset$. When $[\phi_C] \subseteq g$ then $[\phi_C] = g$ which makes $g \cap [\phi_D] = \emptyset$ impossible. When $[\phi_C] \cap g = \emptyset$, then $g \cap [\phi_D] = \emptyset$ is impossible. In either case, $I^{\vee}_{\nu_3}(g)(\alpha) = 0$ cannot be satisfied with any C–exact granule $g$.

Again, the converse is true: when $\alpha$ is not local proper, i.e., $[\phi_C] \cap [\phi_D] = \emptyset$, then $g = [\phi_C]$ does satisfy $I^{\vee}_{\nu_3}(g)(\alpha) = 0$.

A corollary of the same forms follows for *decision rules* in a given decision system $(U, A, d)$, i.e., dependencies of the form $\phi_C \Rightarrow (d = w)$.

## 8 Remarks and Comments

Remark 1. As the reader may not be well–versed in intricate problems of intensional or extensional utterances, be it known that the idea goes back to Gottlob Frege's "Über Sinn und Bedeutung" [6] in which a distinction between *sense* (Sinn), i.e., intension and *reference* (Bedeutung), i.e., extension, had been made. A functional interpretation of the sense/intension had been proposed by Rudolf Carnap, cf., [4] with intensions as functionals on the set of possible states/worlds and extensions as their values; please consult [17] and [7] for a modern treatment of these topics.

Remark 2. The reader may wish as well to peruse [1] with an adaptation of Gallin's scheme to Montague's intensional logic with a three–valued intensional logic as the upshot.

## 9 Conclusions

Intensional logics $grm_\nu$ capture the basic aspects of reasoning in rough set theory as the construction of such logic is oriented toward logical dependency between premises and conclusions of an implicative rule. Let us mention that an application of those logics in a formalization of the idea of *perception calculus* of Zadeh was proposed in [29].

## References


1. Alves, E.H., Guerzoni, J.A.D.: Extending Montague's system: A three–valued intensional logic. Studia Logica 49, 127–132 (1990)
2. Arnold, V.: On functions of three variables. Amer. math. Soc. transl. 28, 51–54 (1963)
3. van Benthem, J.: A Manual of Intensional Logic. CSLI Stanford University (1988)
4. Carnap, R.: Necessity and Meaning. Chicago Univ. Press, Chicago (1947)
5. Chang, C.C.: Proof of an axiom of Łukasiewicz. Trans. Amer. Math. Soc. 87, 55–56 (1958)
6. Frege, G.: Über Sinn und Bedeutung. Zeitschrift für Philosophie und philosophische Kritik, NF 100, 25–50 (1892)
7. Gallin, D.: Intensional and higher–order modal logic. North Holland, Amsterdam (1975)
8. Hájek, P.: Metamathematics of Fuzzy Logic. Kluwer Academic Publ., Dordrecht (2001)
9. Leśniewski, S.: On the foundations of set theory. Topoi 2, 7–52 (1982)
10. Lin, T.Y.: From rough sets and neighborhood systems to information granulation and computing with words. In: Proceedings of the European Congress on Intelligent Techniques and Soft Computing, pp. 1602–1606. Verlag Mainz, Aachen (1997)
11. Ling, C.-H.: Representation of asociative functions. Publ. Math. Debrecen 12, 189–212 (1965)
12. Łukasiewicz, J.: Die Logischen grundlagen der Wahrscheinlichtkeitsrechnung, Cracow (1913)
13. Łukasiewicz, J.: Farewell lecture by professor Jan Łukasiewicz (Warsaw University Lecture Hall. March 7) (1918)
14. Łukasiewicz, J.: On three–valued logic. Ruh Filozoficzny 5, 170–171 (1920)
15. Łukasiewicz, J., Tarski, A.: Untersuchungen ueber den Aussagenkalkuels. C.R. Soc. Sci. Lettr. Varsovie 23, 39–50 (1930)
16. Meredith, C.A.: The dependence of an axiom of Łukasiewicz. Trans. Amer. Math. Soc. 87, 54 (1958)
17. Montague, R.: Formal Philosophy. In: Thomason, R. (ed.). Yale University Press, New Haven (1974)
18. Mostert, P.S., Shields, A.L.: On the structrure of semigroups on a compact manifold with boundary. Ann. Math. 65, 117–143 (1957)
19. Pawlak, Z.: Rough sets. Intern. J. Comp. Inform. Sci. 11, 341–366 (1982)

20. Pawlak, Z.: Rough Sets: Theoretical Aspects of Reasoning about Data. Kluwer, Dordrecht (1991)
21. Polkowski, L.: Rough Sets. In: Mathematical Foundations. Physica Verlag, Heidelberg (2002)
22. Polkowski, L.: A note on 3–valued rough logic accepting decision rules. Fundamenta Informaticae 61, 37–45 (2004)
23. Polkowski, L.: Formal granular calculi based on rough inclusions (a feature talk). In: Hu, X., Liu, Q., Skowron, A., Lin, T.Y., Yager, R.R., Zhang, B. (eds.) Proceedings IEEE GrC 2005, pp. 57–62. IEEE Press, Piscataway (2005)
24. Polkowski, L.: Rough mereology in analysis of vagueness. In: Wang, G., Li, T., Grzymala-Busse, J.W., Miao, D., Skowron, A., Yao, Y. (eds.) RSKT 2008. LNCS (LNAI), vol. 5009, pp. 197–204. Springer, Heidelberg (2008)
25. Polkowski, L.: A unified approach to granulation of knowledge and granular computing based on rough mereology: a survey. In: Pedrycz, W., Skowron, A., Kreinovich, V. (eds.) Handbook of Granular Computing, pp. 375–400. John Wiley and Sons Ltd., Chichester (2008)
26. Polkowski, L.: Granulation of Knowledge: Similarity Based Approach in Information and Decision Systems. In: Springer Encyclopedia of Complexity and System Sciences (2009)
27. Polkowski, L., Artiemjew, P.: On classifying mappings induced by granular structures. In: Peters, J.F., Skowron, A., Rybiński, H. (eds.) Transactions on Rough Sets IX. LNCS, vol. 5390, pp. 264–286. Springer, Heidelberg (2008)
28. Polkowski, L., Semeniuk-Polkowska, M.: On rough set logics based on similarity relations. Fundamenta Informaticae 64, 379–390 (2005)
29. Polkowski, L., Semeniuk-Polkowska, M.: A formal approach to perception calculus of Zadeh by means of rough mereological logic. In: Actes 11th International Conference on Information Processing and Management in Knowledge–Based Systems IPMU 2006, pp. 1468–1473. Univ. Marie Curie, Paris (2006)
30. Polkowski, L., Skowron, A.: Rough mereology. In: Raś, Z.W., Zemankova, M. (eds.) ISMIS 1994. LNCS (LNAI), vol. 869, pp. 85–94. Springer, Heidelberg (1994)
31. Post, E.: Introduction to a general theory of elementary propositions. Amer. J. Math. 43, 163–185 (1921)
32. Wajsberg, M.: Beitraege zum Metaaussagenkalkuel I. Monat. Math. Phys. 42, 221–242 (1935)
33. Zadeh, L.A.: Fuzzy sets and information granularity. In: Gupta, M., Ragade, R., Yager, R.R. (eds.) Advances in Fuzzy Set Theory and Applications, pp. 3–18. North-Holland, Amsterdam (1979)

# On Topological Dominance-based Rough Set Approach

Salvatore Greco[1], Benedetto Matarazzo[1], and Roman Słowiński[2]

[1] Faculty of Economics, University of Catania, Corso Italia, 55, 95129 Catania, Italy
salgreco@unict.it, matarazz@unict.it

[2] Institute of Computing Science, Poznań University of Technology, 60-965 Poznań, and Systems Research Institute, Polish Academy of Sciences, 01-447 Warsaw, Poland
roman.slowinski@cs.put.poznan.pl

**Abstract.** In this article, we characterize the Dominance-based Rough Set Approach (DRSA) from the point of view of its topological properties. Using the concept of a bitopological space, we extend to DRSA the classical results known for the original rough set approach. Moreover, we introduce a topological approach to ordinal classification. These considerations intend to strengthen theoretical foundations of DRSA, giving a deeper knowledge of its specific characteristics.

## 1 Introduction

Dominance-based Rough Set Approach (DRSA) was introduced as a generalization of the rough set approach for dealing with Multiple Criteria Decision Analysis (MCDA), which requires taking into account a preference order in data, as well as a monotonic relationship between values of condition and decision attributes (for a recent state-of-the-art in MCDA, see [6], and for description of DRSA, see [9,10,13,14,15,16,22,32,33]). The ordering and the monotonic relationship are also important, however, in many other problems of data analysis where preferences are not considered. Even when the ordering seems irrelevant, the presence or the absence of a property has an ordinal interpretation, because if two properties are related, the presence rather than the absence of one property should make more (or less) probable the presence of the other property. This is even more apparent when the presence or the absence of a property is graded or fuzzy, because in this case, the more credible the presence of a property, the more (or less) probable the presence of the other property. Since the presence of properties, possibly fuzzy, is the basis of any granulation, DRSA can be seen as a general framework for granular computing [17,18,21,23].

Classical rough set approach has been thoroughly investigated from a topological approach (see [30] for a survey). In this paper, we propose a topological approach to DRSA. There are two main features of DRSA having important consequences for the topological approach. The first feature is that in DRSA there are two types of rough approximation operators, i.e., upward and downward

J.F. Peters et al. (Eds.): Transactions on Rough Sets XII, LNCS 6190, pp. 21–45, 2010.

rough approximation operators. Upward approximation operators deal with the presence of properties which are true to at least some degree, for example, "at least medium students", "at least good students", and so on. Downward approximation operators deal with the presence of properties which are true to at most some degree, for example, "at most bad students", "at most medium students", and so on. The second feature is the graduality of the presence of properties, such that we have nested sets of objects with an increasing degree of the presence of a property, for example, the set of "at least medium students" includes the set of "at least good students", and so on. Similarly, we have nested sets of objects with a decreasing degree of the presence of a property, for example, the set of "at most bad students" is included in the set of "at most medium students", and so on. The first feature related to upward and downward approximation operators implies that topological models of DRSA have to deal with two topologies of a bitopological space, such that the upward lower approximation is the interior operator with respect to one topology, and the downward lower approximation is the interior with respect to the other topology. The second feature related to the graduality of properties implies approximation of a family of upward nested sets, such as "at least medium students", "at least good students", and so on, or approximation of a family of downward nested sets, such as "at most bad students", "at most medium students" and so on, instead of single sets.

The article is organized as follows. In the next section, we sketch main lines of DRSA in the context of decision making. In section 3, a topological approach to DRSA is presented. In section 4, the topological approach to DRSA applied to ordinal classifications is presented. Section 5 groups conclusions.

## 2 Dominance-based Rough Set Approach

This section presents the main concepts of the Dominance-based Rough Set Approach (DRSA) (for a more complete presentation see, e.g., [10,13,16,32,33].

Information about objects is represented in the form of an information table. The rows of the table are labelled by objects, whereas columns are labelled by attributes and entries of the table are attribute-values. Formally, an information table (system) is the 4-tuple $\boldsymbol{S} =< U, Q, V, \phi >$, where $U$ is a finite set of objects, $Q$ is a finite set of attributes, $V = \bigcup_{q \in Q} V_q$ and $V_q$ is the value set of the attribute $q$, and $\phi : U \times Q \to V_q$ is a total function such that $\phi(x, q) \in V_q$ for every $q \in Q$, $x \in U$, called an information function [29]. The set $Q$ is, in general, divided into set $C$ of condition attributes and set $D$ of decision attributes.

Condition attributes with value sets ordered according to decreasing or increasing preference of a decision maker are called *criteria*. For criterion $q \in Q$, $\succeq_q$ is a *weak preference* relation on $U$ such that $x \succeq_q y$ means "$x$ is at least as good as $y$ with respect to criterion $q$". It is supposed that $\succeq_q$ is a complete preorder, i.e., a strongly complete and transitive binary relation, defined on $U$ on the basis of evaluations $\phi(\cdot,q)$. Without loss of generality, the preference is supposed to increase with the value of $\phi(\cdot,q)$ for every criterion $q \in C$, such that for all $x, y \in U$, $x \succeq_q y$ if and only if $\phi(x, q) \geq \phi(y, q)$.

Furthermore, it is supposed that the set of decision attributes $D$ is a singleton $\{d\}$. Values of decision attribute $d$ make a partition of $U$ into a finite number of decision classes, $\boldsymbol{Cl}$=$\{Cl_t, t = 1, ..., n\}$, such that each $x \in U$ belongs to one and only one class $Cl_t \in \boldsymbol{Cl}$. It is supposed that the classes are preference-ordered, i.e., for all $r$,$s \in \{1, ..., n\}$, such that $r > s$, the objects from $Cl_r$ are preferred to the objects from $Cl_s$. More formally, if $\succeq$ is a *comprehensive weak preference relation* on $U$, i.e., if for all $x$,$y \in U$, $x \succeq y$ means "$x$ is comprehensively at least as good as $y$", it is supposed: $[x \in Cl_r, y \in Cl_s, r>s] \Rightarrow [x \succeq y$ and *not* $y \succeq x]$. The above assumptions are typical for consideration of *ordinal classification problems* (also called *multiple criteria sorting problems*).

The sets to be approximated are called *upward union* and *downward union* of classes, respectively:

$$Cl_t^{\geq} = \bigcup_{s \geq t} Cl_s, \quad Cl_t^{\leq} = \bigcup_{s \leq t} Cl_s, \quad t = 1, ..., n.$$

The statement $x \in Cl_t^{\geq}$ means "$x$ belongs to *at least* class $Cl_t$", while $x \in Cl_t^{\leq}$ means "$x$ belongs to *at most* class $Cl_t$". Let us remark that $Cl_1^{\geq} = Cl_n^{\leq} = U$, $Cl_n^{\geq}$=$Cl_n$ and $Cl_1^{\leq}$=$Cl_1$. Furthermore, for $t = 2, \ldots, n$,

$$Cl_t^{\geq} = U - Cl_{t-1}^{\leq} \text{ and } Cl_{t-1}^{\leq} = U - Cl_t^{\geq}.$$

The key idea of the rough set approach is representation (approximation) of knowledge generated by decision attributes, using "*granules of knowledge*" generated by condition attributes.

In DRSA, where condition attributes are criteria and decision classes are preference ordered, the represented knowledge is a collection of upward and downward unions of classes and the "granules of knowledge" are sets of objects defined using a dominance relation.

*$x$ dominates $y$* with respect to $P \subseteq C$ (shortly, *$x$ P-dominates $y$*), denoted by $xD_Py$, if for every criterion $q \in P$, $\phi(x, q) \geq \phi(y, q)$. The relation of $P$-dominance is reflexive and transitive, i.e., it is a partial preorder.

Given a set of criteria $P \subseteq C$ and $x \in U$, the "granules of knowledge" used for approximation in DRSA are:

- a set of objects dominating $x$, called *P-dominating set*, $D_P^+(x)$=$\{y \in U\colon yD_Px\}$,
- a set of objects dominated by $x$, called *P-dominated set*, $D_P^-(x)$=$\{y \in U\colon xD_Py\}$.

Remark that the "granules of knowledge" defined above have the form of upward (positive) and downward (negative) *dominance cones* in the evaluation space.

Let us recall that the *dominance principle* (or Pareto principle) requires that an object $x$ dominating object $y$ on all considered criteria (i.e., $x$ having evaluations at least as good as $y$ on all considered criteria) should also dominate $y$ on the decision (i.e., $x$ should be assigned to at least as good decision class as $y$). This principle is the only objective principle that is widely agreed upon in the multiple criteria comparisons of objects.

Given $P \subseteq C$, the inclusion of an object $x \in U$ to the upward union of classes $Cl_t^{\geq}$, $t = 2, \ldots, n$, is *inconsistent with the dominance principle* if one of the following conditions holds:

- $x$ belongs to class $Cl_t$ or better but it is $P$-dominated by an object $y$ belonging to a class worse than $Cl_t$, i.e., $x \in Cl_t^{\geq}$ but $D_P^+(x) \cap Cl_{t-1}^{\leq} \neq \emptyset$,
- $x$ belongs to a worse class than $Cl_t$ but it $P$-dominates an object $y$ belonging to class $Cl_t$ or better, i.e., $x \notin Cl_t^{\geq}$ but $D_P^-(x) \cap Cl_t^{\geq} \neq \emptyset$.

If, given a set of criteria $P \subseteq C$, the inclusion of $x \in U$ to $Cl_t^{\geq}$, where $t = 2, \ldots, n$, is inconsistent with the dominance principle, then $x$ belongs to $Cl_t^{\geq}$ *with some ambiguity*. Thus, $x$ belongs to $Cl_t^{\geq}$ *without any ambiguity* with respect to $P \subseteq C$, if $x \in Cl_t^{\geq}$ and there is no inconsistency with the dominance principle. This means that all objects $P$-dominating $x$ belong to $Cl_t^{\geq}$, i.e., $D_P^+(x) \subseteq Cl_t^{\geq}$.

Furthermore, $x$ *possibly belongs to* $Cl_t^{\geq}$ with respect to $P \subseteq C$ if one of the following conditions holds:

- according to decision attribute $d$, $x$ belongs to $Cl_t^{\geq}$,
- according to decision attribute $d$, $x$ does not belong to $Cl_t^{\geq}$, but it is inconsistent in the sense of the dominance principle with an object $y$ belonging to $Cl_t^{\geq}$.

In terms of ambiguity, $x$ possibly belongs to $Cl_t^{\geq}$ with respect to $P \subseteq C$, if $x$ belongs to $Cl_t^{\geq}$ with or without any ambiguity. Due to the reflexivity of the dominance relation $D_P$, the above conditions can be summarized as follows: $x$ *possibly belongs* to class $Cl_t$ or better, with respect to $P \subseteq C$, if among the objects $P$-dominated by $x$ there is an object $y$ belonging to class $Cl_t$ or better, i.e., $D_P^-(x) \cap Cl_t^{\geq} \neq \emptyset$.

The *P-lower approximation* of $Cl_t^{\geq}$, denoted by $\underline{P}(Cl_t^{\geq})$, and the *P-upper approximation* of $Cl_t^{\geq}$, denoted by $\overline{P}(Cl_t^{\geq})$, are defined as follows ($t = 1, \ldots, n$):

$$\underline{P}(Cl_t^{\geq}) = \{x \in U : D_P^+(x) \subseteq Cl_t^{\geq}\},$$
$$\overline{P}(Cl_t^{\geq}) = \{x \in U : D_P^-(x) \cap Cl_t^{\geq} \neq \emptyset\}.$$

Analogously, one can define the *P-lower approximation* and the *P-upper approximation* of $Cl_t^{\leq}$ as follows ($t = 1, \ldots, n$):

$$\underline{P}(Cl_t^{\leq}) = \{x \in U : D_P^-(x) \subseteq Cl_t^{\leq}\},$$
$$\overline{P}(Cl_t^{\leq}) = \{x \in U : D_P^+(x) \cap Cl_t^{\leq} \neq \emptyset\}.$$

The $P$-lower and $P$-upper approximations so defined satisfy the following *inclusion property* for each $t \in \{1, \ldots, n\}$ and for all $P \subseteq C$:

$$\underline{P}(Cl_t^{\geq}) \subseteq Cl_t^{\geq} \subseteq \overline{P}(Cl_t^{\geq}), \quad \underline{P}(Cl_t^{\leq}) \subseteq Cl_t^{\leq} \subseteq \overline{P}(Cl_t^{\leq}).$$

The $P$-lower and $P$-upper approximations of $Cl_t^{\geq}$ and $Cl_t^{\leq}$ have an important *complementarity property*, according to which,

$$\underline{P}(Cl_t^{\geq}) = U - \overline{P}(Cl_{t-1}^{\leq}) \text{ and } \overline{P}(Cl_t^{\geq}) = U - \underline{P}(Cl_{t-1}^{\leq}),\ t = 2, \ldots, n,$$

$$\underline{P}(Cl_t^{\leq}) = U - \overline{P}(Cl_{t+1}^{\geq}) \text{ and } \overline{P}(Cl_t^{\leq}) = U - \underline{P}(Cl_{t+1}^{\geq}),\ t = 1, \ldots, n-1.$$

The *P-boundaries* of $Cl_t^{\geq}$ and $Cl_t^{\leq}$, denoted by $Bn_P(Cl_t^{\geq})$ and $Bn_P(Cl_t^{\leq})$ respectively, are defined as follows ($t = 1, \ldots, n$):

$$Bn_P(Cl_t^{\geq}) = \overline{P}(Cl_t^{\geq}) - \underline{P}(Cl_t^{\geq}), \quad Bn_P(Cl_t^{\leq}) = \overline{P}(Cl_t^{\leq}) - \underline{P}(Cl_t^{\leq}).$$

Due to complementarity property, $Bn_P(Cl_t^{\geq}) = Bn_P(Cl_{t-1}^{\leq})$, for $t = 2, \ldots, n$.

For every $P \subseteq C$, the *quality of approximation* of the ordinal classification $\boldsymbol{Cl}$ by a set of criteria $P$ is defined as the ratio of the number of objects $P$-consistent with the dominance principle and the number of all the objects in $U$. Since the $P$-consistent objects are those which do not belong to any $P$-boundary $Bn_P(Cl_t^{\geq})$, $t = 2, \ldots, n$, or $Bn_P(Cl_t^{\leq})$, $t = 1, \ldots, n-1$, the quality of approximation of the ordinal classification $\boldsymbol{Cl}$ by a set of criteria $P$, can be written as

$$\gamma_P(\boldsymbol{Cl}) = \frac{\left| U - \left( \bigcup_{t=2,\ldots,n} Bn_P(Cl_t^{\geq}) \right) \right|}{|U|} = \frac{\left| U - \left( \bigcup_{t=1,\ldots,n-1} Bn_P(Cl_t^{\leq}) \right) \right|}{|U|}.$$

$\gamma_P(\boldsymbol{Cl})$ can be seen as a degree of consistency of the objects from $U$, where $P$ is the set of criteria and $\boldsymbol{Cl}$ is the considered ordinal classification.

Each minimal (with respect to inclusion) subset $P \subseteq C$ such that $\gamma_P(\boldsymbol{Cl}) = \gamma_C(\boldsymbol{Cl})$ is called a *reduct* of $\boldsymbol{Cl}$, and is denoted by $RED_{\boldsymbol{Cl}}$. Let us remark that, for a given set $U$, one can have more than one reduct. The intersection of all reducts is called the *core*, and is denoted by $CORE_{\boldsymbol{Cl}}$. Criteria in $CORE_{\boldsymbol{Cl}}$ cannot be removed from consideration without deteriorating the quality of approximation. This means that, in set $C$, there are three categories of criteria:

- *indispensable* criteria included in the core,
- *exchangeable* criteria included in some reducts, but not in the core,
- *redundant* criteria, neither indispensable nor exchangeable, and thus not included in any reduct.

The dominance-based rough approximations of upward and downward unions of classes can serve to induce "*if..., then...*" decision rules. It is meaningful to consider the following five types of decision rules:

1) Certain $D_{\geq}$-decision rules:
*if* $x_{q1} \succeq_{q1} r_{q1}$ *and* $x_{q2} \succeq_{q2} r_{q2}$ *and* $\ldots$ $x_{qp} \succeq_{qp} r_{qp}$,
*then* $x$ *certainly belongs to* $Cl_t^{\geq}$,
where,
for each $w_q$,$z_q \in X_q$, "$w_q \succeq_q z_q$" means "$w_q$ is at least as good as $z_q$", and $P = \{q_1, \ldots, q_p\} \subseteq C$, $(r_{q1}, \ldots, r_{qp}) \in V_{q1} \times \ldots \times V_{qp}$, $t \in \{2, \ldots, n\}$.

2) Possible $D_{\geq}$-decision rules:
*if* $x_{q1} \succeq_{q1} r_{q1}$ *and* $x_{q2} \succeq_{q2} r_{q2}$ *and* ... $x_{qp} \succeq_{qp} r_{qp}$,
*then* $x$ *possibly belongs to* $Cl_t^{\geq}$,
where
$P = \{q_1, \ldots, q_p\} \subseteq C$, $(r_{q1}, \ldots, r_{qp}) \in V_{q1} \times \ldots \times V_{qp}$, $t \in \{2, \ldots, n\}$.

3) Certain $D_{\leq}$-decision rules:
*if* $x_{q1} \preceq_{q1} r_{q1}$ *and* $x_{q2} \preceq_{q2} r_{q2}$ *and* ... $x_{qp} \preceq_{qp} r_{qp}$,
*then* $x$ *certainly belongs to* $Cl_t^{\leq}$,
where,
for each $w_q, z_q \in X_q$, "$w_q \preceq_q z_q$" means "$w_q$ is at most as good as $z_q$", and
$P = \{q_1, \ldots, q_p\} \subseteq C$, $(r_{q1}, \ldots, r_{qp}) \in V_{q1} \times \ldots \times V_{qp}$, $t \in \{1, \ldots, n-1\}$.

4) Possible $D_{\leq}$-decision rules:
*if* $x_{q1} \preceq_{q1} r_{q1}$ *and* $x_{q2} \preceq_{q2} r_{q2}$ *and* ... $x_{qp} \preceq_{qp} r_{qp}$,
*then* $x$ *possibly belongs to* $Cl_t^{\leq}$,
where
$P = \{q_1, \ldots, q_p\} \subseteq C$, $(r_{q1}, \ldots, r_{qp}) \in V_{q1} \times \ldots \times V_{qp}$, $t \in \{1, \ldots, n-1\}$.

5) Approximate $D_{\geq\leq}$-decision rules:
*if* $x_{q1} \succeq_{q1} r_{q1}$ *and* ... $x_{qk} \succeq_{qk} r_{qk}$ *and*
$x_{q(k+1)} \preceq_{q(k+1)} r_{q(k+1)}$ *and* ... $x_{qp} \preceq_{qp} r_{qp}$, *then* $x \in Cl_s^{\geq} \cap Cl_t^{\leq}$,
where
$O' = \{q_1, \ldots, q_k\} \subseteq C$, $O'' = \{q_{k+1}, \ldots, q_p\} \subseteq C, P = O' \cup O''$,
$O'$ and $O''$ not necessarily disjoint, $(r_{q1}, \ldots, r_{qp}) \in V_{q1} \times \ldots \times V_{qp}$,
and $s, t \in \{1, \ldots, n\}$, such that $s < t$.

The rules of type 1) and 3) represent certain knowledge extracted from the decision table, while the rules of type 2) and 4) represent possible knowledge. Rules of type 5) represent doubtful knowledge.

Let us observe that, in general, the DRSA approximations can be thought of as operators on the powerset of the universe $2^U$. In this abstract context, given a partial preorder $\succeq$ on $U$ (corresponding to the dominance relation considered above) we can define, for any $x \in U$,

$$D^+(x) = \{y \in U : y \succeq x\}, \quad D^-(x) = \{y \in U : x \succeq y\}.$$

For each set $X \subseteq U$, we can then define its *upward lower approximation* $\underline{X}^{>}$ and its *upward upper approximation* $\overline{X}^{>}$ as:

$$\underline{X}^{>} = \left\{x \in U : D^+(x) \subseteq X\right\}, \quad \overline{X}^{>} = \left\{x \in U : D^-(x) \cap X \neq \emptyset\right\}.$$

Analogously, for each $X \subseteq U$, we can define its *downward lower approximation* $\underline{X}^{<}$ and its *downward upper approximation* $\overline{X}^{<}$ as:

$$\underline{X}^{<} = \left\{x \in U : D^-(x) \subseteq X\right\}, \quad \overline{X}^{<} = \left\{x \in U : D^+(x) \cap X \neq \emptyset\right\}.$$

The following properties, corresponding to properties of classical rough sets, hold for DRSA rough approximations: for any $X, Y \subseteq U$,

1. $\underline{X}^{>} \subseteq X \subseteq \overline{X}^{>}, \quad \underline{X}^{<} \subseteq X \subseteq \overline{X}^{<}$,
2. $\underline{\emptyset}^{>} = \overline{\emptyset}^{>} = \emptyset, \quad \underline{\emptyset}^{<} = \overline{\emptyset}^{<} = \emptyset$,
3. $\overline{X \cup Y}^{>} = \overline{X}^{>} \cup \overline{Y}^{>}, \quad \overline{X \cup Y}^{<} = \overline{X}^{<} \cup \overline{Y}^{<}$,
4. $\underline{X \cap Y}^{>} = \underline{X}^{>} \cap \underline{Y}^{>}, \quad \underline{X \cap Y}^{<} = \underline{X}^{<} \cap \underline{Y}^{<}$,
5. $X \subseteq Y \Rightarrow \underline{X}^{>} \subseteq \underline{Y}^{>}, \quad X \subseteq Y \Rightarrow \underline{X}^{<} \subseteq \underline{Y}^{<}$,
6. $X \subseteq Y \Rightarrow \overline{X}^{>} \subseteq \overline{Y}^{>}, \quad X \subseteq Y \Rightarrow \overline{X}^{<} \subseteq \overline{Y}^{<}$,
7. $\underline{X \cup Y}^{>} \supseteq \underline{X}^{>} \cup \underline{Y}^{>}, \quad \underline{X \cup Y}^{<} \supseteq \underline{X}^{<} \cup \underline{Y}^{<}$,
8. $\overline{X \cap Y}^{>} \subseteq \overline{X}^{>} \cap \overline{Y}^{>}, \quad \overline{X \cap Y}^{<} \subseteq \overline{X}^{<} \cap \overline{Y}^{<}$,
9. $\underline{U - X}^{>} = U - \overline{X}^{<}, \quad \underline{U - X}^{<} = U - \overline{X}^{>}$,
10. $\overline{U - X}^{>} = U - \underline{X}^{<}, \quad \overline{U - X}^{<} = U - \underline{X}^{>}$,
11. $\underline{(\underline{X}^{>})}^{>} = \overline{(\underline{X}^{>})}^{>} = \underline{X}^{>}, \quad \underline{(\underline{X}^{<})}^{<} = \overline{(\underline{X}^{<})}^{<} = \underline{X}^{<}$,
12. $\overline{(\overline{X}^{>})}^{>} = \underline{(\overline{X}^{>})}^{>} = \overline{X}^{>}, \quad \overline{(\overline{X}^{<})}^{<} = \underline{(\overline{X}^{<})}^{<} = \overline{X}^{<}$.

## 3 Topological Dominance-based Rough Set Approach

We start this section by recalling some basic notions relative to the topological interpretation of the classical rough set approach (for a survey see Chapter 7 in [30]). By a topological space $(X, \tau)$ we mean a set $X$ along with a topology $\tau$, defined as a family $\tau \subseteq 2^X$ of sets in $X$ (let us remember that $2^X$ is the power set of $X$, i.e., the family of all subsets of $X$) which satisfies the following conditions:

1) $\emptyset \in \tau$ and $X \in \tau$,
2) for any finite sub-family $\tau' \subseteq \tau$, we have $\bigcap \tau' \in \tau$,
3) for any sub-family $\tau' \subseteq \tau$, we have $\bigcup \tau' \in \tau$.

Thus, $\tau$ is closed with respect to operations of finite intersection and arbitrary unions. The elements of $\tau$ are called *open sets*. For the sake of simplicity, in the following we assume that $X$ is a finite set.

For any $A \subseteq X$, the interior is defined as:

$$I(A) = \bigcup \{\alpha \in \tau : \alpha \subseteq A\}.$$

Operator $I$ satisfies the following properties:

(i1) $I(A \cap B) = I(A) \cap I(B)$,
(i2) $I(A) \subseteq A$,
(i3) $I(I(A)) = I(A)$,
(i4) $I(X) = X$.

By duality, we may define the closure operator $C$ as follows: for all $A \subseteq X$,

$$C(A) = X - I(X - A).$$

From properties (i1)-(i4) of the interior operator, it follows that the closure operator satisfies the following properties:

(c1) $C(A \cup B) = C(A) \cup C(B)$,
(c2) $C(A) \supseteq A$,
(c3) $C(C(A)) = C(A)$,
(c4) $C(\emptyset) = \emptyset$.

Each $A \subseteq X$, such that $C(A) = A$, is called a *closed set.* Let us denote by $\mathcal{C}(X)$ the family of all the closed sets. By duality with respect to open sets, this family is closed with respect to finite union and arbitrary intersection. Both operators $I$ and $C$ are monotone, i.e., for all $A \subseteq B \subseteq X$,

$$I(A) \subseteq I(B) \text{ and } C(A) \subseteq C(B).$$

A family **B** of open sets in the topological space $(X, \tau)$ is said to be a base if every open set of $X$ is the union of some sets belonging to **B**.

A topological space satisfies the *clopen sets property* (also called zero-dimensional property) if the following property is satisfied: for any $A \subseteq X$,

$$C(I(A)) = I(A) \text{ and } I(C(A)) = C(A).$$

**Theorem 1.** [34] If $(X, I)$ is a topological space having clopen sets property, then there exists an equivalence relation $R$ in $X$, such that the family composed of the quotient set $X/R$ (the set of equivalence classes of $R$) and the empty set $\emptyset$ is the base of the topology. □

Since the indiscernibility relation of rough sets is an equivalence relation, this means that lower and upper approximations of classical rough sets can be interpreted as interior and closure operators of a topological space $(X, I)$, where $X$ is the universe and $I$ an interior operator induced by the equivalence classes of the indiscernibility relation. Indeed, lower and upper approximations of classical rough sets correspond to Vietoris operators in a topological space [25]. Now, a question arises quite naturally: is it possible to state a similar result with respect to DRSA? The answer is positive, and it is based on the concept of bitopological space [24].

By a bitopological space $(X, \tau_1, \tau_2)$ we mean a set $X$ along with two topologies $\tau_1$ and $\tau_2$. On the basis of $\tau_1$ and $\tau_2$ we can define two interior operators, $I_1$ and $I_2$, and two closure operators, $C_1$ and $C_2$.

A bitopological space can also be defined as a triple $(X, I_1, I_2)$, where $X$ is a set and $I_1$ and $I_2$ are operators satisfying properties (i1)-(i4). By duality, we can define the closure operator $C_1$ and $C_2$ as follows: for all $A \subseteq X$,

$$C_1(A) = X - I_2(X - A),\ C_2(A) = X - I_1(X - A).$$

A bitopological space satisfies the *bi-clopen sets property* (also called pairwise zero-dimensional property) if the following property is satisfied: for any $A \subseteq X$,

$$C_1(I_1(A)) = I_1(A),\ C_2(I_2(A)) = I_2(A),$$

$$I_1(C_1(A)) = C_1(A),\ I_2(C_2(A)) = C_2(A).$$

**Theorem 2.** [1] If $(X, I_1, I_2)$ is a bitopological space having bi-clopen sets property, then there exists a partial preorder $\succeq$ in $X$, such that the set

$$\{\{y \in X : y \succeq x\} : x \in X\} \cup \{\emptyset\}$$

is a base for $\tau_1$ and the set

$$\{\{y \in X : x \succeq y\} : x \in X\} \cup \{\emptyset\}$$

is a base for $\tau_2$. □

Remembering that the dominance relation is a partial preorder, Theorem 2 can be considered as an extension to DRSA of Theorem 1. This means that lower and upper approximations of DRSA can be interpreted as interior and closure operators of a bitopological space $(X, I_1, I_2)$, where $X$ is the universe and $I_1$ and $I_2$ are interior operators induced by the dominance relation.

More precisely, given a bitopological space $(X, I_1, I_2)$ and partial preorder $\succeq$ corresponding to it on the basis of Theorem 2, for any $A \subseteq X$,

$$\underline{A}^{>} = I_1(A), \quad \underline{A}^{<} = I_2(A)$$

and, consequently,

$$\overline{A}^{>} = C_1(A), \quad \overline{A}^{<} = C_2(A).$$

Theorem 2 can also be expressed in terms of Priestley topological space [31]. Let $(X, \succeq)$ be a partially ordered set (poset). We recall that $A \subseteq X$ is an *upset* if $x \in A$ and $y \succeq x$ imply $y \in A$, and that $A$ is a *downset* if $x \in A$ and $x \succeq y$ imply $y \in A$. Let $Up(X)$ denote the set of upsets and $Do(X)$ denote the set of downsets of $(X, \succeq)$. Let $(X, \tau, \succeq)$ be an ordered topological space. We denote by $OpUp(X)$ the family of open upsets, by $ClUp(X)$ the family of closed upsets, and by $CpUp(X)$ the family of clopen upsets of $(X, \tau, \succeq)$. Similarly, let $OpDo(X)$ denote the family of open downsets, $ClDo(X)$ denote the family of closed downsets, and $CpDo(X)$ denote the family of clopen downsets of $(X, \tau, \succeq)$. The next definition is well-known.

An ordered topological space $(X, \tau, \succeq)$ is a *Priestley space* [31] if, whenever *not* $y \succeq x$, there exists a clopen upset $A$ such that $x \in A$ and $y \notin A$.

In the original definition of Priestley space it is supposed that $(X, \tau)$ is compact, but in our case this property is trivially verified because $X$ is a finite set.

Assuming that topology $\tau$ has the base

$$\{\{y \in X : y \succeq x\} : x \in X\} \cup \{\{y \in X : x \succeq y\} : x \in X\} \cup \{\emptyset\},$$

where $\succeq$ corresponds to the bitopological space $(X, I_1, I_2)$ in the sense of Theorem 2, the family of the open upsets of Priestley space $(X, \tau, \succeq)$, $OpUp(X)$, is equivalent to the family of open sets of $(X, \tau_1)$, $\mathcal{O}_1(X)$, as well as, the family of the open downsets of Priestley space $(X, \tau, \succeq)$, $OpDo(X)$, is equivalent to the family of open sets of $(X, \tau_2)$, $\mathcal{O}_2(X)$, where $\tau_1$ and $\tau_2$ are the two topologies of $(X, I_1, I_2)$. Remembering that for clopen sets property, the family of open sets of

$(X, \tau_1)$ is equivalent to the family of closed sets of $(X, \tau_2)$, $\mathcal{C}_2(X)$, as well as, the family of open sets of $(X, \tau_2)$ is equivalent to the family of closed sets of $(X, \tau_1)$, $\mathcal{C}_1(X)$, we can write

$$OpUp(X) = \mathcal{O}_1(X) = \mathcal{C}_2(X),$$

$$OpDo(X) = \mathcal{O}_2(X) = \mathcal{C}_1(X).$$

**Theorem 3.** A bitopological space satisfies the bi-clopen sets property iff there exists a partial preorder $\succeq$ in $X$, such that, for any $A \subseteq X$,

$$I_1(A) = \bigcup \{\{y \in X : y \succeq x\} : \{y \in X : y \succeq x\} \subseteq A\},$$

$$I_2(A) = \bigcup \{\{y \in X : x \succeq y\} : \{y \in X : x \succeq y\} \subseteq A\},$$

$$C_1(A) = \bigcup \{\{y \in X : y \succeq x\} : \{y \in X : x \succeq y\} \cap A \neq \emptyset\},$$

$$C_2(A) = \bigcup \{\{y \in X : x \succeq y\} : \{y \in X : y \succeq x\} \cap A \neq \emptyset\}.$$

**Proof.** Observe that $\tau_1 = \{I_1(A) : A \subseteq X\}$ and $\tau_2 = \{I_2(A) : A \subseteq X\}$. For Theorem 2, there exists $\succeq$ in $X$, such that the set

$$\{\{y \in X : y \succeq x\} : x \in X\} \cup \{\emptyset\}$$

is a base for $\tau_1$ and the set

$$\{\{y \in X : x \succeq y\} : x \in X\} \cup \{\emptyset\}$$

is a base for $\tau_2$. Therefore, each open set $I_1(A)$ with respect to $\tau_1$ is the union of subsets from

$$\{\{y \in X : y \succeq x\} : x \in X\} \cup \{\emptyset\},$$

i.e.,

$$I_1(A) = \bigcup \{\{y \in X : y \succeq x\} : \{y \in X : y \succeq x\} \subseteq A\}.$$

Analogously, each open set $I_2(A)$ with respect to $\tau_2$ is the union of subsets from

$$\{\{y \in X : x \succeq y\} : x \in X\} \cup \{\emptyset\},$$

i.e.,

$$I_2(A) = \bigcup \{\{y \in X : x \succeq y\} : \{y \in X : x \succeq y\} \subseteq A\}.$$

Let us consider $x \in X$ such that

$$\{y \in X : x \succeq y\} \cap A \neq \emptyset. \qquad (i)$$

For contradiction, let us suppose that $\{y \in X : y \succeq x\}$ is not included in $C_1(A)$. This would mean that there exists $z \in X$ such that

$$\{y \in X : z \succeq y\} \subseteq X - A \qquad (ii)$$

and

$$\{y \in X : z \succeq y\} \cap \{y \in X : y \succeq x\} \neq \emptyset. \qquad (iii)$$

From $(iii)$, by the transitivity of $\succeq$, we get $z \succeq x$ and

$$\{y \in X : x \succeq y\} \subseteq \{y \in X : z \succeq y\}. \qquad (iv)$$

From $(ii)$ and $(iv)$ we get

$$\{y \in X : x \succeq y\} \subseteq X - A$$

i.e.

$$\{y \in X : x \succeq y\} \cap A = \emptyset,$$

which contradicts $(i)$. Therefore,

$$C_1(A) \supseteq \bigcup \{\{y \in X : y \succeq x\} : \{y \in X : x \succeq y\} \cap A \neq \emptyset\}. \qquad (v)$$

Let us consider $x \in X$ such that

$$\{y \in X : x \succeq y\} \cap A = \emptyset. \qquad (vi)$$

For contradiction, let us suppose that $\{y \in X : y \succeq x\}$ is included in $C_1(A)$. By the reflexivity of $\succeq$, we get $x \in \{y \in X : y \succeq x\}$ and thus $x \in C_1(A)$. Observe that $(vi)$ means

$$\{y \in X : x \succeq y\} \subseteq X - A,$$

i.e.,

$$\{y \in X : x \succeq y\} \subseteq I_2(X - A). \qquad (vii)$$

By the reflexivity of $\succeq$, we get $x \in \{y \in X : x \succeq y\}$, and, therefore, by $(vii)$ $x \in I_2(X - A)$. But since $C_1(A) = X - I_2(X - A)$, $x$ cannot belong to $C_1(A)$ and $I_2(X - A)$. Therefore $\{y \in X : y \succeq x\}$ cannot be included in $C_1(A)$, such that

$$C_1(A) \subseteq \bigcup \{\{y \in X : y \succeq x\} : \{y \in X : x \succeq y\} \cap A \neq \emptyset\}. \qquad (viii)$$

From $(v)$ and $(viii)$ we get

$$C_2(A) = \bigcup \{\{y \in X : x \succeq y\} : \{y \in X : y \succeq x\} \cap A \neq \emptyset\}. \qquad \square$$

Given a bitopological space $(X, \tau_1, \tau_2)$ and a set $A \subseteq X$, we say that

- $A$ is $\tau_2 - \tau_1$ *regularly open* set if $A = I_1(C_1(A))$,
- $A$ is $\tau_1 - \tau_2$ *regularly open* set if $A = I_2(C_2(A))$,
- $A$ is $\tau_1 - \tau_2$ *regularly closed* set if $A = C_1(I_1(A))$,
- $A$ is $\tau_2 - \tau_1$ *regularly closed* set if $A = C_2(I_2(A))$.

These sets are related to each other by duality in the sense that $A$ is $\tau_2 - \tau_1$ regularly open if and only if $X - A$ is $\tau_2 - \tau_1$ regularly closed, and $A$ is $\tau_1 - \tau_2$ regularly open if and only if $X - A$ is $\tau_1 - \tau_2$ regularly closed. In fact, $A$ is $\tau_2 - \tau_1$ regularly open if $A = I_1(C_1(A))$, such that $X - A = X - I_1(C_1(A))$, from which, by the definition of $I_1$ and $C_1$, we get

$$X - A = X - (X - C_2(X - (X - I_2(X - A)))) = C_2(I_2(X - A)),$$

which means that $X - A$ is $\tau_2 - \tau_1$ regularly closed. Analogously, $A$ is $\tau_1 - \tau_2$ regularly open if $A = I_2(C_2(A))$, such that $X - A = X - I_2(C_2(A))$, from which, by the definition of $I_2$ and $C_2$, we get

$$X - A = X - (X - C_1(X - (X - I_1(X - A)))) = C_1(I_1(X - A)),$$

which means that $X - A$ is $\tau_1 - \tau_2$ regularly closed. Let us introduce now two new auxiliary operations: for any $A \subseteq X$, let $A^{\perp_1} = X - C_1(A)$ and $A^{\perp_2} = X - C_2(A)$. $A^{\perp_1}$ and $A^{\perp_2}$ satisfy the properties shown by the following Theorem 4.

**Theorem 4**

1) $A^{\perp_1} = I_2(X - A)$ and $A^{\perp_2} = I_1(X - A)$.
2) $A$ is $\tau_2 - \tau_1$ *regularly open* if and only if $A = A^{\perp_1\perp_2}$ and $A$ is $\tau_1 - \tau_2$ *regularly open* if and only if $A = A^{\perp_2\perp_1}$.
3) if $A \subseteq B \subseteq X$, then $B^{\perp_1} \subseteq A^{\perp_1}$ and $B^{\perp_2} \subseteq A^{\perp_2}$,
4) if $A$ is an open set with respect to $\tau_1$, then $A \subseteq A^{\perp_1\perp_2}$, and if $A$ is an open set with respect to $\tau_2$, then $A \subseteq A^{\perp_2\perp_1}$,
5) if $A$ is an open set with respect to $\tau_1$, then $A^{\perp_1} = A^{\perp_1\perp_2\perp_1}$, hence $A^{\perp_1\perp_2} = A^{\perp_1\perp_2\perp_1\perp_2}$, and if $A$ is an open set with respect to $\tau_2$, then $A^{\perp_2} = A^{\perp_2\perp_1\perp_2}$, hence $A^{\perp_2\perp_1} = A^{\perp_2\perp_1\perp_2\perp_1}$,
6) For any $A, B \subseteq X$

$$(A \cap B)^{\perp_1\perp_2} \subseteq A^{\perp_1\perp_2} \cap B^{\perp_1\perp_2},$$

$$(A \cap B)^{\perp_2\perp_1} \subseteq A^{\perp_2\perp_1} \cap B^{\perp_2\perp_1},$$

7) For any $A, B \subseteq X$

$$(A \cup B)^{\perp_1} = A^{\perp_1} \cap B^{\perp_1} \text{ and } (A \cup B)^{\perp_2} = A^{\perp_2} \cap B^{\perp_2}.$$

**Proof**

1) By the definition of $C_1(A)$ and $C_2(A)$ we get

$$A^{\perp_1} = X - C_1(A) = X - (X - I_2(X - A)) = I_2(X - A), \text{ and}$$

$$A^{\perp_2} = X - C_2(A) = X - (X - I_1(X - A)) = I_1(X - A).$$

2) Since $A$ is $\tau_2 - \tau_1$ regularly open if $A = I_1(C_1(A))$, by 1) we have $A = I_1(X - I_2(X - A)) = I_1(X - A^{\perp_1}) = A^{\perp_1\perp_2}$. Analogous proof holds for the case of $\tau_1 - \tau_2$ regularly open sets.

3) $A \subseteq B$ implies $C_1(A) \subseteq C_1(B)$. Consequently $X - C_1(B) \subseteq X - C_1(A)$, which means $B^{\perp_1} \subseteq A^{\perp_1}$; analogous proof holds for $B^{\perp_2} \subseteq A^{\perp_2}$.
4) We have $A \subseteq C_1(A)$ by the monotonicity of closure operators, from which we get $I_1(A) \subseteq I_1(C_1(A))$ by the monotonicity of the interior operators. Since $A$ is open with respect to $\tau_1$, $I_1(A) = A$, such that $A \subseteq I_1(C_1(A))$. By the definition of closure operators, we get $I_1(C_1(A)) = X - C_2(X - C_1(A))$. Since by the definition of $^{\perp_1}$ and $^{\perp_2}$ we have $X - C_2(X - C_1(A)) = A^{\perp_1\perp_2}$, finally we get $A \subseteq A^{\perp_1\perp_2}$. Analogous proof holds for $A \subseteq A^{\perp_2\perp_1}$ in case $A$ is an open set with respect to $\tau_2$.
5) Observe that since by 1) $A^{\perp_1} = I_2(X - A)$, then $A^{\perp_1}$ is an open set with respect to $\tau_2$. Therefore we can apply 4) on $A^{\perp_1}$ and we get

$$A^{\perp_1} \subseteq A^{\perp_1\perp_2\perp_1}. \qquad (i)$$

Moreover, since $A$ is an open set with respect to $\tau_1$, by property 4) $A \subseteq A^{\perp_1\perp_2}$, such that, applying property 3),

$$A^{\perp_1\perp_2\perp_1} \subseteq A^{\perp_1}. \qquad (ii)$$

From $(i)$ and $(ii)$ we get

$$A^{\perp_1} = A^{\perp_1\perp_2\perp_1}. \qquad (iii)$$

An analogous proof holds for

$$A^{\perp_2} = A^{\perp_2\perp_1\perp_2} \qquad (iv)$$

in case $A$ is an open set with respect to $\tau_2$. Since we have seen that $A^{\perp_1}$ is an open set with respect to $\tau_2$, we can apply $(iv)$ on it obtaining

$$A^{\perp_1\perp_2} = A^{\perp_1\perp_2\perp_1\perp_2}.$$

Analogous proof holds for

$$A^{\perp_2\perp_1} = A^{\perp_2\perp_1\perp_2\perp_1}.$$

6) Let us consider that $A \cap B \subseteq A$ and $A \cap B \subseteq B$. Applying 3) two times we get

$$(A \cap B)^{\perp_1\perp_2} \subseteq A^{\perp_1\perp_2} \text{ and } (A \cap B)^{\perp_1\perp_2} \subseteq B^{\perp_1\perp_2}, \qquad (v)$$

and

$$(A \cap B)^{\perp_2\perp_1} \subseteq A^{\perp_2\perp_1} \text{ and } (A \cap B)^{\perp_2\perp_1} \subseteq B^{\perp_2\perp_1}. \qquad (vi)$$

From $(v)$ we get

$$(A \cap B)^{\perp_1\perp_2} \subseteq A^{\perp_1\perp_2} \cap B^{\perp_1\perp_2},$$

and from $(vi)$ we get

$$(A \cap B)^{\perp_2\perp_1} \subseteq A^{\perp_2\perp_1} \cap B^{\perp_2\perp_1}.$$

Analogous proof holds for

$$(A \cap B)^{\perp_1\perp_2} \subseteq A^{\perp_1\perp_2} \cap B^{\perp_1\perp_2}.$$

7) By 1)

$$(A \cup B)^{\perp_1} = I_2(X - (A \cup B))$$
$$=$$
$$I_2((X - A) \cap (X - B)) = I_2(X - A) \cap I_2(X - B) = A^{\perp_1} \cap B^{\perp_1}.$$

Analogous proof holds for $(A \cup B)^{\perp_2} = A^{\perp_2} \cap B^{\perp_2}$. □

A set $A \in X$ is said to be $\tau_1 - \tau_2$ biclopen if $A$ is open with respect to $\tau_1$, i.e., $I_1(A) = A$, and closed with respect to $\tau_2$, i.e., $C_1(A) = A$.

A set $A \in X$ is said to be $\tau_2 - \tau_1$ biclopen if $A$ is open with respect to $\tau_2$, i.e., $I_2(A) = A$, and closed with respect to $\tau_1$, i.e., $C_2(A) = A$.

Remembering that $A$ is $\tau_2 - \tau_1$ regularly open if and only if $X - A$ is $\tau_2 - \tau_1$ regularly closed, and $A$ is $\tau_1 - \tau_2$ regularly open if and only if $X - A$ is $\tau_1 - \tau_2$ regularly closed, we can conclude that $A$ is $\tau_1 - \tau_2$ biclopen iff $X - A$ is $\tau_2 - \tau_1$ biclopen.

**Theorem 5.** $A \subseteq X$ is $\tau_1 - \tau_2$ biclopen iff it is $\tau_2 - \tau_1$ regularly open and $\tau_1 - \tau_2$ regularly closed; $A \subseteq X$ is $\tau_2 - \tau_1$ biclopen iff it is $\tau_1 - \tau_2$ regularly open and $\tau_2 - \tau_1$ regularly closed.

**Proof.** Let us suppose that $A \subseteq X$ is $\tau_1 - \tau_2$ biclopen. Therefore,

a) $I_1(A) = A$, and
b) $C_1(A) = A$.

From a) we get $C_1(I_1(A)) = C_1(A)$, which by b) becomes $C_1(I_1(A)) = A$, i.e. $A$ is $\tau_1 - \tau_2$ regularly closed. From b) we get $I_1(C_1(A)) = I_1(A)$, which by a) becomes $I_1(C_1(A)) = A$, i.e. $A$ is $\tau_2 - \tau_1$ regularly open.

Thus we proved that, if $A \subseteq X$ is $\tau_1 - \tau_2$ biclopen, then it is $\tau_2 - \tau_1$ regularly open and $\tau_1 - \tau_2$ regularly closed.

Let us suppose now that $A$ is $\tau_2 - \tau_1$ regularly open and $\tau_1 - \tau_2$ regularly closed. Since $A$ is $\tau_1 - \tau_2$ regularly closed, then $C_1(I_1(A)) = A$, in which, since $I_1(C_1(A)) = A$ because $A$ is $\tau_2 - \tau_1$ regularly open, we can replace the $A$ argument of $I_1$ with $I_1(C_1(A))$. We get

$$C_1(I_1(I_1(C_1(A)))) = A.$$

For property $(i3)$ of interior operators we have $I_1(I_1(C_1(A))) = I_1(C_1(A))$, such that $C_1(I_1(C_1(A))) = A$. Remembering again that $A$ is $\tau_2 - \tau_1$ regularly open, we get $C_1(A) = A$. Analogously we get $I_1(A) = A$.

Thus we proved that if $A \subseteq X$ is $\tau_2 - \tau_1$ regularly open and $\tau_1 - \tau_2$ regularly closed, then it is $\tau_1 - \tau_2$ biclopen.

Analogously, we can prove that $A \subseteq X$ is $\tau_2 - \tau_1$ biclopen iff it is $\tau_1 - \tau_2$ regularly open and $\tau_2 - \tau_1$ regularly closed. □

Let us denote by $CO^{\tau_1,\tau_2}(X)$ the family of $\tau_1 - \tau_2$ biclopen sets and by $CO^{\tau_2,\tau_1}(X)$ the family of $\tau_2 - \tau_1$ biclopen sets.

We introduce now the concept of bipolar Boolean algebra, which is a generalization of Boolean algebra useful to represent properties of above topological concepts. A system $\langle \Sigma, \Sigma^+, \Sigma^-, \wedge, \vee, '^+, '^-, 0, 1\rangle$ is a *bipolar Boolean algebra* if the following properties hold:

1) $\Sigma$ is a distributive lattice with respect to the join and the meet operations $\vee$ and $\wedge$, i.e. for all $a, b, c \in \Sigma$

$$a \vee b = b \vee a, \quad a \wedge b = b \wedge a,$$

$$a \vee (b \vee c) = (a \vee b) \vee c, \quad a \wedge (b \wedge c) = (a \wedge b) \wedge c,$$

$$(a \vee b) \wedge b = b, \quad (a \wedge b) \vee a = a,$$

$$a \wedge (b \vee c) = (a \wedge b) \vee (a \wedge c), \quad a \vee (b \wedge c) = (a \vee b) \wedge (a \vee c);$$

2) $\Sigma^+, \Sigma^- \subseteq \Sigma$ are distributive lattices with respect to the join and the meet operations $\vee$ and $\wedge$; $\Sigma$ is bounded by the least element 0 and the greatest element 1, which implies that also $\Sigma^+$ and $\Sigma^-$ are bounded.
3) for every element $a \in \Sigma^+$ there is an element $a'^+ \in \Sigma^-$ and for every element $b \in \Sigma^-$ there is an element $b'^- \in \Sigma^+$, such that

$$a \vee a'^+ = 1, \quad b \vee b'^- = 1,$$

$$a \wedge a'^+ = 0, \quad b \wedge b'^- = 0.$$

**Theorem 6.** The structure

$$\langle X, CO^{\tau_1,\tau_2}(X), CO^{\tau_2,\tau_1}(X), \cap, \cup, -, -, \emptyset, X\rangle,$$

where, for any $A \in CO^{\tau_1,\tau_2}(X)$ and $B \in CO^{\tau_2,\tau_1}(X)$,

$$-A = X - A, \quad -B = X - B$$

is a bipolar Boolean algebra.

**Proof.** Observe that $X$ and $\emptyset$ are $\tau_1 - \tau_2$ clopen and $\tau_2 - \tau_1$ clopen because

- by the definition of a topology both of them belong to $\tau_1$ and $\tau_2$ and therefore are open with respect to $\tau_1$ and $\tau_2$,
- $X$ and $\emptyset$ are complement each other and therefore both of them are closed with respect to $\tau_1$ and $\tau_2$.

Since by definition of topology the finite unions and intersections of open sets are open sets as well as the finite unions and intersections of closed sets are closed sets, then the finite unions and intersections of $\tau_1 - \tau_2$ clopen sets are $\tau_1 - \tau_2$ clopen sets as well as the finite unions and intersections of $\tau_2 - \tau_1$ clopen sets are $\tau_2 - \tau_1$ clopen sets. Finally, remember that the complement of a $\tau_1$ open set is a $\tau_1$ closed set (i.e. for all $A \subseteq X$, $I_1(A) = A$ is equivalent to $C_2(X - A) = X - A$) as well as the complement of a $\tau_2$ open set is a $\tau_2$ closed set (i.e. for all $A \subseteq X$, $I_2(A) = A$ is equivalent to $C_1(X - A) = X - A$). From this observation we can conclude that the complement of $\tau_1 - \tau_2$ clopen sets is a $\tau_2 - \tau_1$ clopen set and viceversa. □

## 4 A Topological Approach to Ordinal Classification

We consider a universe $X$. $n^X$ is the set of all ordered partitions of $X$ into $n$ classes, i.e.,

$$n^X = \left\{ \langle A_1,\ldots,A_n\rangle : A_1,\ldots,A_n \subseteq X,\ \bigcup_{i=1}^{n} A_i = X, A_i \cap A_j = \emptyset\ \forall i,j = 1,\ldots,n \right\}.$$

Observe that $\langle A_1,\ldots,A_n\rangle$ can be identified with the set of decision classes $\boldsymbol{Cl}=\{Cl_1,\ldots,Cl_n\}$.

Given $\langle A_1,...,A_n\rangle \in n^X$, we define upward unions $A^{\geq i}$ and downward unions $A^{\leq i}$, $i = 1,\ldots,n$, as follows:

$$A^{\geq i} = \bigcup_{j=i}^{n} A_j, \quad A^{\leq i} = \bigcup_{j=1}^{i} A_j.$$

Let us consider the sets

$$X^{\geq} = \left\{ \langle A^{\geq 2},\ldots,A^{\geq n}\rangle : \langle A_1,\ldots,A_n\rangle \in n^X \right\};$$

$$X^{\leq} = \left\{ \langle A^{\leq 1},\ldots,A^{\leq n-1}\rangle : \langle A_1,\ldots,A_n\rangle \in n^X \right\}.$$

We define the following operations on $X^{\geq}$ and $X^{\leq}$: given $\langle A_1,\ldots,A_n\rangle$, $\langle B_1,\ldots,B_n\rangle \in n^X$,

$$\langle A^{\geq 2},\ldots,A^{\geq n}\rangle \cap \langle B^{\geq 2},\ldots,B^{\geq n}\rangle = \langle A^{\geq 2}\cap B^{\geq 2},\ldots,A^{\geq n}\cap B^{\geq n}\rangle,$$

$$\langle A^{\leq 1},\ldots,A^{\leq n-1}\rangle \cap \langle B^{\leq 1},\ldots,B^{\leq n-1}\rangle = \langle A^{\leq 1}\cap B^{\leq 1},\ldots,A^{\leq n-1}\cap B^{\leq n-1}\rangle,$$

$$\langle A^{\geq 2},\ldots,A^{\geq n}\rangle \cup \langle B^{\geq 2},\ldots,B^{\geq n}\rangle = \langle A^{\geq 2}\cup B^{\geq 2},\ldots,A^{\geq n}\cup B^{\geq n}\rangle,$$

$$\langle A^{\leq 1},\ldots,A^{\leq n-1}\rangle \cup \langle B^{\leq 1},\ldots,B^{\leq n-1}\rangle = \langle A^{\leq 1}\cup B^{\leq 1},\ldots,A^{\leq n-1}\cup B^{\leq n-1}\rangle.$$

Now, let us consider a bitopological space $(X,\tau_1,\tau_2)$ on the basis of which we can define two interior operators, $I_1$ and $I_2$, and two closure operators, $C_1$ and $C_2$ on $X$. Let us also introduce operators $I^{\geq}$ and $C^{\geq}$ on $X^{\geq}$, and operators $I^{\leq}$ and $C^{\leq}$ on $X^{\leq}$, as follows: given $\langle A_1,\ldots,A_n\rangle, \langle B_1,\ldots,B_n\rangle \in n^X$,

$$I^{\geq}(\langle A^{\geq 2},\ldots,A^{\geq n}\rangle) = \langle I_1(A^{\geq 2}),\ldots,I_1(A^{\geq n})\rangle,$$

$$C^{\geq}(\langle A^{\geq 2},\ldots,A^{\geq n}\rangle) = \langle C_1(A^{\geq 2}),\ldots,C_1(A^{\geq n})\rangle,$$

$$I^{\leq}(\langle A^{\leq 1},\ldots,A^{\leq n-1}\rangle) = \langle I_2(A^{\leq 1}),\ldots,I_2(A^{\leq n-1})\rangle,$$

$$C^{\leq}(\langle A^{\leq 1},\ldots,A^{\leq n-1}\rangle) = \langle C_2(A^{\leq 1}),\ldots,C_2(A^{\leq n-1})\rangle.$$

If $\langle A_1,\ldots,A_n\rangle$ are identified with the set of decision classes $\boldsymbol{Cl}=\{Cl_1,\ldots,Cl_n\}$ in an information table $\boldsymbol{S} =< U,Q,V,\phi >$, for a fixed subset $P \subseteq Q$,

- $I_1(A^{\geq t})$ represents the lower approximation $\underline{P}(Cl_t^{\geq})$, $t = 2,\ldots,n$,
- $C_1(A^{\geq t})$ represents the upper approximation $\overline{P}(Cl_t^{\geq})$, $t = 2,\ldots,n$,

- $I_2(A^{\leq t})$ represents the lower approximation $\underline{P}(Cl_t^{\leq})$, $t = 1, \ldots, n-1$,
- $C_2(A^{\leq t})$ represents the upper approximation $\overline{P}(Cl_t^{\leq})$, $t = 1, \ldots, n-1$.

Therefore,

$$I^{\geq}(\langle A^{\geq 2}, \ldots, A^{\geq n}\rangle) = \left\langle \underline{P}(Cl_2^{\geq}), \ldots, \underline{P}(Cl_n^{\geq})\right\rangle,$$

$$C^{\geq}(\langle A^{\geq 2}, \ldots, A^{\geq n}\rangle) = \left\langle \overline{P}(Cl_2^{\geq}), \ldots, \overline{P}(Cl_n^{\geq})\right\rangle,$$

$$I^{\leq}(\langle A^{\leq 1}, \ldots, A^{\leq n-1}\rangle) = \left\langle \underline{P}(Cl_1^{\leq}), \ldots, \underline{P}(Cl_{n-1}^{\leq})\right\rangle,$$

$$C^{\leq}(\langle A^{\leq 1}, \ldots, A^{\leq n-1}\rangle) = \left\langle \overline{P}(Cl_1^{\leq}), \ldots, \overline{P}(Cl_{n-1}^{\leq})\right\rangle.$$

We can introduce the following inclusion binary relations on $X^{\geq}$ and $X^{\leq}$: given $\langle A_1, ..., A_n\rangle, \langle B_1, \ldots, B_n\rangle \in n^X$,

$$\langle A^{\geq 2}, \ldots, A^{\geq n}\rangle \subseteq \langle B^{\geq 2}, \ldots, B^{\geq n}\rangle$$

$$\Leftrightarrow$$

$$A^{\geq i} \subseteq B^{\geq i}, \quad i = 2, \ldots, n,$$

$$\langle A^{\leq 1}, \ldots, A^{\leq n-1}\rangle \subseteq \langle B^{\leq 1}, \ldots, B^{\leq n-1}\rangle$$

$$\Leftrightarrow$$

$$A^{\leq i} \subseteq B^{\leq i}, \quad i = 1, \ldots, n-1.$$

Observe that operators $I^{\geq}$ and $I^{\leq}$ are interior operators. In fact,

(i1$^{\geq}$) $I^{\geq}(\langle A^{\geq 2}, \ldots, A^{\geq n}\rangle \cap \langle B^{\geq 2}, \ldots, B^{\geq n}\rangle) =$
$I^{\geq}(\langle A^{\geq 2}, \ldots, A^{\geq n}\rangle) \cap I^{\geq}(\langle B^{\geq 2}, \ldots, B^{\geq n}\rangle)$,
(i2$^{\geq}$) $I^{\geq}(\langle A^{\geq 2}, \ldots, A^{\geq n}\rangle) \subseteq \langle A^{\geq 2}, \ldots, A^{\geq n}\rangle)$,
(i3$^{\geq}$) $I^{\geq}(I^{\geq}(\langle A^{\geq 2}, \ldots, A^{\geq n}\rangle)) = I^{\geq}(\langle A^{\geq 2}, \ldots, A^{\geq n}\rangle)$,
(i4$^{\geq}$) $I^{\geq}(\langle X, \ldots, X\rangle) = \langle X, \ldots, X\rangle$,

and

(i1$^{\leq}$) $I^{\leq}(\langle A^{\leq 1}, \ldots, A^{\leq n-1}\rangle \cap \langle B^{\leq 1}, \ldots, B^{\leq n-1}\rangle) =$
$I^{\leq}(\langle A^{\leq 1}, \ldots, A^{\leq n-1}\rangle) \cap I^{\leq}(\langle B^{\leq 1}, \ldots, B^{\leq n-1}\rangle)$,
(i2$^{\leq}$) $I^{\leq}(\langle A^{\leq 1}, \ldots, A^{\leq n-1}\rangle) \subseteq \langle A^{\leq 1}, \ldots, A^{\leq n-1}\rangle)$,
(i3$^{\leq}$) $I^{\leq}(I^{\leq}(\langle A^{\leq 1}, \ldots, A^{\leq n-1}\rangle)) = I^{\leq}(\langle A^{\leq 1}, \ldots, A^{\leq n-1}\rangle)$,
(i4$^{\leq}$) $I^{\leq}(\langle X, \ldots, X\rangle) = \langle X, \ldots, X\rangle$.

For any $\alpha \in \tau_1$ and for any $h = 2, \ldots, n$, we can define the following set

$$Z_h^{\geq}(\alpha) = \left\langle Z_h^{\geq 2}(\alpha), \ldots, Z_h^{\geq n}(\alpha)\right\rangle$$

such that $Z_h^{\geq k}(\alpha) = \alpha$ if $k \leq h$, and $Z_h^{\geq k}(\alpha) = \emptyset$ otherwise. Analogously, for any $\alpha \in \tau_2$ and for any $h = 1, \ldots, n-1$, we can define the following set

$$Z_h^{\leq}(\alpha) = \left\langle Z_h^{\leq 1}(\alpha), \ldots, Z_h^{\leq n-1}(\alpha) \right\rangle$$

such that $Z_h^{\leq k}(\alpha) = \alpha$ if $k \geq h$, and $Z_h^{\geq k}(\alpha) = \emptyset$ otherwise.

Observe that for any $\alpha \in \tau_1$ and for any $h = 2, \ldots, n$, $Z_h^{\geq}(\alpha) \in X^{\geq}$, as well as for any $\alpha \in \tau_2$ and for any $h = 1, \ldots, n-1$, $Z_h^{\leq}(\alpha) \in X^{\leq}$.

**Theorem 7.** If $(X, I_1, I_2)$ is a bitopological space, then for any $\langle A_1, ..., A_n \rangle \in n^X$

$$I^{\geq}(\langle A^{\geq 2}, \ldots, A^{\geq n} \rangle)$$

$$=$$

$$\bigcup \{Z_h^{\geq}(\alpha), \alpha \in \tau_1, h = 2, \ldots, n : Z_h^{\geq}(\alpha) \subseteq \langle A^{\geq 2}, \ldots, A^{\geq n} \rangle\},$$

$$I^{\leq}(\langle A^{\leq 1}, \ldots, A^{\leq n-1} \rangle)$$

$$=$$

$$\bigcup \{Z_h^{\leq}(\alpha), \alpha \in \tau_2, h = 1, \ldots, n-1 : Z_h^{\leq}(\alpha) \subseteq \langle A^{\leq 1}, \ldots, A^{\leq n-1} \rangle\}.$$

**Proof.** Observing that for all $h, k = 2, \ldots, n$ with $h \geq k$, we have $A^{\geq h} \subseteq A^{\geq k}$, and, consequently, for any $\alpha \in \tau_1$, $\alpha \subseteq A^{\geq h}$ implies $\alpha \subseteq A^{\geq k}$. Let

$$\bigcup \{Z_h^{\geq}(\alpha), \alpha \in \tau_1, h = 2, \ldots, n : Z_h^{\geq}(\alpha) \subseteq \langle A^{\geq 2}, \ldots, A^{\geq n} \rangle\}$$

$$=$$

$$\langle \mathcal{A}_2, \ldots, \mathcal{A}_n \rangle .$$

For any $h$=2,. . ., $n$ and for all $\alpha \in \tau_1$, we have that $\alpha \subseteq \mathcal{A}_h$ iff $Z_h^{\geq}(\alpha) \subseteq \langle A^{\geq 2}, \ldots, A^{\geq n} \rangle$. Remembering the definition of $Z_h^{\geq}(\alpha)$ and $I_1(A^{\geq h})$, for any $h = 2, \ldots, n$, we get

$$\mathcal{A}_h = \bigcup \left\{ \alpha \in \tau_1 : Z_h^{\geq}(\alpha) \subseteq \langle A^{\geq 2}, \ldots, A^{\geq n} \rangle \right\}$$

$$=$$

$$\bigcup \{\alpha \in \tau_1 : \alpha \subseteq A^{\geq h}\} = I_1(A^{\geq h}).$$

Therefore,

$$\langle \mathcal{A}_2, \ldots, \mathcal{A}_n \rangle = I^{\geq}(\langle A^{\geq 2}, \ldots, A^{\geq n} \rangle),$$

i.e.,

$$I^{\geq}(\langle A^{\geq 2}, \ldots, A^{\geq n} \rangle)$$

$$=$$

$$\bigcup \{Z_h^{\geq}(\alpha), \alpha \in \tau_1, h = 2, \ldots, n : Z_h^{\geq}(\alpha) \subseteq \langle A^{\geq 2}, \ldots, A^{\geq n} \rangle\}.$$

Analogous proof holds for

$$I^{\leq}(\langle A^{\leq 1}, \ldots, A^{\leq n-1}\rangle)$$

$$=$$

$$\bigcup\{Z_h^{\leq}(\alpha), \alpha \in \tau_2, h = 1, \ldots, n-1 : Z_h^{\leq}(\alpha) \subseteq \langle A^{\leq 1}, \ldots, A^{\leq n-1}\rangle\}. \qquad \square$$

Theorem 7 gives a formulation of $I^{\geq}$ in terms of a Vietoris interior operator in the topological space

$$\{X^{\geq}, \{Z_h^{\geq}(\alpha) : \alpha \in \tau_1, h = 2, \ldots, n\}\}$$

as well as a formulation of $I^{\leq}$ in terms of a Vietoris interior operator in the topological space

$$\{X^{\leq}, \{Z_h^{\leq}(\alpha) : \alpha \in \tau_2, h = 1, \ldots, n-1\}\}.$$

If the bitopological space $(X, I_1, I_2)$ satisfies the *bi-clopen sets property*, then for any $\langle A_1, \ldots, A_n\rangle \in n^X$,

$$C^{\geq}(I^{\geq}(\langle A^{\geq 2}, \ldots, A^{\geq n}\rangle)) = I^{\geq}(\langle A^{\geq 2}, \ldots, A^{\geq n}\rangle),$$

$$C^{\leq}(I^{\leq}(\langle A^{\leq 1}, \ldots, A^{\leq n-1}\rangle)) = I^{\leq}(\langle A^{\leq 1}, \ldots, A^{\leq n-1}\rangle),$$

$$I^{\geq}(C^{\geq}(\langle A^{\geq 2}, \ldots, A^{\geq n}\rangle)) = C^{\geq}(\langle A^{\geq 2}, \ldots, A^{\geq n}\rangle),$$

$$I^{\leq}(C^{\leq}(\langle A^{\leq 1}, \ldots, A^{\leq n-1}\rangle)) = C^{\leq}(\langle A^{\leq 1}, \ldots, A^{\leq n-1}\rangle).$$

Let us remember that, according to Theorem 2 of the previous section, if $(X, I_1, I_2)$ is a bitopological space having bi-clopen sets property, then there exists a partial preorder $\succeq$ in $X$, such that the set

$$U^{\uparrow} = \{\{y \in X : y \succeq x\} : x \in X\} \cup \{\emptyset\}$$

is a base for $\tau_1$ and the set

$$U^{\downarrow} = \{\{y \in X : x \succeq y\} : x \in X\} \cup \{\emptyset\}$$

is a base for $\tau_2$.

For any $x \in X$ and for any $h = 2, \ldots, n$, we can define the following set

$$W_h^{\geq}(x) = \left\langle W_h^{\geq 1}(x), \ldots, W_h^{\geq n}(x)\right\rangle$$

such that $W_h^{\geq 1}(x) = X$ and $W_h^{\geq k}(x) = \{y \in X : y \succeq x\}$ if $k \leq h$ and $W_h^{\geq k}(x) = \emptyset$ otherwise. Analogously, for any $x \in X$ and for any $h = 1, \ldots, n-1$, we can define the following set

$$W_h^{\leq}(x) = \left\langle W_h^{\leq 1}(x), \ldots, W_h^{\leq n-1}(x)\right\rangle$$

such that $W_h^{\leq n}(x) = X$ and $W_h^{\leq k}(x) = \{y \in X : x \succeq y\}$ if $k \geq h$ and $W_h^{\geq k}(x) = \emptyset$ otherwise.

Observe that, for any $x \in X$ and for any $h = 2, \ldots, n$, $W_h^{\geq}(x) \in X^{\geq}$, as well as for any $x \in X$ and for any $h = 1, \ldots, n-1$, $W_h^{\leq}(x) \in X^{\leq}$.

**Theorem 8.** If $(X, I_1, I_2)$ is a bitopological space having the bi-clopen sets property, then for any $\langle A_1, \ldots, A_n \rangle \in n^X$

$$I^{\geq}(\langle A^{\geq 2}, \ldots, A^{\geq n} \rangle)$$

$$=$$

$$\bigcup \{W_h^{\geq}(x), x \in X, h = 2, \ldots, n : W_h^{\geq}(x) \subseteq \langle A^{\geq 2}, \ldots, A^{\geq n} \rangle\},$$

$$I^{\leq}(\langle A^{\leq 1}, \ldots, A^{\leq n-1} \rangle)$$

$$=$$

$$\bigcup \{W_h^{\leq}(x), x \in X, h = 1, \ldots, n-1 : W_h^{\leq}(x) \subseteq \langle A^{\leq 1}, \ldots, A^{\leq n} \rangle\},$$

$$C^{\geq}(\langle A^{\geq 2}, \ldots, A^{\geq n} \rangle)$$

$$=$$

$$\bigcup \{W_h^{\geq}(x), x \in X, h = 2, \ldots, n : W_h^{\leq}(x) \cap \langle A^{\geq 2}, \ldots, A^{\geq n} \rangle) \neq \langle \emptyset, \ldots, \emptyset \rangle\},$$

$$C^{\leq}(\langle A^{\leq 1}, \ldots, A^{\leq n-1} \rangle)$$

$$=$$

$$\bigcup \{W_h^{\leq}(x), x \in X, h = 1, \ldots, n-1 : W_h^{\geq}(x) \cap \langle A^{\leq 1}, \ldots, A^{\leq n-1} \rangle) \neq \langle \emptyset, \ldots, \emptyset \rangle\}.$$

**Proof.** From Theorem 7 we know that, if $(X, I_1, I_2)$ is a bitopological space, then for any $\langle A_1, \ldots, A_n \rangle \in n^X$

$$I^{\geq}(\langle A^{\geq 2}, \ldots, A^{\geq n} \rangle)$$

$$=$$

$$\bigcup \{Z_h^{\geq}(\alpha), \alpha \in \tau_1, h = 2, \ldots, n : Z_h^{\geq}(\alpha) \subseteq \langle A^{\geq 1}, \ldots, A^{\geq n} \rangle\}.$$

Since $(X, I_1, I_2)$ is a bitopological space having the bi-clopen sets property, from Theorem 2 we know that

$$\{\{y \in X : y \succeq x\} : x \in X\} \cup \{\emptyset\}$$

is a base for $\tau_1$. Therefore we can replace vectors $Z_h^{\geq}(\alpha), \alpha \in \tau_1$, with vectors $W_h^{\geq}(x), x \in X$, such that

$$I^{\geq}(\langle A^{\geq 2},\ldots,A^{\geq n}\rangle)$$

$$=$$

$$\bigcup\{W_h^{\geq}(x), x \in X, h=2,\ldots,n : W_h^{\geq}(x) \subseteq \langle A^{\geq 2},\ldots,A^{\geq n}\rangle\}.$$

Observing that, for all $h, k = 2,\ldots,n$ with $h \geq k$, we have $A^{\geq h} \subseteq A^{\geq k}$, and, consequently, for any $x \in X$, $\{y \in X : x \succeq y\} \cap A^{\geq h} \neq \emptyset$ implies $\{y \in X : x \succeq y\} \cap A^{\geq k} \neq \emptyset$.

Let

$$\bigcup\{W_h^{\geq}(x), x \in X, h=2,\ldots,n : W_h^{\leq}(x) \cap \langle A^{\geq 2},\ldots,A^{\geq n}\rangle) \neq \langle\emptyset,\ldots,\emptyset\rangle\}$$

$$=$$

$$\langle \mathcal{A}_2,\ldots,\mathcal{A}_n\rangle .$$

For any $h = 2,\ldots,n$ and for all $x \in X$, we have that $\{y \in X : x \succeq y\} \cap \mathcal{A}_h \neq \emptyset$ iff $W_h^{\leq}(x) \cap \langle A^{\geq 2},\ldots,A^{\geq n}\rangle \neq \langle\emptyset,\ldots,\emptyset\rangle$. Remembering the definition of $W_h^{\geq}(x)$, for any $h = 2,\ldots,n$, we get

$$\mathcal{A}_h = \bigcup\left\{\{y \in X : y \succeq x\} : W_h^{\leq}(x) \cap \langle A^{\geq 2},\ldots,A^{\geq n}\rangle \neq \emptyset\right\}$$

$$=$$

$$\bigcup\left\{\{y \in X : y \succeq x\} : \{y \in X : x \succeq y\} \cap A^{\geq h} \neq \emptyset\right\}.$$

From Theorem 3, we know that, if $(X, I_1, I_2)$ is a bitopological space having the bi-clopen sets property, then

$$C_1(A) = \bigcup\{\{y \in X : y \succeq x\} : \{y \in X : x \succeq y\} \cap A \neq \emptyset\}.$$

Therefore,

$$\bigcup\left\{\{y \in X : y \succeq x\} : \{y \in X : x \succeq y\} \cap A^{\geq h} \neq \emptyset\right\} = C_1(A^{\geq h}),$$

such that

$$\langle \mathcal{A}_2,\ldots,\mathcal{A}_n\rangle$$

$$=$$

$$\langle C_1(A^{\geq 2},\ldots,C_1(A^{\geq n}\rangle$$

$$=$$

$$C^{\geq}(\langle A^{\geq 2},\ldots,A^{\geq n}\rangle),$$

i.e.,

$$C^{\geq}(\langle A^{\geq 2},\ldots,A^{\geq n}\rangle)$$

$$=$$

$$\bigcup\{W_h^{\geq}(x), x \in X, h=2,\ldots,n : W_h^{\leq}(x) \cap \langle A^{\geq 2},\ldots,A^{\geq n}\rangle) \neq \langle\emptyset,\ldots,\emptyset\rangle\}.$$

Analogous proof holds for

$$I^{\leq}(\langle A^{\leq 1},\dots,A^{\leq n-1}\rangle)$$
$$=$$
$$\bigcup\{W_h^{\leq}(x), x\in X, h=1,\dots,n-1 : W_h^{\leq}(x)\subseteq\langle A^{\leq 1},\dots,A^{\leq n}\rangle\},$$

$$C^{\leq}(\langle A^{\leq 1},\dots,A^{\leq n-1}\rangle)$$
$$=$$
$$\bigcup\{W_h^{\leq}(x), x\in X, h{=}1,\dots,n-1 : W_h^{\geq}(x)\cap\langle A^{\leq 1},\dots,A^{\leq n-1}\rangle\neq\langle\emptyset,\dots,\emptyset\rangle\}.\ \square$$

Theorem 8 is a specialization of above Theorem 7 to the case of bitopological spaces having the bi-clopen sets property. Therefore, it gives a formulation of interior and closure operators $I^{\geq}$ and $C^{\geq}$ in terms of Vietoris operators in the bitopological space

$$\{X^{\geq},\mathcal{W}^{\geq},\mathcal{W}^{\leq}\},$$

where $\mathcal{W}^{\geq}$ is the family of open sets having as base $\{W_h^{\geq}(x) : x\in X, h=2,\dots,n\}\cup\{\langle\emptyset,\dots,\emptyset\rangle\}$ and $\mathcal{W}^{\leq}$ is the family of open sets having as base $\{W_h^{\leq}(x) : x\in X, h=1,\dots,n-1\}\cup\{\langle\emptyset,\dots,\emptyset\rangle\}$, as well as a formulation of interior and closure operators $I^{\leq}$ and $C^{\leq}$ in terms of Vietoris operators in the bitopological space

$$\{X^{\leq},\mathcal{W}^{\leq},\mathcal{W}^{\geq}\}.$$

Let us consider now the set

$$\mathcal{X}=\{\langle A_1,\dots,A_{n-1}\rangle : A_1,\dots,A_{n-1}\subseteq X\}.$$

We define the following operations on $\mathcal{X}$: given $\langle A_1,\dots,A_{n-1}\rangle,\langle B_1,\dots,B_{n-1}\rangle\in\mathcal{X}$,

$$\langle A_1,\dots,A_{n-1}\rangle\cap\langle B_1,\dots,B_{n-1}\rangle=\langle A_1\cap B_1,\dots,A_{n-1}\cap B_{n-1}\rangle,$$
$$\langle A_1,\dots,A_{n-1}\rangle\cup\langle B_1,\dots,B_{n-1}\rangle=\langle A_1\cup B_1,\dots,A_{n-1}\cup B_{n-1}\rangle.$$

Taking into account the bitopological space $\{X,\tau_1,\tau_2\}$, let us denote by $\mathcal{CO}^{\tau_1,\tau_2}(X)$ the family of vectors $\langle A^{\geq 2},\dots,A^{\geq n}\rangle\in X^{\geq}$ such that $A^{\geq h}$ is $\tau_1-\tau_2$ biclopen set for all $h=2,\dots,n$ and by $\mathcal{CO}^{\tau_2,\tau_1}(X)$ the family of vectors $\langle A^{\leq 1},\dots,A^{\leq n-1}\rangle\in X^{\leq}$ such that $A^{\leq h}$ is $\tau_2-\tau_1$ biclopen set for all $h=1,\dots,n-1$.

We consider also the negations $'^{+-} : \mathcal{CO}^{\tau_1,\tau_2}(X)\rightarrow\mathcal{CO}^{\tau_2,\tau_1}(X)$ and $'^{-+} : \mathcal{CO}^{\tau_2,\tau_1}(X)\rightarrow\mathcal{CO}^{\tau_1,\tau_2}(X)$ defined as follows: for all $\langle A^{\geq 2},\dots,A^{\geq n}\rangle\in\mathcal{CO}^{\tau_1,\tau_2}$, $\langle A^{\leq 1},\dots,A^{\leq n-1}\rangle\in\mathcal{CO}^{\tau_2,\tau_1}$,

$$\langle A^{\geq 2},\dots,A^{\geq n}\rangle'^{+-}=\langle X-A^{\geq 2},\dots,X-A^{\geq n}\rangle,$$
$$\langle A^{\leq 1},\dots,A^{\leq n-1}\rangle'^{-+}=\langle X-A^{\leq 1},\dots,X-A^{\leq n-1}\rangle.$$

**Theorem 9.** The structure

$$\langle\mathcal{X},\mathcal{CO}^{\tau_1,\tau_2}(X),\mathcal{CO}^{\tau_2,\tau_1}(X),\cap,\cup,'^{+-},'^{-+},\langle\emptyset,\dots,\emptyset\rangle,\langle X,\dots,X\rangle\rangle,$$

is a bipolar Boolean algebra.

**Proof.** It is a consequence of Theorem 6. □

## 5 Conclusions

In this article, we characterized the Dominance-based Rough Set Approach from the point of view of its topological properties. Using the concept of a bitopological space, we extended to DRSA the classical results known for the original rough set approach. Moreover, we introduced a topological approach to ordinal classification. We believe that these theoretical results will contribute to proving solid foundations of DRSA, giving a deeper knowledge of its specific characteristics. We also trust that this deeper knowledge will increase applications of DRSA to reasoning about data concerning real life problems.

## Acknowledgment

The third author wishes to acknowledge financial support from the Ministry of Science and Higher Education, grant N N519 314435.

## References

1. Bezhanishvili, G., Bezhanishvili, N., Gabelaia, D., Kurz, A.: Bitopological Duality for Distributive Lattices and Heyting Algebras. Mathematical Structures in Computer Science 20, 359–393 (2010)
2. Cattaneo, G., Ciucci, D.: Algebraic structures for rough sets. In: Peters, J.F., Skowron, A., Dubois, D., Grzymała-Busse, J.W., Inuiguchi, M., Polkowski, L. (eds.) Transactions on Rough Sets II. LNCS, vol. 3135, pp. 208–252. Springer, Heidelberg (2004)
3. Cattaneo, G., Nisticó, G.: Brouwer-Zadeh posets and three valued Łukasiewicz posets. Fuzzy Sets and Systems 33, 165–190 (1989)
4. Dubois, D., Prade, H.: Rough fuzzy sets and fuzzy rough sets. Internat. J. General Systems 17, 191–209 (1990)
5. Dubois, D., Prade, H.: Putting rough sets and fuzzy sets together. In: Słowiński, R. (ed.) Intelligent Decision Support - Handbook of Applications and Advances of the Rough Sets Theory, pp. 203–232. Kluwer, Dordrecht (1992)
6. Figueira, J., Greco, S., Ehrgott, M. (eds.): Multiple Criteria Decision Analysis: State-of-the-Art Surveys. Springer, Berlin (2005)
7. Fodor, J., Roubens, M.: Fuzzy Preference Modelling and Multicriteria Decision Support. Kluwer, Dordrecht (1994)
8. Greco, S., Inuiguchi, M., Słowiński, R.: Dominance-based rough set approach using possibility and necessity measures. In: Alpigini, J.J., Peters, J.F., Skowron, A., Zhong, N. (eds.) RSCTC 2002. LNCS (LNAI), vol. 2475, pp. 85–92. Springer, Heidelberg (2002)
9. Greco, S., Matarazzo, B., Słowiński, R.: A new rough set approach to evaluation of bankruptcy risk. In: Zopounidis, C. (ed.) Operational Tools in the Management of Financial Risks, pp. 121–136. Kluwer, Dordrecht (1998)
10. Greco, S., Matarazzo, B., Słowiński, R.: The use of rough sets and fuzzy sets in MCDM. In: Gal, T., Stewart, T., Hanne, T. (eds.) Advances in Multiple Criteria Decision Making, ch. 14, pp. 14.1–14.59. Kluwer, Boston (1999)

11. Greco, S., Matarazzo, B., Słowiński, R.: Rough set processing of vague information using fuzzy similarity relations. In: Calude, C., Paun, G. (eds.) From Finite to Infinite, pp. 149–173. Springer, Berlin (2000)
12. Greco, S., Matarazzo, B., Słowiński, R.: A fuzzy extension of the rough set approach to multicriteria and multiattribute sorting. In: Fodor, J., De Baets, B., Perny, P. (eds.) Preferences and Decisions under Incomplete Information, pp. 131–154. Physica-Verlag, Heidelberg (2000)
13. Greco, S., Matarazzo, B., Słowiński, R.: Rough sets theory for multicriteria decision analysis. European Journal of Operational Research 129, 1–47 (2001)
14. Greco, S., Matarazzo, B., Słowiński, R.: Dominance-Based Rough Set Approach to Knowledge Discovery (I) - General Perspective. In: Zhong, N., Liu, J. (eds.) Intelligent Technologies for Information Analysis, ch. 20, pp. 513–552. Springer, Berlin (2004)
15. Greco, S., Matarazzo, B., Słowiński, R.: Dominance-Based Rough Set Approach to Knowledge Discovery (II) - Extensions and Applications. In: Zhong, N., Liu, J. (eds.) Intelligent Technologies for Information Analysis, ch. 21, pp. 553–612. Springer, Berlin (2004)
16. Greco, S., Matarazzo, B., Słowiński, R.: Decision rule approach. In: Figueira, J., Greco, S., Ehrgott, M. (eds.) Multiple Criteria Decision Analysis: State of the Art Surveys, ch. 13, pp. 507–563. Springer, Berlin (2005)
17. Greco, S., Matarazzo, B., Słowiński, R.: Dominance-based Rough Set Approach as a proper way of handling graduality in rough set theory. In: Peters, J.F., Skowron, A., Marek, V.W., Orłowska, E., Słowiński, R., Ziarko, W.P. (eds.) Transactions on Rough Sets VII. LNCS (LNAI), vol. 4400, pp. 36–52. Springer, Heidelberg (2007)
18. Greco, S., Matarazzo, B., Słowiński, R.: Fuzzy set extensions of the Dominance-based Rough Set Approach. In: Bustince, H., Herrera, F., Montero, J. (eds.) Fuzzy Sets and their Extensions, Representation, Aggregation and Models, pp. 239–261. Springer, Berlin (2007)
19. Greco, S., Matarazzo, B., Słowiński, R.: Algebraic structures for Dominance-based Rough Set Approach. In: Wang, G., Li, T., Grzymala-Busse, J.W., Miao, D., Skowron, A., Yao, Y. (eds.) RSKT 2008. LNCS (LNAI), vol. 5009, pp. 252–259. Springer, Heidelberg (2008)
20. Greco, S., Matarazzo, B., Słowiński, R.: Dominance-based Rough Set Approach and Bipolar Abstract Approximation Spaces. In: Chan, C.-C., Grzymala-Busse, J.W., Ziarko, W.P. (eds.) RSCTC 2008. LNCS (LNAI), vol. 5306, pp. 31–40. Springer, Heidelberg (2008)
21. Greco, S., Matarazzo, B., Słowiński, R.: Granular computing for reasoning about ordered data: the Dominance-based Rough Set Approach. In: Pedrycz, W., Skowron, A., Kreinovich, V. (eds.) Handbook of Granular Computing, ch. 15, pp. 347–373. John Wiley & Sons, Chichester (2008)
22. Greco, S., Matarazzo, B., Słowiński, R.: Dominance-based Rough Set Approach to Interactive Multiobjective Optimization. In: Branke, J., Deb, K., Miettinen, K., Słowiński, R. (eds.) Multiobjective Optimization: Interactive and Evolutionary Approaches, ch. 5, pp. 121–155. Springer, Berlin (2008)
23. Greco, S., Matarazzo, B., Słowiński, R.: Granular Computing and Data Mining for Ordered Data - the Dominance-based Rough Set Approach. In: Meyers, R.A. (ed.) Encyclopedia of Complexity and Systems Science, pp. 4283–4305. Springer, New York (2009)
24. Kelly, J.C.: Bitopological spaces. Proc. London Math. Soc. 13, 71–89 (1963)
25. Michael, E.: Topologies on spaces of subsets. Trans. Amer. Math. Soc. 71, 152–182 (1951)

26. Pagliani, P.: Rough set and Nelson algebras. Fundamenta Informaticae 18, 1–25 (1993)
27. Pagliani, P.: Rough set theory and logic-algebraic structures. In: Orłowska, E. (ed.) Incomplete Information: Rough Set Analysis, pp. 109–190. Physica, Heidelberg (1998)
28. Pawlak, Z.: Rough Sets. International Journal of Computer and Information Sciences 11, 341–356 (1982)
29. Pawlak, Z.: Rough Sets. Kluwer, Dordrecht (1991)
30. Polkowski, L.: Rough Set: Mathematical Foundations. Physica-Verlag, Heidelberg (2002)
31. Priestley, H.A.: Representation of distributive lattices by means of ordered Stone spaces. Bull. London Math. Soc. 2, 186–190 (1970)
32. Słowiński, R., Greco, S., Matarazzo, B.: Rough set based decision support. In: Burke, E.K., Kendall, G. (eds.) Search Methodologies: Introductory Tutorials in Optimization and Decision Support Techniques, ch. 16, pp. 475–527. Springer, New York (2005)
33. Słowiński, R., Greco, S., Matarazzo, B.: Rough Sets in Decision Making. In: Meyers, R.A. (ed.) Encyclopedia of Complexity and Systems Science, pp. 7753–7786. Springer, New York (2009)
34. Wiweger, A.: On topological rough sets. Bull. Polish Academy of Sciences, ser. Mathematics 37, 89–93 (1988)

# A Study of Multiple-Source Approximation Systems

Md. Aquil Khan and Mohua Banerjee

Department of Mathematics and Statistics,
Indian Institute of Technology,
Kanpur 208 016, India
{mdaquil,mohua}@iitk.ac.in

**Abstract.** The article continues an investigation of multiple-source approximation systems ($MSAS$s) [1,2]. These are collections of Pawlak approximation spaces over the same domain, and embody the situation where information arrives from a collection of sources. Notions of strong/weak lower and upper approximations of a subset of the domain were introduced in [1]. These result in a division of the domain into five mutually disjoint sets. Different kinds of definability of a set are then defined. In this paper, we study further properties of all these concepts in a structure called multiple-source approximation system with group knowledge base ($MSAS^G$), where we also have equivalence relations representing the combined knowledge base of each group of sources. Some of the properties of combined knowledge base are presented and its relationship with the strong/weak lower and upper approximation is explored. Specifically, ordered structures that arise from these concepts are studied in some detail. In this context, notions of dependency, that reflect how much the information provided by a $MSAS^G$ depends on an individual source or group of sources, are introduced. Membership functions for $MSAS$s were investigated in [2]. These are also studied afresh here.

## 1 Introduction

In this paper, we are interested in a multiple-source scenario for *information systems* [3] – where information is obtained from different sources about the same set of objects. In such a situation, different sources may consider different sets of attributes to study the same set of objects, or they may even assign different attribute values to the objects for the same attribute. For this purpose, a *distributed* information system [4] becomes relevant, where we have an index set equipped with a symmetric relation on it and each element of the index set is associated with an information system. The domain of these information systems could be different. The situation of our interest could be expressed by a special type of distributed information system, where the index set represents the set of sources (with the identity relation), and the information systems corresponding to the sources are based on the same set of objects. Thus information from each source determines a partition on the domain, and a family of Pawlak approximation spaces [5,3] is obtained. Work on a collection of equivalence relations on

J.F. Peters et al. (Eds.): Transactions on Rough Sets XII, LNCS 6190, pp. 46–75, 2010.

the same domain may be found, e.g. in [3,6,7,8]. Multi-agent systems (sources interpreted as 'agents') are discussed, for instance, in [9,10,11,12,13]. Some are also surveyed in [14]. Continuing the investigations, a *multiple-source approximation system* ($MSAS$) is defined in [1,2], as a tuple $\mathfrak{F} := (U, \{R_i\}_{i\in N})$, where $U$ is a non-empty set, $N$ an initial segment of the set $\mathbb{N}$ of positive integers and for each $i \in N$, $R_i$ is an equivalence relation on $U$. $R_i$ may be said to represent the knowledge base of the $i^{th}$ source (cf. e.g. [11]). We would like to extend this study here, to the combined knowledge base of a group of sources.

**Definition 1.** *A* multiple-source approximation system with group knowledge base ($MSAS^G$) *is a tuple $\mathfrak{F} := (U, \{R_P\}_{P\subseteq N})$, where $U$ is a non-empty set, $N$ an initial segment of the set $\mathbb{N}$ of positive integers and for each $P \subseteq N$, $R_P$ is a binary relation on $U$ satisfying the following:*

**(M1)** *$R_P$ is an equivalence relation;*
**(M2)** *$R_P = \bigcap_{i\in P} R_i$, for each $P \subseteq N$.*

*$|N|$ is referred to as the* cardinality of $\mathfrak{F}$ *and is denoted by $|\mathfrak{F}|$.*

For $i \in N$, we shall write $R_i$ instead of $R_{\{i\}}$. For each $P \subseteq N$, $R_P$ represents the combined knowledge base of the group $P$ of sources such that two objects are distinguishable from the information provided by $R_P$, if and only if any of the sources belonging to $P$ can distinguish it with its knowledge base. Note that the above notion of combined knowledge base is also discussed by Rauszer [11] as *distributed knowledge*. $MSAS^G$s are different from *dynamic spaces* considered by Pagliani in [15], which are of the form $(U, \{R_i\}_{i\in N})$, where $\{R_i\}_{i\in N}$ is a collection of binary, or equivalence relations in a special case, on $U$. $MSAS^G$s are also different from the information structures introduced by Orłowska [16], which are tuples of the form $(U, PAR, \{R_P\}_{P\subseteq PAR})$, where we have conditions (M1), but (M2) is replaced by (M3), (M4) where,

**(M3)** $R_\emptyset = U \times U$ and
**(M4)** $R_{P\cup Q} = R_P \cap R_Q$.

Observe that (M2) implies (M3), (M4) and so every $MSAS^G$ is an information structure. In fact, if the relations are indexed over the power set of finite sets, then these two notions coincide. But in general, we can have an information structure which is not a $MSAS^G$.

When we have a number of (equivalence) relations on the domain instead of just one, a natural question would be about counterparts of the standard concepts such as approximations of sets, definability of sets, membership functions. For instance, suppose we are provided with a $MSAS^G$ and asked whether an object $x$ is an element of a set $X$ or not. Under the assumption that we do not prefer one source over another, our natural decision should be based on the number of sources which consider the object to be a positive/negative element of the set. Thus, based on practical situations, we may decide to call $x$ a positive element of $X$ based on any of the following principles:

- if all the sources of the system consider $x$ to be a positive element of $X$,
- if there is at least one source that considers $x$ to be a positive element of $X$,
- the number of sources which consider $x$ to be a positive element of $X$ is greater than the number of sources which consider $x$ to be a negative element of $X$,
- if the ratio of the number of sources which consider $x$ to be a positive element of $X$ with the total number of sources in the system, exceeds some (*a priori*) fixed threshold.

One may also wish to use the combined knowledge base of some group of sources to decide the membership of the objects.

Our study of $MSAS^G$ is motivated from these observations. In [1], the *strong lower (upper) approximation* and *weak lower (upper) approximation* of a rough set in a $MSAS$ were defined, which are generalizations of the standard lower and upper approximations. Moreover, different notions of definability were also introduced. In Section 2, we present these in the context of a set of sources. Lattice structures with varying definitions of 'complements', arise from these generalized concepts defined on $MSAS^G$s. These are studied in Section 3. In Section 4, we give some expressions to dependency [3] in this context, which reflect how much the information provided by a $MSAS^G$ depends on that of a source or a group of sources.

In Section 5, we come to the issue of membership functions for $MSAS^G$. In a $MSAS^G$, it could happen that an object $y$ has greater possibility than $x$ to be an element of a set $X$ for one source of the $MSAS^G$, but the situation is reversed for the other. So we need to define the membership functions for $MSAS^G$ considering the information provided by all the sources of the $MSAS^G$. In this direction, one may define membership functions, for instance, (i) which grade the objects depending only on the number of sources that consider the objects positive/negative element of the set, or, (ii) which depend on the distinguishability of the objects in the approximation spaces corresponding to sources. A study of this issue was done in [2]. We explore some possible candidates for such functions here as well. Section 6 concludes the article.

## 2 Notions of Lower and Upper Approximations in $MSAS^G$

Let $\mathfrak{F} := (U, \{R_P\}_{P \subseteq N})$ be a $MSAS^G$, and $P, Q \subseteq N$, $X \subseteq U$. Corresponding to each group of sources in $\mathfrak{F}$, we have an approximation space representing how the group perceives the objects with their combined knowledge base. Thus we obtain lower and upper approximations of sets corresponding to each group. Note that if $P \subseteq Q$, we have $R_Q \subseteq R_P$ and so $[x]_{R_Q} \subseteq [x]_{R_P}$. In particular, $[x]_{R_Q} \subseteq [x]_{R_i}$ for $i \in Q$. Using this fact, it is not difficult to obtain the proposition below, which shows how these lower and upper approximations are related.

**Proposition 1.** *1.* $\underline{X}_{R_P} \subseteq \underline{X}_{R_Q}$*, if* $P \subseteq Q$*.*
*2.* $\overline{X}_{R_Q} \subseteq \overline{X}_{R_P}$*, if* $P \subseteq Q$*.*

*3.* $Bn_{R_Q}(X) \subseteq Bn_{R_Q}(X)$ *for* $P \subseteq Q$*, where* $Bn_R(X)$ *represents the boundary of the set* $X$ *in the approximation space* $(U, R)$*.*

*4.* $\underline{\underline{X}_{R_Q}}_{R_P} \subseteq \underline{X}_{R_P}$.

*5.* $\underline{\underline{X}_{R_Q}}_{R_P} \subseteq \underline{X}_{R_Q}$.

*Proof.* We only prove 4. The rest may be done in the same way.
$x \in \underline{\underline{X}_{R_Q}}_{R_P}$ implies that $[x]_{R_P} \subseteq \underline{X}_{R_Q} \subseteq X$. Thus $x \in \underline{X}_{R_P}$. □

From 1 and 2, it follows that if an object is a positive or negative element of a set with respect to the combined knowledge base of a group $P$, then it remains so with respect to the combined knowledge base of any other group which contains $P$. In other words, with the increase of the number of sources, the knowledge about the objects of the domain at least does not decrease. Note that equality in 4 does not hold in general. Moreover, we may not have any set inclusion between $\underline{\underline{X}_{R_Q}}_{R_P}$ and $\underline{\underline{X}_{R_P}}_{R_Q}$. But in a special case when $R_P \subseteq R_Q$, i.e. when the group $P$ has 'finer' knowledge than the group $Q$, we obtain the following.

**Proposition 2.** *1.* $R_P \subseteq R_Q$*, if and only if* $\underline{\underline{X}_{R_Q}}_{R_P} = \underline{X}_{R_Q}$.

*2.* $R_P \subseteq R_Q$ *implies* $\underline{\underline{X}_{R_Q}}_{R_P} = \underline{\underline{X}_{R_P}}_{R_Q}$.

*Proof.* 1. From the property of lower approximation, we have $\underline{\underline{X}_{R_Q}}_{R_P} \subseteq \underline{X}_{R_Q}$. So, we prove the reverse inclusion. Let $x \in \underline{X}_{R_Q}$. Let $y, z \in U$ such that $(x, y) \in R_P$ and $(y, z) \in R_Q$. We need to prove $z \in X$. Since $(x, y) \in R_P$ and $R_P \subseteq R_Q$, we also have $(x, y) \in R_Q$ and hence, using the transitivity of $R_Q$, we obtain $(x, z) \in R_Q$. This gives $z \in X$ as $x \in \underline{X}_{R_Q}$.
Conversely, let $(x, y) \in R_P$ and $(x, y) \notin R_Q$. Let $X := \{z \in U : (x, z) \in R_Q\}$. Then $x \in \underline{X}_{R_Q}$ but $x \notin \underline{\underline{X}_{R_Q}}_{R_P}$.

2. Since $\underline{X}_{R_P} \subseteq X$, we have $\underline{\underline{X}_{R_P}}_{R_Q} \subseteq \underline{X}_{R_Q} = \underline{\underline{X}_{R_Q}}_{R_P}$ (by 1).
So, it remains to show $\underline{\underline{X}_{R_Q}}_{R_P} \subseteq \underline{\underline{X}_{R_P}}_{R_Q}$. Let $x \in \underline{\underline{X}_{R_Q}}_{R_P}$. Then we obtain

$$[x]_{R_Q} \subseteq X. \tag{1}$$

Let $y, z \in U$ such that $(x, y) \in R_Q$ and $(y, z) \in R_P$. We need to show $z \in X$. Using the transitivity of $R_Q$ and the fact that $R_P \subseteq R_Q$, we obtain $(x, z) \in R_Q$. Therefore, from (1), we obtain $z \in X$. □

Dually, one may obtain results similar to Propositions 1 and 2 for upper approximation. It follows from these propositions that for $i \in P$,

$$\underline{X}_{R_i} \subseteq \underline{X}_{R_P}, \overline{X}_{R_P} \subseteq \overline{X}_{R_i}, \underline{\underline{X}_{R_P}}_{R_i} = \underline{X}_{R_P} \text{ and } \underline{\underline{X}_{R_P}}_{R_i} = \underline{\underline{X}_{R_i}}_{R_P}.$$

## 2.1 Strong, Weak Lower and Upper Approximations for $MSAS^G$

Let us consider the following example [2].

*Example 1.* Suppose we have information regarding the attribute set {transport facilities(Tra), law and order(LO), literacy(Li)} for the cities Calcutta(C), Mumbai(M), Delhi(D), Chennai(Ch), Bangalore(B) and Kanpur(K) from four different agencies $M_1, M_2, M_3$ and $M_4$:

| | $M_1$ | | | $M_2$ | | | $M_3$ | | | $M_4$ | | |
|---|---|---|---|---|---|---|---|---|---|---|---|---|
| | Tra | LO | Li | Tra | LO | Li | Tra | LO | Li | Tra | LO | Li |
| C | a | a | g | a | a | a | a | a | g | g | g | a |
| M | g | g | a | a | a | g | a | g | p | p | a | g |
| D | g | g | a | a | a | a | a | g | p | a | a | g |
| Ch | g | p | g | a | a | g | a | p | g | p | p | g |
| B | a | a | g | a | a | g | a | g | p | p | a | g |
| K | p | g | g | p | g | p | a | p | g | p | p | g |

Here g, a, p stand for *good*, *average* and *poor*. Let $U := \{C, M, D, Ch, B, K\}$. Each $M_i$, $1 \leq i \leq 4$, then gives rise to an equivalence relation $R_i$ on $U$
$R_1 := \{\{C, B\}, \{M, D\}, \{Ch\}, \{K\}\}$;
$R_2 := \{\{C, D\}, \{M, Ch, B\}, \{K\}\}$;
$R_3 := \{\{C\}, \{M, D, B\}, \{Ch, K\}\}$;
$R_4 := \{\{C\}, \{M, B\}, \{D\}, \{Ch, K\}\}$.

We thus have a $MSAS^G$ $(U, \{R_P\}_{P \subseteq N})$, $N = \{1, 2, 3, 4\}$, where $R_P := \bigcap_{i \in P} R_i$. Many questions may be raised now. For example, for a given $X (\subseteq U)$, we may ask the following.

**(Q1)** Which cities are considered to be a positive element of $X$ by every agency?
**(Q2)** Take any particular city, say, Calcutta. Is it the case that Calcutta is a boundary element of $X$ for every agency?
**(Q3)** Is it the case that if a city is not a boundary element of $X$ for some agency, then it will also not be a boundary element of $X$ with respect to the combined knowledge base ($R_N$) of all the agencies?
**(Q4)** Is it the case that if a city is not a boundary element of $X$ with respect to $R_N$, then there is some agency, for which it will also not be a boundary element of the set?
**(Q5)** Is there any agency which considers a city to be a positive element of $X$, but some other agency considers the same city to be a negative element of $X$?
**(Q6)** Is there a city which is a boundary element of $X$ for each agency, but is not a boundary element of the set with respect to $R_N$?
**(Q7)** Let $P \subseteq N$, and take the collection $S$ of cities that are considered to be positive elements of $X$ by at least one agency of $P$. Now with respect to the combined knowledge base $R_P$ of $P$, will the cities of $S$ also be positive elements of $S$ itself? This question points to a kind of 'iteration' in knowledge, that will become clear in the sequel.

These kind of questions lead us to generalize the definitions in [1] in an obvious way to the following.
Let $\mathfrak{F} := (U, \{R_P\}_{P \subseteq N})$ be a $MSAS^G$, $X \subseteq U$ and $P$ a non-empty subset of $N$.

**Definition 2.** *The* strong lower approximation $\underline{X}_{s(P)_{\mathfrak{F}}}$, weak lower approximation $\underline{X}_{w(P)_{\mathfrak{F}}}$, strong upper approximation $\overline{X}_{s(P)_{\mathfrak{F}}}$, *and* weak upper approximation $\overline{X}_{w(P)_{\mathfrak{F}}}$ *of* $X$ *with respect to* $P$, *respectively, are defined as follows.*

$$\underline{X}_{s(P)_{\mathfrak{F}}} := \bigcap_{i\in P} \underline{X}_{R_i};\ \underline{X}_{w(P)_{\mathfrak{F}}} := \bigcup_{i\in P} \underline{X}_{R_i}.$$
$$\overline{X}_{s(P)_{\mathfrak{F}}} := \bigcap_{i\in P} \overline{X}_{R_i};\ \overline{X}_{w(P)_{\mathfrak{F}}} := \bigcup_{i\in P} \overline{X}_{R_i}.$$

If there is no confusion, we shall omit $\mathfrak{F}$ as the subscript in the above definition. If $P = N$, then these are just the strong/weak lower and upper approximations $\underline{X}_s$, $\underline{X}_w$, $\overline{X}_s$ and $\overline{X}_w$ defined in [1]. We observe that $x \in \underline{X}_{s(P)}$, provided $x$ is a positive element of $X$ for every source in $P$. On the other hand, $x \in \underline{X}_{w(P)}$, provided $x$ is a positive element of $X$ for some source in $P$. Similarly, $x \in \overline{X}_{s(P)}$, if $x$ is a possible element of $X$ for every source in $P$, and $x \in \overline{X}_{w(P)}$, if $x$ is a possible element of $X$ for some source in $P$. The relationship between the defined sets is:

(*) $$\underline{X}_{s(P)} \subseteq \underline{X}_{w(P)} \subseteq X \subseteq \overline{X}_{s(P)} \subseteq \overline{X}_{w(P)}.$$

Moreover, like lower and upper approximations, we obtain:

$$\underline{\emptyset}_{w(P)} = \underline{\emptyset}_{s(P)} = \overline{\emptyset}_{s(P)} = \overline{\emptyset}_{w(P)} = \emptyset;\ \underline{U}_{w(P)} = \underline{U}_{s(P)} = \overline{U}_{s(P)} = \overline{U}_{w(P)} = U.$$

If $\mathfrak{F} := (U, \{R\})$ then there is only one source and $\underline{X}_s = \underline{X}_w = \underline{X}_R$, $\overline{X}_s = \overline{X}_w = \overline{X}_R$. So in the special case of a single approximation space, the weak/strong lower and upper approximations are just the standard lower and upper approximations respectively.

As observed in [1], based on the information provided by a group $P$ of sources of $MSAS^G$ $\mathfrak{F}$, the domain $U$ is divided into five disjoint sets (cf. Fig. 1), viz. $\underline{X}_{s(P)}$, $\underline{X}_{w(P)} \setminus \underline{X}_{s(P)}$, $\overline{X}_{s(P)} \setminus \underline{X}_{w(P)}$, $\overline{X}_{w(P)} \setminus \overline{X}_{s(P)}$, and $(\underline{X}_{w(P)})^c$. Moreover, the possibility of an element $x \in U$ to belong to $X$ on the basis of information provided by the group $P$ of sources of $\mathfrak{F}$, reduces as we go from $\underline{X}_{s(P)}$ to $(\underline{X}_{w(P)})^c$. If $x \in \underline{X}_{s(P)}$, then we are certain that $x$ is an element of $X$. On the other hand, if $x \in (\overline{X}_{w(P)})^c$, then we are certain that $x$ is not an element of $X$.

**Definition 3.** $x \in U$ *is said to be a*

certain positive element *of* $X$ *for* $P$, *if* $x \in \underline{X}_{s(P)}$,
possible positive element *of* $X$ *for* $P$, *if* $x \in \underline{X}_{w(P)} \setminus \underline{X}_{s(P)}$,
certain negative element *of* $X$ *for* $P$, *if* $x \in (\overline{X}_{w(P)})^c$,
possible negative element *of* $X$ *for* $P$, *if* $x \in \overline{X}_{w(P)} \setminus \overline{X}_{s(P)}$, *and*
certain boundary element *of* $X$ *for* $P$, *if* $x \in \overline{X}_{s(P)} \setminus \underline{X}_{w(P)}$.

So if $x$ is certain positive element of $X$ for $P$, then every source belonging to $P$ consider it a positive element of $X$. If $x$ is a possible positive element, then there is no source in $P$ which considers it a negative element of $X$; moreover, there must be at least one source in $P$ which considers it as a positive element,

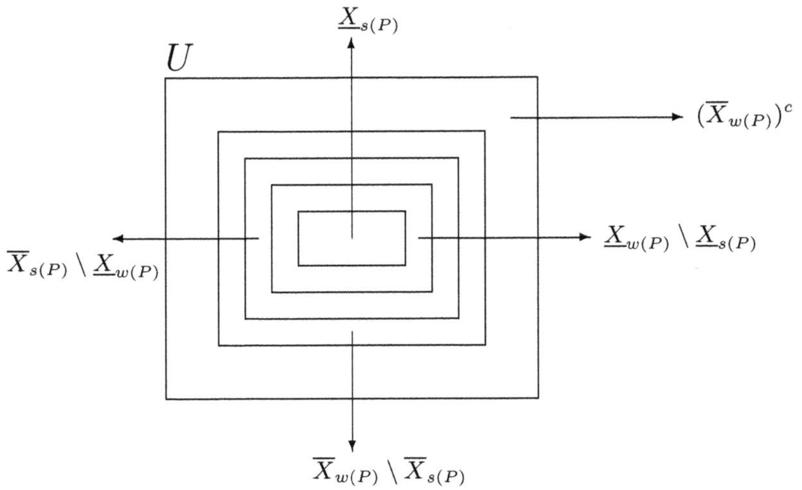


**Fig. 1.** A partition of $U$ based on the information provided by group $P$ of sources

and one for which $x$ is boundary element of $X$. Similar is the case with negative elements. If $x$ is a certain boundary element of $X$ for $P$, it is clearly a boundary element of $X$ for each source in $P$.

Consider a family of subsets $\{X_i\}_{i\in\Delta}$ of $U$, $\Delta$ being the index set.

**Proposition 3.**

1. *(a)* $\underline{\bigcap_{i\in\Delta}X_i}_{s(P)} = \bigcap_{i\in\Delta}\underline{X_i}_{s(P)}$; *(b)* $\underline{\bigcup_{i\in\Delta}X_i}_{s(P)} \supseteq \bigcup_{i\in\Delta}\underline{X_i}_{s(P)}$;
   *(c)* $\underline{\bigcap_{i\in\Delta}X_i}_{w(P)} \subseteq \bigcap_{i\in\Delta}\underline{X_i}_{w(P)}$; *(d)* $\underline{\bigcup_{i\in\Delta}X_i}_{w(P)} \supseteq \bigcup_{i\in\Delta}\underline{X_i}_{w(P)}$.
2. *(a)* $\overline{\bigcap_{i\in\Delta}X_i}_{w(P)} \subseteq \bigcap_{i\in\Delta}\overline{X_i}_{w(P)}$; *(b)* $\overline{\bigcup_{i\in\Delta}X_i}_{w(P)} = \bigcup_{i\in\Delta}\overline{X_i}_{w(P)}$.
   *(c)* $\overline{\bigcap_{i\in\Delta}X_i}_{s(P)} \subseteq \bigcap_{i\in\Delta}\overline{X_i}_{s(P)}$; *(d)* $\overline{\bigcup_{i\in\Delta}X_i}_{s(P)} \supseteq \bigcup_{i\in\Delta}\overline{X_i}_{s(P)}$.
3. *(a)* $\underline{X^c}_{s(P)} = (\overline{X}_{w(P)})^c$ ; *(b)* $\underline{X^c}_{w(P)} = (\overline{X}_{s(P)})^c$ ;
   *(c)* $\overline{X^c}_{s(P)} = (\underline{X}_{w(P)})^c$ ; *(d)* $\overline{X^c}_{w(P)} = (\underline{X}_{s(P)})^c$.
4. *If* $X \subseteq Y$ *then* $\underline{X}_{s(P)} \subseteq \underline{Y}_{s(P)}$, $\underline{X}_{w(P)} \subseteq \underline{Y}_{w(P)}$, $\overline{X}_{s(P)} \subseteq \overline{Y}_{s(P)}$ *and* $\overline{X}_{w(P)} \subseteq \overline{Y}_{w(P)}$.
5. *(a)* $\underline{X}_{w(P)} = \underline{(\underline{X}_{w(P)})}_{w(P)}$; *(b)* $\overline{X}_{s(P)} = \overline{(\overline{X}_{s(P)})}_{s(P)}$; *(c)* $X \subseteq \underline{\overline{X}_{w}}_{s(P)}$;
   *(d)* $\overline{X}_{w(P)} = \underline{(\overline{X}_{w(P)})}_{w(P)} = \overline{(\overline{X}_{s(P)})}_{w(P)}$; *(e)* $\overline{(\underline{X}_{s(P)})}_{w(P)} \subseteq \underline{X}_{w(P)}$.

*Proof.* We prove some of the items here.

1(a). $x \in \underline{\bigcap_{i\in\Delta}X_i}_{s(P)}$

$\Leftrightarrow [x]_{R_j} \subseteq \bigcap_{i\in\Delta} X_i$ for all $j \in P$

$\Leftrightarrow [x]_{R_j} \subseteq X_i$ for all $j \in P$ and $i \in \Delta$
$\Leftrightarrow x \in \underline{X_i}_{s(P)}$ for all $i \in \Delta$
$\Leftrightarrow x \in \bigcap_{i\in\Delta} \underline{X_i}_{s(P)}$.

1(b). $x \in \bigcup_{i\in\Delta} \underline{X_i}_{s(P)}$ implies that for some $i \in \Delta$, $[x]_{R_j} \subseteq X_i$ for all $j \in P$. This means $[x]_{R_j} \subseteq \bigcup_{i\in\Delta} X_i$, for all $j \in P$. Thus $x \in \underline{\bigcup_{i\in\Delta} X_i}_{s(P)}$.

3(a). $x \in \underline{X^c}_{s(P)}$
$\Leftrightarrow [x]_{R_i} \subseteq X^c$ for all $i \in P$
$\Leftrightarrow [x]_{R_i} \cap X = \emptyset$ for all $i \in P$
$\Leftrightarrow x \notin \overline{X}_{w(P)}$
$\Leftrightarrow x \in (\overline{X}_{w(P)})^c$.
Thus $\underline{X^c}_{s(P)} = (\overline{X}_{w(P)})^c$. The other part can be done similarly.

5(a). Since $\underline{X}_{w(P)} \subseteq X$, we have

$$\underline{(\underline{X}_{w(P)})}_{w(P)} \subseteq \underline{X}_{w(P)}. \tag{2}$$

Let $x \in \underline{X}_{w(P)}$, then

$$[x]_{R_i} \subseteq X \text{ for some } i \in P. \tag{3}$$

Now, $y \in [x]_{R_i}$ implies $[y]_{R_i} = [x]_{R_i} \subseteq X$ (by (3)) and hence $y \in \underline{X}_{w(P)}$. Thus $[x]_{R_i} \subseteq \underline{X}_{w(P)}$ and so $x \in \underline{(\underline{X}_{w(P)})}_{w(P)}$. This implies

$$\underline{X}_{w(P)} \subseteq \underline{(\underline{X}_{w(P)})}_{w(P)}. \tag{4}$$

From (2) and (4), we have $\underline{X}_{w(P)} = \underline{(\underline{X}_{w(P)})}_{w(P)}$.

5(d). First note that

$$\underline{(\overline{X}_{w(P)})}_{w(P)} \subseteq \overline{X}_{w(P)}. \tag{5}$$

Let $x \in \overline{X}_{w(P)}$. Then $[x]_{R_i} \cap X \neq \emptyset$ for some $i \in P$.
So there exists a $z$ such that

$$z \in [x]_{R_i} \cap X. \tag{6}$$

Now, $y \in [x]_{R_i}$
$\Rightarrow z \in [x]_{R_i} \cap X = [y]_{R_i} \cap X$ (by (6))
$\Rightarrow [y]_{R_i} \cap X \neq \emptyset$
$\Rightarrow y \in \overline{X}_{w(P)}$.

So $[x]_{R_i} \subseteq \overline{X}_{w(P)}$. This implies that $x \in \underline{(\overline{X}_{w(P)})}_{w(P)}$. Thus we have shown $\underline{(\overline{X}_{w(P)})}_{w(P)} \supseteq \overline{X}_{w(P)}$. This together with (5) gives $\underline{(\overline{X}_{w(P)})}_{w(P)} = \overline{X}_{w(P)}$.
Next, we show $\overline{X}_{w(P)} = \overline{(\overline{X}_{s(P)})}_{w(P)}$. Obviously $\overline{X}_{w(P)} \subseteq \overline{(\overline{X}_{s(P)})}_{w(P)}$ (by 4). So it remains to show reverse inclusion. Now,
$x \in \overline{(\overline{X}_{s(P)})}_{w(P)}$

$\Rightarrow [x]_{R_i} \cap \overline{X}_{s(P)} \neq \emptyset$ for some $i \in P$
$\Rightarrow$ there exists a $y$ such that $(x,y) \in R_i$ and $[y]_{R_j} \cap X \neq \emptyset$ for all $j \in P$
$\Rightarrow [y]_{R_i} \cap X \neq \emptyset$
$\Rightarrow [x]_{R_i} \cap X \neq \emptyset$ ($\because (x,y) \in R_i$)
$\Rightarrow x \in \underline{X}_{w(P)}$. □

We observe from Proposition 3 (3) that $\overline{X}_{w(P)}$ is the dual of $\underline{X}_{s(P)}$, while $\overline{X}_{s(P)}$ is the dual of $\underline{X}_{w(P)}$. Also note that strong lower and weak upper approximations behave like the Pawlak's lower and upper approximation with respect to set theoretic intersection and union. But it is not the case with weak lower and strong upper approximations. In fact, reverse inclusion in the item 1(c) and 2(d) does not hold as shown in the Example 2. Moreover, from item 5, it follows that, like Pawlak's lower/upper approximations, the weak lower and strong upper approximations are also *idempotent.* But the strong lower and weak upper approximations do not have this property.

*Example 2.* Recall the $MSAS^G$ of Example 1. For $P := \{2, 3, 4\}$, different lower and upper approximations of some sets are given by the following table.

| $Z$ | $\underline{Z}_{s(P)}$ | $\underline{Z}_{w(P)}$ | $\overline{Z}_{s(P)}$ | $\overline{Z}_{w(P)}$ |
|---|---|---|---|---|
| $\{Ch, M, B\}$ | $\emptyset$ | $\{Ch, M, B\}$ | $\{M, B, Ch\}$ | $\{K, M, B, D, Ch\}$ |
| $\{Ch, K\}$ | $\{K\}$ | $\{Ch, K\}$ | $\{Ch, K\}$ | $\{M, B, K, Ch\}$ |
| $\{Ch\}$ | $\emptyset$ | $\emptyset$ | $\emptyset$ | $\emptyset$ |
| $\{C, K, D\}$ | $\{C\}$ | $\{C, D, K\}$ | $\{C, K, D\}$ | $\{C, M, B, D, Ch, K\}$ |
| $\{C, D, B, M\}$ | $\{C, D\}$ | $\{C, D, B, M, K\}$ | $\{C, M, B, D\}$ | $\{M, B, Ch, D, C\}$ |
| $\{C, M, D, B, K\}$ | $\{C, D\}$ | $\{C, B, M, K, D\}$ | $\{C, K, M, B, Ch, D\}$ | $\{K, M, B, Ch, D, C\}$ |
| $\{C, M, D\}$ | $\{C\}$ | $\{C, D\}$ | $\{C, M, B, D\}$ | $\{C, M, B, D, Ch\}$ |

From the table it follows that for $X_1 := \{Ch, M, B\}$ and $X_2 := \{Ch, K\}$, we have $\underline{X_1}_{w(P)} \cap \underline{X_2}_{w(P)} \nsubseteq \underline{X_1 \cap X_2}_{w(P)}$. Similarly, for $Y_1 := \{C, K, D\}$, $Y_2 := \{C, D, B, M\}$, we have $Ch \in \overline{Y_1 \cup Y_2}_{s(P)}$, but $Ch \notin \overline{Y_1}_{s(P)} \cup \overline{Y_2}_{s(P)}$.
Observe that $C$ is a certain positive element of $X$ for the group $P$. Although $D$ is a positive element of $X$ for both the second and third sources, it is only a possible positive element of $X$ and not a certain positive element for $P$. Similarly, $K$ and $Ch$ are respectively certain and possible negative elements of $X$ for $P$. Moreover, $M$ and $B$ are both certain boundary elements of $X$ for the group $P$. The next proposition lists some more properties of strong, weak lower and upper approximations.

**Proposition 4.** *1.* $\underline{X}_{s(P \cup Q)} \subseteq \underline{X}_{s(P)} \cup \underline{X}_{s(Q)}$ *(reverse inclusion may not hold).*
*2.* $\underline{X}_{w(P \cup Q)} = \underline{X}_{w(P)} \cup \underline{X}_{w(Q)}$.
*3.* $\underline{\underline{X}_{s(P)}}_{R_i} \subseteq \underline{X}_{R_i}$ *(reverse inclusion may not hold).*
*4.* $\underline{\underline{X}_{w(P)}}_{R_j} \subseteq \underline{X}_{R_j}$ *and* $\underline{X}_{R_i} \subseteq \underline{\underline{X}_{w(P)}}_{R_i}$, *when* $i \in P$
*5.* $\underline{\underline{X}_{R_j}}_{w(P)} \subseteq \underline{X}_{R_j}$ *and* $\underline{X}_{R_i} \subseteq \underline{\underline{X}_{R_i}}_{w(P)}$, *when if* $i \in P$. *(Note that* $\underline{X}_{R_i} \subseteq \underline{\underline{X}_{R_i}}_{s(P)}$, $i \in P$ *does not hold).*

6. $\underline{\underline{X}_{s(P)}}_{R_i} \subseteq \underline{X}_{s(P)}$ *(reverse inclusion does not hold even if* $i \in P$*).*
7. $\underline{X}_{s(P)} \subseteq \underline{\underline{X}_{w(P)}}_{R_i}$, $i \in P$.

*Proof.* We only prove 4.
$x \in \underline{X}_{R_i}$
$\Rightarrow [y]_{R_i} \subseteq X$ for all $y \in [x]_{R_i}$
$\Rightarrow y \in \underline{X}_{w(P)}$ for all $y \in [x]_{R_i}$ as $i \in P$
$\Rightarrow [x]_{R_i} \subseteq \underline{X}_{w(P)}$
$\Rightarrow x \in \underline{\underline{X}_{w(P)}}_{R_i}$.
Thus $\underline{X}_{R_i} \subseteq \underline{\underline{X}_{w(P)}}_{R_i}$. Also since, $\underline{X}_{w(P)} \subseteq X$, we obtain $\underline{\underline{X}_{w(P)}}_{R_i} \subseteq \underline{X}_{R_i}$. Thus $\underline{\underline{X}_{w(P)}}_{R_i} \subseteq \underline{X}_{R_i}$. □

From 1, it follows that if $x$ is certain positive element of $X$ for a group $P$, then it remains so for every subset of the group. Properties 3-7 give us results about 'iterations' of knowledge.

*Example 3.* Let us recall the $MSAS^G$ of Example 2. Note that for $P := \{2,3,4\}$ and $X := \{Ch\}$, we have $Ch \in \underline{X}_{R_P}$, but $\underline{X}_{w(P)} = \emptyset$. This shows that lower approximation of a set with respect to combined knowledge base $R_P$ of a group of sources may be different from strong/weak lower approximation of the set. The next proposition shows how the strong/weak lower and upper approximations corresponding to a group $P$ of sources are related with the lower and upper approximations with respect to their combined knowledge base $R_P$.

**Proposition 5.** *1.* $\underline{X}_{w(P)} \subseteq \underline{X}_{R_P} \subseteq \overline{X}_{R_P} \subseteq \overline{X}_{s(P)}$.
2. $\underline{X}_{w(P)} \setminus \underline{X}_{s(P)} \subseteq \underline{X}_{R_P}$.
3. $\overline{X}_{R_P} \setminus \underline{X}_{R_P} \subseteq \overline{X}_{s(P)} \setminus \underline{X}_{w(P)}$.
4. $\underline{X}_{s(P)} = \underline{\underline{X}_{s(P)}}_{R_P}$
5. $\underline{X}_{w(P)} = \underline{\underline{X}_{w(P)}}_{R_P}$ *(but* $\underline{X}_{w(P)} \subseteq \underline{\underline{X}_{s(P)}}_{R_P}$ *does not hold).*
6. $\underline{\underline{X}_{w(P)}}_{R_P} \subseteq \underline{X}_{R(P)}$ *(reverse inclusion does not hold)*

*Proof.* We only prove 1 and 4.
1. It is enough to show $\underline{X}_{w(P)} \subseteq \underline{X}_{R_P}$ and $\overline{X}_{R_P} \subseteq \overline{X}_{s(P)}$.

Now, $x \in \underline{X}_{w(P)}$
$\Rightarrow [x]_{R_i} \subseteq X$ for some $i \in P$
$\Rightarrow [x]_{R_P} \subseteq X$, ($\because [x]_{R_P} \subseteq [x]_{R_i}$ for all $i \in P$)
$\Rightarrow x \in \underline{X}_{R_P}$.

Again, $x \in \overline{X}_{R_P}$
$\Rightarrow [x]_{R_P} \cap X \neq \emptyset$
$\Rightarrow [x]_{R_i} \cap X \neq \emptyset$ for all $i \in P$ ($\because [x]_{R_P} \subseteq [x]_{R_i}$ for all $i \in P$)
$\Rightarrow x \in \overline{X}_{s(P)}$.

4. $x \in \underline{X}_{s(P)}$
$\Rightarrow [x]_{R_i} \subseteq X$ for all $i \in P$
$\Rightarrow$ for all $y \in [x]_{R_P}$, $[y]_{R_j} \subseteq X$, for all $j \in P$ ($\because [x]_{R_j} = [y]_{R_j}$ for all $j \in P$)
$\Rightarrow$ for all $y \in [x]_{R_P}$, $y \in \underline{X}_{s(P)}$
$\Rightarrow [x]_{R_P} \subseteq \underline{X}_{s(P)}$
$\Rightarrow x \in \underline{\underline{X}_{s(P)}}_{R_P}$.

We also have $\underline{\underline{X}_{s(P)}}_{R_P} \subseteq \underline{X}_{s(P)}$ from the property of lower approximation. Thus we obtain $\underline{\underline{X}_{s(P)}}_{R_P} = \underline{X}_{s(P)}$. □

From 2, it follows that if $x$ is a possible positive element of a set $X$ for a group $P$, then $x$ is also a positive element of $X$ with respect to the combined knowledge base $R_P$. 3 says that if an object is a boundary element of a set with respect to $R_P$, then it will also be a certain boundary element of the set for $P$.

Let us return to the Example 1. (Q1) and (Q2) can be rephrased respectively as: What are the elements of the set $\underline{X}_s$? Does $C \in \overline{X}_s \setminus \underline{X}_w$? (Q3)-(Q7) reduces to checking the set inclusions (i) $\overline{X}_{R_N} \setminus \underline{X}_{R_N} \subseteq \overline{X}_s \setminus \underline{X}_w$, (ii) $\overline{X}_s \setminus \underline{X}_w \subseteq \overline{X}_{R_N} \setminus \underline{X}_{R_N}$, (iii) $\underline{X}_w \not\subseteq \overline{X}_w \cap (\overline{X}_w \setminus \overline{X}_s)^c$, (iv) $\overline{X}_s \setminus \underline{X}_w \not\subseteq \overline{X}_{R_N} \setminus \underline{X}_{R_N}$ and (v) $\underline{X}_{w(P)} \subseteq \underline{\underline{X}_{w(P)}}_{R_P}$ respectively for all $Y \subseteq U$.

Note that for $X := \{C, M, D\}$, $\underline{X}_s = \emptyset$ and $\overline{X}_s \setminus \underline{X}_w = \{B\}$ and hence in this case answer to Q2 is no. From Proposition 5 (1), it follows that we always have the inclusion of (i), whatever $MSAS^G$ and set is considered, and thus answer to Q3 is yes. Reverse of inclusion in (i), i.e. inclusion in (ii) may not always hold. For instance, for the $MSAS^G$ of Example 1, if $Y := \{B\}$, then $B \in \overline{Y}_s \setminus \underline{Y}_w$, but $B \notin \overline{Y}_{R_N} \setminus \underline{Y}_{R_N}$. Thus answer to Q4 is no. Moreover, if $X := \{B\}$, then answer to Q6 is yes. Since, we always have $\underline{Y}_w \subseteq \overline{Y}s$, the answer to Q5 is no. Also, from Proposition 5 (5), it follows that answers to questions like Q7 are always yes.

### 2.2 $MSAS^G$ and Tolerance Approximation Spaces

A tolerance approximation space (cf. [17], [13]) can be thought of as a generalization of a Pawlak approximation space where the relation is only a tolerance i.e. reflexive and symmetric (not necessarily an equivalence). In this section, we give the relationship between the strong lower and weak upper approximations with the lower and upper approximations for tolerance approximation spaces.

Let us consider a $MSAS^G$ $\mathfrak{F} := (U, \{R_P\}_{P \subseteq N})$ and the relation $R_{\mathfrak{F}} := \bigcup_{i \in N} R_i$. $R_{\mathfrak{F}}$ is a tolerance relation and so we have the tolerance approximation space $(U, R_{\mathfrak{F}})$. On the other hand, we have

**Proposition 6.** *Let $(U, R)$ be a tolerance approximation space with finite $U$. Then there exists a $MSAS^G$ $\mathfrak{F}$ such that $R_{\mathfrak{F}} = R$.*

*Proof.* For each $x, y \in U$ such that $xRy$, consider the set $A_{xy} = \{(x, y)\} \subseteq U \times U$. Let $A_1, A_2, \ldots, A_n$ be an enumeration of all such $A_{xy}$.
Define equivalence relations $R_1, R_2, \ldots, R_n$ on $U$ such that

$$U/R_i = \{\{x, y\} : \ (x, y) \in A_i\} \cup \{\{z\} : z \neq x, z \neq y\}.$$

Obviously, $R = R_1 \cup R_2 \cup \ldots \cup R_n$. Consider $N := \{1, 2, \ldots, n\}$ and the $MSAS^G$ $\mathfrak{F} := (U, \{R_P\}_{P \subseteq N})$ with $R_P := \bigcap_{i \in P} R_i$. As already observed $R_{\mathfrak{F}} = R$. □

**Observation 1.** *In the tolerance approximation space $(U, R_{\mathfrak{F}})$, we have for any $X \subseteq U$, the standard definitions:*

$$\underline{X}_{R_{\mathfrak{F}}} := \{x : R_{\mathfrak{F}}(x) \subseteq X\}; \quad \overline{X}_{R_{\mathfrak{F}}} := \{x : R_{\mathfrak{F}}(x) \cap X \neq \emptyset\},$$

*where $R_{\mathfrak{F}}(x) = \{y : (x, y) \in R_{\mathfrak{F}}\}$. It is not difficult to show that*

$$\underline{X}_{R_{\mathfrak{F}}} = \underline{X}_s; \quad \overline{X}_{R_{\mathfrak{F}}} = \overline{X}_w.$$

From Proposition 6 and Observation 1, it is clear that the strong lower and weak upper approximations corresponding to $MSAS^G$ with finite domains behave like the lower and upper approximations corresponding to tolerance approximation spaces with finite domain. In fact, this restriction of 'finite domain' can be removed. We can do so with the help of methods involving logic, but that is beyond the scope of this article.

## 3 Posets Based on $MSAS^G$

In this section, we explore what kind of ordered structures are formed using the concepts defined on $MSAS^G$. But before going to that, we state an important result related to lattice theory [18] which we shall use.

**Theorem 1.** *1. Let $L$ be an ordered set such that for each $\mathcal{H} \subseteq L$, g.l.b. $\dot{\bigwedge} \mathcal{H}$ of $\mathcal{H}$ exists. Then $L$ is a complete lattice in which l.u.b. of each $\mathcal{H} \subseteq L$ is given by:*

$$\dot{\bigvee} \mathcal{H} := \dot{\bigwedge} \{x \in L : (\forall a \in \mathcal{H}) a \leq x\}.$$

*2. Let $L$ be an ordered set such that for each $\mathcal{H} \subseteq L$, l.u.b. $\dot{\bigvee} \mathcal{H}$ of $\mathcal{H}$ exists. Then $L$ is a complete lattice in which g.l.b. of each $\mathcal{H} \subseteq L$ is given by:*

$$\dot{\bigwedge} \mathcal{H} := \dot{\bigvee} \{x \in L : (\forall a \in \mathcal{H}) x \leq a\}.$$

### 3.1 Posets Corresponding to Different Notions of Definable Sets

In [1], different notions of definability of sets was presented. We present these now with respect to a set of sources. Let $\mathfrak{F} := (U, \{R_P\}_{P \subseteq N})$, $P \subseteq N$ and $X \subseteq U$.

**Definition 4.** *$X$ is said to be*

- $P$-lower definable *in $\mathfrak{F}$, if $\underline{X}_{w(P)} = \underline{X}_{s(P)}$,*
- $P$-upper definable *in $\mathfrak{F}$, if $\overline{X}_{w(P)} = \overline{X}_{s(P)}$,*
- $P$-strong definable *in $\mathfrak{F}$, if $\overline{X}_{w(P)} = \underline{X}_{s(P)}$,* i.e. every element of $U$ is either certain positive or certain negative for $P$,
- $P$-weak definable *in $\mathfrak{F}$, if $\overline{X}_{s(P)} = \underline{X}_{w(P)}$,* i.e. $X$ does not have any certain boundary element for $P$.

So, if $X$ is $P$-lower definable, then the sets of positive elements in all the approximation spaces $(U, R_i), i \in N$ are identical. Similarly, if $X$ is $P$-upper definable, then the sets of negative elements in all the approximation spaces $(U, R_i), i \in N$ are identical. Let us denote the set of all $P$-strong and $P$-weak definable sets in $\mathfrak{F}$ by $SD^P_{\mathfrak{F}}$ and $WD^P_{\mathfrak{F}}$ respectively. While, $UD^P_{\mathfrak{F}}$, $LD^P_{\mathfrak{F}}$ denote the set of $P$-upper and $P$-lower definable sets in $\mathfrak{F}$.

*Example 4.* Consider the $MSAS^G$ of Example 2. Observe that the set $\{Ch, K\}$ is $Q$-strong definable, where $Q := \{1, 3, 4\}$. Moreover, for $P := \{2, 3, 4\}$, the set $U \setminus \{Ch\}$ is $P$-upper definable, but it is none of weak, strong or lower $P$-definable. In fact, it is even not definable in any of the approximation spaces corresponding to sources 2, 3 and 4. Similarly, $\{Ch, M, B\}$ is $P$-weak definable but it is none of strong, upper or lower $P$-definable.

Here is the the exact relationship among these notions of definability.

**Proposition 7.** *The following are equivalent.*
*1. $X$ is $P$-strong definable.*
*2. $X$ is $P$-weak, lower and upper definable.*

### Strong and weak definable sets

Let us first consider the collection $SD^P_{\mathfrak{F}}$ of strong definable sets. This collection is the most well-behaved among the collection of definable sets given above, as is clear from the following proposition.

**Proposition 8.** $(SD^P_{\mathfrak{F}}, \bigcup, \bigcap, ^c)$ *forms a complete field of sets [19].*

*Proof.* We note that $U, \emptyset$ are both P-strong definable. Also from Proposition 3(3), it follows that if $X$ is P-strong definable, then so is $X^c$.
Now, let $\{X_i : i \in \Delta\}$ be an arbitrary family of strong definable sets.
Let $Y := \bigcup_{i \in \Delta} X_i$ and $Z := \bigcap_{i \in \Delta} X_i$. To show $\underline{Y}_{s(P)} = \overline{Y}_{w(P)}$ and $\underline{Z}_{s(P)} = \overline{Z}_{w(P)}$.
Obviously, $\underline{Y}_{s(P)} \subseteq \overline{Y}_{w(P)}$ and $\underline{Z}_{s(P)} \subseteq \overline{Z}_{w(P)}$. Now, using Proposition 3 and the fact that each $X_i$ is strong definable, we obtain

$$\overline{Y}_{w(P)} = \bigcup_{i \in \Delta} \overline{X_i}_{w(P)} = \bigcup_{i \in \Delta} \underline{X_i}_{s(P)} \subseteq \underline{\bigcup_{i \in \Delta} X_i}_{s(P)} = \underline{Y}_{s(P)},$$

$$\overline{Z}_{w(P)} \subseteq \bigcap_{i \in \Delta} \overline{X_i}_{w(P)} = \bigcap_{i \in \Delta} \underline{X_i}_{s(P)} = \underline{\bigcap_{i \in \Delta} X_i}_{s(P)} = \underline{Z}_{s(P)}.$$

□

We do not have such nice properties for $P$-weak definable sets. Union or intersection of two $P$-definable sets may not be $P$-definable. In fact, the collection of $P$-weak definable sets does not form a lattice with respect to set inclusion, as we see from the following example.

*Example 5.* Let us consider an $MSAS^G$ $\mathfrak{F} := (U, \{R_P\}_{P \subseteq \{1,2\}})$ where $U := \{a,b,c,d,e\}$, $U/R_1 := \{\{e,a,c\},\{b\},\{d\}\}$ and $U/R_2 := \{\{e,b,d\},\{a\},\{c\}\}$. The subsets $\{c\},\{d\},\{a,c,d,e\}$ and $\{b,c,d,e\}$ are all weak definable, but $\{c,d\}$ and $\{c,d,e\}$ are not. Note that if $X$ is a l.u.b. of $\{c\}$ and $\{d\}$ with respect to set inclusion, then it must contain both $\{c\}$ and $\{d\}$; moreover, it must be contained in both $\{a,c,d,e\}$ and $\{b,c,d,e\}$. But we do not have such a weak definable $X$ and hence, $\{c\}$ and $\{d\}$ do not have a l.u.b. in the collection of all weak definable sets with respect to set inclusion.

**Lower and upper definable sets**

Let us note the following fact.

**Proposition 9.** *1. Arbitrary intersection of P-lower definable sets is also P-lower definable.*
*2. Arbitrary union of P-upper definable sets is also P-upper definable.*

*Proof.* Let $\{X_i\}_{i\in\Delta}$ be a family of P-lower definable sets. So $\underline{X_i}_{s(P)} = \underline{X_i}_{w(P)}$, for all $i \in \Delta$. We need to prove $\underline{\bigcap_{i\in\Delta} X_i}_{s(P)} = \underline{\bigcap_{i\in\Delta} X_i}_{w(P)}$.
Obviously, $\underline{\bigcap_{i\in\Delta} X_i}_{s(P)} \subseteq \underline{\bigcap_{i\in\Delta} X_i}_{w(P)}$. So we need to prove the reverse inclusion. Here,

$$\begin{aligned}
\underline{\bigcap_{i\in\Delta} X_i}_{w(P)} &\subseteq \bigcap_{i\in\Delta} \underline{X_i}_{w(P)} \text{ (by Proposition 3(4))}\\
&= \bigcap_{i\in\Delta} \underline{X_i}_{s(P)} (\because X_i \text{ are P-lower definable})\\
&= \underline{\bigcap_{i\in\Delta} X_i}_{s(P)} \quad \text{(by Proposition 3).}
\end{aligned}$$

One can similarly prove 2. □

Note that intersection (union) of two $P$-upper ($P$-lower) definable set may not be $P$-upper ($P$-lower) definable as shown in Example 6.

*Example 6.* Let us consider an $MSAS^G$ $\mathfrak{F} := (U, \{R_P\}_{P \subseteq \{1,2\}})$ where $U := \{a,b,c,d\}$, $U/R_1 := \{\{a,c\},\{b\},\{d\}\}$ and $U/R_2 := \{\{a,b\},\{c,d\}\}$. The subsets $Y_1 := \{a\}$, $Y_2 := \{c\}$ of $U$ are lower definable, but their union, i.e. the set $Y_1 \cup Y_2 = \{a,c\}$, is not lower definable. Similarly, the subsets $Z_1 := \{a,b,d\}$, $Z_2 := \{b,c,d\}$ are upper definable, but the set $Z_1 \cap Z_2 = \{b,d\}$ is not upper definable.

However, due to Theorem 1 and Proposition 9, we have the following.

**Proposition 10.** *1. $UD^P_{\mathfrak{F}}$ forms a complete lattice with respect to set inclusion such that for all $\mathcal{H} \subseteq UD^P_{\mathfrak{F}}$, the l.u.b. and g.l.b. of $\mathcal{H}$ is given by*

$$\dot{\bigvee}\mathcal{H} := \bigcup_{X\in\mathcal{H}} X \text{ and}$$

$$\dot{\wedge}\mathcal{H} := \bigcup\{X \in UD^P_{\mathfrak{F}} : (\forall Y \in \mathcal{H})\ X \subseteq Y\},$$

*and* $\emptyset$, $U$ *are the zero and unit elements of this lattice.*

2. $LD^P_{\mathfrak{F}}$ *forms a complete lattice with respect to set inclusion such that for all* $\mathcal{H} \subseteq LD^P_{\mathfrak{F}}$, *the g.l.b. and l.u.b. of* $\mathcal{H}$ *is given by*

$$\dot{\wedge}\mathcal{H} := \bigcap_{X \in \mathcal{H}} X \text{ and}$$

$$\dot{\vee}\mathcal{H} := \bigcap\{X \in LD^P_{\mathfrak{F}} : (\forall Y \in \mathcal{H})\ Y \subseteq X\},$$

*and* $\emptyset$, $U$ *are the zero and unit elements of this lattice.*

### 3.2 Posets for Strong/Weak Lower and Upper Approximations

Let us consider a $MSAS^G$ $\mathfrak{F} := (U, \{R_P\}_{P \subseteq N})$. In this section, we are interested in the collection of $P$-strong/weak lower and upper approximations, $P \subseteq N$, i.e in the following sets:

$U^{s(P)} := \{\overline{X}_{s(P)} : X \subseteq U\}$.
$U_{w(P)} := \{\underline{X}_{w(P)} : X \subseteq U\}$.
$U^{w(P)} := \{\overline{X}_{w(P)} : X \subseteq U\}$.
$U_{s(P)} := \{\underline{X}_{s(P)} : X \subseteq U\}$.

**Posets for strong upper and weak lower approximations**

The following proposition will be used to show that $U^{s(P)}$ and $U_{w(P)}$ form complete lattices with respect to set inclusions.

**Proposition 11.** *1.* $\overline{X}_{s(P)} = \bigcap\{\overline{Y}_{s(P)} : X \subseteq \overline{Y}_{s(P)}\}$.

*2.* $\underline{X}_{w(P)} = \bigcup\{\underline{Y}_{w(P)} : \underline{Y}_{w(P)} \subseteq X\}$.

*Proof.* 1 follows from the facts that (a) $X \subseteq \overline{X}_{s(P)}$ and that (b) $X \subseteq \overline{Y}_{s(P)}$ gives $\overline{X}_{s(P)} \subseteq \overline{Y}_{s(P)}$ (by Proposition 3). □

**Proposition 12.** *1.* $U^s := \{\overline{X}_s : X \subseteq U\}$ *forms a complete lattice with respect to set inclusion such that for all* $\mathcal{H} \subseteq U^s$, *the g.l.b. and l.u.b. of* $\mathcal{H}$ *is given by*

$$\dot{\wedge}\mathcal{H} := \bigcap_{\overline{X}_s \in \mathcal{H}} \overline{X}_s \text{ and}$$

$$\dot{\vee}\mathcal{H} := \overline{\left(\bigcup_{\overline{X}_s \in \mathcal{H}} \overline{X}_s\right)}_s$$

*and* $\emptyset$, $U$ *are the zero and unit elements of this lattice.*

*2.* $U_w := \{\underline{X}_w : X \subseteq U\}$ *forms a complete lattice with respect to set inclusion such that for all* $\mathcal{H} \subseteq U_w$, *the l.u.b. and g.l.b. of* $\mathcal{H}$ *is given by*

$$\dot{\vee}\mathcal{H} := \bigcup_{\underline{X}_w \in \mathcal{H}} \underline{X}_w \text{ and}$$

$$\dot{\wedge}\mathcal{H} := \underline{\left(\bigcap_{\underline{X}_w \in \mathcal{H}} \underline{X}_w\right)}_w$$

*and* $\emptyset$, $U$ *are the zero and unit elements of this lattice.*

*Proof.* We only prove 1.
First we prove that $\bigcap_{\overline{X}_s \in \mathcal{H}} \overline{X}_s \in U^s$. Here,

$$\begin{aligned}\bigcap_{\overline{X}_s \in \mathcal{H}} \overline{X}_s &\subseteq \overline{\left(\bigcap_{\overline{X}_s \in \mathcal{H}} \overline{X}_s\right)}_s \quad (\because Z \subseteq \overline{Z}_s \text{ for all } Z \subseteq U) \\ &\subseteq \overline{(\overline{Y}_s)}_s \quad \text{for all } \overline{Y}_s \in \mathcal{H} \text{ (by Proposition 3(4))} \\ &= \overline{Y}_s \quad \text{for all } \overline{Y}_s \in \mathcal{H} \text{ (by Proposition 3(5))}\ .\end{aligned}$$

Therefore, $\bigcap_{\overline{X}_s \in \mathcal{H}} \overline{X}_s \subseteq \overline{\left(\bigcap_{\overline{X}_s \in \mathcal{H}} \overline{X}_s\right)}_s \subseteq \bigcap_{\overline{X}_s \in \mathcal{H}} \overline{X}_s$, whereby

$\bigcap_{\overline{X}_s \in \mathcal{H}} \overline{X}_s = \overline{\left(\bigcap_{\overline{X}_s \in \mathcal{H}} \overline{X}_s\right)}_s$ . Therefore $\bigcap_{\overline{X}_s \in \mathcal{H}} \overline{X}_s \in U^s$.
Obviously $\bigcap_{\overline{X}_s \in \mathcal{H}} \overline{X}_s$ is the g.l.b. of $\mathcal{H}$.
Therefore, from Theorem 1, it follows that the set

$$A := \bigcap\{B \in U^s : \overline{X}_s \subseteq B \text{ for all } \overline{X}_s \in \mathcal{H}\},$$

is the l.u.b. of $\mathcal{H}$. Now,

$$\begin{aligned}A &= \bigcap\{B \in U^s : \overline{X}_s \subseteq B \text{ for all } \overline{X}_s \in \mathcal{H}\} \\ &= \bigcap\{B \in U^s : \bigcup_{\overline{X}_s \in \mathcal{H}} \overline{X}_s \subseteq B\} \\ &= \overline{\left(\bigcup_{\overline{X}_s \in \mathcal{H}} \overline{X}_s\right)}_s \quad \text{(by Proposition 11)}\end{aligned}$$

Also note that $\emptyset, U \in U^s$. □

**Posets for weak upper and strong lower approximations**

Let us now consider the strong lower and weak upper approximations of sets. From Proposition 3(1), it follows that if $\mathcal{H} \subseteq U^w$ and $\mathcal{K} \subseteq U_s$, then l.u.b. of $\mathcal{H}$ and g.l.b. of $\mathcal{K}$ both exists in $U^w$ and $U_s$ respectively and is given by $\bigcup_{\overline{X}_w \in \mathcal{H}} \overline{X}_w$ and $\bigcap_{\underline{X}_s \in \mathcal{H}} \underline{X}_s$ respectively. Moreover, $U, \emptyset \in U^w \cap U_s$. Thus again using Theorem 1, we obtain the following.

**Proposition 13.** *1. $U^w := \{\overline{X}_w : X \subseteq U\}$ forms a complete lattice with respect to set inclusion such that for all $\mathcal{H} \subseteq U^w$, the l.u.b. and g.l.b. of $\mathcal{H}$ is given by*

$$\dot{\vee}\mathcal{H} := \bigcup_{\overline{X}_w \in \mathcal{H}} \overline{X}_w \text{ and } \dot{\wedge}\mathcal{H} := \bigcup\{X \in U^w : (\forall\, Y \in \mathcal{H})\ X \subseteq Y\}$$

*and* $\emptyset$, $U$ *are the zero and unit elements of this lattice.*

2. $U_s := \{\underline{X}_s : X \subseteq U\}$ *forms a complete lattice with respect to set inclusion such that for all* $\mathcal{H} \subseteq U_s$, *the g.l.b. and l.u.b. of* $\mathcal{H}$ *is given by*

$$\dot{\wedge}\mathcal{H} := \bigcap_{\underline{X}_s \in \mathcal{H}} \underline{X}_s \text{ and } \dot{\vee}\mathcal{H} := \bigcap\{X \in U_s : (\forall\, Y \in \mathcal{H})\ Y \subseteq X\}$$

*and* $\emptyset$, $U$ *are the zero and unit elements of this lattice.*

### Equivalence and inclusion through strong/weak lower and upper approximations

Notions of equality and inclusion are generalized [3] in rough set theory using lower and upper approximations . For instance, in an approximation space $(U, R)$, two sets $X, Y \subseteq U$ are said to be *bottom* $R$*-equal* if $\underline{X}_R = \underline{Y}_R$. Similarly, $X$ is said to be *bottom* $R$*-included* in $Y$ if $\underline{X}_R \subseteq \underline{Y}_R$. In this section, we define inclusion and equality using the strong/weak lower and upper approximations and we check what kind of lattice structures they generate.

**Definition 5.** *Given a* $MSAS^G$ $\mathfrak{F} := (U, \{R_P\}_{P \subseteq N})$ *and* $P \subseteq N$, *we define binary relations* $\overset{\mathfrak{S}}{\approx}_P, \underset{\mathfrak{S}}{\approx}_P, \overset{\mathfrak{W}}{\approx}_P$ *and* $\underset{\mathfrak{W}}{\approx}_P$ *on* $\mathcal{P}(U)$ *as follows.*

$X \overset{\mathfrak{S}}{\approx}_P Y$ *if and only if* $\overline{X}_{s(P)} = \overline{Y}_{s(P)}$.

$X \underset{\mathfrak{S}}{\approx}_P Y$ *if and only if* $\underline{X}_{s(P)} = \underline{Y}_{s(P)}$.

$X \overset{\mathfrak{W}}{\approx}_P Y$ *if and only if* $\overline{X}_{w(P)} = \overline{Y}_{w(P)}$.

$X \underset{\mathfrak{W}}{\approx}_P Y$ *if and only if* $\underline{X}_{w(P)} = \underline{Y}_{w(P)}$.

It is immediate that $\overset{\mathfrak{S}}{\approx}_P, \underset{\mathfrak{S}}{\approx}_P, \overset{\mathfrak{W}}{\approx}_P$ and $\underset{\mathfrak{W}}{\approx}_P$ are equivalence relations on $\mathcal{P}(U)$. Thus, the relation $\overset{\mathfrak{S}}{\approx}_P$ collects together all the sets which have same strong upper approximations with respect to $P$. Now, we turn to inclusions; clearly the following partial orders on the quotient sets $\mathcal{P}(U)/\overset{\mathfrak{S}}{\approx}_P$, $\mathcal{P}(U)/\underset{\mathfrak{S}}{\approx}_P$, $\mathcal{P}(U)/\overset{\mathfrak{W}}{\approx}_P$, $\mathcal{P}(U)/\underset{\mathfrak{W}}{\approx}_P$ are induced.

$[X]_{\overset{\mathfrak{S}}{\approx}_P} \leq^s [Y]_{\overset{\mathfrak{S}}{\approx}_P}$ if and only if $\overline{X}_{s(P)} \subseteq \overline{Y}_{s(P)}$,

$[X]_{\underset{\mathfrak{S}}{\approx}_P} \leq_s [Y]_{\underset{\mathfrak{S}}{\approx}_P}$ if and only if $\underline{X}_{s(P)} \subseteq \underline{Y}_{s(P)}$,

$[X]_{\overset{\mathfrak{W}}{\approx}_P} \leq^w [Y]_{\overset{\mathfrak{W}}{\approx}_P}$ if and only if $\overline{X}_{w(P)} \subseteq \overline{Y}_{w(P)}$,

$[X]_{\underset{\mathfrak{W}}{\approx}_P} \leq_w [Y]_{\underset{\mathfrak{W}}{\approx}_P}$ if and only if $\underline{X}_{w(P)} \subseteq \underline{Y}_{w(P)}$.

Let us consider the posets $(\mathcal{P}(U)/\stackrel{\mathfrak{S}}{\approx}_P, \leq^s), (\mathcal{P}(U)/\underset{\mathfrak{s}}{\approx}_P, \leq_s), (\mathcal{P}(U)/\stackrel{\mathfrak{W}}{\approx}_P, \leq^w)$ and $(\mathcal{P}(U)/\underset{\mathfrak{w}}{\approx}_P, \leq_w)$. Then we have

**Proposition 14.** *1.* $(\mathcal{P}(U)/\stackrel{\mathfrak{S}}{\approx}_P, \leq^s) \cong (U^s, \subseteq))$.

*2.* $(\mathcal{P}(U)/\underset{\mathfrak{s}}{\approx}_P, \leq_s) \cong (U_s, \subseteq))$.

*3.* $(\mathcal{P}(U)/\stackrel{\mathfrak{W}}{\approx}_P, \leq^w) \cong (U^w, \subseteq))$.

*4.* $(\mathcal{P}(U)/\underset{\mathfrak{w}}{\approx}_P, \leq^w) \cong (U_w, \subseteq))$.

*Proof.* Consider the mappings $f_a([X]_{\stackrel{\mathfrak{S}}{\approx}_P}) = \overline{X}_{s(P)}$, $f_b([X]_{\underset{\mathfrak{s}}{\approx}_P}) = \underline{X}_{s(P)}$, $f_c([X]_{\stackrel{\mathfrak{W}}{\approx}_P}) = \overline{X}_{w(P)}$ and $f_d([X]_{\underset{\mathfrak{w}}{\approx}_P}) = \underline{X}_{w(P)}$ for 1, 2, 3 and 4 respectively. □

From Propositions 12 and 13, we have that $(\mathcal{P}(U)/\stackrel{\mathfrak{S}}{\approx}_P, \leq^s)$, $(\mathcal{P}(U)/\underset{\mathfrak{s}}{\approx}_P, \leq_s)$, $(\mathcal{P}(U)/\stackrel{\mathfrak{W}}{\approx}_P, \leq^w)$ and $(\mathcal{P}(U)/\underset{\mathfrak{w}}{\approx}_P, \leq_w)$ are all complete lattices.

### 3.3 'Complements' on the Posets

So far, we have seen that the structures $(UD_{\mathfrak{F}}^P, \subseteq), (LD_{\mathfrak{F}}^P, \subseteq), (U^{s(P)}, \subseteq), (U_{s(P)}, \subseteq), (U^{w(P)}, \subseteq)$ and $(U_{w(P)}, \subseteq)$ are all complete lattices. A natural question would be about complements in these ordered structures. Unlike the set $SD_{\mathfrak{F}}^P$ consisting of strong definable sets which forms a complete field of sets, candidates for a 'complement' for the collections $UD_{\mathfrak{F}}^P$, $LD_{\mathfrak{F}}^P$, $U^{s(P)}$, $U_{s(P)}$, $U^{w(P)}$ and $U_{w(P)}$ are not obvious. However, here are some possibilities.

**Definition 6.** *Unary operators* $*_1, *_2, *_3, *_4, *_5, *_6$ *on* $(UD_{\mathfrak{F}}^P, \subseteq)$, $(LD_{\mathfrak{F}}^P, \subseteq)$, $(U^{s(P)}, \subseteq), (U_{s(P)}, \subseteq), (U^{w(P)}, \subseteq), (U_{w(P)}, \subseteq)$ *respectively, are given as follows.*

$$X^{*_1} := \underline{X^c}_{s(P)},\ X \in UD_{\mathfrak{F}}^P.$$

$$X^{*_2} := \overline{X^c}_{w(P)},\ X \in LD_{\mathfrak{F}}^P.$$

$$(\overline{X}_{s(P)})^{*_3} := \overline{(\underline{X^c}_{w(P)})}_{s(P)}.$$

$$(\underline{X}_{s(P)})^{*_4} := \underline{(\overline{X^c}_{w(P)})}_{s(P)}.$$

$$(\overline{X}_{w(P)})^{*_5} := \overline{(\underline{X^c}_{s(P)})}_{w(P)}.$$

$$(\underline{X}_{w(P)})^{*_6} := \underline{(\overline{X^c}_{s(P)})}_{w(P)}.$$

Note that , we have $\underline{X^c}_{s(P)} \in UD_{\mathfrak{F}}^P$ and $\overline{Y^c}_{w(P)} \in LD_{\mathfrak{F}}^P$, provided $X \in UD_{\mathfrak{F}}^P$ and $Y \in LD_{\mathfrak{F}}^P$. The propositions below give some properties of these operators.

**Proposition 15.** *For $j \in \{1,4,6\}$, we have the following.*

1. $X \dot{\wedge} X^{*_j} = \emptyset$.
2. $\emptyset^{*_j} = U$ *and* $U^{*_j} = \emptyset$.
3. *If* $X \subseteq Y$, *then* $Y^{*_j} \subseteq X^{*_j}$.
4. *If* $X \dot{\vee} Y = U$, *then* $X^{*_j} \subseteq Y$.
5. $X \subseteq X^{*_j *_j}$.
6. $X^{*_j *_j *_j} = X^{*_j}$.
7. $X^{*_1} \dot{\vee} X^{*_1 *_1} = U$.

*Proof.* 1. ($*_1$): Since $\bigcup\{X \in UD : (\forall Y \in \mathcal{H}) X \subseteq Y\} \subseteq \bigcap_{Y \in \mathcal{H}} Y$, it is enough to show that $X \cap X^{*_1} = \emptyset$, i.e, $X \cap \underline{X^c}_{s(P)} = \emptyset$. But this is obvious as $x \in \underline{X^c}_{s(P)}$ implies $[x]_{R_i} \in X^c$ for all $i \in N$ and hence $x \in X^c$.
($*_4$): Here, we need to prove $\underline{X}_{s(P)} \cap \underline{(\overline{X^c}_{w(P)})}_{s(P)} = \emptyset$. This follows as

$$\underline{(\overline{X^c}_{w(P)})}_{s(P)} \subseteq \overline{X^c}_{w(P)} = (\underline{X}_{s(P)})^c (\text{using Proposition 3(3)}).$$

($*_6$): Here, we need to prove $\underline{(\underline{X}_{w(P)} \cap \underline{(\overline{X^c}_{s(P)})}_{w(P)})}_{w(P)} = \emptyset$. This follows as $\underline{(\underline{X}_{w(P)} \cap \underline{(\overline{X^c}_{s(P)})}_{w(P)})}_{w(P)} \subseteq \underline{X}_{w(P)} \cap \underline{(\overline{X^c}_{s(P)})}_{w(P)}$ and $\underline{(\overline{X^c}_{s(P)})}_{w(P)} \subseteq \overline{X^c}_{s(P)} = (\underline{X}_{w(P)})^c$.

2. Follows from the the fact that $\underline{\emptyset}_{w(P)} = \underline{\emptyset}_{s(P)} = \overline{\emptyset}_{s(P)} = \overline{\emptyset}_{w(P)} = \emptyset$; $\underline{U}_{w(P)} = \underline{U}_{s(P)} = \overline{U}_{s(P)} = \overline{U}_{w(P)} = U$.

3. Follows from the Proposition 3(4).

4. Let $X \cup Y = U$. Then $X^c \subseteq Y$. Using it and Proposition 3, we obtain $X^{*_1} = \underline{X^c}_{s(P)} \subseteq X^c \subseteq Y$.
For, $X = \underline{Z}_{s(P)}$, $X^{*_4} = (\underline{Z}_{s(P)})^{*_4} = \underline{(\overline{Z^c}_{w(P)})}_{s(P)} \subseteq \overline{Z^c}_{w(P)} = (\underline{Z}_{s(P)})^c \subseteq Y$.
Similarly, we can prove for $*_6$.

5. $X^{*_1} = \underline{X^c}_{s(P)} = (\overline{X}_{w(P)})^c$.
So, $X^{*_1 *_1} = \underline{((\overline{X}_{w(P)})^c)^c}_{s(P)} = \underline{(\overline{X}_{w(P)})}_{s(P)}$.
Therefore, using Proposition 3(5), we obtain the desired result.
Here, $(\underline{X}_{s(P)})^{*_4} = \underline{(\overline{X^c}_{w(P)})}_{s(P)} = \underline{(\overline{(\underline{X}_{s(P)})^c})}_{s(P)} = (\overline{\underline{X}_{s(P)}}_{w(P)})^c$.
Therefore, $(\underline{X}_{s(P)})^{*_4 *_4} = \underline{(\overline{(\overline{\underline{X}_{s(P)}}_{w(P)})}_{w(P)})}_{s(P)}$.

Now, $x \in \underline{X}_{s(P)} \subseteq \overline{\underline{X}_{s(P)}}_{w(P)}$

$\Rightarrow$ for all $i \in N$, $[x]_{R_i} \subseteq \overline{(\overline{\underline{X}_{s(P)}}_{w(P)})}_{w(P)}$

$\Rightarrow x \in \underline{(\overline{(\overline{\underline{X}_{s(P)}}_{w(P)})}_{w(P)})}_{s(P)}$.

Thus we obtain the desired result. Similarly, we can prove for $*_6$.

6. Follows from (c) and (e).

7. We need to prove $X^{*_1 *_1} \subseteq (X^{*_1})^c$.
Here, $X^{*_1} = \underline{X^c}_{w(P)} = (\overline{X}_{s(P)})^c$.
Therefore, $X^{*_1 *_1} = \underline{(\overline{X}_{s(P)})}_{w(P)} \subseteq \overline{X}_{s(P)} = (X^{*_1})^c$. □

Similarly, we have

**Proposition 16.** *For $j \in \{2, 3, 5\}$, we have the following.*

1. $X \dot{\vee} X^{*_j} = U$.
2. $\emptyset^{*_j} = U$ *and* $U^{*_j} = \emptyset$.
3. *If* $X \subseteq Y$, *then* $Y^{*_j} \subseteq X^{*_j}$.
4. *If* $X \cap Y = \emptyset$, *then* $X \subseteq Y^{*_j}$.
5. $X^{*_j *_j} \subseteq X$.
6. $X^{*_j *_j *_j} = X^{*_j}$.
7. $X^{*_2} \dot{\wedge} X^{*_2 *_2} = \emptyset$

We note that an element $x^*$ is a *pseudocomplement* [18] of an element $x$ of a lattice $L$ with a least element 0, if it satisfies the following two conditions:

(a) $x \cap x^* = 0$;
(b) for all $a \in L$, $x \cap a = 0$ implies $a \leq x^*$.

Although the operators $*_j$, $j \in \{1, 4, 6\}$ satisfy the condition (a), but these fail to satisfy (b). On the other hand, the complements $*_j$, $j \in \{2, 3, 5\}$ satisfy the condition (b), but fail to satisfy the other one.

## 4 Dependency of Information

Dependency [3] of knowledge bases represented by equivalence relations is a very important aspect of rough set theory. Given two equivalence relations $R, Q$ over the same domain $U$, representing two knowledge bases, we say $Q$ depends on $R$ if and only if $R \subseteq Q$, i.e. $[x]_R \subseteq [x]_Q$ for all $x \in U$. Usually a knowledge base does not depend wholly but partially on other knowledge bases. A measure of this is given [3] by the expression $\frac{|POS_R(Q)|}{|U|}$, assuming $U$ is finite, where $POS_R(Q) = \bigcup_{X \in U/Q} \underline{X}_R$. In this case, we say $Q$ depends in a degree $\frac{|POS_R(Q)|}{|U|} = k$, $(0 \leq k \leq 1)$ on $R$, and denote it as $R \Rightarrow_k Q$. Note that this measure behaves like a function: given any two equivalence relations $Q, R$ over the same finite domain $U$, we always have a $k \in [0, 1]$ such that $R \Rightarrow_k Q$.

Let us closely observe the set $POS_R(Q)$. It is not difficult to show that this set is actually the set $\{x \in U : [x]_R \subseteq [x]_Q\}$. In fact, $POS_R(Q)$ consists of precisely all those objects $x$ such that: *if $Q$ considers $x$ to be positive or negative element of a set, then $R$ also considers it positive or negative element of the set accordingly; in other words, if $R$ is not able to decide whether $x$ is an element* *of a set $X$ or not (i.e. $x$ is boundary element of $X$ with respect to $R$), then $Q$ is also not able to do so.*

In this section, our aim is to come up with some dependency functions which determine how much the information provided by a $MSAS^G$ depends on the information provided by an individual source or a group of sources. In this direction, motivated from the set $POS_R(Q)$ as described above, we find that the following sets emerge. Let $\mathfrak{F} := (U, \{R_P\}_{P \subseteq N})$ and $P \subseteq N$.

**Definition 7.** $A_1(P) := \{x \in U : [x]_{R_P} \subseteq [x]_{R_N}\}$.

$A_2(P) := \{x \in U : [x]_{R_{N \setminus P}} \not\subseteq [x]_{R_N}\}$.

$A_3(P) := \{x \in U : [x]_{R_Q} \subseteq [x]_{R_P}$ *for some* $Q \subseteq N \setminus P\}$.

$A_4(P) := \{x \in U :$ *for all* $i \in P$ *there exists* $j \in N \setminus P$ *such that* $[x]_{R_j} \subseteq [x]_{R_i}\}$.

$A_5(P) := \{x \in U : [x]_{R_{N \setminus P}} \subseteq [x]_{R_N}\}$.

$A_6(P) := \{x \in U :$ *for all* $j \in N \setminus P$ *there exists* $i \in P$ *such that* $[x]_{R_i} \subseteq [x]_{R_j}\}$.

$A_7(P) := \{x \in U :$ *for all* $Q \subseteq N \setminus P$ *there exists* $i \in P$ *such that* $[x]_{R_i} \subseteq [x]_{R_Q}\}$.

As we know, $\overline{X}_R \setminus \underline{X}_R$ is the set of all boundary elements of $X$ with respect to the equivalence relation $R$. Now, we note that $A_1(P)$ is actually the set $POS_{R_P}(R_N)$ and so it consists of precisely those objects $x$ such that if $x$ is a boundary element of a set $X$ with respect to the combined knowledge base of the group $P$, then $x$ will also be the boundary element of $X$ with respect to the combined knowledge base of all the sources, i.e. for all $X \subseteq U$, $x \notin \overline{X}_{R_N} \setminus \underline{X}_{R_N}$ implies $x \notin \overline{X}_{R_P} \setminus \underline{X}_{R_P}$. Similarly, $A_2(P)$ consists of precisely those objects $x$ for which there exists a set $X$ such that $x$ is a boundary element of $X$ with respect to combined knowledge base of the group $N \setminus P$, but $x$ is not a boundary element of $X$ with respect to combined knowledge base of all the sources. In other words, $A_2(P)$ collects all those objects $x$ such that if we dismiss the sources belonging to $P$ from the system, then we will lose some information about these objects– information whether these are positive or negative elements of some set.

$A_3(P)$ consists of precisely those objects $x$ such that if $x$ is not a boundary element of some set $X$ with respect to $R_P$, then it is also not a boundary element of $X$ with respect to combined knowledge base of some group consisting of elements which are not in $P$; in other words, we have *replacement of the group* $P$ in the system regarding the information about these objects in the sense that if combined knowledge base of $P$ can determine that $x$ is positive or negative element of a set, then we have some other group not involving the sources of $P$ which can also do so. We get the following proposition giving the interpretation of all the sets defined above.

**Proposition 17.** *1.* $x \in A_1(P)$ *if and only if for all* $X \subseteq U$, $x \in \overline{X}_{R_P} \setminus \underline{X}_{R_P}$ *implies* $x \in \overline{X}_{R_N} \setminus \underline{X}_{R_N}$.

2. $x \in A_2(P)$ *if and only if there exists a* $X \subseteq U$ *such that,* $x \in (\overline{X}_{R_{N\setminus P}} \setminus \underline{X}_{R_{N\setminus P}}) \cap (\overline{X}_{R_N} \setminus \underline{X}_{R_N})^c$.
3. $x \in A_3(P)$ *if and only if for all* $X \subseteq U$, $x \notin \overline{X}_{R_P} \setminus \underline{X}_{R_P}$ *implies* $x \notin \overline{X}_{R_Q} \setminus \underline{X}_{R_Q}$ *for some* $Q \subseteq N \setminus P$.
4. $x \in A_4(P)$ *if and only if for all* $i \in P$ *and for all* $X \subseteq U$, $x \notin \overline{X}_{R_i} \setminus \underline{X}_{R_i}$ *implies* $x \notin \overline{X}_{R_j} \setminus \underline{X}_{R_j}$ *for some* $j \in N \setminus P$.
5. $x \in A_5(P)$ *if and only if for all* $X \subseteq U$, $x \notin \overline{X}_{R_N} \setminus \underline{X}_{R_N}$ *implies* $x \notin \overline{X}_{R_{N\setminus P}} \setminus \underline{X}_{R_{R_{N\setminus P}}}$.
6. $x \in A_6(P)$ *if and only if for all* $j \in N \setminus P$ *and for all* $X \subseteq U$, $x \notin \overline{X}_{R_j} \setminus \underline{X}_{R_j}$ *implies* $x \notin \overline{X}_{R_i} \setminus \underline{X}_{R_i}$ *for some* $i \in P$.
7. $x \in A_7(P)$ *if and only if for all* $Q \subseteq N \setminus P$ *and for all* $X \subseteq U$, $x \notin \overline{X}_{R_j} \setminus \underline{X}_{R_j}$ *implies* $x \notin \overline{X}_{R_i} \setminus \underline{X}_{R_i}$ *for some* $i \in P$.

The proposition below expresses some of the the sets given in Definition 7 in terms of weak lower approximations.

**Proposition 18.** *1.* $A_4(P) = \bigcap_{i\in P}(\bigcup_{X\in U/R_i} \underline{X}_{w(N\setminus P)})$.

2. $A_6(P) = \bigcap_{i\in N\setminus P}(\bigcup_{X\in U/R_i} \underline{X}_{w(P)})$.
3. $A_7(P) = \bigcap_{Q\subseteq N\setminus P}(\bigcup_{X\in U/R_Q} \underline{X}_{w(P)})$.

*Proof.* We only prove 1.
$x \in A_4(P)$
$\Leftrightarrow$ for all $i \in P$, there exists $j \in N \setminus P$ such that $[x]_{R_j} \subseteq [x]_{R_i}$
$\Leftrightarrow$ for all $i \in P$, $x \in \underline{[x]_{R_i}}_{w(N\setminus P)}$
$\Leftrightarrow$ for all $i \in P$, $x \in \underline{X}_{w(N\setminus P)}$ for some $X \in U/R_i$
$\Leftrightarrow$ for all $i \in P$, $x \in \bigcup_{X\in U/R_i} \underline{X}_{w(N\setminus P)}$
$\Leftrightarrow x \in \bigcap_{i\in P}(\bigcup_{X\in U/R_i} \underline{X}_{w(N\setminus P)})$. □

Now, just as in standard rough set theory, we define some dependency functions $D : 2^N \to [0,1]$ for the $MSAS^G$ $\mathfrak{F} := (U, \{R_P\}_{P\subseteq N})$, using the sets given in Definition 7. Note that $D(P) = k$ will signify that information provided by the $MSAS^G$ depends in a degree $k$ on the information of the group $P$.

**Definition 8.** *Let* $\mathfrak{F} := (U, \{R_P\}_{P\subseteq N})$ *be an* $MSAS^G$ *with finite* $U$. *For each* $i \in \{1, 2, \ldots, 7\}$, *we define a dependency function* $D_i : 2^N \to [0,1]$ *as follows:*

$$D_i(P) := \frac{|A_i(P)|}{|U|} \text{ for } i \in \{1,2,6,7\}, \text{ and}$$

$$D_i(P) := 1 - \frac{|A_i(P)|}{|U|} \text{ for } i \in \{3,4,5\}.$$

*Example 7.* Let us consider the $MSAS^G$ $\mathfrak{F}$ of Example 1. Let $P := \{M_1, M_2\}$ and $Q := \{M_3, M_4\}$. Then we have,

$D_1(P) = 1$ and $D_1(Q) = \frac{1}{3}$; $D_2(P) = \frac{2}{3}$ and $D_2(Q) = 0$;
$D_3(P) = \frac{2}{3}$ and $D_2(Q) = 0$; $D_4(P) = \frac{5}{6}$ and $D_4(Q) = \frac{5}{6}$;
$D_5(P) = \frac{2}{3}$ and $D_5(Q) = 0$; $D_6(P) = \frac{1}{2}$ and $D_2(Q) = \frac{1}{6}$;
$D_7(P) = \frac{1}{3}$ and $D_7(Q) = \frac{1}{6}$.

We note the following facts about this example.
(i) According to every dependency function except $D_4$, information provided by $\mathfrak{F}$ depends on $P$ more than $Q$.

(ii) According to dependency function $D_4$, information provided by $\mathfrak{F}$ depends equally on both $P$ and $Q$. Moreover, $A_4(P) = \{K\}$ and $A_4(Q) = \{C\}$. Therefore, every city except $K$ ($C$) is such that it is a positive or negative element of some city with respect to knowledge base of some source belonging to $P$ ($Q$), but it is a boundary element of the set for every source not belonging to $P$ ($Q$).

(iii) $D_1$: the information provided by $\mathfrak{F}$ totally depends on $P$. So if any object is a positive or negative element of a set with respect to combined knowledge base of all sources, then it is also so with respect to combined knowledge base of the group $P$.

(iv) $D_6$: the information provided by $\mathfrak{F}$ is neither totally dependent nor independent of $P$. So there are objects (in fact exactly three) such that if any of these is a positive or negative element of some set with respect to knowledge base of some source not belonging to $P$, then it is also not a boundary element of the set with respect to knowledge base of some source belonging to $P$.

(v) $D_5$: the information provided by $\mathfrak{F}$ is totally independent of $Q$. So if any object is a positive or negative element of a set with respect to combined knowledge base of all the sources, then it remains so with respect to the combined knowledge base of the group $N \setminus Q$.

Lastly, the proposition below lists some properties of dependency functions.

**Proposition 19.** *1. If $Q \subseteq P$, then $D_i(Q) \leq D_i(P)$, $i \in \{1, 2, \ldots, 7\}$.*

2. *If $R_P \subseteq R_Q$, then $D_i(P) \geq D_i(Q)$, $i \in \{1, 3\}$.*
3. *If $R_{N\setminus P} \subseteq R_{N\setminus Q}$, then $D_i(P) \leq D_i(Q)$, $i \in \{2, 5\}$.*
4. *$D_i(P \cup Q) \geq D_i(Q)$, $i \in \{1, 2, \ldots, 7\}$.*
5. *If $D_7(P) = 1$, then $D_6(P) = D_1(P) = 1$.*
6. *If $D_6(P) = 1$, then $D_1(P) = 1$.*
7. *If $D_1(P) = 0$, then $D_7(P) = D_6(P) = 0$.*
8. *If $D_6(P) = 0$, then $D_7(P) = 0$.*
9. *$D_1(P) = 1$ if and only if $D_2(N \setminus P) = 0$.*
10. *$D_1(P) = 0$ if and only if $D_2(N \setminus P) = 1$.*

*Proof.* 1. Note that for $Q \subseteq P$, we have $A_i(Q) \subseteq A_i(P)$ for $i \in \{1,2,6,7\}$ and $A_i(P) \subseteq A_i(Q)$ for $i \in \{3,4,5\}$. This gives $D_i(Q) \leq D_i(P)$ for $i \in \{1,2,\ldots,7\}$.

2 follows from the fact that when $R_P \subseteq R_Q$, then $A_3(P) \subseteq A_3(Q)$ and $A_1(Q) \subseteq A_1(P)$.

3 follows from the fact that when $R_{N\setminus P} \subseteq R_{N\setminus Q}$, then $A_2(P) \subseteq A_2(Q)$ and $A_5(Q) \subseteq A_5(P)$.

4 follows from (1).

5- 8 follows from the fact that $A_7(P) \subseteq A_6(P) \subseteq A_1(P)$.

9 follows from the fact that $A_1(P) = U$ if and only if $A_2(N \setminus P) = \emptyset$.

10 follows from the fact that$A_1(P) = \emptyset$ if and only if $A_2(N \setminus P) = U$. □

## 5 Membership Functions

In this section, our aim is to define some membership functions for $MSAS^G$. As a $MSAS^G$ involves information from multiple sources, membership functions must grade the objects considering the information provided by all the sources. Here, we shall define membership functions based on

**(i)** number of sources which consider objects positive or negative element of the sets; and
**(ii)** the distinguishability of the objects in the approximation spaces corresponding to sources.

**(i)** Let us recall that given an approximation space $(U, R)$ and $X \subseteq U$, the domain $U$ is divided into three disjoint sets namely $\underline{X}_R$, $\overline{X}_R \setminus \underline{X}_R$ and $(\overline{X}_R)^c$. The membership function $f$ satisfies $1 = f(x_1, X) > f(x_2, X) > f(x_3, X) = 0$ for $x_1 \in \underline{X}_R$, $x_2 \in \overline{X}_R \setminus \underline{X}_R$ and $x_3 \in (\overline{X}_R)^c$. Similarly, as shown in Section 2, given a $MSAS^G$ $\mathfrak{F}$ and $X \subseteq U$, the domain $U$ is divided into five disjoint sets namely, $\underline{X}_s$, $\underline{X}_w \setminus \underline{X}_s$, $\overline{X}_s \setminus \underline{X}_w$, $\overline{X}_w \setminus \overline{X}_s$, and $(\underline{X}_w)^c$. So, we would like to have a membership function $G$ which satisfies the following:

$$1 = G(x_1, X) > G(x_2, X) > G(x_3, X) > G(x_4, X) > G(x_5, X) = 0,$$

for $x_1 \in \underline{X}_s$, $x_2 \in \underline{X}_w \setminus \underline{X}_s$, $x_3 \in \overline{X}_s \setminus \underline{X}_w$, $x_4 \in \overline{X}_w \setminus \overline{X}_s$ and $x_5 \in (\underline{X}_{w(P)})^c$. Since, the above division is made based on the number of sources which consider objects positive or negative element of $X$, the definition of $G$ should also be on the same line. In [2], a mathematical formulation of such a membership function is given. It is defined as

$$G(X, x) := \frac{\sum_{i \in N} \mu_i(X, x)}{|N|},$$

assuming $MSAS^G$ is of finite cardinality, where each $\mu_i$ is a three valued function which takes the value 1, $\frac{1}{2}$ and 0 for $x$ according as $x$ belongs to $\underline{X}_{R_i}$, or $\overline{X}_{R_i} \setminus \underline{X}_{R_i}$ or $(\overline{X}_{R_i})^c$.

Note that the inclusion $\underline{X}_w \subseteq \overline{X}_s$ guarantees that there cannot be two sources such that one considers an object to be a positive element of a set but the other considers it a negative element of the same set. Due to this fact, we obtain the following.

**Proposition 20.** *$G(x, X) > G(y, X)$, if and only if more sources consider $x$ to be a positive element of $X$ than $y$, or less sources consider $x$ to be a negative element of $X$ than $y$.*

**(ii)** Let us recall the definition of the membership function in an approximation space $(U, R)$ [20]. It is given as

$$f(x, X) := \frac{|[x]_R \cap X|}{|[x]_R|}.$$

So, $f$ depends on the number of objects which are indistinguishable from $x$. But in the case of $MSAS^G$, an object $y$ may be indistinguishable from $x$ for some source, but may not be so for the other. In order to handle this situation, one can give some *threshold* $k$ such that if the ratio of the number of sources which cannot distinguish $y$ with $x$ with the total number of sources exceeds that threshold, then we take $y$ indistinguishable from $x$. For instance, if we get information, say from five sources, then we can put the threshold to be $\frac{2}{5}$ and so if at least two sources can distinguish $x$ and $y$, then we take $x$ and $y$ distinguishable. Similarly, one may wish to make two objects distinguishable if any of the sources can distinguish it. Now, using this distinguishability relation, one can define the membership function in the same way as in the case of a Pawlak approximation space. We formally present this below.

Let us consider here $MSAS^G$ $\mathfrak{F} := (U, \{R_P\}_{P \subseteq N})$ with finite $N$ and $U$. For basic notions on fuzzy sets, we refer to [21].

**Definition 9.** *The* degree of indiscernibility *of objects $x, y \in U$ in $\mathfrak{F}$, is given by a fuzzy relation $R$ on $U \times U$ defined as:*

$$R(x, y) = \frac{|rel(x, y)|}{|N|}, \quad x, y \in U,$$

*where $rel(x, y) := \{R_i : (x, y) \in R_i, i \in N\}$ (so that $0 \leq |rel(x, y)| \leq |N|$).*

$R$ is clearly a (fuzzy) reflexive and symmetric relation. Now, consider the fuzzy set $R_x$, where

$$R_x(y) := R(x, y), \ x, y \in U.$$

Note that (i) $R_x(y) = R_y(x)$ and (ii) $R_x(x) = 1$.
Consider the $\alpha - cut$, $\alpha \in (0, 1]$ of the set $R_x$, i.e.

$$R_x^\alpha := \{y \in U : R_x(y) \geq \alpha\}$$

and for $\alpha = 0$, consider the strict $\alpha - cut$, i.e.

$$R_x^{0^+} := \{y \in U : R_x(y) > 0\}.$$

*Remark 1.* Note that in the case when $MSAS^G$ consists of only one relation, say $R_1$, then $R_x$ is actually the characteristic function of the set $[x]_{R_1}$; moreover, $R_x^\alpha = R_x^{0+} = [x]_{R_1}$.

Some simple properties of $R_x^\alpha$ and $R_x^{0+}$ are given by the following proposition.

**Proposition 21.** *1.* $R_x^\alpha \subseteq R_x^\beta$ *for* $\alpha \geq \beta$.
*2.* $x \in R_x^\alpha \cap R_x^{0+}$.
*3. If* $y \in R_x^\alpha$, *then* $x \in R_y^\alpha$.
*4.* $R_x^1 = [x]_{R_N}$.
*5.* $x \in R_y^1$ *and* $y \in R_z^1$, *then* $x \in R_z^1$.
*6.* $R_x^{0^+} = [x]_{R_\mathfrak{F}}$, *where* $R_\mathfrak{F} := \bigcup_{i \in N} R_i$ *(cf. Section 2.2).*

Given the above notion of $R_x^\alpha$ and $R_x^{0+}$, a natural generalization of membership function would be the following.

$$f_\alpha(X, x) := \frac{|R_x^\alpha \cap X|}{|R_x^\alpha|}, \text{ for } \alpha \in (0, 1] \text{ and}$$

$$f_0(X, x) := \frac{|R_x^{0^+} \cap X|}{|R_x^{0^+}|}.$$

From Remark 1, it follows that in the case when $MSAS^G$ consists of a single relation $R_1$, then $f_\alpha$ and $f_0$ reduce to the membership function on the approximation space $(U, R_1)$.
Let us consider the boundary cases of the above membership function. First, consider the case when $\alpha$ is 1. From Proposition 21(4), it follows that $f_1$ is nothing but the membership function on the approximation space $(U, R_N)$. In the case when $\alpha$ is 0, Proposition 21(6) shows that $f_0$ is actually the membership function on the tolerance approximation space $(U, R_\mathfrak{F})$ (cf. Section 2.2). Note that $[x]_{R_\mathfrak{F}}$ consists of all those objects which cannot be distinguished from $x$ by at least one source. Another interesting case is when $\alpha$ is $\frac{1}{2}$. In this case, we note that $R_x^{\frac{1}{2}}$ consists of those objects which are indistinguishable from $x$ by at least half or more of the total number of sources.

Apart from the membership functions discussed above, one may think of some natural and simple membership functions. For instance, for the $MSAS^G$ $\mathfrak{F} := (U, \{R_P\}_{P \subseteq N})$ with finite $N$, one can consider the mean of the membership functions on the approximation spaces $(U, R_i), i \in N$. That is, we define a membership function $M : 2^U \times U \to [0, 1]$ as

$$M(X, x) := \frac{\sum_{i \in N} f_i(X, x)}{|N|},$$

where $f_i$ is the membership function on the approximation space $(U, R_i)$, $i \in N$. In some applications, a 'preference' may need to be accorded to information provided by one source over that by another. In such cases, one may define membership functions by considering weighted mean. We would also like to mention that all the membership functions defined in this section consider the information from all sources. But, if one so wishes, then these definitions can be trivially modified for any subset of sources.

*Example 8.* Let us again consider the $MSAS^G$ of Example 1 and the set $X := \{C, M, D\}$. Calculating the membership functions $f_0, f_{\frac{1}{2}}, f_1$ and $G$, we obtain the following ordering. We do not mention $X$, as it is understood.

$$0 = G(K) < G(Ch) < G(B) < G(M) < G(D) < G(C) < 1.$$

$$0 = f_1(K) = f_1(Ch) = f_1(B) < f_1(M) = f_1(D) = f_1(C) = 1.$$

$$0 = f_{\frac{1}{2}}(K) = f_{\frac{1}{2}}(Ch) < f_{\frac{1}{2}}(B) < f_{\frac{1}{2}}(M) < f_{\frac{1}{2}}(D) = f_{\frac{1}{2}}(C) = 1.$$

$$0 = f_0(K) < f_0(Ch) < f_0(M) < f_0(B) = f_0(D) = f_0(C) = 1.$$

From the above table we note the following facts.
(i) According to agency $M_1$, $M$ is a positive element of $X$ and $C$ is a boundary element of $X$. But when we consider the information provided by each source, it appears that $C$ has a greater possibility to be an element of $X$ than $M$.

(ii) $M$ and $D$ are both possible positive elements of $X$. But $G$ makes it clear that $D$ has a greater possibility to be an element of $X$ than $M$. Moreover, $f_0$ and $f_{\frac{1}{2}}$ also strengthen this observation.

(iii) $B$ is a certain boundary element of $X$ and $M$ is a possible positive element. But according to $f_0$, $B$ has a greater possibility to be an element of $X$ than $M$.

(iv) $f_1(B) = 0$, but $f_0(B) \neq 0$ as well as $f_{\frac{1}{2}}(B) \neq 0$. In fact, we have $f_0(B) > f_0(M)$, where $M$ is a possible positive element of $X$.

The next proposition lists some of the properties of the membership functions given above. These can be derived easily.

**Proposition 22.** *1. For $g \in \{G, f_\alpha, M\}$, we have the following.*

*(a) If $(x, y) \in R_i$ for all $i \in N$, then $g(X, x) = g(Y, x)$.*

*(b) If $X \subseteq Y$, then $g(X, x) \leq g(Y, x)$ for all $x \in U$.*

*2. If $A = \{X_1, X_2, \ldots, X_m\}$ is a family of pairwise disjoint subsets of $U$, then $g(\bigcup_{i=1}^n X_i, x) = \sum_{i=1}^m g(X_i, x)$, for $g \in \{f_\alpha, M\}$.*

*3. For $g \in \{G, f_\alpha, M\}$, $g(X^c, x) = 1 - g(X, x)$.*

*4. $g(X, x) = 1$ if and only if $x \in \underline{X}_s$, for $g \in \{G, f_0, M\}$.*

*5. $g(X, x) = 0$ if and only if $x \notin \overline{X}_w$, for $g \in \{G, f_0, M\}$.*

*6. $g(X \cap Y, x) = 1$ if and only if $g(X, x) = 1$ and $g(Y, x) = 1$, for $g \in \{G, f_\alpha, M\}$.*

*7. $g(X \cup Y, x) = 0$ if and only if $g(X, x) = 0$ and $g(Y, x) = 0$, for $g \in \{G, f_\alpha, M\}$.*

*8. $0 < G(X, x) < \frac{1}{2}$ if and only if $x \in \overline{X}_w \setminus \overline{X}_s$.*

*9.* $\frac{1}{2} < G(X,x) < 1$ *if and only if* $x \in \underline{X}_w \setminus \underline{X}_s$.

*10.* $G(X,x) = \frac{1}{2}$ *if and only if* $x \in \overline{X}_s \setminus \underline{X}_w$.

We end this section with the remark that the properties given in items 2-7 are standard properties of the rough membership function with $\underline{X}_s$ and $\overline{X}_w$ replaced with Pawlak's lower and upper approximation respectively. Thus with respect to these properties, the functions $f_0$ and $M$ behave in a manner similar to the standard rough membership function.

## 6 Conclusions

We continue our study of rough sets in a multiple-source situation and define multiple-source approximation systems with group knowledge base ($MSAS^G$), where we also have equivalence relations representing the combined knowledge base of groups of sources. Standard notions, such as approximation of sets, dependency, membership functions related with Pawlak approximation spaces are generalized to define these notions on $MSAS^G$. In this direction, the notions of strong/weak lower and upper approximation introduced in [1], are defined with respect to a group of sources and relationships with combined knowledge base of groups are also presented. Moreover, lattice structures arising from the concepts defined on $MSAS^G$ are also studied – the paper presents an investigation which needs to be pursued further.

In [1], a quantified propositional modal logic $LMSAS$ is defined to facilitate formal reasoning with rough sets in multiple-source approximation systems ($MSAS$). The language of $LMSAS$ is not strong enough to express set approximations with respect to combined knowledge base of group of sources and hence not suitable for $MSAS^G$. Thus, we need to extend $LMSAS$ suitably to obtain a logic for $MSAS^G$, where one can express the properties and concepts related to $MSAS^G$.

In this paper, we have restricted our study to equivalence relations only. But depending on the application, the Pawlak approximation space is generalized in several ways (cf. [22,23,24,25,26,27]). For generalized approximation spaces, there has been much work, for instance, on exploring membership functions [28,20,29,30,31,23,32,22]. Some basic differences from the cases considered here may arise in the generalized models. For instance, we observe that covering and neighbourhood based approximation spaces would admit a situation where one source considers an object to be a positive element of a set, but another considers that object to be a negative element of the same set. So a multiple-source extension of these generalized spaces would differ from $MSAS^G$s. It seems worth investigating, how the basic ideas of this work may be extended to these formalisms.

We have made the assumption here that each source is equally preferred as far as deciding membership of an object is concerned. But one may extend the notion of a $MSAS^G$ to include a preference order on the set of sources. Notions such as approximations of sets or membership functions would then depend on the knowledge base of the sources of the system as well as on the position of the sources in the hierarchy giving the preference of sources.

## References


1. Khan, M.A., Banerjee, M.: Formal reasoning with rough sets in multiple-source approximation systems. International Journal of Approximate Reasoning 49(2), 466–477 (2008)
2. Khan, M.A., Banerjee, M.: Multiple-source approximation systems: membership functions and indiscernibility. In: Wang, G., Li, T., Grzymala-Busse, J.W., Miao, D., Skowron, A., Yao, Y. (eds.) RSKT 2008. LNCS (LNAI), vol. 5009, pp. 80–87. Springer, Heidelberg (2008)
3. Pawlak, Z.: Rough Sets. Theoretical Aspects of Reasoning about Data. Kluwer Academic Publishers, Dordrecht (1991)
4. Ras, Z., Dardzinska, A.: Ontology-based distributed autonomous knowledge systems. Information Systems 29, 47–58 (2004)
5. Pawlak, Z.: Rough sets. International Journal of Computer and Information Science 11(5), 341–356 (1982)
6. Farinas Del Cerro, L., Orłowska, E.: *DAL* – a logic for data analysis. Theoretical Computer Science 36, 251–264 (1997)
7. Wong, S.K.M.: A rough set model for reasoning about knowledge. In: Polkowski, L., Skowron, A. (eds.) Rough Sets in Knowledge Discovery 1: Methodology and Applications, pp. 276–285. Physica, Heidelberg (1998)
8. Liau, C.J.: An overview of rough set semantics for modal and quantifier logics. International Journal of Uncertainty, Fuzziness and Knowledge-Based Systems 8(1), 93–118 (2000)
9. Rasiowa, H.: Mechanical proof systems for logic of reaching consensus by groups of intelligent agents. Int. J. Approximate Reasoning 5(4), 415–432 (1991)
10. Rauszer, C.M.: Knowledge representation systems for groups of agents. In: Woleński, J. (ed.) Philosophical Logic in Poland, pp. 217–238. Kluwer, Dordrecht (1994)
11. Rauszer, C.M.: Rough logic for multiagent systems. In: Masuch, M., Polos, L. (eds.) Logic at Work 1992. LNCS (LNAI), vol. 808, pp. 161–181. Springer, Heidelberg (1994)
12. Pagliani, P., Chakraborty, M.K.: Information quanta and approximation spaces I: non-classical approximation operators. In: Proc. 2005 IEEE Conf. on Granular Computing, pp. 605–610. IEEE Press, Los Alamitos (2005)
13. Skowron, A.: Approximate reasoning in *MAS*: rough set approach. In: Proc. 2006 IEEE/WIC/ACM Conf. on Intelligent Agent Technology, pp. 12–18. IEEE Computer Society, Washington (2006)
14. Banerjee, M., Khan, M.A.: Propositional logics from rough set theory. Transactions on Rough Sets VI, 1–25 (2007)
15. Pagliani, P.: Pretopologies and dynamic spaces. Fundamenta Informaticae 59(2-3), 221–239 (2004)
16. Orłowska, E.: Kripke semantics for knowledge representation logics. Studia Logica XLIX, 255–272 (1990)
17. Komorowski, J., Pawlak, Z., Polkowski, L., Skowron, A.: Rough sets: a tutorial. In: Pal, S.K., Skowron, A. (eds.) Rough Fuzzy Hybridization: A New Trend in Decision-Making, pp. 3–98. Springer, Singapore (1999)

18. Birkhoff, G.: Lattice theory. American Mathematical Society Colloquium Publications, New York (1967)
19. Rasiowa, H.: An Algebraic Approach to Non-classical Logics. North Holland, Amsterdam (1974)
20. Wong, S.K.M., Ziarko, W.: Comparison of the probabilistic approximate classification and fuzzy set model. Fuzzy Sets and Systems 21, 357–362 (1986)
21. Klir, G.J., Yuan, B.: Fuzzy Sets and Fuzzy Logic: Theory and Applications. Prentice-Hall, Englewood Cliffs (1995)
22. Skowron, A., Stepaniuk, J.: Tolerance approximation spaces. Fundamenta Informaticae 27, 245–253 (1996)
23. Yao, Y.Y., Wong, S.K.M., Lin, T.Y.: A review of rough set models. In: Lin, T.Y., Cercone, N. (eds.) Rough Sets and Data Mining: Analysis for Imprecise Data, pp. 47–75. Kluwer, Boston (1997)
24. Ślęzak, D., Ziarko, W.: The investigation of the Bayesian rough set model. International Journal of Approximate Reasoning 40, 81–91 (2005)
25. Yao, Y.Y.: Probabilistic rough set approximations. International Journal of Approximate Reasoning 49(2), 255–271 (2008)
26. Ziarko, W.: Probabilistic approach to rough sets. International Journal of Approximate Reasoning 49(2), 272–284 (2008)
27. Greco, S., Matarazzo, B., Słowiński, R.: Parametrized rough set model using rough membership and Bayesian confirmation measures. International Journal of Approximate Reasoning 49(2), 285–300 (2008)
28. Wong, S.K.M., Ziarko, W., Li Ye, R., Li, R.: Comparison of rough-set and statistical methods in inductive learning. International Journal of Man-Machine Studies 24, 53–72 (1986)
29. Pawlak, Z., Wong, S.K.M., Ziarko, W.: Rough sets: probabilistic versus deterministic approach. International Journal of Man-Machine Studies 29, 81–95 (1988)
30. Ziarko, W.: Variable precision rough set model. Journal of Computer and System Sciences 46, 39–59 (1993)
31. Pawlak, Z., Skowron, A.: Rough membership functions. In: Yager, R., Fedrizzi, M., Kacprzyk, J. (eds.) Advances in the Dempster-Shafer Theory of Evidence, pp. 251–271. John Wiley and Sons, Inc., New York (1994)
32. Intan, R., Mukaidono, M.: Generalization of rough membership function based on $\alpha-$coverings of the universe. In: Pal, N.R., Sugeno, M. (eds.) AFSS 2002. LNCS (LNAI), vol. 2275, pp. 129–135. Springer, Heidelberg (2002)

# A New Knowledge Reduction Algorithm Based on Decision Power in Rough Set

Jiucheng Xu and Lin Sun

College of Computer and Information Technology
Henan Normal University, Xinxiang Henan 453007, China
xjch3701@sina.com, slinok@126.com

**Abstract.** Many researchers are working on developing fast data mining methods for processing huge data sets efficiently, but some current reduction algorithms based on rough sets still have some disadvantages. In this paper, we indicated their limitations for reduct generation, then a new measure to knowledge was introduced to discuss the roughness of rough sets, and we developed an efficient algorithm for knowledge reduction based on rough sets. So, we modified the mean decision power, and proposed to use the algebraic definition of decision power. To select optimal attribute reduction, the judgment criterion of decision with an inequality was presented and some important conclusions were obtained. A complete algorithm for the attribute reduction was designed. Finally, through analyzing the given example, it is shown that the proposed heuristic information is better and more efficient than the others, and the presented method in the paper reduces time complexity and improves the performance. We report experimental results with several data sets from UCI Machine Learning Repository, and we compare the results with some other methods. The results prove that the proposed method is promising, which enlarges the application areas of rough sets.

**Keywords:** Rough set, knowledge reduction, decision table, conditional entropy, positive region, algebra view, decision power, heuristic information.

## 1 Introduction

The theory of rough sets, proposed by Pawlak [1] in the 1980s, as a valid mathematical tool that deals with imprecise, uncertain, vague or incomplete knowledge of a decision system, is an extension of set theory for the study of intelligent systems characterized by insufficient and incomplete information, and it is established on the basis of classification mechanism, which takes classification according to equivalence relation. Because knowledge holds granularity, then the finer the data classification, the smaller the information granularity, and the more precise the knowledge will be, so it is known that data classification and knowledge granulation are associated with equivalence relation [2-4]. Therefore, rough set theory has been applied to many areas successfully including expert systems, machine learning, pattern recognition, decision analysis, process control,

J.F. Peters et al. (Eds.): Transactions on Rough Sets XII, LNCS 6190, pp. 76–89, 2010.

and knowledge discovery in databases [5-7]. In recent years, rough set theory has become one of flash point in the research area of information science, and then it has been widely discussed and used in attribute reduction and feature selection. Attribute reduction, also called feature subset selection, is usually employed as a preprocessing step to select part of the features and focuses the learning algorithm on the relevant information, and then reduct means a minimal attribute subset with the same approximating power as the whole set in an information system [8-11], that is to say, the reduct should have the least redundant information and not loss the classification ability of the raw data set. What's more, the attributes in a reduct should not only be strongly relevant to the learning task, but also be not redundant with each other.

Up to now, reduction of knowledge has been one of the most important topics, thus attribute reduction is so significant in lots of applications that many experts major in researching on a series of approaches to search reducts [4, 8-30], and then many types of reducts have been proposed, each of which aimed at some basic requirements. A fundamental notion supporting such applications is the concept of attribute reduction [1]. The objective of reduct construction is to reduce the number of attributes, and at the same time, preserve a certain property that we want. In the following, we briefly review these relevant literatures.

Pawlak first proposed attribute reduction from the algebraic point of view. In [12, 13], Wang proposed the reducts from the algebra and information views, which are equivalent in a consistent decision table, while the reduct based on the algebra view is included in the information view for an inconsistent decision table, and then Wang introduced two novel heuristic algorithms of knowledge reduction with the time complexity $O(|C||U|^2)+O(|U|^3)$ and $O(|C|^2|U|)+O(|C||U|^3)$, respectively, where $|C|$ and $|U|$ denote the number of condition attributes and objects in $U$ respectively. In [14], based on the mutual information, the heuristic algorithm costing the time complexity $O(|C||U|^2) + O(|U|^3)$ was constructed. In [15, 16], the discernibility matrix studied roundly is also a good method to design attribute reduction algorithms, but it has been proved that attribute reduction based on the discernibility matrix of Skowron is not equal to that based on the information view. In [10], matrices were introduced to store the features to distinguish the corresponding pair of objects. What's more, Boolean operations were conducted on the matrices to search all of the reducts. However, the main disadvantage of this method is much space and time cost. Based on the indiscernibility relation and positive region, a complete algorithm for reduction of attributes with time complexity $O(|C|^2|U|\log|U|)$ and space complexity $O(|C||U|)$ was introduced in [17]. Recently, in [18], a new and relatively reasonable formula was designed for an efficient attribute reduction algorithm, whose worst time complexity was cut down to $\mathrm{Max}(O(|C||U|), O(|C|^2|U/C|))$. At present, the best idea of many algorithms for computing $U/C$ is based on quick sorting, and its time complexity is $O(|C||U|\log|U|)$, which is not very good. However, in [18], a new algorithm based on radix sorting for computing $U/C$ is provided, and its complexity is cut down to $O(|C||U|)$, so far, which is fortunate. In [20], by using the lower and upper approximations in rough set theory, knowledge hidden in information systems could be unravelled and expressed in the form of decision rules.

In the above studies, the whole approaches is inspired by the notion of inadequacy of available information to perform complete classification of objects belonging to a specified category. However, these presented reduction algorithms have still their some own limitations, such as sensitivity to noises, relatively high complexities and so on. What's more, Hu [19] proposed also some problems in dependency based attribute reduction. The dependency function in rough set approaches is the ratio of sizes of the positive region over the sample space, which is the sample set which can be undoubtedly classified into a certain class according to the existing attributes. So it is seen that the dependency function ignores the influence of boundary samples, which may be belong to more than one class. Then in classification learning, the boundary samples also exert an influence on the learned results. Liu [26] proposed some drawbacks in dealing with knowledge reduction for an inconsistent decision table. In [21, 22], it is known that reliability and coverage of a decision rule are all the most important standards for estimating the decision quality. Thus, through detail analyses, we have the proposition that these algorithms [5-7, 12-18, 23] can't reflect objectively the change of decision quality as fully as possible.

To compensate for their limitations, we construct a new method for separating consistent objects from inconsistent objects, the corresponding judgment criterion with an inequality, and many propositions, which are used in searching for the minimal or optimal reducts. Then we design a new heuristic reduction algorithm with relatively lower time complexity. In a large decision table, since usually $|U| \gg |C|$, by modifying the mean decision power, the proposed reduction algorithm will be more efficient than them discussed above. Finally, six data sets from UCI Machine Learning Repository are used to illustrate the performance of the proposed algorithm and the comparative result with the existing methods is reported in detail.

## 2 Rough Set Theory Preliminaries

An information system is usually denoted by a quintuplet $S = (U, C, D, V, f)$, called a decision table, where $U$ is the nonempty finite set of objects, called the universe, $C$ is the set of condition attributes, $D$ is the set of decision attributes, and $C \cap D = \emptyset$. $V = \cup\{V_a | a \in C \cup D\}$, where $V_a$ is the value range of attribute $a$. And $f$: $U \times (C \cup D) \rightarrow V$ is an information function, which an information value for each attribute of object holds, i. e., $\forall a \in C \cup D$, $x \in U$, $f(x, a) \in V_a$.

In a decision table $S = (U, C, D, V, f)$, every attribute subset $P \subseteq C \cup D$ determines a binary discernibility relation, denoted by $IND(P) = \{(x, y) \in U \times U | f(x, a) = f(y, a), \forall a \in P\}$. $IND(P)$ is called an equivalence relation, and partitions $U$ into a family of a disjoint subsets $U/IND(P)$ (in short $U/P$) called a quotient set of $U$, denoted by $U/P = \{[x]_P | \forall x \in U\}$, where $[x]_P$ denotes the equivalence class determined by $x$ with respect to $P$, i. e., $[x]_P = \{y \in U \,|(x, y) \in IND(P)\}$. The sets $\underline{P}Y = \cup\{[x]_P | [x]_P \subseteq Y\}$ means the $P$-lower approximation of $Y$, and $\overline{P}Y = \cup\{[x]_P | [x]_P \cap Y \neq \emptyset\}$ means the $P$-upper approximation of $Y$, where $Y \subseteq U$.

In a decision table $S = (U, C, D, V, f)$, let $P, Q \subseteq C \cup D$ be two equivalent relations on $U$, $U/P = \{P_1, P_2, \cdots, P_t\}$, $U/Q = \{Q_1, Q_2, \cdots, Q_s\}$, and the partition $U/Q$ is coarser than the partition $U/P$, in symbols $U/P \leq U/Q$ if and only if $\forall P_i \in U/P$, and $\exists Q_j \in U/Q$ such that $P_i \subseteq Q_j$. If $U/P \leq U/Q$ and $U/P \geq U/Q$, then $U/P = U/Q$. If $U/P \leq U/Q$ and $U/P \neq U/Q$, then $U/Q$ is strictly coarser than $U/P$, i.e., $U/P < U/Q$.

Hence, we have the proposition that if $P, Q \subseteq C \cup D$ are two equivalent relations on $U$, and $Q \subseteq P$, then $U/P \leq U/Q$ always holds.

In a decision table $S = (U, C, D, V, f)$, let $P, Q \subseteq C \cup D$ be two equivalent relations on $U$, $U/P = \{X_1, X_2, \cdots, X_n\}$, $U/Q = \{Y_1, Y_2, \cdots, Y_m\}$, then the conditional entropy of knowledge $Q$ with respect to $P$ is denoted by

$$H(Q|P) = -\sum_{i=1}^{n} p(X_i) \sum_{j=1}^{m} p(Y_j|X_i) \log(p(Y_j|X_i)), \tag{1}$$

where $p(X_i) = \frac{|X_i|}{|U|}$, $p(Y_j|X_i) = \frac{|X_i \cap Y_j|}{|X_i|}$.

Hence, if $U/C = \{C_1, C_2, \cdots, C_m\}$ , and $U/D = \{D_1, D_2, \cdots, D_k\}$, then the conditional entropy of decision attributes set $D$ with respect to condition attributes set $C$ is denoted by

$$H(D|C) = -\sum_{i=1}^{m} p(C_i) \sum_{j=1}^{k} p(D_j|C_i) \log(p(D_j|C_i)). \tag{2}$$

In a decision table $S = (U, C, D, V, f)$, let $P, Q \subseteq C \cup D$ be two equivalent relations on $U$, $U/P=\{X_1, X_2, \ldots, X_r\}$, and $U/Q=\{X_1, X_2, \ldots, X_{i-1}, X_{i+1}, \ldots, X_{j-1}, X_{j+1}, \ldots, X_r, X_i \cup X_j\}$ be a new partition formed by unifying $X_i$ and $X_j$ in $U/P$ to $X_i \cup X_j$, then we have $U/P \leq U/Q$.

Meanwhile, let $B \subseteq C \cup D$ be an equivalent relation on $U$, suppose that $U/B = \{Y_1, Y_2, \ldots, Y_s\}$, then for the conditional entropy, $H(B|P) \leq H(B|Q)$ holds, and the equation holds if and only if $\frac{|X_i \cap Y_k|}{|X_i|} = \frac{|X_j \cap Y_k|}{|X_j|}$, where $\forall Y_k \in U/B$, and $\forall X_i, X_j \in U/P$. Hence, we have $H(D|P) \leq H(D|Q)$, and the equation holds if and only if $\frac{|X_i \cap D_r|}{|X_i|} = \frac{|X_j \cap D_r|}{|X_j|}$, where $\forall D_r \in U/D$, and $\forall X_i, X_j \in U/P$.

If $r \in P$, and $H(D|P) = H(D|P - \{r\})$, then $r$ in $P$ is unnecessary for $D$, else $r$ is necessary. If every element in $P$ is necessary for $D$, then $P$ is independent relative to $D$.

So in a decision table $S = (U, C, D, V, f)$, we have the proposition that if $\forall A_1 \subseteq A_2 \subseteq C$, then $H(D|A_1) \geq H(D|A_2)$ always holds, and then the necessary and sufficient condition of equation is that for any $X_i, X_j \in U/A_2$, $X_i \neq X_j$, if $X_i \cup X_j \subseteq U/A_1$, then $\frac{|X_i \cap D_r|}{|X_i|} = \frac{|X_j \cap D_r|}{|X_j|}$ always holds, where $\forall D_r \in U/D$.

Let $P$, $Q \subset C \cup D$ be two equivalent relations on $U$, then the $P$ positive region of $Q$ is defined as

$$POS_P(Q) = \cup\{\underline{P}Y|Y \in U/Q\}. \tag{3}$$

Thus, if $X \subseteq U$, and $P \subseteq C$, then we have the positive region of $P$ with respect to $D$ denoted by

$$POS_P(D) = \cup\{\underline{P}X|X \in U/D\}. \tag{4}$$

Hence, $P$-quality of approximation classification with respect to $D$ in $U$ is denoted by $\gamma_P(D) = \frac{|POS_P(D)|}{|U|}$.

Therefore, in a decision table $S = (U, C, D, V, f)$, on the basis of the definition of positive region, we have the proposition that if $POS_C(D) = U$, then the decision table is called a consistent one, otherwise an inconsistent one. Thus, we have the definition that the set $POS_C(D)$ is also called the (positive region) consistent object set, and $U - POS_C(D)$ is called the inconsistent object set [4].

# 3 The Proposed Approach

## 3.1 Limitations of Current Reduction Algorithms

Thus, we analyze algorithms based on the positive region and the conditional entropy deeply. Firstly, in a decision table $S = (U, C, D, V, f)$, if any $P \subseteq C$ is an equivalent relation on $U$, the $P$-quality of approximation with respect to $D$ is equal to the $C$-quality of approximation with respect to $D$, i.e., $\gamma_P(D) = \gamma_C(D)$, and there doesn't exist $P^* \subset P$ such that $\gamma_{P*}(D) = \gamma_C(D)$, then $P$ is called the reduct of $C$ with respect to $D$ [7, 15-18, 23]. Take these algorithms for example, whether or not any condition attributes is redundant depends on whether the lower approximation quality corresponding to decision set is changed or not, after the attribute is deleted from $C$. Accordingly if new inconsistent objects are added to the decision table, it is not taken into account whether the conditional probability distribution of the primary inconsistent objects is changed in their corresponding decision class. Hence, if the generated deterministic decision rules are the same, they will support the same important standards for estimating decision quality. Suppose that the generated deterministic decision rules are the same, that is to say, the prediction of these rules has not change. Thus it is seen that these presented algorithms only take into account whether or not the prediction of deterministic decision rules has change after reduction.

Secondly, in a decision table $S = (U, C, D, V, f)$, if any $P \subseteq C$ is an equivalent relation on $U$, $H(D|P) = H(D|C)$, and $P$ is independent relative to $D$, then $P$ is called the reduct of $C$ with respect to $D$ [12, 13, 24, 25]. Hence, whether any condition attributes is redundant or not depends on whether the conditional entropy value of decision table is changed or not, after the attribute is deleted from $C$. It is known that the conditional entropy value generated by $POS_C(D)$ is 0, thus $U - POS_C(D)$ can lead to a change of conditional entropy value. Due to the new added and primary inconsistent objects in their corresponding decision class, if their conditional probability distribution is changed, then it will cause the change of conditional entropy value of the whole decision table. Therefore, we have the proposition that the main criterions of these algorithms in estimating decision quality include two aspects, the invariability of the deterministic decision rules, the invariability of the reliability of nondeterministic decision rules.

Therefore, some researchers above only think about the change of reliability for all decision rules after reduction. However, in decision application, besides the reliability of decision rules, the object coverage of decision rules is also one of the most important standards of estimating decision quality. Hence, we have the proposition that these current reduction algorithms above can't reflect the change of decision quality objectively. Then it is necessary to seek for a new kind of measure to search for the precise reducts effectively.

### 3.2 Representation of Decision Power on Decision Table

Let $P$, $Q$ be two equivalent relations on $U$, then $U/(P \cup Q) = U/P \cap U/Q$ is obtained in [23].

Thus, in a decision table $S = (U, C, D, V, f)$, if $U/C = \{X_1, X_2, \cdots, X_n\}$, $U/D = \{Y_1, Y_2, \cdots, Y_m\}$, and $U/(C \cup D) = U/C \cap U/D = \{Z_1, Z_2, \cdots, Z_k\}$, then

$$\sigma_S = \frac{1}{k} \sum_{i=1}^{n} \sum_{j=1}^{m} \left( \frac{|X_i \cap Y_j|}{|X_i|} \times \frac{|X_i \cap Y_j|}{|U|} \right) \tag{5}$$

is called the mean decision power of $S$, where $k$ is the radix of partition for equivalent classes $\{Z_1, Z_2, \cdots, Z_k\}$ [28].

From (5), the $\frac{|X_i \cap Y_j|}{|X_i|}$ represents the reliability of any decision rule generated by decision table, the $\frac{|X_i \cap Y_j|}{|U|}$ represents the objects coverage corresponding to the decision rule above, and $k$ is the number of the generated decision rules corresponding to attribute sets before reduction.

The most concise decision rules set that satisfies condition of the mean decision power discussed in the paper [28] is regarded as the final reduction result in the new reduction model. Some experiments show that the mean decision power can acquire good standards. However, at all points of attribute reduction on decision table based on the mean decision power, we suppose that a new measure of knowledge reduction is presented without the number of original decision rules, compared with classical reduction algorithms. Thus, it not only occupies much smaller storage space, requires much lower computational costs, and implementation complexity, but also has no effects on helping to get the minimal or optimal reducts.

From the algebra view in [7, 15-17, 23], through analyzing the algorithms based on the positive region, we have the proposition that this significance of attribute is regarded as the quantitative computation of radix about positive region, which merely describes the subsets of certain classes in $U$, while from the viewpoint of information view in [12, 13, 24, 25], this significance of attribute is only obtained by detaching objects in different decision classes from equivalent classes of condition attributes subset. However, due to inconsistent objects, we have the proposition that the current defined measures for reducts to rough set, which are based on positive region and information view, when used as heuristic information, are still lacking in dividing $U$ into consistent objects sets and inconsistent objects sets in an inconsistent decision table. Therefore, based on a series of theoretical analyses above, we have the further conclusion that these heuristic reduction algorithms above will not be equivalent in the representation of concepts and operations for the inconsistent decision table.

Meanwhile, in a decision table $S = (U, C, D, V, f)$, if $A \subseteq C$ is an equivalent relation on $U$, then $POS_A(D) = POS_C(D)$ if and only if the $A$-lower approximation of $D_i$ is equal to the $C$-lower approximation of $D_i$, i.e., $\underline{A}D_i = \underline{C}D_i$, for any $D_i \in U/D$ [26].

Then suppose that $D_0 = U - POS_C(D)$, we have $\underline{C}D_0 = D_0$. If all subsets from $\{\underline{A}D_0, \underline{A}D_1, \underline{A}D_2, \ldots, \underline{A}D_m\}$ are nonempty, then the whole sets is created a new partition of $U$. Thus, we assume that $\underline{A}D_i$ is empty, for any $i$ ($i$=1, 2, ..., $m$), then $\underline{A}D_i$ is called a redundant set of the new partition. After all redundant sets are taken out, we can obtain another new partition of $U$, i. e., $\{\underline{A}D_0, \underline{A}D_1, \underline{A}D_2, \ldots, \underline{A}D_m\}$.

Thus, suppose that $A \subset C$, then in the partition $\{\underline{A}D_0, \underline{A}D_1, \underline{A}D_2, \ldots, \underline{A}D_m\}$, and all inconsistent objects are concentrated as the whole set $\underline{A}D_0$. What's more, another new partition of condition attribute set $C$ on $U$ is $\{\underline{C}D_0, \underline{C}D_1, \underline{C}D_2, \ldots, \underline{C}D_m\}$, then we have the new equivalence relation generated by this new partition, denoted by $R_D$, i.e., $U/R_D=\{\underline{C}D_0, \underline{C}D_1, \underline{C}D_2, \ldots, \underline{C}D_m\}$. Accordingly the decision partition $U/R_D$ not only covers consistent objects from the different decision classes in $U$, but also separates the consistent objects from the inconsistent objects, while $U/D$ is gained through extracting objects from different decision classes with respect to equivalence classes.

Thereby, based on $U/R_D$, the mean decision power is introduced and modified to discuss the roughness and attribute reduction in rough sets. Thus we propose the algebra definition of decision power, which has effects not only on the subsets of the certain classes, but also on the subsets of the uncertain (relative to the decision) classes in $U$.

**Definition 1.** In a decision table $S = (U, C, D, V, f)$, if $P \subseteq C$ is an equivalent relation on $U$, $U/P = \{X_1, X_2, \ldots, X_t\}$, $D = \{d\}$, $U/D = \{Y_1, Y_2, \ldots, Y_m\}$, and $U/R_D=\{\underline{C}Y_0, \underline{C}Y_1, \underline{C}Y_2, \ldots, \underline{C}Y_m\}$, then the decision power of equivalence relation $R_D$ with respect to $P$ is denoted by $S(R_D; P)$, defined as

$$S\left(R_D; P\right) = \sum_{i=1}^{t}\sum_{j=0}^{m}\left(\frac{|X_i \cap \underline{C}Y_j|}{|X_i|} \times \frac{|X_i \cap \underline{C}Y_j|}{|U|}\right) = \sum_{i=1}^{t}\sum_{j=0}^{m}\left(\frac{|X_i \cap \underline{C}Y_j|^2}{|X_i||U|}\right). \quad (6)$$

From (6), it is easily seen that any $X_i$ is defined by $C$-lower approximation, $X_i$ is a union of some equivalent classes of $C$-lower approximation, and $Y_j$ is also defined by $C$-lower approximation. Hence, the $C$-lower approximation of $Y_j$ is also a union of some equivalent classes of $C$-lower approximation. Thus, we obtain that the intersection of $X_i$ and $C$-lower approximation of $Y_j$ is equal to the union of all equivalent classes of $C$-lower approximation, which are included in the $C$-lower approximation of $Y_j$ and $X_i$. What's more, through the analyses above, it is known that $\frac{|X_i \cap \underline{C}Y_j|}{|X_i|}$ and $\frac{|X_i \cap \underline{C}Y_j|}{|U|}$ represent the reliability of a decision rule and the object coverage corresponding to that rule, respectively. Then the definition of decision power cover not only the reliability of decision rules, but also the object coverage of decision rules in a decision table, so that it can objectively reflect the change of decision quality as fully as possible.

**Theorem 1.** In a decision table $S = (U, C, D, V, f)$, for any $r \in P \subseteq C$, then we have $S(R_D; P) \geq S(R_D; P - \{r\})$.

*Proof.* To simplify notation in the proof, at the beginning, we consider only a special case. The proof in the general case goes in the analogous way. We assume that $U/P = \{X_1, \ldots, X_t\}$ and $U/(P - \{r\})$ contains the same classes as in $U/P$ with the only one exception that $X_p$ and $X_q$ are joined, i.e., $X_p \cup X_q$ is a class of $U/(P - \{r\})$, and $X_p$, $X_q$ are not. If many classes in $U/P$ will be also joined, after $r$ in $P$ is deleted, then the coalition may be considered as automatically comprising more two partitions continually. Thus, from (6), we have the formulas as follows:

$$
\begin{aligned}
S(R_D; P) &= \sum_{i=1}^{t} \sum_{j=0}^{m} \left(\frac{|X_i \cap \underline{C}Y_j|^2}{|X_i||U|}\right), \\
S(R_D; P - \{r\}) &= S(R_D; P) - \sum_{j=0}^{m} \left(\frac{|X_p \cap \underline{C}Y_j|^2}{|X_p||U|}\right) \\
&\quad - \sum_{j=0}^{m} \left(\frac{|X_q \cap \underline{C}Y_j|^2}{|X_q||U|}\right) + \sum_{j=0}^{m} \left(\frac{|(X_p \cup X_q) \cap \underline{C}Y_j|^2}{|X_p \cup X_q||U|}\right), \\
S_{\triangle} &= S(R_D; P) - S(R_D; P - \{r\}) \\
&= \sum_{j=0}^{m} \left(\frac{|X_p \cap \underline{C}Y_j|^2}{|X_p||U|}\right) + \sum_{j=0}^{m} \left(\frac{|X_q \cap \underline{C}Y_j|^2}{|X_q||U|}\right) \\
&\quad - \sum_{j=0}^{m} \left(\frac{|(X_p \cap \underline{C}Y_j) \cup (X_q \cap \underline{C}Y_j)|^2}{|X_p \cup X_q||U|}\right).
\end{aligned}
$$

Suppose that $|X_p| = x$, $|X_q| = y$, $|X_p \cap \underline{C}Y_j| = ax$, $|X_q \cap \underline{C}Y_j| = by$, there must exist $x > 0$, $y > 0$, $0 \leq a \leq 1$, and $0 \leq b \leq 1$. Thus, we get the formula below

$$
\begin{aligned}
S_{\triangle} &= \sum_{j=0}^{m} \frac{(ax)^2}{x|U|} + \sum_{j=0}^{m} \frac{(by)^2}{y|U|} - \sum_{j=0}^{m} \frac{(ax + by)^2}{(x + y)|U|} \\
&= \frac{1}{|U|} \sum_{j=0}^{m} \frac{xy(a - b)^2}{x + y}.
\end{aligned}
$$

Assume that a function $f_j = \dfrac{xy(a-b)^2}{x+y}$, for any $j$ ($j = 0, 1, \ldots, m$), thus, it is obviously true that when $a = b$, then $\dfrac{|X_p \cap \underline{C}Y_j|}{|X_p|} = \dfrac{|X_q \cap \underline{C}Y_j|}{|X_q|}$ always holds, i.e., $f_j = 0$.

Therefore, when any attribute $r$ in $P \subseteq C$ is deleted from the decision table $S$, there must exist $S_\triangle \geq 0$. Thus, the proposition $S(R_D; P) \geq S(R_D; P - \{r\})$ is true.

Hence, we obtain the conclusion that the decision power of knowledge decreases non-monotonously as the information granularities become finer.

**Theorem 2.** In a decision table $S = (U, C, D, V, f)$, let $P \subseteq C$ be an equivalent relation on $U$, $U/P = \{X_1, X_2, \ldots, X_t\}$, $D = \{d\}$, $U/D = \{Y_1, Y_2, \ldots, Y_m\}$, and $U/R_D$=$\{\underline{C}Y_0$, $\underline{C}Y_1$, $\underline{C}Y_2$, $\ldots, \underline{C}Y_m\}$. If the decision table $S$ is a consistent one, we have $\underline{C}Y_0 = \emptyset$, and $U/R_D = U/D$. We assume that $\dfrac{|X_p \cap \underline{C}Y_j|}{|X_p|} = \dfrac{|X_q \cap \underline{C}Y_j|}{|X_q|}$ such that $\dfrac{|X_p \cap Y_j|}{|X_p|} = \dfrac{|X_q \cap Y_j|}{|X_q|}$, then $S(R_D; P) = S(R_D; P - \{r\})$ $\Leftrightarrow H(D|P) = H(D|P - \{r\}) \Leftrightarrow \gamma_P(D) = \gamma_{p-\{r\}}(D)$, where $\forall X_p, X_q \in U/P$, $\underline{C}Y_j \in U/R_D$, and $Y_j \in U/D$.

*Proof.* One can prove Theorem 2 easily from Lemma 1 in [13].

**Theorem 3.** In a decision table $S = (U, C, D, V, f)$, let $P \subseteq C$ be an equivalent relation on $U$, $U/P = \{X_1, X_2, \ldots, X_t\}$, $D = \{d\}$, $U/D = \{Y_1, Y_2, \ldots, Y_m\}$, and $U/R_D$=$\{\underline{C}Y_0, \underline{C}Y_1, \underline{C}Y_2, \ldots, \underline{C}Y_m\}$. If the decision table $S$ is an inconsistent one, we have $\underline{C}Y_0 = Y_0 \neq \emptyset$. We assume that $\dfrac{|X_p \cap \underline{C}Y_0|}{|X_p|} = \dfrac{|X_q \cap \underline{C}Y_0|}{|X_q|}$ such that $\dfrac{|X_p \cap Y_0|}{|X_p|} = \dfrac{|X_q \cap Y_0|}{|X_q|}$, then $S(R_D; P) = S(R_D; P - \{r\}) \Leftrightarrow \gamma_P(D) = \gamma_{p-\{r\}}(D)$, where $\forall X_p, X_q \in U/P$, $\underline{C}Y_0 \in U/R_D$, and $Y_0 \in U/D$.

*Proof.* One can also prove Theorem 3 from Lemma 1 in [13].

**Corollary 1.** In a decision table $S = (U, C, D, V, f)$, let $P \subseteq C$ be an equivalent relation on $U$, then any attribute $r$ in P is said to be unnecessary with respect to $D$ if and only if $S(R_D; P) = S(R_D; P - \{r\})$.

**Definition 2.** In a decision table $S = (U, C, D, V, f)$, if $P \subseteq C$ is an equivalent relation on $U$, then the significance of any attribute $r \in C - P$ with respect to $D$ is denoted by $SGF(r, P, D)$, defined as

$$SGF(r, P, D) = S(R_D; P \cup \{r\}) - S(R_D; P). \tag{7}$$

Notice that when $P = \emptyset$, then we have $SGF(r, \emptyset, D) = S(R_D; \{r\})$.

From Lemma 1 and (7), we know that if $SGF(r, P, D) = 0$, then this significance of attribute based on positive region is also 0, on the other hand, if the radix of positive region fills out after adding any attribute, then, that significance of attribute isn't 0. What's more, we also have $SGF(r, P, D) \neq 0$. Therefore, $SGF(r, P, D)$ can not only include more information than that based on positive region, but also compensate for some limitations of significance of attribute based on the algebra point of view and the information point of view.

**Definition 3.** In a decision table $S = (U, C, D, V, f)$, if $P \subseteq C$ is an equivalent relation on $U$, $S(R_D; P) = S(R_D; C)$, and $S(R_D; P^*) < S(R_D; P)$, for any $P^* \subset P$, then $P$ is an attribute reduction of $C$ with respect to $D$.

Definition 3 indicates that reduct of condition attributes set $C$ is a subset, which can discern decision classes with the same discriminating capability as $C$, and none of the attributes in the reduct can be eliminated without decreasing its discriminating capability.

### 3.3 Design of Reduction Algorithm Based on Decision Power

We know that the calculated $S(R_D; P)$ is the same every time, then calculating any attribute $r$ with the maximum of $SGF(r, P, D)$, used as heuristic information, is in fact to calculate that with the maximum of $S(R_D; P \cup \{r\})$. Therefore, we only need to calculate $S(R_D; P \cup \{r\})$ except $S(R_D; P)$. However, calculating $S(R_D; P \cup \{r\})$ is in fact to calculate corresponding to partitions and positive region principally, and we then make full use of the effective computational methods of equivalence (indiscernibility) classes, positive region and attribute importance in [18], attribute core in [27, 30]. Thus it will help to reduce the quantity of computation, the time and space of search.

The processes of reduction implementation consists of the following steps, first, detaching objects from equivalent classes step by step, then determining attribute core of the whole objects, finally, obtaining the minimum relative reducts through adding attributes bottom-up.

**Algorithm1. Knowledge Reduction Based on Decision Power**
**Input:** Decision table $S = (U, C, D, V, f)$
**Output:** A relative attribute reduction $P$
(1) Compute $POS_C(D)$, and $U - POS_C(D)$ for the partition $U/R_D$
(2) Compute $S(R_D; C)$, and $CORE_D(C)$
(3) Initialize $P = CORE_D(C)$
(4) If $P=\emptyset$
then turn to (6)
(5) If $S(R_D; P) = S(R_D; C)$
then turn to (11)
(6) Compute $S(R_D; P \cup \{r\})$ for any $r \in C - P$
(7) Select an attribute $r$ with the maximum of $S(R_D; P \cup \{r\})$
(8) If this $r$ is not only
then select that with the maximum of $|U/(P \cup \{r\})|$
else select the front
(9) Let $P = P \cup \{r\}$
(10) If $S(R_D; P) \neq S(R_D; C)$
then turn to (6)
else let $P^* = P - CORE_D(C)$
Let $t = |P^*|$

For $i$=1 to $t$ begin
Let $r_i \in P^*$
Let $P^* = P^* - \{r_i\}$
If $S(R_D; P^* \cup CORE_D(C)) < S(R_D; P)$
then let $P^* = P^* \cup \{r_i\}$
End
Let $P = P^* \cup CORE_D(C)$
(11) Output $P$ called a minimum relative attribute reduction
(12) End

It is clear that this algorithm is complete, that is to say, none of the attributes in $P$ can be eliminated again without decreasing its discriminating quality, whereas many algorithms are incomplete, which can't ensure that the final reducts will be obtained [17]. Thus, the algorithms in [4, 17, 23, 26, 29] are complete, but the algorithms in [13, 14] are not.

In the following, we discuss and analyze its time complexity, and space complexity of search. The worst time complexity of the step 1 is $O(|C||U|)$. From step 2 to step 6, the most numbers is $|C|$, thus, the time complexity of this algorithm is $O(|C||U|) + O((|C|-1)|U|) + O((|C|-2)|U|) + \ldots + O(|U|)=O(|C|^2|U|)$. Hence, we can easily see that the time complexity of algorithm is $O(|C|^2|U|)$ in the worst case, which is less than that of [12-17, 22, 23, 25, 26, 28, 29], and the worst space complexity of algorithm is $O(|C||U|)$.

## 4 Experimental Results

In Table 1, we report an inconsistent decision table $S = (U, C, D, V, f)$, where $U = \{x_1, x_2, \ldots, x_{10}\}$, $C = \{a_1, a_2, \ldots, a_5\}$, and $D = \{d\}$.

**Table 1.** Inconsistent Decision Table $S$

| $U$ | $a_1$ | $a_2$ | $a_3$ | $a_4$ | $a_5$ | $d$ |
|---|---|---|---|---|---|---|
| $x_1$ | 0 | 1 | 0 | 0 | 1 | 0 |
| $x_2$ | 0 | 0 | 0 | 1 | 0 | 0 |
| $x_3$ | 0 | 1 | 0 | 0 | 1 | 1 |
| $x_4$ | 0 | 0 | 1 | 0 | 0 | 0 |
| $x_5$ | 1 | 0 | 1 | 1 | 1 | 1 |
| $x_6$ | 0 | 1 | 1 | 0 | 0 | 0 |
| $x_7$ | 0 | 0 | 0 | 0 | 1 | 1 |
| $x_8$ | 0 | 0 | 1 | 0 | 0 | 1 |
| $x_9$ | 1 | 0 | 0 | 1 | 1 | 1 |
| $x_{10}$ | 0 | 0 | 0 | 0 | 1 | 0 |

In Table 2, there is the significance of attribute with respect to the core $\{a_2\}$, and the relative reducts, thus, the Algorithm in [23], CEBARKCC in [13], Algorithm 2 in [26], and the proposed Algorithm in this paper are denoted by Algorithm_$a$, Algorithm_$b$, Algorithm_$c$, and Algorithm_$d$, respectively. And $m, n$ are the number of primal attributes and universe, respectively.

**Table 2.** Comparison of Experimental Results

| Algorithm | Relative to $\{a_2\}$ | | | | Reduction Result | Time Complexity |
|---|---|---|---|---|---|---|
| | $a_1$ | $a_3$ | $a_4$ | $a_5$ | | |
| Algorithm_$a$ | 0.200 | 0.100 | 0 | 0.100 | $\{a_1,a_2,a_3,a_5\}$ | $O(m^3n^2)$ |
| Algorithm_$b$ | 0.204 | 0.089 | 0.014 | 0.165 | $\{a_1,a_2,a_3,a_5\}$ | $O(mn^2)+O(n^3)$ |
| Algorithm_$c$ | 0.604 | 0.365 | 0.689 | 0.565 | $\{a_2,a_4,a_5\}$ | $O(m^2n\log(n))$ |
| Algorithm_$d$ | 0.240 | 0.200 | 0.267 | 0.200 | $\{a_2,a_4,a_5\}$ | $O(m^2n)$ |

From Table 2, the significance of attribute $a_4$ is relatively minimum [13, 23], then their reducts are $\{a_1, a_2, a_3, a_5\}$, rather than the minimum relative reduct $\{a_2, a_4, a_5\}$. However, the $SGF(a_4, \{a_2\}, D)$ is relatively maximum. Thus we get the minimum relative reduction $\{a_2, a_4, a_5\}$ generated by Algorithm_$c$ and Algorithm_$d$,. Compared with Algorithm_$a$ and Algorithm_$b$, the new proposed algorithm in this paper does not need much mathematical computation, logarithm computation in particular. What's more, we know that the general schema of adding attributes is typical for old approaches to forward selection of attributes, although they are using different evaluation measures, but it is clear that on the basis of $U/R_D$, the proposed decision power in this paper is feasible to discuss the roughness of rough set theory. Hence, the new heuristic information can compensate for the proposed limitations of those current algorithms. Therefore, this algorithm's effects on reduction of knowledge are well remarkable.

Here we choose six discrete data sets from UCI Machine Learning Repository and five algorithms to do more experiments on PC (P4 2.6G, 256M RAM, WINXP) under JDK1.4.2 in Table 3 below, where T or F indicates that the data sets are consistent or not, $m, n$ are the number of primal attributes and after reduction respectively, $t$ is the time of operation, and Algorithm_$e$ denotes the algorithm in [22]. Then, in Table 3, all five algorithms (Algorithm_$a$, Algorithm_$b$, Algorithm_$c$, Algorithm_$d$, and Algorithm_$e$) are also in short $a$, $b$, $c$, $d$, and $e$ respectively.

**Table 3.** Comparison of Reducts for Data Sets

| Database | T or F | Objects Radix | m | $a$ | | $b$ | | $c$ | | $d$ | | $e$ | |
|---|---|---|---|---|---|---|---|---|---|---|---|---|---|
| | | | | n | t | n | t | n | t | n | t | n | t |
| Balloon(1) | T | 20 | 4 | 2 | 0.20 | 2 | 0.06 | 2 | 0.04 | 2 | 0.02 | 2 | 0.12 |
| Zoo | F | 101 | 17 | 10 | 0.36 | 11 | 0.29 | 10 | 0.14 | 10 | 0.09 | 10 | 5.83 |
| Voting-records | T | 435 | 16 | 10 | 0.98 | 9 | 0.51 | 9 | 0.27 | 9 | 0.26 | 9 | 6.75 |
| Tic-tac-toe | T | 958 | 9 | 8 | 0.95 | 8 | 1.38 | 8 | 0.56 | 8 | 0.52 | 8 | 9.65 |
| Chess end-game | T | 3196 | 36 | 29 | 275.27 | 29 | 23.15 | 29 | 5.56 | 29 | 5.25 | 29 | 32.28 |
| Mushroom | T | 8124 | 22 | 5 | 486.36 | 4 | 16.58 | 4 | 6.68 | 4 | 6.56 | 4 | 29.87 |

## 5 Conclusion

Nowadays, the development of rough computational method is one of the most important research tasks. In this paper, to reflect the change of decision quality objectively, a measure for reduction of knowledge, its judgment theorems with an inequality, and many propositions are established by introducing the decision power from the algebraic point of view. To compensate for these current disadvantages of classical algorithms, we design an efficient complete algorithm for reduction of knowledge with the worst time complexity reduced to $O(|C|^2|U|)$ (In preprocessing, the complexity for computing $U/C$ based on radix sorting is cut down to $O(|C||U|)$, and the complexity for measuring attribute importance based on the positive region is descended to $O(|C-P||U^{'}-U_P^{'}|)$ in [18]), and the result of this method is very objective. Furthermore, to exert rough set theory in many more applied fields, the further researches will be to use massive database to test the results in the experiments and find more effective and efficient feature subsets.

**Acknowledgment.** This work was supported by the National Natural Science Foundation of China under Grant (No. 60873104), New Century Excellence Genius Support Plan of Henan Province of China (No. 2006HANCET-19), and Natural Science Foundation of Educational Department of Henan Province of China (No. 2008B520019).

## References

1. Pawlak, Z.: Rough sets. Int. J. of Comp. Inf. Sci. 11, 341–356 (1982)
2. Xu, J.C., Shen, J.Y., Wang, G.Y.: Rough set theory analysis on decision subdivision. In: Tsumoto, S., Słowiński, R., Komorowski, J., Grzymała-Busse, J.W. (eds.) RSCTC 2004. LNCS (LNAI), vol. 3066, pp. 340–345. Springer, Heidelberg (2004)
3. Xu, J.C., Shen, J.Y., An, Q.S., Li, N.Q.: Study on decision subdivision based on information granularity and rough sets. Journal of Xi'an Jiaotong University 39(4), 335–338 (2005)
4. Xu, J.C., Sun, L.: Knowledge reduction and its rough entropy representation of decision tables in rough set. In: Proceedings of the 2007 IEEE International Conference on Granular Computing, Silicon Valley, California, pp. 249–252 (2007)
5. Pawlak, Z.: Rough set theory and its application to data analysis. Cybernetics and Systems 29, 661–668 (1998)
6. Polkowski, L., Tsumoto, S., Lin, T.Y. (eds.): Rough Set Methods and Applications. Physica-Verlag, Berlin (2000)
7. Pawlak, Z.: Rough sets and intelligent data analysis. Inf. Sci. 147, 1–12 (2002)
8. Bhatt, R.B., Gopal, M.: On fuzzy-rough sets approach to feature selection. Pattern Recognition Letters 26, 965–975 (2005)
9. Dash, M., Liu, H.: Consistency-based search in feature selection. Artificial Intelligence 151, 155–176 (2003)
10. Swiniarski, R.W., Skowron, A.: Rough set methods in feature selection and recognition. Pattern Recognition Letters 24, 833–849 (2003)
11. Guyon, I., Elisseeff, A.: An introduction to variable and feature selection. Journal of Machine Learning Research 3, 1157–1182 (2003)
12. Wang, G.Y.: Rough reduction in algebra view and information view. International Journal of Intelligent System 18, 679–688 (2003)
13. Wang, G.Y., Yu, H., Yang, D.C.: Decision table reduction based on conditional information entropy. Journal of Computers 25(7), 759–766 (2002)

14. Miao, D.Q., Hu, G.R.: A heuristic algorithm for reduction of knowledge. Journal of Computer Research and Development 36(6), 681–684 (1999)
15. Wang, J., Wang, J.: Reduction algorithms based on discernibility matrix: the ordered attributes method. J. of Comp. Sci. and Tech. 16(6), 489–504 (2001)
16. Guan, J.W., Bell, D.A., Guan, Z.: Matrix computation for information systems. Information Sciences 131, 129–156 (2001)
17. Liu, S.H., Sheng, Q.J., Wu, B., et al.: Research on efficient algorithms for rough set methods. Journal of Computers 26(5), 524–529 (2003)
18. Xu, Z.Y., Liu, Z.P., et al.: A quick attribute reduction algorithm with complexity of Max($O(|C||U|)$,$O(|C|^2|U/C|)$). Journal of Computers 29(3), 391–399 (2006)
19. Hu, Q.H., Zhao, H., et al.: Consistency based attribute reduction. In: Zhou, Z.-H., Li, H., Yang, Q. (eds.) PAKDD 2007. LNCS (LNAI), vol. 4426, pp. 96–107. Springer, Heidelberg (2007)
20. Mi, J.S., Wu, W.Z., Zhang, W.X.: Approaches to knowledge reductions based on variable precision rough sets model. Information Sciences 159(3-4), 255–272 (2004)
21. Liang, J.Y., Shi, Z.Z., Li, D.Y.: Applications of inclusion degree in rough set theory. International Journal of Computationsl Cognition 1(2), 67–68 (2003)
22. Jiang, S.Y., Lu, Y.S.: Two new reduction definitions of decision table. Mini-Micro Systems 27(3), 512–515 (2006)
23. Guan, J.W., Bell, D.A.: Rough computational methods for information systems. International Journal of Artificial Intelligences 105, 77–103 (1998)
24. Ślęzak, D.: Approximate entropy Reducts. Fundamenta Informaticae 53, 365–390 (2002)
25. Ślęzak, D., Wróblewski, J.: Order based genetic algorithms for the search of approximate entropy reducts. In: Wang, G.Y., Liu, Q., Yao, Y.Y., Skowron, A. (eds.) Proc. Conference Rough Sets, Fuzzy Sets, Data Mining, and Granular Computing. LNCS (LNAI), vol. 2639, pp. 308–311. Springer, Heidelberg (2003)
26. Liu, Q.H., Li, F., et al.: An efficient knowledge reduction algorithm based on new conditional information entropy. Control and Decision 20(8), 878–882 (2005)
27. Wang, G.Y.: Calculation methods for core attributes of decision table. Journal of Computers 26(5), 611–615 (2003)
28. Jiang, S.Y.: An incremental algorithm for the new reduction model of decision table. Computer Engineering and Applications 28, 21–25 (2005)
29. Ślęzak, D.: Various approaches to reasoning with frequency-based decision reducts: A survey. In: Polkowski, L., Lin, T.Y., Tsumoto, S. (eds.) Rough Set Methods and Applications: New Developments in Knowledge Discovery in Information Systems, vol. 56, pp. 235–285. Springer, Heidelberg (2000)
30. Han, J.C., Hu, X.H., Lin, T.Y.: An efficient algorithm for computing core attributes in database systems. In: Zhong, N., Raś, Z.W., Tsumoto, S., Suzuki, E. (eds.) ISMIS 2003. LNCS (LNAI), vol. 2871, pp. 663–667. Springer, Heidelberg (2003)

# Comparison of Some Classification Algorithms Based on Deterministic and Nondeterministic Decision Rules

Paweł Delimata[1], Barbara Marszał-Paszek[2], Mikhail Moshkov[3], Piotr Paszek[2], Andrzej Skowron[4], and Zbigniew Suraj[1]

[1] Chair of Computer Science, University of Rzeszów
Rejtana 16A, 35-310 Rzeszów, Poland
pdelimata@wp.pl, zsuraj@univ.rzeszow.pl

[2] Institute of Computer Science, University of Silesia
Będzińska 39, 41-200 Sosnowiec, Poland
{bpaszek,paszek}@us.edu.pl

[3] Division of Mathematical and Computer Sciences and Engineering
King Abdullah University of Science and Technology
P.O. Box 55455, Jeddah 21534, Saudi Arabia
mikhail.moshkov@kaust.edu.sa

[4] Institute of Mathematics, Warsaw University
Banacha 2, 02-097 Warsaw, Poland
skowron@mimuw.edu.pl

**Abstract.** We discuss two, in a sense extreme, kinds of nondeterministic rules in decision tables. The first kind of rules, called as inhibitory rules, are blocking only one decision value (i.e., they have all but one decisions from all possible decisions on their right hand sides). Contrary to this, any rule of the second kind, called as a bounded nondeterministic rule, can have on the right hand side only a few decisions. We show that both kinds of rules can be used for improving the quality of classification. In the paper, two lazy classification algorithms of polynomial time complexity are considered. These algorithms are based on deterministic and inhibitory decision rules, but the direct generation of rules is not required. Instead of this, for any new object the considered algorithms extract from a given decision table efficiently some information about the set of rules. Next, this information is used by a decision-making procedure. The reported results of experiments show that the algorithms based on inhibitory decision rules are often better than those based on deterministic decision rules. We also present an application of bounded nondeterministic rules in construction of rule based classifiers. We include the results of experiments showing that by combining rule based classifiers based on minimal decision rules with bounded nondeterministic rules having confidence close to 1 and sufficiently large support, it is possible to improve the classification quality.

**Keywords:** rough sets, classification, decision tables, deterministic decision rules, inhibitory decision rules, lazy classification algorithms (classifiers), nondeterministic decision rules, rule based classifiers.

J.F. Peters et al. (Eds.): Transactions on Rough Sets XII, LNCS 6190, pp. 90–105, 2010.

# 1 Introduction

Over the years many methods based on rule induction (called also as covering methods [26]) and rule based classification systems were developed (see, e.g., different versions of AQ system by Ryszard Michalski [17], several other systems such as 1R [15], systems based on the PRISM algorithm [9,8] or different methods based on rough sets [24,21], [35,37], [34], and software systems such as LERS [14], RSES [30], ROSETTA [29], and others [25]). In this paper, we show that still there is a room for improving the rule based classification systems. We discuss two methods for rule inducing. The first one is based on classification of new objects only on the basis of the rule votes against decisions and the second one is based on searching for strong rules for union of a few relevant decision classes.

In the rough set approach, decision rules belong to the most important objects of study. The rule induction can be treated as searching for neighborhoods of the approximation spaces which are relevant for classification approximation on extensions of the training sample. One of the commonly used methods is based on tuning the balance between the quality of classification approximation and the description length of rule based models (classifiers) [32,6]. Hence, this approach is strongly related to the minimum length principle (MLD) [27,28].

We discuss two kinds of, in a sense extreme, nondeterministic rules in decision tables. The first kind of rules, called as inhibitory rules, are blocking only one decision value (i.e., they have all but one decisions on their right hand sides). Contrary to this, any rule of the second kind, called as a bounded nondeterministic rule, can have on the right hand side only a few decisions. Bounds on nondeterminism in decision rules were introduced to reduce the complexity of searching for relevant nondeterministic rules. We show that both kinds of rules can be used for improving the quality of classification.

In the paper, the following classification problem is considered: for a given decision table $T$ [23,24] and a new object $v$ generate a value of the decision attribute on $v$ using values of conditional attributes on $v$.

First, we compare two lazy [1,21] classification algorithms based on deterministic and inhibitory decision rules of the forms

$$a_1(x) = b_1 \wedge \ldots \wedge a_t(x) = b_t \Rightarrow d(x) = b,$$

$$a_1(x) = b_1 \wedge \ldots \wedge a_t(x) = b_t \Rightarrow d(x) \neq b,$$

respectively, where $a_1, \ldots, a_t$ are conditional attributes, $b_1, \ldots, b_t$ are values of these attributes, $d$ is the decision attribute and $b$ is a value of $d$. By $V_d(T)$ we denote the set of values of the decision attribute $d$.

The first algorithm (D-algorithm) was proposed and studied by J.G. Bazan [3,4,5]. This algorithm is based on the deterministic decision rules. For any new object $v$ and each decision $b \in V_d(T)$ we find (using polynomial-time algorithm) the number $D(T, b, v)$ of objects $u$ from the decision table $T$ such that there exists a deterministic decision rule $r$ satisfying the following conditions: (i) $r$ is true for the decision table $T$, (ii) $r$ is realizable for $u$ and $v$, and (iii) $r$ has the equality $d(x) = b$ on the right hand side. For the new object $v$ we choose a decision

$b \in V_d(T)$ for which the value $D(T, b, v)$ is maximal. Note that this approach was generalized by J.G. Bazan [3,4,5] to the case of approximate decision rules, and by A. Wojna [36] to the case of decision tables with not only nominal but also numerical attributes.

The second algorithm (I-algorithm) is based on the inhibitory decision rules. For any new object $v$ and each decision $b \in V_d(T)$ using a polynomial-time algorithm it is computed the number $I(T, b, v)$ of objects $u$ from the decision table $T$ such that there exists an inhibitory decision rule $r$ satisfying the following conditions: (i) $r$ is true for the decision table $T$, (ii) $r$ is realizable for $u$ and $v$, and (iii) $r$ has the relation $d(x) \neq b$ on the right hand side. For the new object $v$ we choose a decision $b \in V_d(T)$ for which the value $I(T, b, v)$ is minimal. Hence, for $v$ we vote, in a sense, for the decision $b$ for which there are weakest arguments "against". Note that in [22] the dissimilarity measures are used for obtaining arguments against classification of a given handwritten digit to some decision classes. These arguments can be interpreted as inhibitory rules.

Instead of the consideration of inhibitory rules for the original decision table $T$ we can study decision rules for new decision tables $T_b$, $b \in V_d(T)$. The decision tables $T$ and $T_b$ coincide with the exception of decision attributes $d$ and $d_b$:

$$d_b(x) = \begin{cases} b, \text{ if } \ d(x) = b, \\ c, \text{ if } \ d(x) \neq b, \end{cases}$$

where $c \neq b$. One can show that, for a new object $v$, $I(T, b, v) = I(T_b, b, v)$ and $I(T_b, b, v) = D(T_b, c, v)$ (see Remark 1 at the end of Section 6).

Results of experiments show that the algorithm based on inhibitory decision rules is, often, better than the algorithm based on deterministic decision rules.

This work was inspired by results of comparison of deterministic and inhibitory rules for information system $S = (U, A)$ [23,24], where $U$ is a finite set of objects and $A$ is a finite set of attributes (functions defined on $U$). The considered rules are of the following form:

$$a_1(x) = b_1 \wedge \ldots \wedge a_t(x) = b_t \Rightarrow a_{t+1}(x) = b_{t+1},$$
$$a_1(x) = b_1 \wedge \ldots \wedge a_t(x) = b_t \Rightarrow a_{t+1}(x) \neq b_{t+1},$$

where $a_1, \ldots, a_{t+1}$ are attributes from $A$ and $b_1, \ldots, b_{t+1}$ are values of these attributes. We consider only true and realizable rules. *True* means that the rule is true for any object from $U$. *Realizable* means that the left hand side of the rule is true for at least one object from $U$. We identify objects from $U$ and tuples of values of attributes from $A$ on these objects. Let $V$ be the set of all tuples of known values of attributes from $A$. We say that the set $U$ can be described by deterministic (inhibitory) rules if there exists a set $Q$ of true and realizable deterministic (inhibitory) rules such that the set of objects from $V$, for which all rules from $Q$ are true, is equal to $U$.

In [31,33] it was shown that there exist information systems $S = (U, A)$ such that the set $U$ can not be described by deterministic rules. In [18,19] it was shown that for any information system $S = (U, A)$ the set $U$ can be described by inhibitory rules. It means that the inhibitory rules can express essentially more information encoded in information systems than the deterministic rules. This fact is a motivation for a wider use of inhibitory rules, in particular, in classification algorithms and in algorithms for synthesis of concurrent systems [31]. To compare experimentally the classification quality based on inhibitory and deterministic rules, we create two similar families of lazy classification algorithms based on deterministic and inhibitory rules, respectively [11]. Results of experiments show that the algorithms based on inhibitory rules are noticeably better than those based on deterministic rules.

We also present an application of bounded nondeterministic rules in construction of rule based classifiers. These are the rules of the following form:

$$a_1(x) = b_1 \wedge \ldots \wedge a_t(x) = b_t \Rightarrow d(x) = c_1 \vee \ldots \vee d(x) = c_s,$$

where $a_1, \ldots, a_t$ are conditional attributes of the decision table $T$, $d$ is the decision of $T$, and $\emptyset \neq \{c_1, \ldots, c_s\} \subseteq V_d(T)$, where $V_d(T)$ is the value set of the decision $d$ [24]. We consider nondetermninistic rules with cardinality $|\{c_1, \ldots, c_s\}|$ small in comparison with $|V_d(T)|$.

We include the results of experiments showing that by combining rule based classifiers based on minimal decision rules [24] with the second kind of rules having the confidence [2] close to 1 and the sufficiently large support [2] (more precisely, normalized support, in which we include some penalty measure relative to $|\{c_1, \ldots, c_s\}|$) it is possible to improve the classification quality.

This paper is an essentially extended version of [12].

The paper consists of nine sections. In Sect. 2, we recall the notion of decision table. In Sects. 3, 4 and 5, we describe notions of deterministic, inhibitory and nondeterministic decision rules. Sect. 6 contains definitions of two lazy classification algorithms. Results of experiments are discussed in Sect. 7. In Sect. 8, we present the main steps in construction of classifiers enhanced by bounded nondeterministic rules as well as results of performed computer experiments showing that such rules may lead to increasing the classification quality. Section 9 contains short conclusions.

## 2 Decision Tables

Let $T = (U, A, d)$ be a *decision table*, where $U = \{u_1, \ldots, u_n\}$ is a finite nonempty set of *objects*, $A = \{a_1, \ldots, a_m\}$ is a finite nonempty set of *conditional attributes* (functions defined on $U$), and $d$ is the *decision attribute* (function defined on $U$). We assume that for each $u_i \in U$ and each $a_j \in A$ the value $a_j(u_i)$ and the value $d(u_i)$ belong to $\omega$, where $\omega = \{0, 1, 2, \ldots\}$ is the set of nonnegative integers. By $V_d(T)$ we denote the set of values of the decision attribute $d$ on objects from $U$.

For $i = 1, \ldots, n$ we identify the object $u_i \in U$ with the tuple

$$(a_1(u_i), \ldots, a_m(u_i)).$$

Besides objects from $U$ we consider also objects from $\mathcal{U}(T) = \omega^m$. The set $\mathcal{U}(T)$ is called the *universe* for the decision table $T$. For any object (tuple) $v \in \mathcal{U}(T)$ and any attribute $a_j \in A$ the value $a_j(v)$ is equal to $j$-th integer component of $v$.

## 3 Deterministic Decision Rules

Let us consider a rule

$$a_{j_1}(x) = b_1 \wedge \ldots \wedge a_{j_t}(x) = b_t \Rightarrow d(x) = b, \tag{1}$$

where $t \geq 0$, $a_{j_1}, \ldots, a_{j_t} \in A$, $b_1, \ldots, b_t \in \omega$, $b \in V_d(T)$ and numbers $j_1, \ldots, j_t$ are pairwise different. Such rules are called *deterministic decision* rules. The rule (1) is called *realizable for an object* $u \in \mathcal{U}(T)$ if $a_{j_1}(u) = b_1, \ldots, a_{j_t}(u) = b_t$ or $t = 0$. The rule (1) is called *true for an object* $u_i \in U$ if $d(u_i) = b$ or (1) is not realizable for $u$. The rule (1) is called *true for* $T$ if it is true for any object from $U$. The rule (1) is called *realizable for* $T$ if it is realizable for at least one object from $U$. By $Det(T)$ we denote the set of all deterministic decision rules which are true for $T$ and realizable for $T$. The rule (1) true and realizable for $T$ is called *minimal* if a rule obtained from this rule by drooping any descriptor from its left hand side is no longer true for $T$.

Our aim is to recognize, for given objects $u_i \in U$ and $v \in \mathcal{U}(T)$, and given value $b \in V_d(T)$ if there exists a rule from $Det(T)$ which is realizable for $u_i$ and $v$ and has $d(x) = b$ in the right hand side. Such a rule "supports" the assignment of the decision $b$ to the new object $v$.

Let $M(u_i, v) = \{a_j : a_j \in A, a_j(u_i) = a_j(v)\}$ and $P(u_i, v) = \{d(u) : u \in U, a_j(u) = a_j(v) \text{ for any } a_j \in M(u_i, v)\}$. Note that if $M(u_i, v) = \emptyset$, then $P(u_i, v) = \{d(u) : u \in U\} = V_d(T)$.

**Proposition 1.** *Let $T = (U, A, d)$ be a decision table, $u_i \in U$, $v \in \mathcal{U}(T)$, and $b \in V_d(T)$. Then in $Det(T)$ there exists a rule, which is realizable for $u_i$ and $v$ and has $d(x) = b$ in the right hand side, if and only if $P(u_i, v) = \{b\}$.*

*Proof.* Let $P(u_i, v) = \{b\}$. In this case, the rule

$$\bigwedge_{a_j \in M(u_i, v)} a_j(x) = a_j(v) \Rightarrow d(x) = b \tag{2}$$

belongs to $Det(T)$, is realizable for $u_i$ and $v$, and has $d(x) = b$ in the right hand side.

Let us assume that there exists a rule (1) from $Det(T)$, which is realizable for $u_i$ and $v$, and has $d(x) = b$ in the right hand side. Since (1) is realizable for $u_i$ and $v$, we have $a_{j_1}, \ldots, a_{j_t} \in M(u_i, v)$. Since (1) is true for $T$, the rule (2) is true for $T$. Therefore, $P(u_i, v) = \{b\}$. □

From Proposition 1 it follows that there exists a polynomial algorithm recognizing, for a given decision table $T = (U, A, d)$, given objects $u_i \in U$ and $v \in \mathcal{U}(T)$, and a given value $b \in V_d(T)$, if there exists a rule from $Det(T)$, which is realizable for $u_i$ and $v$, and has $d(x) = b$ in the right hand side. This algorithm constructs the set $M(u_i, v)$ and the set $P(u_i, v)$. The considered rule exists if and only if $P(u_i, v) = \{b\}$.

## 4 Inhibitory Decision Rules

Let us consider a rule

$$a_{j_1}(x) = b_1 \wedge \ldots \wedge a_{j_t}(x) = b_t \Rightarrow d(x) \neq b, \tag{3}$$

where $t \geq 0$, $a_{j_1}, \ldots, a_{j_t} \in A$, $b_1, \ldots, b_t \in \omega, b \in V_d(T)$, and numbers $j_1, \ldots, j_t$ are pairwise different. Such rules are called *inhibitory decision* rules. The rule (3) is called *realizable for an object* $u \in \mathcal{U}(T)$ if $a_{j_1}(u) = b_1, \ldots, a_{j_t}(u) = b_t$ or $t = 0$. The rule (3) is called *true for an object* $u_i \in U$ if $d(u_i) \neq b$ or (3) is not realizable for $u_i$. The rule (3) is called *true for* $T$ if it is true for any object from $U$. The rule (3) is called *realizable for* $T$ if it is realizable for at least one object from $U$. By $Inh(T)$ we denote the set of all inhibitory decision rules which are true for $T$ and realizable for $T$.

Our aim is to recognize for given objects $u_i \in U$ and $v \in \mathcal{U}(T)$, and given value $b \in V_d(T)$ if there exists a rule from $Inh(T)$, which is realizable for $u_i$ and $v$, and has $d(x) \neq b$ in the right hand side. Such a rule "contradicts" the assignment of the decision $b$ to the new object $v$.

**Proposition 2.** *Let* $T = (U, A, d)$ *be a decision table,* $u_i \in U$*,* $v \in \mathcal{U}(T)$*, and* $b \in V_d(T)$*. Then in* $Inh(T)$ *there exists a rule, which is realizable for* $u_i$ *and* $v$*, and has* $d(x) \neq b$ *in the right hand side, if and only if* $b \notin P(u_i, v)$*.*

*Proof.* Let $b \notin P(u_i, v)$. In this case, the rule

$$\bigwedge_{a_j \in M(u_i,v)} a_j(x) = a_j(v) \Rightarrow d(x) \neq b \tag{4}$$

belongs to $Inh(T)$, is realizable for $u_i$ and $v$, and has $d(x) \neq b$ in the right hand side.

Let us assume that there exists a rule (3) from $Inh(T)$, which is realizable for $u_i$ and $v$, and has $d(x) \neq b$ in the right hand side. Since (3) is realizable for $u_i$ and $v$, we have $a_{j_1}, \ldots, a_{j_t} \in M(u_i, v)$. Since (3) is true for $T$, the rule (4) is true for $T$. Therefore, $b \notin P(u_i, v)$. □

From Proposition 2 it follows that there exists a polynomial algorithm recognizing for a given decision table $T = (U, A, d)$, given objects $u_i \in U$ and $v \in \mathcal{U}(T)$, and a given value $b \in V_d(T)$ if there exists a rule from $Inh(T)$, which is realizable for $u_i$ and $v$, and has $d(x) \neq b$ in the right hand side. This algorithm constructs the set $M(u_i, v)$ and the set $P(u_i, v)$. The considered rule exists if and only if $b \notin P(u_i, v)$.

## 5 Nondeterministic Decision Rules

Inhibitory decision rules are examples of nondeterministic decision rules. In general, nondeterministic decision rules in a given decision table $T$ are of the form

$$a_{j_1}(x) = b_1 \wedge \ldots \wedge a_{j_t}(x) = b_t \Rightarrow d(x) = c_1 \vee \ldots \vee d(x) = c_s, \tag{5}$$

where $a_{j_1}, \ldots, a_{j_t} \in A$, $b_1, \ldots, b_t \in \omega$, numbers $j_1, \ldots, j_t$ are pairwise different, and $\emptyset \neq \{c_1, \ldots, c_s\} \subseteq V_d(T)$. Let $r$ be the rule (5). In the sequel, we write $d(x) \in V(r)$, where $V(r) = \{c_1, \ldots, c_s\} \subset V_d(T)$, instead of $d(x) = c_1 \vee \ldots \vee d(x) = c_s$.

The rule (5) is called *realizable for an object* $u \in U$ if $a_{j_p}(u) = b_p$ for all $p \in \{1, \ldots, t\}$ or $t = 0$. The rule (5) is called *true for an object* $u_j \in U$ if $d(u_i) = c_j$ for at least one $j \in \{1, \ldots, s\}$ or (5) is not realizable for $u_i$. The rule (5) is called *true for* $T$ if it is true for any object from $U$. The rule (5) is called *realizable for* $T$ if it is realizable for at least one object from $U$. By $NON\_DET(T)$ we denote the set of all nondeterministic decision rules which are true for $T$ and realizable for $T$.

Let us introduce some notation. If $r$ is the nondeterministic rule (5) then by $lh(r)$ we denote its left hand side, i.e., the formula $a_{j_1}(x) = b_1 \wedge \ldots \wedge a_{j_t}(x) = b_t$, and by $rh(r)$ its right hand side, i.e., the formula $d(x) \in V(r)$. If $\alpha$ is a boolean combination of descriptors, i.e., formulas of the form $a(x) = v$, where $a \in A \cup \{d\}$ and $v \in \omega$ then by $\|\alpha\|_T$ (or $\|\alpha\|$, for short) we denote all objects from $U$ satisfying $\alpha$ [24]. Hence, the nondeterministic rule (5) is true in $T$ if and only if $\|lh(r)\| \subseteq \|rh(r)\|$.

In this paper, we also consider nondeterministic rules which are partially true in a given decision table. To measure the quality of such rules we use coefficients called the *support* and the *confidence* [2]. They are defined as follows. If $r$ is a nondeterministic rule of the form (5) then the support of this rule in the decision table $T$ is defined by

$$supp_T(r) = \frac{|\|lh(r)\|_T \cap \|rh(r)\|_T|}{|U|}, \tag{6}$$

and the confidence of $r$ in $T$ is defined by

$$conf_T(r) = \frac{|\|lh(r)\|_T \cap \|rh(r)\|_T|}{|\|lh(r)\|_T|}. \tag{7}$$

We also use a normalized support of $r$ in $T$ defined by

$$norm\_supp_T(r) = \frac{supp_T(r)}{\sqrt{|V(r)|}}. \tag{8}$$

For any two nondeterministic decision rules $r_1, r_2$ with the same support in $T$, we give a higher priority to a rule $r_1$ over a given rule $r_2$ if $r_2$ has more decision values on the right hand side than the rule $r_1$.

Now we can define a set of nondeterministic decision rules which are used in Section 8 for enhancing the quality of classification of rule based classifiers. This set is defined relative to the following three parameters:

1. $con \in (0, 1]$ – a threshold used as the lower bound for the confidence of rules; in our method we assume that $con$ is close to 1;
2. $n\_sup \in (0, 1]$ – a threshold used as the lower bound for the normalized support of rules;

3. $l$ – a threshold used as an upper bound on the number of decision values on the right hand sides of rules; in our heuristic method $l$ is assumed to be small.

The set of nondeterministic rules $RULE_T(con, n\_sup, l)$ is defined as the set of all nondeterministic rules $r$ (over attributes in $T$) such that

1. $conf_T(r) \geq con$;
2. $norm\_supp_T(r) \geq n\_sup$ and;
3. $|V(r)| \leq l$.

The algorithm presented in Section 8 is searching for bounded nondetermninistic rules $r$ with possibly large (normalized) support and with sets $\|lh(r)\|_T$ *almost* included is sums of *a few* decision classes. Such rules are combined with minimal rules [24] for increasing the classification quality. The details are presented in Section 8.

## 6 Classification Algorithms Based on Deterministic and Inhibitory Rules

Let $T = (U, A, d)$ be a decision table. We consider the following classification problem: for an object $v \in \mathcal{U}(T)$ predict the value of the decision attribute $d$ on $v$ using only values of attributes from $A$ on $v$. To this end, we use the D-classification algorithm (D-algorithm) and the I-classification algorithm (I-algorithm).

D-algorithm is based on the use of the parameter $D(T, b, v)$, $b \in V_d(T)$. This parameter is equal to the number of objects $u_i \in U$ for which there exists a rule from $Det(T)$, that is realizable for $u_i$ and $v$, and has $d(x) = b$ in the right hand side. From Proposition 1 it follows that there exists a polynomial algorithm which for a given decision table $T = (U, A, d)$, a given object $v \in \mathcal{U}(T)$ and a given value $b \in V_d(T)$ computes the value $D(T, b, v) = |\{u_i : u_i \in U, P(u_i, v) = \{b\}\}|$.

*D-algorithm:*
For given object $v$ and each $b \in V_d(T)$ we find the value of the parameter $D(T, b, v)$. As the value of the decision attribute for $v$ we choose $b \in V_d(T)$ such that $D(T, b, v)$ has the maximal value. If more than one such $b$ exists then we choose the minimal $b$ for which $D(T, b, v)$ has the maximal value.

I-algorithm is based on the use of the parameter $I(T, b, v)$, $b \in V_d(T)$. This parameter is equal to the number of objects $u_i \in U$ for which there exists a rule from $Inh(T)$, that is realizable for $u_i$ and $v$, and has $d(x) \neq b$ in the right hand side. From Proposition 2 it follows that there exists a polynomial algorithm which for a given decision table $T = (U, A, d)$, a given object $v \in \mathcal{U}(T)$ and a given value $b \in V_d(T)$ computes the value $I(T, b, v) = |\{u_i : u_i \in U, b \notin P(u_i, v)\}|$.

*I-algorithm:*
For given object $v$ and each $b \in V_d(T)$ we find the value of the parameter $I(T, b, v)$. As the value of the decision attribute for $v$ we choose $b \in V_d(T)$ such that $I(T, b, v)$ has the minimal value. If more than one such $b$ exists then we choose the minimal $b$ for which $I(T, b, v)$ has the minimal value.

*Example 1.* Let us consider decision table $T$ with three conditional attributes $a_1$, $a_2$ and $a_3$, and five objects (rows) $u_1$, $u_2$, $u_3$, $u_4$ and $u_5$ (see Figure 1). The set $V_d(T)$ of values of the decision attribute $d$ is equal to $\{1, 2, 3\}$.

$T =$

| | $a_1$ | $a_2$ | $a_3$ | $d$ |
|---|---|---|---|---|
| $u_1$ | 0 | 0 | 1 | 1 |
| $u_2$ | 1 | 0 | 0 | 1 |
| $u_3$ | 1 | 1 | 1 | 2 |
| $u_4$ | 0 | 1 | 0 | 3 |
| $u_5$ | 1 | 1 | 0 | 3 |

**Fig. 1.** Decision table $T$

Let $v = (0, 1, 1)$ be a new object. One can show that

$$\begin{aligned}
&M(u_1, v) = \{a_1, a_3\}, M(u_2, v) = \emptyset, M(u_3, v) = \{a_2, a_3\},\\
&M(u_4, v) = \{a_1, a_2\}, M(u_5, v) = \{a_2\},\\
&P(u_1, v) = \{1\}, P(u_2, v) = \{1, 2, 3\}, P(u_3, v) = \{2\},\\
&P(u_4, v) = \{3\}, P(u_5, v) = \{2, 3\}.
\end{aligned}$$

Let us consider outputs returned by D-algorithm and I-algorithm.

D-algorithm. One can show that $D(T, 1, v) = 1$, $D(T, 2, v) = 1$, $D(T, 3, v) = 1$. Therefore, we assign to $v$ the decision 1.

I-algorithm. One can show that $I(T, 1, v) = 3$, $I(T, 2, v) = 2$, $I(T, 3, v) = 2$. Therefore, we assign to $v$ the decision 2.

*Remark 1.* Let the decision attribute $d$ of a decision table $T$ has exactly two values $b$ and $c$. Let $v$ be a new object. One can show that $I(T, c, v) = D(T, b, v)$ and $I(T, b, v) = D(T, c, v)$. Therefore D-algorithm and I-algorithm return for $v$ the same output.

## 7 Results of Experiments

We have performed experiments with D-algorithm and I-algorithm and decision tables from [20] using DMES system [10]. Some attributes in tables are discretized, and missing values are filled by algorithms from RSES2 [30]. We removed attributes of the kind "name" that are distinct for each instance. To evaluate the accuracy of an algorithm on a decision table (the percent of correctly classified objects) we use either train-and-test method or cross-validation method.

Let us, for some algorithm and some table, use $n$-fold cross-validation method for the estimation of the accuracy. Then for this table we obtain $n$ accuracies $x_1, \ldots, x_n$. As the final accuracy we use the value $\bar{x} = \frac{1}{n}\sum_{i=1}^{n} x_i$ which is the arithmetic mean of $x_1, \ldots, x_n$. *Maximal relative deviation* for $x_1, \ldots, x_n$ is equal to $\max\left\{\frac{|x_i - \bar{x}|}{\bar{x}} : i = 1, \ldots, n\right\}$. This value characterizes algorithm stability.

Table 1 contains results of experiments for D-algorithm and I-algorithm and initial decision tables from [20]. Columns "D ac" and "I ac" contain accuracy of D-algorithm and I-algorithm. Columns "D mrd" and "I mrd" contain maximal relative deviations of accuracies in the performed cross-validation.

**Table 1.** Results of experiments with initial decision tables

| Decision table | D ac | I ac | D mrd | I mrd | Decision table | D ac | I ac | D mrd | I mrd |
|---|---|---|---|---|---|---|---|---|---|
| monk1 | 89.8 | 89.8 | | | lenses | 71.6 | 76.6 | 0.534 | 0.565 |
| monk2 | 80.0 | 80.0 | | | soybean-small | 57.5 | 57.5 | 0.652 | 0.652 |
| monk3 | 93.9 | 93.9 | | | soybean-large | 85.6 | 85.9 | | |
| lymphography | 78.5 | 79.2 | 0.272 | 0.279 | zoo | 85.1 | 94.0 | 0.177 | 0.063 |
| diabetes | 75.2 | 75.2 | 0.168 | 0.168 | post-operative | 65.5 | 64.4 | 0.152 | 0.310 |
| breast-cancer | 76.4 | 76.4 | 0.121 | 0.121 | hayes-roth | 92.8 | 85.7 | | |
| primary-tumor | 35.7 | 37.5 | 0.322 | 0.290 | lung-cancer | 40.0 | 40.0 | 0.665 | 0.665 |
| balance-scale | 78.7 | 76.6 | 0.099 | 0.097 | solar-flare | 97.8 | 97.8 | | |

For 3 decision tables the accuracy of D-algorithm is greater than the accuracy of I-algorithm, for 5 decision tables the accuracy of I-algorithm is greater than the accuracy of D-algorithm, and for 8 decision tables D-algorithm and I-algorithm have the same accuracy. The considered algorithms are not stable.

Table 2 contains results of experiments for D-algorithm and I-algorithm and modified decision tables from [20]. For each initial table from [20] we choose a number of many-valued (with at least three values) attributes different from the decision attribute, and consider each such attribute as new decision attribute. As the result we obtain the same number of new decision tables as the number of chosen attributes (this number can be found in the column "New"). The column "D opt" contains the number of new tables for which the accuracy of D-algorithm is greater than the accuracy of I-algorithm. The column "I opt" contains the number of new tables for which the accuracy of I-algorithm is greater than the accuracy of D-algorithm. The columns "D aac" and "I aac" contain the average accuracy of D-algorithm and I-algorithm for new tables. We have chosen many-valued attributes since for two-valued ones D-algorithm and I-algorithm have the same accuracy for new decision tables (see Remark 1).

For 8 new decision tables the accuracy of D-algorithm is greater than the accuracy of I-algorithm, for 24 new decision tables the accuracy of I-algorithm is greater than the accuracy of D-algorithm, and for 16 new decision tables D-algorithm and I-algorithm have the same accuracy. Note also that for each of the considered initial tables the average accuracy of I-algorithm for new tables corresponding to the initial one is greater than the average accuracy of D-algorithm.

**Table 2.** Results of experiments with modified decision tables

| Decision table | New | D opt | I opt | D aac | I aac |
|---|---|---|---|---|---|
| lymphography | 9 | 4 | 3 | 54.9 | 55.3 |
| primary-tumor | 3 | 1 | 2 | 64.8 | 65.7 |
| balance-scale | 4 | 0 | 4 | 19.5 | 23.9 |
| soybean-large | 5 | 0 | 3 | 86.2 | 86.4 |
| zoo | 1 | 0 | 1 | 73.0 | 74.0 |
| post-operative | 8 | 1 | 4 | 56.4 | 57.4 |
| hayes-roth | 4 | 1 | 3 | 45.5 | 49.9 |
| lung-cancer | 7 | 0 | 1 | 58.7 | 59.1 |
| solar-flare | 7 | 1 | 3 | 67.0 | 67.8 |

## 8 Classification Algorithms Based on Combination of Deterministic and Bounded Nondeterministic Rules

In this section, we present an application of bounded nondeterministic rules (i.e., rules with a few decisions on their right hand sides) for classification of objects.

An $ND$-heuristic was developed for generation of such rules from minimal decision rules [24] with the confidence parameters close to 1 and with sufficiently large normalized supports. The main steps of the $ND$-heuristic are as follows.
Input: $T$ – decision table;
$con$ – confidence threshold;
$n\_sup$ – support threshold;
$l$ – bound on nondeterminism in decision rules.
Output: a set of rules from $RULE_T(con, n\_sup, l)$.
*Step 1.* Generate a set $X$ of minimal decision rules of $T$ such that $X$ covers objects in $T$.
*Step 2.* Search for rules from $RULE_T(con, n\_sup, l)$ which can be obtained by dropping some descriptors from the left hand sides of rules from $X$.

In above $ND$-heuristic, we use heuristics from RSES for generation of sets of minimal rules (e.g., a genetic algorithm) because of the high computational complexity of the generation of all minimal rules [13]. In searching for relevant nondeterministic decision rules, we use bounds on nondeterminism to reduce the complexity of searching for relevant nondeterministic rules based on dropping of descriptors from minimal decision rules.

The above procedure is repeated for different parameters $con, n\_sup, l$. The final set of rules is selected from sets of rules corresponding to maximal values of $(con, n\_sup)$ (relative to the natural partial order on the set of pairs of real numbers).

In our experiments, we used classification algorithms constructed by the combination of two auxiliary classification algorithms. The first one is the classification algorithm based on minimal rules generated by system RSES [7]. This algorithm uses the standard voting procedure. The second classification algorithm is based on nondeterministic rules generated by the $ND$-heuristic. In the latter case, for any new object a decision value set is predicted instead of single decision value. This decision value set is generated as follows. First, for any new object, all nondeterministic rules matching this object are extracted. Next, from these matched rules, a rule with the largest (normalized) support is selected. In the case when several rules have the same support, the decision value set $V(r)$ of the nondeterministic rule $r$ with the smallest decision value set is selected. If still several nondeterministic rules with the above property exist then one of them is selected randomly. In this way, for any new object we obtain a decision value $c$ (given by the first auxiliary classification algorithm) and a decision value set $V(r)$, where $r$ is the rule selected from the set of nondeterministic rules generated by the $ND$-heuristic. It should be noted that each of the considered auxiliary classification algorithms can leave the new object unclassified (if there are no rules matching this object).

In our experiments, we used two kinds $C_1$ and $C_2$ of the classification algorithm. For the construction of the first version $C_1$, the classification algorithm from RSES based on all minimal decision rules was used. The second version $C_2$ was realized using the classification algorithm from RSES based on (some) minimal decision rules generated by genetic algorithm.

The final decision for a given new object is obtained from the decision $c$ and decision value set $V(r)$ (where $r$ is the nondeterministic rule selected by the second auxiliary classification algorithm from the set of rules generated by the $ND$-heuristic). This decision is defined by the *excluding classification method.*

The excluding classification method is based on the following steps [16]:

1. If for a given new object the auxiliary classification algorithm based on minimal rules predicts the decision value $c$ and $c \in V(r)$, (i.e., no conflict arises) then we take as the final decision the single decision $c$.
2. If for a given new object the auxiliary classification algorithm based on minimal rules predicts the decision value $c$ and $c \notin V(r)$ (i.e., conflict arises) then we take as the final decision value the single decision value $c$ provided the minimal rule support (i.e., the support of the rule $lh(r_1) \vee \ldots \vee lh(r_k) \Rightarrow d(x) = c$ in the considered decision table $T$, where $r_1, \ldots, r_k$ are all minimal rules matched by the new object) is larger than the normalized support of the decision rule $r$ generated by the $ND$-heuristic and selected for the given new object. In the opposite case, we take as the final decision a single decision value from the set $V(r)$, with the largest support in $T$ among decisions from $V(r)$.
3. If for a new object, the auxiliary classification algorithm based on minimal rules predicts the decision value $c$ and this object does not match any rule generated by the $ND$-heuristic then we assign the decision $c$ as the final decision.
4. If a given new object does not match any of the minimal rules used for construction the auxiliary classification algorithm based on these rules then we assign as the final decision the single decision from $V(r)$ with the largest

**Table 3.** Results of experiments with nondeterministic rules

| Decision table | Classification factor | Classification algorithm | | | | | |
|---|---|---|---|---|---|---|---|
| | | $Alg_i$ | $C_i$ | | | | |
| | | | $con^{(3)}$ | | | | |
| | | | 1.0 | 0.9 | 0.8 | 0.7 | 0.6 |
| Balance Scale $^{(1)}$ | acc × cover | 7.75 | **73.00** | 68.00 | 66.12 | | |
| | mrd | 0.177 | 0.051 | 0.049 | 0.059 | | |
| Dermatology$^{(2)}$ | acc × cover | 95.17 | **95.26** | 91.35 | 87.07 | 86.61 | 82.88 |
| | mrd | 0.036 | 0.035 | 0.026 | 0.018 | 0.025 | 0.054 |
| Ecoli $^{(1)}$ | acc × cover | 53.35 | 59.45 | **60.61** | 60.27 | 60.42 | 56.25 |
| | mrd | 0.043 | 0.026 | 0.031 | 0.047 | 0.049 | 0.037 |
| Lymphography $^{(1)}$ | acc × cover | 55.11 | 55.11 | 55.11 | 55.11 | | |
| | mrd | 0.741 | 0.741 | 0.741 | 0.741 | | |
| Primary Tumor $^{(1)}$ | acc × cover | 65.29 | 65.49 | **66.08** | 66.08 | 66.08 | |
| | mrd | 0.188 | 0.185 | 0.174 | 0.174 | 0.174 | |
| Vehicle$^{(2)}$ | acc × cover | 62.41 | 62.86 | **63.26** | 61.27 | 59.30 | 55.79 |
| | mrd | 0.049 | 0.049 | 0.043 | 0.038 | 0.035 | 0.038 |

$^{(1)}$ $i = 1$ – in column marked by $Alg_i$ classification is defined by the classification algorithm $Alg_1$ from RSES based on all minimal rules;
in column marked by $C_i$ classification is defined by the classification algorithm $C_1$ (with $Alg_1$).

$^{(2)}$ $i = 2$ – in column marked by $Alg_i$ classification is defined by the classification algorithm $Alg_2$ from RSES based on decision rules generated by genetic algorithm;
in column marked by $C_i$ classification is defined by the classification algorithm $C_2$ (with $Alg_2$).

$^{(3)}$ confidence of nondeterministic rules generated by the $ND$-heuristic is not smaller than the parameter $con$.

support among decisions from $V(r)$, where $r$ is the rule selected by the second auxiliary classification algorithm.

5. In the remaining cases, a given new object is not classified.

We have performed experiments on decision tables from [20] using two classification algorithms $C_1, C_2$. The classification algorithm $C_1$ is obtained by the described above combination of the auxiliary classification algorithm from RSES based on all minimal decision rules with the auxiliary classification algorithm based on nondeterministic rules generated by the $ND$-heuristic. The classification algorithm $C_2$ is obtained by the described above combination of the auxiliary classification algorithm from RSES based on minimal decision rules generated by a genetic algorithm with the auxiliary classification algorithm based on nondeterministic rules generated by the $ND$-heuristic.

Some attributes in decision tables used for experiments were discretized, and missing values were filled by algorithms from RSES. We removed attributes of the kind "name" discerning any two objects (cases). In evaluation of the accuracy of classification algorithms on decision table (i.e., the percentage of correctly classified objects) the cross-validation method was used.

For any considered data table, we used one of the classification algorithms $C_1$ or $C_2$ for different values of parameter *con*. On testing sets the accuracy and the coverage factor were calculated. Also the *maximal relative deviation* (mrd) was calculated.

Table 3 includes the results of our experiments.

For five decision tables the classification quality measured by *accuracy* $\times$ *coverage* was better for the classification algorithm $C_1$ (or $C_2$) than in the case of the classification algorithm from RSES based only on minimal rules with standard voting.

For one of the decision tables, the classification quality measured by *accuracy* $\times$ *coverage* was the same for the classification algorithm based on a combination of the two auxiliary classification algorithms and for the classification algorithm from RSES based only on minimal rules with standard voting.

For four decision tables, the mrd was no greater then 5% in the case when we used the classification algorithm $C_1$ (or $C_2$). Hence, using the classification algorithm $C_1$ (or $C_2$) may lead to more stable classification.

For obtaining those better results, it was necessary to optimize the threshold *con* for each data table. This means that the parameter *con* should be tuned to the data tables.

## 9 Conclusions

Results of experiments show that the algorithm based on inhibitory decision rules is, often, better than the algorithm based on deterministic decision rules. It means that inhibitory decision rules are as relevant to classification algorithms as deterministic decision rules. There is an additional (intuitive) motivation for the use of inhibitory decision rules in classification algorithms: the inhibitory decision rules have much more chance to have larger support than the deterministic ones. Results of experiments with bounded nondeterministic rules generated by the $ND$-heuristic are showing that also the nondeterministic rules can lead to improving the classification quality. We have demonstrated this using a combination of classification algorithms based on minimal decision rules and bounded nondeterministic rules.

We plan to extend our approach on classification algorithms constructed by hybridization of the rough set methods with the Dempster–Shafer approach.

## Acknowledgements

The research has been partially supported by the grant N N516 368334 from Ministry of Science and Higher Education of the Republic of Poland. The authors are indebted to anonymous reviewers for helpful suggestions.

## References

1. Aha, D.W. (ed.): Lazy Learning. Kluwer Academic Publishers, Dordrecht (1997)
2. Agrawal, R., Imielinski, T., Swami, A.: Mining Association Rules Between Sets of Items in Large Databases. In: Buneman, P., Jajodia, S. (eds.) Proceedings of the 1993 ACM SIGMOD International Conference on Management of Data, Washington, D.C., May 26-28, pp. 207–216. ACM Press, New York (1993)
3. Bazan, J.G.: Discovery of Decision Rules by Matching New Objects Against Data Tables. In: Polkowski, L., Skowron, A. (eds.) RSCTC 1998. LNCS (LNAI), vol. 1424, pp. 521–528. Springer, Heidelberg (1998)
4. Bazan, J.G.: A Comparison of Dynamic and Non-Dynamic Rough Set Methods for Extracting Laws from Decision Table. In: Polkowski, L., Skowron, A. (eds.) Rough Sets in Knowledge Discovery, pp. 321–365. Physica-Verlag, Heidelberg (1998)
5. Bazan, J.G.: Methods of Approximate Reasoning for Synthesis of Decision Algorithms. Ph.D. Thesis. Warsaw University (1998) (in Polish)
6. Bazan, J., Skowron, A., Swiniarski, R.: Rough Sets and Vague Concept Approximation: From Sample Approximation to Adaptive Learning. In: Peters, J.F., Skowron, A. (eds.) Transactions on Rough Sets V. LNCS, vol. 4100, pp. 39–62. Springer, Heidelberg (2006)
7. Bazan, J.G., Szczuka, M.S., Wojna, A., Wojnarski, M.: On the Evolution of Rough Set Exploration System. In: Tsumoto, S., Słowiński, R., Komorowski, J., Grzymała-Busse, J.W. (eds.) RSCTC 2004. LNCS (LNAI), vol. 3066, pp. 592–601. Springer, Heidelberg (2004)
8. Bramer, M.A.: Automatic Induction of Classification Rules from Examples Using NPRISM. In: Research and Development in Intelligent Systems XVI, pp. 99–121. Springer, Heidelberg (2000)
9. Cendrowska, J.: PRISM: An Agorithm for Inducing Modular Rules. International Journal of Man-Machine Studies 27(4), 349–370 (1987)
10. Data Mining Exploration System (Software), `http://www.univ.rzeszow.pl/rspn`
11. Delimata, P., Moshkov, M., Skowron, A., Suraj, Z.: Two Families of Classification Algorithms. In: An, A., Stefanowski, J., Ramanna, S., Butz, C.J., Pedrycz, W., Wang, G. (eds.) RSFDGrC 2007. LNCS (LNAI), vol. 4482, pp. 297–304. Springer, Heidelberg (2007)
12. Delimata, P., Moshkov, M.J., Skowron, A., Suraj, Z.: Comparison of Lazy Classification Algorithms Based on Deterministic and Inhibitory Decision Rules. In: Wang, G., Li, T., Grzymala-Busse, J.W., Miao, D., Skowron, A., Yao, Y. (eds.) RSKT 2008. LNCS (LNAI), vol. 5009, pp. 55–62. Springer, Heidelberg (2008)
13. Delimata, P., Moshkov, M.J., Skowron, A., Suraj, Z.: Inhibitory Rules in Data Analysis: A Rough Set Approach. In: Studies in Computational Intelligence, vol. 163. Springer, Heidelberg (2009)
14. Grzymala-Busse, J.W.: LERS - A Data Mining System. In: Maimon, O., Rokach, L. (eds.) The Data Mining and Knowledge Discovery Handbook, pp. 1347–1351. Springer, New York (2005)
15. Holte, R.: Very Simple Classification Rules Perform Well on Most Commonly Used Data Sets. Machine Learning 11, 63–91 (1993)
16. Marszał-Paszek, B., Paszek, P.: Minimal Templates and Knowledge Discovery. In: Kryszkiewicz, M., Peters, J.F., Rybiński, H., Skowron, A. (eds.) RSEISP 2007. LNCS (LNAI), vol. 4585, pp. 411–416. Springer, Heidelberg (2007)
17. Ryszard Michalski, `http://www.mli.gmu.edu/michalski/`

18. Moshkov, M., Skowron, A., Suraj, Z.: On Maximal Consistent Extensions of Information Systems. In: Proceedings of the Conference Decision Support Systems, Zakopane, Poland, December 2006, vol. 1, pp. 199–206. University of Silesia, Katowice (2007)
19. Moshkov, M., Skowron, A., Suraj, Z.: Maximal Consistent Extensions of Information Systems Relative to Their Theories. Information Sciences 178, 2600–2620 (2008)
20. UCI Repository of Machine Learning Databases, University of California, Irvine, `http://www.ics.uci.edu/~mlearn/MLRepository.html`
21. Nguyen, H.S.: Scalable Classification Method Based on Rough Sets. In: Alpigini, J.J., Peters, J.F., Skowron, A., Zhong, N. (eds.) RSCTC 2002. LNCS (LNAI), vol. 2475, pp. 433–440. Springer, Heidelberg (2002)
22. Nguyen, T.T.: Handwritten Digit Recognition Using Adaptive Classifier Construction Techniques. In: Pal, S.K., Polkowski, L., Skowron, A. (eds.) Rough-Neural Computing: Techniques for Computing with Words, pp. 573–585. Springer, Heidelberg (2003)
23. Pawlak, Z.: Rough Sets – Theoretical Aspects of Reasoning about Data. Kluwer Academic Publishers, Dordrecht (1991)
24. Pawlak, Z., Skowron, A.: Rudiments of Rough Sets. Information Sciences 177, 3–27 (2007); Rough Sets: Some Extensions. Information Sciences 177, 28–40 (2007); Rough Sets and Boolean Reasoning. Information Sciences 177, 41–73 (2007)
25. Polkowski, L., Skowron, A. (eds.): Rough Sets in Knowledge Discovery 2: Applications, Case Studies and Software Systems. Studies in Fuzziness and Soft Computing, vol. 19. Physica-Verlag, Heidelberg (1998)
26. Pulatova, S.: Covering (Rule-Based) Algorithms. In: Berry, M.W., Browne, M. (eds.) Lecture Notes in Data Mining, pp. 87–97. World Scientific, Singapore (2006)
27. Rissanen, J.: Modeling by Shortest Data Description. Automatica 14, 465–471 (1978)
28. Rissanen, J.: Information and Complexity in Statistical Modeling. In: Springer Sciences and Business Media, LLC, New York (2007)
29. Rosetta, `http://www.lcb.uu.se/tools/rosetta/`
30. Rough Set Exploration System, `http://logic.mimuw.edu.pl/~rses`
31. Skowron, A., Suraj, Z.: Rough Sets and Concurrency. Bulletin of the Polish Academy of Sciences 41, 237–254 (1993)
32. Skowron, A., Swiniarski, R., Synak, P.: Approximation Spaces and Information Granulation. In: Peters, J.F., Skowron, A. (eds.) Transactions on Rough Sets III. LNCS, vol. 3400, pp. 175–189. Springer, Heidelberg (2005)
33. Suraj, Z.: Some Remarks on Extensions and Restrictions of Information Systems. In: Ziarko, W.P., Yao, Y. (eds.) RSCTC 2000. LNCS (LNAI), vol. 2005, pp. 204–211. Springer, Heidelberg (2001)
34. Triantaphyllou, E., Felici, G. (eds.): Data Mining and Knowledge Discovery Approaches Based on Rule Induction Techniques. In: Springer Science and Business Media, LLC, New York (2006)
35. Tsumoto, S.: Modelling Medical Diagnostic Rules Based on Rough Sets. In: Polkowski, L., Skowron, A. (eds.) RSCTC 1998. LNCS (LNAI), vol. 1424, pp. 475–482. Springer, Heidelberg (1998)
36. Wojna, A.: Analogy-Based Reasoning in Classifier Construction (Ph.D. Thesis). In: Peters, J.F., Skowron, A. (eds.) Transactions on Rough Sets IV. LNCS, vol. 3700, pp. 277–374. Springer, Heidelberg (2005)
37. Yao, J.T., Yao, Y.Y.: Induction of Classification Rules by Granular Computing. In: Alpigini, J.J., Peters, J.F., Skowron, A., Zhong, N. (eds.) RSCTC 2002. LNCS (LNAI), vol. 2475, pp. 331–338. Springer, Heidelberg (2002)

# Gene Selection and Cancer Classification: A Rough Sets Based Approach

Lijun Sun, Duoqian Miao, and Hongyun Zhang

Key Laboratory of Embedded System and Service Computing,
Ministry of Education,
Tongji University, Shanghai 201804, P.R.China
Department of Computer Science and Technology,
Tongji University, Shanghai, 201804, P.R.China
Sunlj1028@yahoo.com.cn, Miaoduoqian@163.com, Zhanghongyun583@sina.com

**Abstract.** Indentification of informative gene subsets responsible for discerning between available samples of gene expression data is an important task in bioinformatics. Reducts, from rough sets theory, corresponding to a minimal set of essential genes for discerning samples, is an efficient tool for gene selection. Due to the compuational complexty of the existing reduct algoritms, feature ranking is usually used to narrow down gene space as the first step and top ranked genes are selected . In this paper,we define a novel certierion based on the expression level difference btween classes and contribution to classification of the gene for scoring genes and present a algorithm for generating all possible reduct from informative genes.The algorithm takes the whole attribute sets into account and find short reduct with a significant reduction in computational complexity. An exploration of this approach on benchmark gene expression data sets demonstrates that this approach is successful for selecting high discriminative genes and the classification accuracy is impressive.

**Keywords:** gene selection, cancer classification, rough sets, reduct, feature ranking, bioinformatics, gene expression.

## 1 Introduction

Standard medical classification systems for cancer tumors are based on clinical observations and the microscopical appearances of the tumors, these systems fail to recognize the molecular characteristics of the cancer that often correspond to subtypes that need different treatment. Studying the expression levels of genes in tumor issues may reveal such subtypes and may also diagnose the disease before it manifest itself on a clinical level, tumor classification based on gene expression data analysis is becoming one of the most important research areas in bioinformatics.

Gene expression data often has thousands of genes while not more than a few dozens of tissue samples, with such a huge attribute space, it is almost certain that very accurate classification of tissue samples is difficult. Recent research

J.F. Peters et al. (Eds.): Transactions on Rough Sets XII, LNCS 6190, pp. 106–116, 2010.

has shown that the expression level of fewer than ten genes are often sufficient for accurate diagnosis of most cancers, even though the expression levels of a large number of genes are strongly correlated with the disease[1],[2],[3],[4]. In fact, the use of a much larger set of gene expression levels has been shown to have a deleterious effect on the diagnostic accuracy due to the phenomenon known as the curse of dimensionality. Thus Performing gene selection prior to classification will help to narrowing down the attribute number and improving classification accuracy. More importantly, by identifying a small subset of genes on which to base a diagnostic accuracy, we can gain possibly significant insights into the nature of disease and genetic mechanisms responsible for it. In addition, assays that require very few gene expression levels to be measured in order to make diagnosis are far more likely to be widely deployed in a clinical setting. How to select the most informative genes for cancer classification is becoming a very challenging task.

Rough set theory proposed by Pawlak provides a mathematic tool that can be used to find out all possible feature sets [5],[6].It works by gradually eliminating superfluous or redundant features to find a minimal set of essential features for classification. Reduct, corresponds to such a minimal subset, will be used to instead of all features in the learning process. Thus a feature selection method which is using rough set theory can be regarded as finding such a reduct with respect to the best classification [7].

Gene expression data analysis represents a fascinating and important application area for rough sets-based methods. Recently, researchers have focused their attention on gene subsets selection and cancer classification based on rough sets [8],[9]. Midelfart et al. used rough set-based learning methods implemented with ROSETTA involving GAs and dynamic reducts for gastric tumor classification; rule models for each of six clinical parameters are induced [10]. Valdes et al. investigated an approach using clustering in combination with Rough Sets and neural networks to select high discriminated genes[11] . Momin et al. developed a positive region based algorithm for generating reducts from gene expression data [12], results on benchmark gene expression datasets demonstrate more than 90% reduction of redundant genes. Banerjee et al. adopted an evolutionary rough feature selection algorithm for classifying microarray gene expression patterns. Since the data typically consist of a large number of redundant features, an initial redundancy reduction of the attributes is done to enable faster convergence. Thereafter rough set theory is employed to generate reducts in a multiobjective framework. The effectiveness of the algorithm is demonstrated on three cancer datasets, colon, lymphoma, and leukemia. In case of the leukemia data and lymphoma data, two genes are selected, whereas the colon data results in an eight-gene reduct size [13].

In this paper, we propose a rough set based method for identifying informative genes for cancer classification. We first develop a novel criterion for scoring genes, top ranked k genes are selected with this criterion, and then rough sets attribute reduction is employed to obtain all possible subsets that define the same partition with all gens, rule sets induced from all reducts used as the classifier for labeling the new samples. The rest of the paper is organized as follows: Section 2 provides the basics of rough sets, then, our method is detailed in Section 3. Experimental results demonstrated on benchmark microarray data sets are listed, the discussions of these results are given in section 4. Finally, the conclusion is drawn in Section 5.

## 2 Rough Sets Preliminaries

In rough set theory, a decision table is denoted by $T = (U, C \cup D)$, where $U$ is a nonempty finite set called universe, $C$ is a nonempty finite set called condition attribute sets and $D = \{d\}$ is decision feature. Rows of the decision table correspond to objects, and columns correspond to attributes [7].

Let $B \subseteq C \cup D$, a binary relation $IND(B)$, called the indiscernibility relation, is defined as

$$IND(B) = \{(x, y) \in U \times U : \forall a \in B, a(x) = a(y)\} \tag{1}$$

$IND(B)$ is also an equivalence relationship on $U$. Let $U/IND(B)$ denotes the family of all equivalence classes of the relation $IND(B)$ (or classification of $U$), $U/IND(B)$ is also a definable partition of the universe induced by $IND(B)$.

A subset $X \subseteq U$ can be approximated by a pair of sets, called lower and upper approximation with respect to an attribute subset $B \subseteq C$, the $B-$ lower approximation of $X$ is denoted as $\underline{B}(X) = \{Y \in U/IND(B) : Y \subseteq X\}$ and the $B-$ upper approximation of $X$ is denoted as $\overline{B}(X) = \{Y \in U/IND(B) : Y \cap X \neq \varphi\}$ respectively.

The $C$ -positive region of $D$, denoted by $POS_C(D)$, is defined as:

$$POS_C(D) = \bigcup_{X \in U/IND(D)} \underline{C}(X) \tag{2}$$

For an attribute $c \in C$, if $POS_C(D) \neq POS_{C-\{c\}}(D)$, $c$ is an indispensible feature, delete it from $C$ will decrease the dicernibility between objects. Otherwise, if $POS_C(D) = POS_{C-\{c\}}(D)$, $c$ is a dispensable feature, that is, $c$ is redundancy and can be deleted from $C$ without affecting the discernibility of objects. The subset of attributes $R \subseteq C$ is a reduct of attribute $C$ with respect to $D$ if

$$POS_R(D) = POS_C(D) \tag{3}$$

A reduct $R$ of $C$ is called a minimal reduct if $\forall Q \subset R$, $POS_Q(D) \neq POS_C(D)$ ,thus $R$ represents the minimal set of non-redundant features which capable of discerning objects in a decision table, thus it will be used to instead of $C$ in a rule discovery algorithm. The set of all indispensable features in $C$ is core of $C$ with respect to $D$, denoted by $CORE_D(C)$, we have

$$CORE_D(C) = \cap RED(C) \tag{4}$$

Where $RED(C)$ is the set of all reducts of $C$ with respect to $D$.

Reducts have been nicely characterized in [17] by discernibility matrices and discernibility functions. If $T = (U, C \cup D)$ is a decision table, with $U = \{x_1, x_2, ......, x_n\}$, the disneribility matrix of T, denoted by M(T), mean a $n \times n$ matrix defined as:

$$m_{ij} = \{c \in C : c(x_i) \neq c(x_j) \wedge d(x_i) \neq d(x_j)\}, i = 1, 2, ......, n \quad (5)$$

Thus entry $m_{ij}$ is the set of all attributes that classify object $x_i$ and $x_j$ into different decision class in $U/IND(D)$.The $CORE(C)$ can be defined as the set of all single element entries of the discernibility matrix, that is

$$CORE(C) = \{c \in C : mij = \{c\}\} \quad (6)$$

A discernibility function $f$ is a Boolean function of $m$ boolean variables $a_1, a_2, ......, a_m$ correspond to the attributes $c_1, c_2, ......, c_m$, defined as:

$$f = \wedge\{\vee(m_{ij}) : 1 \leq j < i \leq n, m_{ij} \neq \varphi\} \quad (7)$$

Where $\vee(m_{ij})$ is the disjunction of all the variables $a$ with $c \in m_{ij}$.It is seen that $\{c_{i1}, c_{i2}, ......, c_{ip}\}$ is a reduction in the decision table $T$ if and only if $\{c_{i1} \wedge c_{i2} \wedge ...... \wedge c_{ip}\}$ is a prime implicant of $f$.

In rough set community, most feature subset selection algorithm are attribute reduct-oriented, that is, finding minimum reducts of the conditional attributes of decision tables. Two main approaches to finding attribute reducts are recognized as discernibility function-based and attribute dependency-based [14],[15],[16],[17]. These algorithms, however, suffer from intensive computations of either discernibility functions for the former or positive region for the latter, which limit its application on large scale data sets.

## 3 Informative Genes

As mentioned above, for all genes measured by microarray, only a few of them play major role in the processes that underly the differences between the classes, the expression levels of many other genes maybe irrelevant to the distinction between tissue classes. The goal of this work is to extract those genes that demonstrate high discriminating capabilities between the classes of samples, these genes are called informative genes, an important characteristic of them is that the expression level in different classes has the remarkable difference, that is, the gene that demonstrate high difference in its expression levels in different classes is a good significant genes that is typically highly related with the disease of samples.

For a binary classification problem (assuming class1 vs. class2), when we consider a gene , if $U/IND(g) = \{X_1, X_2\}$ and all the samples in $X_1$ belong to class1 and all the samples in $X_2$ belong to class2, then gene $g$ is most useful gene for classify new samples.

For that, we exam for each gene, all the values of each class, and get the number of equivalence classes.

$a$ =# of equivalence classes of gene $g$ in class1
$b$ =# of equivalence classes of gene $g$ in class2

Thus the most useful gene is the one that has the lowest $a$ and $b$ values and the case $a$ =1 and $b$ =1 gives the least noise, thus the measure $a \times b$ is a good indicator of how much a gene differentiates between two classes ,therefore we compute a $V$ score for each gene

$$V = a \times b \tag{8}$$

In real applications, two or more genes may have same $V$ values, therefore mutual information (MI) is used to distinguish such genes. Given the partition by $D$, $U/IND(D)$ , of $U$, the mutual information based on the partition by $g \in C$ , $U/IND(g)$, of $U$, is given by

$$I(D,g) = H(D) - H(D|\{g\}\} \\ = \frac{1}{|U|} \sum_{X \in U/IND(D)} |X| \log \frac{|X|}{|U|} - \frac{1}{|U|} \sum_{X \in U/IND(D)} \sum_{Y \in U/IND(g)} |X \cap Y| \log \frac{|X \cap Y|}{|Y|} \tag{9}$$

$I(D,g)$ quantifies the relevance of gene $g$ for the classification task. Thus the criterion for scoring genes can be described below:

1. A gene with the lowest $V$ value is the one that have the highest differences in its expression levels between two classes of samples.
2. If two or more genes have same $V$ values, the one has maximal mutual information is more significant.

# 4 Our Proposed Methods for Gene Selection and Cancer Classification

Our learning problem is to select high discriminate genes from gene expression data for cancer classification, which define the same partition as the whole set of genes. We may formalize this problem as a decision table $T = (U, C \cup \{d\})$ , where universe $U = \{x_1, x_2, ..., x_n\}$ is a set of tumors. The conditional attributes set $C = \{g_1, g_2, ..., g_m\}$ contains each gene, the decision attribute d corresponds to class label of each sample. Each attribute $g_i \in C$ is represented by a vector $g_i = \{v_{1,i}, v_{2,i}, ..., v_{n,i}\}$, $i = 1, 2, ...m$ where $v_{k,i}$ is the expression level of gene $i$ at sample $k, k = 1, 2, ..., n$ .

## 4.1 Discretization

The methods of rough sets can only deal with data sets having discrete attributes because real values attribute will lead to large number of equivalence classes, thus lead to large number of rules, thereby making rough sets classifiers inefficient. In our study, a simplest method is used to discretize the data set. It is applied for each continuous attribute independently, for each attribute, it sorts the continuous feature values and gets a cut $c = (V_a(x_i) + V_a(x_{i+1}))/2$ when the adjacent two samples $x_i$ and $x_{i+1}$ have different class labels, namely $d(x_i) \neq d(x_{i+1})$.

### 4.2 Gene Selection

A major obstacle for using rough sets to deal with gene expression data may be the large scale of gene expression data and the comparatively slow computational speed of rough sets algorithms. The computation of discernibility has a time complexity of $O(n^2)$ , which is still much higher than many algorithms in bioinformatics. In order to solve this problem, each gene is measured for correlation with the class according to some criteria, top ranked genes are selected before we employee the existing attribute reduct algorithm on gene expression data sets. The literature discusses several method for scoring genes for relevance, Simple methods based on t-test and mutual information are proved to be effective [18],[19],[20].

Feature sets so obtained have certain redundancy because genes in similar pathways probably all have very similar score and therefore no additional information gain. If several pathways involved in perturbation but one has main influence it is Possible to describe this pathway with fewer genes, therefore Rough sets attribute reduction is used to minimize the feature sets.

Our proposed method uses the criterion described above in section 3 to score genes and starts with choosing genes having highest significance. From these chosen genes, possible subsets are formed and subsets having same classification capability with the entire gene set are chosen, the algorithm is formulated as the following:

Step 1: compute the significance of each gene

Step 2: sort the genes in descending order of significance

Step 3: select top ranked $k$ genes which satisfied with $POS_{top}(D) = POS_C(D)$

Step 4: finding all possible reducts of the top ranked $k$ genes

Where $POS_{top}(D)$ is the positive region of the top ranked $k$ genes, thus each reduct of the top ranked $k$ genes is also a reduct of all genes.

This algorithm finds the most promising $k$ attributes from the whole data set based on the attribute significance and positive region coverage, the reducts are then generated from these promising attributes, thus there is significant reduction in computational complexity and time required for getting reducts as compared to getting reducts from the entire gene space. Moreover, this algorithm takes into account the positive region coverage of all the genes, hence no information lost.

### 4.3 Classification

Decision rules as knowledge representation can be used to express relationship between condition attributes (or the antecedent) and decision attribute (or the consequent). The rough set theory as a mathematical tool provides an efficient mechanism to induce decision rules from decision table. After attribute reduction, each row in the decision table corresponds to a decision rule; rule set induced from the reduct can be used as the classifier to predict the class label of a new sample. In order to simplify the decision rules, value reduction method proposed in [5] is adopted.

In our study, we mix all the rules we obtained from each possible reduct together to be the classifier. When rules conflict, stand voting is used to select the best one. Leave-one out cross-validation (LOOCV), which is a widely used process for gene classification, is employed to evaluate the performance of classification process. With LOOCV, each object in data set will in turn be the test set, and the left are training set. Thus each sample of the data set will be predicted once by classifier trained with the left samples. All iterations are then averaged to obtain an unbiased number of performance estimates.

## 5 Experimental Results and Discussions

In this study, we have focus on four sets of gene expression data as summarized in Table1.They are described as follows:

**Table 1.** Data sets

| Data set | Number of genes | Number of Classes | Number of samples |
|---|---|---|---|
| Acute Leukemia | 7129 | 2(AML vs.ALL) | 72(47ALL:25AML) |
| Colon Cancer | 2000 | 2(tumor vs. normal) | 62(22tumor:40normalL) |
| Lung Cancer | 12533 | 2(MPM vs. ACDA) | 181(31MPM:150ACDA) |
| DLBCL | 4026 | 2(germinal vs. activated) | 47(24germinal:23activated) |

1. The Acute Leukemia data set [21] consists of samples from two different types of acute leukemia, acute lymphoblastic leukemia (ALL) and acute myeloid leukemia (AML). There are 47 ALL and 25 AML samples; each sample has expression patterns of 7129 genes measured.

2. The Colon Cancer data set [22] is a collection of 62 gene expression measurement from biopsy samples generated using high density oligonucleotide microarrays. There are 22 normal and 40 tumor samples; each sample has expression patterns of 2000 genes measured.

3. The Lung Cancer data set [23] provides a collection of 181 gene expression samples, 150 for ACDA, 31 for PMP; each sample has expression patterns of 12533 genes measured.

4. The DLBCL data set [24] provides expression measurement of 47 samples, 24 for germinal, 31 for activated; each sample has expression patterns of 4026 genes measured.

For each data set, we select top ranked $k$ genes($k$ =1,2,3,4,,5,6,7,8,9,10) to form the subset, and search for all possible reducts of all genes from them , rule sets generated from each reduct are mixed together to be the classifier for predicting the class label of new samples.

Table2 lists the top 10 genes we selected in each dataset. Most of them are also identified by the different other methods in the published literatures. For example, Krishnapuram et.al. selects 20 genes in leukemia data set and colon

**Table 2.** top 10 genes selected by our proposed method in each data set

| No. of genes | Acute Leukemia | Colon Cancer | Lung Cancer | DLBCL |
|---|---|---|---|---|
| 1 | M23197 | M76378 | 37157_at | GENE3330X |
| 2 | X95735 | X12671 | 37954_at | GENE3261X |
| 3 | M28791 | X56597 | 36533_at | GENE3327X |
| 4 | D88422_at | M22382 | 33328_at | GENE3967X |
| 5 | M84526_at | R87126 | 37716_at | GENE3329X |
| 6 | M92287 | M16937 | 39640_at | GENE3256X |
| 7 | U46499 | H28711 | 515_s_at | GENE3332X |
| 8 | M31523 | U25138 | 1500_at | GENE3328X |
| 9 | M11722 | D63874 | 38127_at | GENE3315X |
| 10 | L09209_s_at | M63391 | 179_at | GENE1252X |

**Table 3.** the classification accuracy results with top ranked 1 to 10 genes

| No. of genes | Acute Leukemia | | Colon Cancer | |
|---|---|---|---|---|
| | Proposed criterion | MI | Proposed criterion | MI |
| 1 | 98.6% | 98.6% | 93.5% | 74.2% |
| 2 | 100% | 100% | 95.2% | 93.5% |
| 3 | 100% | 100% | 98.4% | 96.8% |
| 4 | 100% | 100% | 98.4% | 96.8% |
| 5 | 100% | 100% | 98.4% | 98.4% |
| 6 | 100% | 100% | 98.4% | 98.4% |
| 7 | 100% | 100% | 98.4% | 98.4% |
| 8 | 100% | 100% | 98.4% | 98.4% |
| 9 | 100% | 100% | 98.4% | 98.4% |
| 10 | 100% | 100% | 98.4% | 98.4% |
| No. of genes | Lung Cancer | | DLBCL | |
| | Proposed criterion | MI | Proposed criterion | MI |
| 1 | - | 86.2% | 93.6% | 78.7% |
| 2 | 98.9% | 90.6% | 100% | 91.5% |
| 3 | 99.4% | 92.3% | 100% | 97.9% |
| 4 | 99.4% | 95.6% | 100% | 100% |
| 5 | 100% | 96.7% | 100% | 100% |
| 6 | 100% | 96.7% | 100% | 100% |
| 7 | 100% | 96.7% | 100% | 100% |
| 8 | 100% | 98.3% | 100% | 100% |
| 9 | 100% | 98.3% | 100% | 100% |
| 10 | 100% | 98.3% | 100% | 100% |

data set respectively and achieved 100% classification of all the samples [25], our selected gene M23197, M27891, M84526_at, M11722 from acute leukemia data set and M76378, R87126 from colon data set are among them. Gene M23197, M95735, M92287, L09209_s_at we selected from leukemia data set is also identified by Deb's work [26]. For DLBCL data set, Gene GENE3330X, GENE3328X are proved to be significant [18].These proves the effectiveness of our proposed method.

Table 3 summarizes the results, which demonstrate that our feature selection method can produces high significant features as the classification accuracy is very impressive. For acute leukemia data set, when using one gene X23197, all ALL samples can be correctly classified and the classification accuracy of AML samples is 96%, when using two or more genes, the overall classification accuracy 100% are achieved. For Lung cancer data set, the first gene selected is gene 37157_at, it can not form a reduct, when using 2 genes, ACDA samples can be fully classified(100%), and when using 5 or more genes, all the samples can be fully classified(100%). For colon data set, normal samples can be fully classified with the top ranked 2 genes, and we achieve the overall classification accuracy 98.4% when 3 or more genes are selected. For DLBCL data set, we achieve 93.6% and 100% classification accuracy respectively with the top ranked 2 genes. For comparison, experiment results when using MI to select the top ranked genes are also list. We can see that, for acute leukemia data set, both methods obtain the same results, for the other 3 data sets, our proposed method can obtain higher classification accuracy with fewer genes on each data set.

## 6 Conclusions

Gene expression data set has very unique characteristics which are very different from all the previous data used for classification. In order to achieve good classification performance, and obtain more useful insight about the biological related issues in cancer classification, gene selection should be well explored to both reduce the noise and avoid overfitting of classification algorithm.

This paper explores feature selection techniques based on rough set theory within the context of gene expression data for sample classification. We define a novel measurement for scoring genes according to the expression level differences between classes of a gene and its Mutual information, and then present a gene selection method based on rough sets theory, the method takes whole attributes set into account and extract all possible gene subsets from the top ranked informative genes that allow for sample classification with high accuracy. Experimental results on benchmark datasets indicate our method has successfully achieved its objectives: obtain high classification accuracy with a small number of high discriminative genes.

## References


1. Frank, A.: A New Branch and Bound Feature Selection Algorithm. M.Sc. Thesis, submitted to Technion, Israel Institute of Technology (2002)
2. Xiong, M., Li, W., Zhao, J., Jin, L., Boerwinkle, E.: Feature (gene) Selection In Gene Expression-based Tumor Classication. Molecular Genetics and Metabolism 73, 239–247 (2001)
3. Wang, L.P., Feng, C., Xie, X.: Accurate Cancer Classifcation Using Expressions of Very Few Genes. EE/ACM Transactions on Computational Biology and Bioinformatics 4, 40–53 (2007)
4. Li, W., Yang, Y.: How Many Genes Are Needed For A Discriminant Microarray Data Analysis? In: Methods of Microarray Data Analysis. Kluwer academic Publisher, Norwell (2002)
5. Pawlak, Z.: Rough Set- Theoretical Aspects of Reasoning about Data. Kluwer Academic Publishers, Dorderecht (1991)
6. Palawk, Z.: Rough Sets. International Journal of Computer and Information Science 11, 341–356 (1982)
7. Zhong, N., Dong, J., Ohsuga, S.: Using rough sets with heruristic for feature selection. Journal of Intelligent Information Systems 16, 119–214 (2001)
8. Mitra, S., Hayashi, Y.: Bioinformatics with Soft Computing. IEEE Transactions on Systems, Man and Cybernetics-Part C: Applications and Reviews 36, 616–635 (2006)
9. Hvidsten, T.R., Komorowski, J.: Rough Sets in Bioinformatics. In: Peters, J.F., Skowron, A., Marek, V.W., Orłowska, E., Słowiński, R., Ziarko, W.P. (eds.) Transactions on Rough Sets VII. LNCS, vol. 4400, pp. 225–243. Springer, Heidelberg (2007)
10. Midelfart, H., Komorowski, J., Nørsett, K., Yadetie, F., Sandvik, A.K., Lægreid, A.: Learning Rough Set Classifiers From Gene Expressions And Clinical Data. Fundamenta Inf. 53, 155–183 (2002)
11. Valdes, J.J., Barton, A.J.: Gene Discovery in Leukemia Revisited: A Computational Intelligence Perspective. In: Orchard, B., Yang, C., Ali, M. (eds.) IEA/AIE 2004. LNCS (LNAI), vol. 3029, pp. 118–127. Springer, Heidelberg (2004)
12. Momin, B.F., Mitra, S., Datta Gupta, R.: Reduct Generation and Classifcation of Gene Expression Data. In: Proceeding of First International Conference on Hybrid Information Technology (ICHICT 2006), pp. 699–708. IEEE Press, New York (2006)
13. Banerjee, M., Mitra, S., Banka, H.: Evolutinary-Rough Feature Selection in Gene Expression Data. IEEE Transaction on Systems, Man, and Cyberneticd, Part C: Application and Reviews 37, 622–632 (2007)
14. Wang, J., Wang, J.: Reduction Algorithms Based on Discernibly Matrix: The Ordered Attributes Method. Journal of Computer Science and Technology 16, 489–504 (2002)
15. Miao, D.Q., Hu, G.R.: A Heuristic Algorithm for Reduction of Knowledge. Journal of Computer Research and Development 36, 681–684 (1999)
16. Shen, Q., Chouchoulas, A.: A modular approach to generating fuzzy rules with reduced attributes for monitoring of complex systems. Engineering Applications of Artificial Intellegence 12, 263–278 (2000)
17. Skowron, A., Rauszer, C.: The discernibility matrices and functions in information systems. In: Intelligent decision Support. Handbook of Applications and Advances of the Rough Sets Theory. Kluwer Academic, Dordrecht (1992)

18. Wang, Y., Tetko, I.V., Hall, M.A., Frank, E., Facius, A., Mayer, K.F.X., Mewes, H.W.: Gene Selection from Microarray Data for Cancer Classification-A Machine Learning Approach. Computational Biology and Chemistry 29, 37–46 (2005)
19. Zhou, W.G., Zhou, C.G., Liu, G.X., Wang, Y.: Artificial Intelligence Applications and Innovations. In: Proceeding of IFIP Intenational Federation for Information, pp. 492–499. Springer, Heidelberg (2006)
20. Ding, C., Peng, H.C.: Minimum Redundancy Feature Selection from Microarray Gene Expression Data. Journal of Bioinformatics and Computational Biology 3, 185–205 (2003)
21. Golub, T.R., Slonim, D.K., Tamayo, P., Huard, C., Gaasenbeek, M., Mesirov, J.P., Coller, H., Loh, M.L., Downing, J.R., Caligiuri, M.A., Bloomfield, C.D., Lander, E.S.: Molecular Classification of Cancer: Class Discovery and Class Prediction by Gene Expression Monitoring. Science 286, 531–537 (1999)
22. Alon, U., Barkai, N., Notterman, D.A.: Broad Patterns of Gene Expression Revealed By Clustering Analysis of Tumor And Normal Colon Tissues Probed By Oligonucleotide Arrays. PNASUSA 96, 6745–6750 (1999)
23. Armstrong, S.A.: MLL Translocations Specify A Distinct Gene Distinguishes A Expression Profile That Unique Leukemia. Nature Genetics 30, 41–47 (2002)
24. Alizadeh, A.A., et al.: Distict types of diffuse large B-cell lymphoma identified by gene expressionprofiling. Nature 403, 503–511 (2000)
25. Krishnapuram, B., et al.: Joint classifier and feature selection optimization for Cancer diagnosis using gene expression Data. In: Proceedings of the Seventh Annual International Conference on Research in Computational Molecular Biology, pp. 167–175. ACM, New York (2003)
26. Deb, K., Reddy, A.R.: Reliable Classifcation of Two Class Cancer Data Using Evolutionary Algorithms. BioSystems 72, 111–129 (2003)

# Evolutionary-Rough Feature Selection for Face Recognition

Debasis Mazumdar[1], Soma Mitra[1], and Sushmita Mitra[2]

[1] CDAC, Kolkata, Plot - E2/1, Block- GP, Sector - V, Salt Lake Electronics Complex, Kolkata - 700091, India
[2] Machine Intelligence Unit, Indian Statistical Institute, Kolkata 700 108, India

**Abstract.** Elastic Bunch Graph Matching is a feature-based face recognition algorithm which has been used to determine facial attributes from an image. However the dimension of the feature vectors, in case of EBGM, is quite high. Feature selection is a useful preprocessing step for reducing dimensionality, removing irrelevant data, improving learning accuracy and enhancing output comprehensibility.

In rough set theory reducts are the minimal subsets of attributes that are necessary and sufficient to represent a correct decision about classification. The high complexity of the problem has motivated investigators to apply various approximation techniques like the multi-objective GAs to find near optimal solutions for reducts.

We present here an application of the evolutionary-rough feature selection algorithm to the face recognition problem. The input corresponds to biometric features, modeled as Gabor jets at each node of the EBGM. Reducts correspond to feature subsets of reduced cardinality, for efficiently discriminating between the faces. The whole process is optimized using MOGA. The simulation is performed on large number of Caucasian and Indian faces, using the FERET and CDAC databases. The merit of clustering and their optimality is determined using cluster validity indices. Successful retrieval of faces is also performed.

**Keywords:** Face recognition, Gabor jets, Softcomputing, Rough sets, Multiobjective genetic algorithms, C-means clustering, Clustering validity index, Elastic Bunch Graph Matching.

## 1 Introduction

Recognition of human face is a complex visual pattern recognition problem. While the human perception system does this as a routine task, the building up of a similar computer system with all possible flexibilities and relaxations is an on-going research issue. Face recognition is a referral problem in the area of deformable pattern recognition. The complexities of the problem are mainly due to significant intra-class variations in appearance pertaining to pose, facial expressions, ageing, illumination, different resolution of imaging devices, and ornamental variations like varying hairstyle, growth of beard, moustache etc.

J.F. Peters et al. (Eds.): Transactions on Rough Sets XII, LNCS 6190, pp. 117–142, 2010.

The earliest research on automatic machine recognition of faces can be traced back to the 1970s in the engineering literature [1,2]. Over the past few decades extensive research has been conducted by psychophysicists, neuroscientists and engineers on various aspects of face recognition by humans and machines. Their comparative studies are also in progress. Many methods and algorithms have been proposed in the area of computerized face recognition during the past 30 years [3]. It has been reported by psychophysicists and neuroscientists [4] that in human perception a face is recognized more as a whole or as interrelationships between different features, than as a group of individual features. The other school of thought believes in feature salience [5]. They found that the central face is vital for recognition; not surprisingly - as this region includes the eyes, nose and mouth. Moreover the general viewpoint seems to be that all the facial features are not of the same importance; rather the eyes are more important than the mouth, which in turn is more significant than the nose [6].

The literature on face recognition is vast and diverse. However the psychological studies of how humans use holistic and local features specifically suggests us a guideline to categorize the methodology and algorithms [3]:

1. Holistic matching methods use the whole face region as the basic input to a recognition system. Face is represented as eigenpictures [7,8], which are based on principal component analysis.
2. Feature-based (structural) matching methods first extract the local features like eyes, nose, eyebrows, mouth, etc. Then their locations, and local statistics of geometric and other appearance-based properties (like texture) are computed. The local feature properties are next fed into a structural classifier.
3. Hybrid methods use both local features as well as the whole facial region to recognize a face.

A face recognition system, developed around any of the above three methods, should have the capabilities to recognize a human face across pose, illumination and expression variation. Over the past 15 years, research has been focused towards the fully automatic face recognition system having the aforementioned capabilities. Holistic approach of face recognition (which are principally appearance-based) have proved to be effective in experiments with large databases. Feature-based graph matching approaches have also been quite successful. Compared to holistic approaches, like PCA, ICA, such structural graph matching approaches are less sensitive to variations in illumination, pose, expression, as well as to any inaccuracy in landmark point location or face localization [3].

Elastic Bunch Graph matching algorithm (EBGM), proposed by Wiskott *et al.* [9], falls under this category. It has been used fairly reliably to determine facial attributes from image. EBGM can even recognize human faces from a large database consisting of only one image per person. Note that this task is difficult because of image variation in terms of position, size, expression, pose and illumination. Holistic approaches, on the other hand, require more than one facial images (in different pose, illumination, expression) of the same person to be present in the database. This is unrealistic since in the real life scenario, particularly when the subjects are non-cooperative, the availability of more than one facial images of the same person in different poses, sizes, expressions, become infeasible.

Faces in EBGM [10,9], are represented by labeled graphs. The algorithm extracts some biometric landmark points (sometimes referred to as fiducial points, in the literature) on the face, like the end points of the eyebrows, inner and outer canthus of both eyes, tip of the nose, end points of the lips, etc. Around each such point in a face, it extracts texture and generates a labeled graph. The edges of the graph represent the distance between the landmark points. The nodes are labeled with a set of complex Gabor Wavelet coefficients, called a jet. The labeled graph of all the facial images is stored systematically in a database called gallery. The values of the Gabor jets are the components of the feature vector representing a face in the original feature space. A query image is also represented by a labeled graph and its computed feature vector is compared with the existing feature vectors (of images) in the gallery. Selection of the matched pair is performed on the highest similarity score.

However the dimensions of the feature vector, in case of EBGM, is quite large. In the real time vision scenario, a face recognition system, has to deal with data streams in the range of several Mbytes/sec. Improving the computational efficiency is therefore of major concern when developing systems for real-world applications. It may be noted that redundant features increase the level of confusion. Since the landmark points do not contain the same amount of information - they contribute with different degrees of importance towards the recognition of a face [6]. Dimensionality reduction is a helpful preprocessing step towards removing irrelevant data, improving learning accuracy and enhancing output comprehensibility.

Many approaches for dimensionality reduction have been reported in literature [11,12,13]. The existing method can be grouped into feature extraction and feature selection. In feature extraction problems [13,14] the original features in the measurement space are transformed into a reduced space via some specified transformation. Significant features are then determined in this projected space. On the contrary, feature selection attempts to seek important subsets of the original features [15,16,17,18,19,20] that preserve the information carried by the complete data. Often the less important features are redundant, and can be ignored [21].

### 1.1 Soft Computing

Soft computing [22,23] involves the synergistic integration of paradigms like fuzzy sets, neural networks, genetic algorithms and rough sets. It helps to generate low-cost, low-precision (approximate), good solutions. Thereby we utilize the learning ability of neural networks for adapting, uncertainty handling capacity of fuzzy sets and rough sets for modeling ambiguity, and the search potential of genetic algorithms for efficiently traversing large search spaces.

Rough set theory(RST) [24] provides an important and mathematically established tool, for dimensionality reduction in large data. RST was developed by Pawlak as a tool to deal with inexact and incomplete data. Over the years, it has become a topic of great interest to researchers and has been applied to many domains, in particular to knowledge databases. This success is due in part to the following aspects of the theory, viz. only the facts hidden in data are analyzed, no additional information about the data is required, and a minimal knowledge representation is generated.

In RST the reducts are minimal subsets of attributes which give the same classification as the whole set of input features. These correspond to the minimal feature sets that are necessary and sufficient to represent a correct decision about classification. The task of finding reducts is reported to be NP-hard [25]. The high complexity of this problem has motivated investigators to apply various approximation techniques like genetic algorithms (GAs) [26] to find near-optimal solutions of reducts [27].

Genetic algorithms provide an efficient search technique, involving a set of evolutionary operators like selection, crossover and mutation. A population of chromosomes is made to evolve over generations by optimizing a fitness function, which provides a quantitative measure of the fitness of individuals in the pool.

When there are two or more conflicting characteristics to be optimized, often the single-objective GA requires an appropriate formulation of the fitness function in terms of an additive combination of the different criteria involved. In such cases multi-objective GAs (MOGAs) [28] provide an alternative, more efficient, approach to search for optimal solutions. The evolutionary-rough feature selection algorithm [29] has been used for classifying gene expression patterns in bioinformatics. Since the data typically consisted of a large number of redundant features, an initial redundance reduction of the attributes was done to enable faster convergence. RST was employed to generate reducts, which represent the minimal sets of non-redundant features capable of discerning between all objects, in a multi-objective framework.

### 1.2 Outline of the Proposed Model

We present here an application of the evolutionary-rough feature selection algorithm [29] to the face recognition problem. The input features are the Gabor jets at each node of the EBGM. Hybridization of RST with MOGA is utilized. The notion of reducts from RST helps to determine the distinction between different faces, involving a feature subset of reduced cardinality. The whole process is optimized using MOGA. Our contribution can be summarized as follows.

1. The selected optimal feature subsets have been analyzed and correlated with the physical biometric properties of the face. The result obtained is in good agreement with the established results of anthropometric and forensic exploration [30].
2. The results indicate more compact and better separated clusters in the reduced feature space. The faces are found to be effectively grouped, based on

their similarity in matching. Therefore texture pattern extracted around the landmark points in a face possess some characteristics which help to distinguish one face from another. Retrieval of faces in the reduced feature space provides promising results.

The simulation has been performed on a large database of Caucasian faces (FERET)[1] and our own database[2] containing large number of faces of different Indian ethnic groups. We have tried to explore the following challenging issues.

1. As biometric landmark how do the different part(s) of the facial features represent a face in a unique and distinguishable way?
2. In a large set of facial image how discriminating are the features?
3. How sensitive is the similarity score on the resolution and the orientation of the Gabor jet?
4. Are the biometric features, extracted as visual cues from facial images and utilized for computerized face recognition system, in tune with the findings from forensics, anthropometrics and psychometrics?

The remaining part of this article is organized as follows. Section 2 describes briefly our face recognition model based on EBGM. Section 3 deals with dimensionality reduction in the soft computing framework, involving the evolutionary-rough feature selection algorithm. The Davies-Bouldin validity index is used to determine the optimal number of clusters. The Adjusted Rank index helps quantify the improvement in partitioning in the reduced space. Section 4 presents the experimental results for efficient clustering and face retrieval. Finally Section 5 concludes the article.

## 2 Elastic Bunch Graph Matching and the Face Manifolds

In tasks like face recognition, involving within-class discrimination of objects, it is necessary to have information specific to the structure common in all objects of the class (say, corresponding to the fiducial points). This is crucial for the extraction, from the image, of those structural traits which are important for discrimination. Normally the class-specific information for each pose is kept in a data structure called the bunch graph. These are stacks of a moderate number of different faces, jet-sampled to an appropriate set of fiducial points (placed over eyes, nasal bridge, tip of the nose, mouth corner, etc.) The bunch graph is a combinatorial entity in which, for each fiducial point, jets from different sample faces can be selected. It enhances the robustness of the system.

It should be noted here that the original EBGM algorithm was proposed to recognize facial images of frontal views. In the real life scenario, however, the recognition across pose variation is essential. Therefore, two data structures are

[1] http://www.itl.nist/iad/feret/feret_master.html

[2] Centre for Development of Advanced Computing (CDAC) database http://www.kolkatacdac.in can be obtained, on request, for non-commercial use.

used to pursue the two different tasks. The bunch graph [see Fig.1(b)] includes a wide range of possible variations in the appearance of faces, such as differently shaped eyes, noses, mouth, different types of beards, variations due to sex, age, etc. Hence the face bunch graph combines a representative set of different possible variations around the fiducial points into a stack-like structure. Its utilization lies in the event of automatic selection of fiducial points in a new (unseen) face. Another data structure used in EBGM is termed the image graph [see Fig.1(a)]. It represents a face in the form of a graph, with each node corresponding to a fiducial point. Here at each node a stack of Gabor jets is stored to represent textural information at different orientations and frequencies.

In the case of face recognition across pose variations one, therefore, needs to develop different bunch graphs for the different poses. This is not a practical solution in the real life scenario. In our proposed face recognition engine we use only a single bunch graph for the frontal view. The information about the locations of the fiducial points across the various non-frontal poses is acquired in terms of a novel feature selection algorithm. All the modules are described in further detail in the subsequent sections.

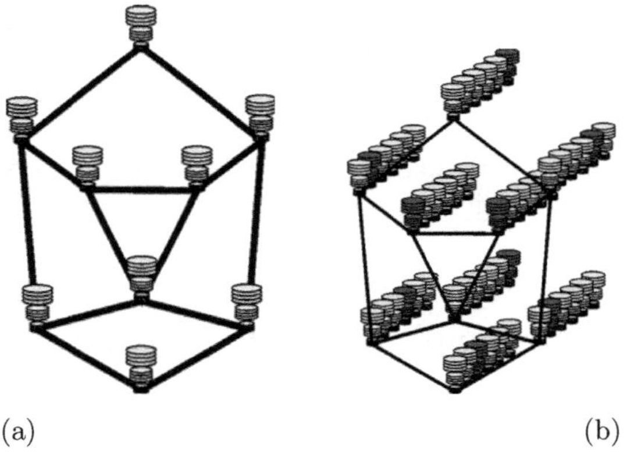


**Fig. 1.** Frontal view of face, representing the (a) image graph of fiducial points, and (b) face bunch graph

### 2.1 EBGM for Face Recognition

A face recognition system is developed based on the Elastic Bunch Graph Matching algorithm first proposed by Wiskott *et al.* [9]. The novelty of the algorithm lies in recognizing human faces from a large database containing only one image per person. The core of the algorithm uses two novel data structures, namely the image graph (IG) and the face bunch graph (FBG). Image graph (IG) is the mapping of a human face into a graph, as depicted in Fig. 1(a). Each node of the IG represents a biometric landmark point, like inner corners of the eyebrows, inner and outer corners of eyes, lip ends, nose tip, in terms of texture information extracted from the corresponding location in the facial image using a set of

Gabor wavelets of different resolutions and orientations. These IGs are stored in the face gallery against each facial image.

The face bunch graph is used to store class specific information. FBG stacks a moderate number (70 in Wiskott *et al.* [9]) of different faces, jet-sampled in an appropriate set of fiducial points (placed over eyes, lip ends, eyebrows), as depicted in Fig. 1(b). Bunch graphs are treated as combinatorial entities in which, corresponding to each fiducial point, a jet from different sample faces can be selected; thus creating a highly adaptable model. This model is matched to new facial images to reliably find fiducial points in the image.

In the case of face recognition across pose variation, a number of bunch graphs have to be generated in EBGM, one corresponding to each pose. However in the practical scenario, it is often very difficult to have even a moderate number of face images at each pose. In order to avoid such difficulties, we modify in the algorithm. A single bunch graph (corresponding to frontal face) is used to locate the landmark points in the frontal facial images. In case of rotated faces, instead of the face bunch graph we use a set of feature extraction algorithms - like lip corner detector [31], eye corner detector [32], to locate the biometric landmark points in the faces at different poses. This eliminates the need for multiple face bunch graphs while recognizing faces across pose variations.

Graphs of different faces are generated by EBGM [9,33,34], and then added into the gallery as new model graphs. Note that a model graph corresponding to a face contains all information relevant for the face discrimination task. The family of Gabor Kernels used are [35]

$$\Psi_j(\overrightarrow{x}) = \frac{k_j^2}{\sigma^2} \exp\left(-\frac{k_j^2 x^2}{2\sigma^2}\right) \left[\exp(i\overrightarrow{k_j}\overrightarrow{x}) - \exp(-\frac{\sigma^2}{2})\right], \tag{1}$$

where $\sigma$ is the standard deviation, $\overrightarrow{x}$ is the spatial variable representing gray scale patch around a given pixel and $\overrightarrow{k_j}$ is the wave vector corresponding to the plane waves restricted by the Gaussian envelop function. These represent the shape of plane waves, with the wave vector $\overrightarrow{k_j}$ being modulated by a Gaussian envelop function. The kernels are self-similar, and are generated from one mother wavelet by dilation and rotation. We employ a discrete set of five different frequencies (representing width of the modulating Gaussian envelop at the base), with indices $\gamma = 0, ...., 4$, and eight orientations in the interval 0 to $\pi$, having indices $\mu = 0, ..., 7$. We have

$$\overrightarrow{k_j} = (k_\gamma \cos\phi_\mu, k_\gamma \sin\phi_\mu), \tag{2}$$

where $k_\gamma = 2^{\frac{-(\gamma+2)}{2}}\pi$, $\phi_\mu = \frac{\mu\pi}{8}$, and the index $j = \mu + 8\gamma$. This sampling evenly covers a band in the frequency space. The choice of five different frequencies is made empirically over a large number of facial images, and the response of the Gabor wavelets are recorded. The wavelets with the chosen set of frequencies have shown the best responses.

For images of size (128 × 128) pixels, the lowest and highest frequencies have corresponding wavelengths of size 16 and 4 pixels respectively. The intermediate wavelengths include $4\sqrt{2}$, 8, $8\sqrt{2}$ pixels respectively. This range is found to be suitable for collecting information from each landmark region, with the upper limit determining that no two landmark points can be covered within the largest window. It is to be noted that orientations from $\pi$ to $2\pi$ would be redundant due to the even/odd symmetry of the wavelets. The width $\frac{\sigma}{\overrightarrow{k_j}}$ of the Gaussian is controlled by the parameter $s = 2\pi$. The term $\exp\left(-\frac{\sigma^2}{2}\right)$ in eqn. (1) makes the kernels DC free, *ie.*, the integral $\int \Psi_j(\overrightarrow{x})d^2\overrightarrow{x}$ vanishes. Fig. 2 schematically represents the process flow of converting an input image into the corresponding image graph [9]. A set of Gabor wavelet kernels are convolved at the landmark points of the input image to obtain its image graph.

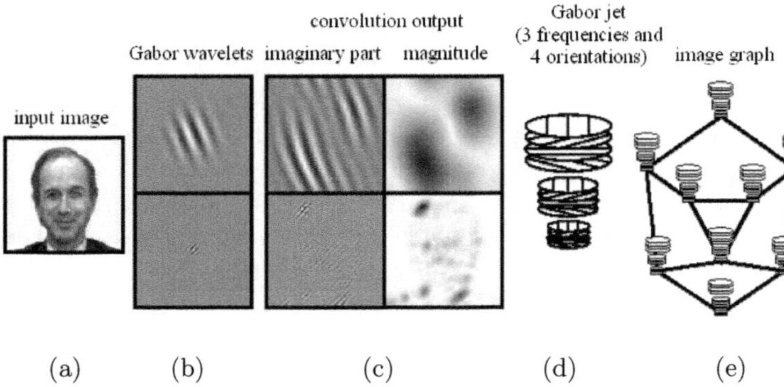


**Fig. 2.** Schematic representation of converting an (a) input image, with (b) Gabor wavelets of lower and higher frequencies (upper and lower half), resulting in (c) convolution of image with Gabor wavelet (left and right halves indicating imaginary part and magnitude). (d) The Gabor jet with 40 coefficients obtained at one landmark point for 3 frequencies and 4 orientations, and (e) the resultant image graph consisting of a collection of Gabor jets at the location of landmark points in the face.

A jet $J$ is defined as the set $\{J_j\}$ of forty complex Gabor wavelet coefficients obtained at each fiducial point. Let $\zeta_j(\overrightarrow{x})$ represent the magnitude of $j$ Gabor coefficients and $\theta_j(\overrightarrow{x})$ be the corresponding phases which rotate with a rate set by the spatial frequency or, wave vector $\overrightarrow{k_j}$ of the kernels. We express eqn. (1) as

$$\Psi_j(\overrightarrow{x}) = \zeta_j \exp(i\theta_j), \tag{3}$$

where $\zeta_j = \left(\frac{k_j}{\sigma}\right)^2 \times$

$$\left[\left\{\exp\left(-\frac{k_j^2x^2}{2\sigma^2}\cos\overline{k_jx}\right) - \exp\left(\overline{-\frac{k_j^2x^2}{2\sigma^2}+\frac{\sigma^2}{2}}\right)\right\}^2 + \left\{\exp\left(-\frac{k_j^2x^2}{2\sigma^2}\sin\overline{k_jx}\right)\right\}^2\right]^{\frac{1}{2}},$$

$$\theta_j = \arctan \frac{\exp\left\{-\frac{k_j^2 x^2}{2\sigma^2}\sin(k_j x)\right\}}{\exp\left\{-\frac{k_j^2 x^2}{2\sigma^2}\cos(k_j x)\right\} - \exp\left\{-\left(\frac{k_j^2 x^2}{2\sigma^2}+\frac{\sigma^2}{2}\right)\right\}}. \tag{4}$$

However, jets taken from fiducial points only a few pixels apart have very different coefficients although corresponding to almost the same set of local features. This can cause severe problems during matching.

Following Wiskott *et al.* [9] we have, therefore, ignored the phase component in our model and chosen the similarity function as

$$S_a(J, J') = \frac{\Sigma_j \zeta_j \zeta_j'}{\sqrt{\Sigma_j \zeta_j^2 \Sigma_j \zeta_j'^2}}. \tag{5}$$

It should be noted here that more Gabor kernels apparently tend to yield better results as more information is extracted. But this effect is found to saturate and result in confusing information if the chosen kernel frequencies are too close; with the corresponding Gabor coefficients becoming highly correlated due to overlap between the kernels. Moreover, the computational effort increases linearly with the number of Gabor kernels. Selection of the optimal number of Gabor kernels is therefore a compromise between recognition performance and speed.

We demonstrate here the utility of dimensionality reduction towards face recognition. The reduced set of features represented by the coefficients of Gabor kernels, computed at different fiducial points, is found to improve the performance of the face recognition task. This is because the less relevant features serve to increase the intra-class variance, and lead to increased computational complexity.

### 2.2 Feature Space

Merits of any face recognition algorithm are evaluated on the basis of the following two factors. The accuracy of the features that are extracted to represent the face pattern and the choice of the feature points to represent the uniqueness of the objects in the feature space. Different studies suggest that both holistic and feature information are crucial for the perception and recognition of faces [36,37]. It also suggests the possibility of global descriptions serving as a front end for finer feature-based perception.

Dominant features like pupil of the eyes and curvature of the eyebrows may be assigned higher priority in terms of weight instead of the holistic description. For example, in face recall studies, humans are found to quickly focus on odd features such as big ears, a crooked nose, a bulging eye, etc [3]. The face inversion effect presents the strongest piece of evidence that the process of face recognition not only involves the feature points, but also uses their configural properties [38].

Choice of fiducial points in creating graphs to represent faces therefore requires ranking of significance of facial features. Studies shows that hair, face outline, eyes, and mouth are more important for perceiving and remembering faces [6,36]. It has also been found that the upper part of the face is more useful for face recognition than the lower part [6].

In the proposed system 14 fiducial points are considered. Each fiducial point in the graph is labeled with forty Gabor jets (five frequencies at eight orientations), as shown in Fig. 1(a). Hence it is a feature space of dimension $N = 14 \times 5 \times 8 = 560$. Any face in this feature space acquires a distribution or manifold, which accounts for all possible variation on facial appearance due to pose, illumination, expression, etc. Properties of these manifolds have been studied by many authors and found to be highly nonlinear and non convex [39]. Designing discriminator between these manifolds is highly complex.

We employ the evolutionary-rough feature selection algorithm [29] to extract the reduced set of Gabor jets at different fiducial points. It has been shown in Section 4 that sub-manifolds in the reduced feature space are better clustered into separable groups to facilitate recognition. The well-known c means clustering [40] is used to evaluate the resultant partitioning of the faces in this reduced space, in terms of the Davies Bouldin validity index [41]. These are elaborated in further detail in the following section.

## 3 Evolutionary-Rough Feature Selection of Gabor Jets

Soft computing tools like genetic algorithms and rough sets have been used to perform the tasks of feature reduction in face recognition [34,42]. We describe here the reduct generation *i.e.,* feature selection procedure, incorporating initial redundancy reduction, in multi-objective framework using rough sets. We focus our analysis to two-class problems. In other words, this implies considering whether an unknown image matches with the face of a particular person or not (*ie.*, it is the face of somebody else).

### 3.1 Feature Selection with Rough Sets

It is a process that selects a minimum subset of $M$ features from an original set of $N$ features ($M \leq N$), so that the feature space is optimally reduced according to an evaluation criterion. Finding the best feature subset is often intractable or NP-hard.

One of the important problems in extracting and analyzing information from large databases is the associated high complexity. Feature selection is helpful as a preprocessing step for reducing dimensionality, removing irrelevant data, improving learning accuracy and enhancing output comprehensibility, while simultaneously decreasing the computational costs.

There are two basic categories of feature selection algorithms, *viz.*, filter and wrapper models. The filter model selects feature subsets independently of any learning algorithm, and relies on various measures of the general characteristics of the training data. The wrapper model uses the predictive accuracy of a predetermined learning algorithm to determine the goodness of the selected subsets, and is computationally expensive.

An information system consisting of a domain $U$ of objects/observations and a set $A$ of attributes/features induces a partitioning (classification) of $U$ by $A$.

A block of the partition would contain those objects of $U$ that share identical feature values, i.e. are *indiscernible* with respect to the given set $A$ of features. But the whole set $A$ may not always be necessary to define the classification/partition of $U$. Many of the attributes may be superfluous, and we may find *minimal* subsets of attributes which give the same classification as the whole set $A$. These subsets are called *reducts* in rough set parlance.

Our system considers fourteen fiducial points, each being labeled with forty Gabor jets having five frequencies at eight orientations. Hence, a face is represented by a feature space of dimension $N = 14 \times 5 \times 8 = 560$. We consider $m$ persons each providing five different poses. This creates an attribute table $A$ of size $5m \times 560$ to be considered for preprocessing. We perform attribute-wise normalization to generate normalized table $A'$. Now $a'_j, \forall j = 1, 2, \ldots, 560$, corresponds to the normalized $j$th feature component in the attribute table, having values lying in the range 0 to 1.

A reduct is the minimal set of attributes that are sufficient to distinguish between the objects (faces of a pair of persons). Its computation is made using a distinction table (or, a discernibility matrix). The objective is to discern across different poses of each person, and to identify those attributes that can do so. Considering $m$ persons, each having five different poses, the number of rows in the distinction table becomes
$[(m-1) + (m-2) + \ldots + 1] \times 5 \times 5 = m(m-1)/2 \times 25$.
Here the distinction is made between each pair of persons in 25 ways.

A distinction table is thus a binary matrix with $\frac{(m^2-m)}{2} \times 25$ rows and 560 columns. An entry $b((k,j),i)$ of the matrix corresponds to the attribute $a'_i$ and pair of objects $(x_k, x_j)$, *ie.* faces, and is given by

$$b((k,j),i) = \begin{cases} 1 & \text{if} \quad a_i'(x_k) - a_i'(x_j) > Th \\ 0 & \text{if} \quad a_i'(x_k) - a_i'(x_j) < Th \\ *(\text{don't care}) & \text{otherwise} \end{cases} \tag{6}$$

where $Th = 0.5$ is a user-supplied threshold.

Find the average occurrences of '$*$' over the entire attribute value table. Choose this as threshold $Th_a$. Remove from the table those attributes for which the number of '$*$'s are $\geq Th_a$. This is the *modified* (reduced) *attribute value table* $A'$. The threshold serves to enhance contrast between the two categories of attributes capable of discerning and not discerning among the object pairs.

### 3.2 Optimization with Multi-objective GA

The presence of an '1' signifies the ability of the attribute $a'_i$ to discern (or distinguish) between the pair of faces $(x_k, x_j)$. The essential properties of reducts here are

1. to classify among all face pairs in the distinction table with the same accuracy as the starting attribute (feature) set, and
2. to be of small cardinality.

The objective is to select a minimal set of ones that are sufficient to cover all face pairs in the distinction table. A close observation reveals that these two characteristics are of a conflicting nature. Hence the determination of reducts is better represented as a bi-objective problem. We employ the multi-objective feature selection algorithm [29] that has been used for classification of benchmark gene expression data in different types of cancer.

It uses, among others, the criteria like non-domination and crowding distance. The solutions are represented by binary strings of length $N$, where $N$ is number of attributes (Gabor jets). In the bit representation '1' means that the attribute is present and '0' means that it is not. Two fitness functions $f_1$ and $f_2$ are considered for each individual (face). We have [29]

$$f_1(\vec{\nu}) = \frac{N - L_{\vec{\nu}}}{N} \tag{7}$$

and

$$f_2(\vec{\nu}) = \frac{C_{\vec{\nu}}}{m_1 * m_2}, \tag{8}$$

where $N$ is the total number of attributes, $\vec{\nu}$ is the reduct candidate, $L_{\vec{\nu}}$ represents the number of 1's in $\vec{\nu}$, $C_{\vec{\nu}}$ indicates the number of object combinations $\vec{\nu}$ can discern between, and $m_1$ and $m_2$ are the number of objects (faces in different poses) in the pair of classes (individual persons) being distinguished. The fitness function $f_1$ gives the candidate credit for containing less attributes (fewer 1's), while the function $f_2$ determines the extent to which the candidate can discern among objects. We outline the main steps of the algorithm [29].

1. Initialize the population randomly.
2. Calculate the multi-objective fitness functions.
3. Rank the population using dominance criteria, and calculate crowding distance.
4. Do selection using crowding selection operator.
5. Do crossover and mutation (as in conventional GA) to generate children population.
6. Combine parent and children population.
7. Replace the parent population by the best members of the combined population. Initially, members of lower fronts replace the parent population. When it is not possible to accommodate all the members of a particular front, then that front is sorted according to the crowding distance. Selection of individuals is done on the basis of higher crowding distance.

Thus, by generating a reduct we are focusing on that minimal set of attributes (Gabor jets) which can essentially distinguish between all patterns (faces of individuals in different poses) in the given set. In this manner, a reduct is mathematically more meaningful as the most appropriate set of non-redundant features selected from a high-dimensional data. One-point crossover is employed with probability $p_c = 0.7$. Probability $p_m$ of mutation on a single position of individual was taken as 0.05. Mutation of one position means replacement of '1' by '0', or '0' by '1'.

# 4 Experimental Results

We present here the investigations demonstrating the benefit of dimensionality reduction on the performance of EBGM-based face recognition algorithm.

## 4.1 Data Description

All experiments were conducted on the following facial image databases.

1. The benchmark database - FERET, which mostly includes Caucasian faces. There are 14,051 eight-bit gray scale images of human heads with views ranging from frontal to left and right profiles.
2. In order to generalize the results on different non-Caucasian faces, particularly with those from different Indian ethnic groups, the experiments were repeated on our facial image database CDAC. This database includes colored facial images of 200 persons from different Indian ethnic groups. For each person a total of 57 facial images at different poses and orientations were recorded. This resulted in a total of 11,400 images. Out of these 57 views we have chosen 5 images per person to include the frontal face and other facial views (rotated upto $\pm 20°$) at $5°$ intervals. These are termed as "poses" in the sequel.

A portion of the FERET and CDAC databases are depicted in Fig. 3.

## 4.2 Feature Extraction

Fig. 4 illustrates the preprocessing and feature extraction procedures on a sample CDAC image. The major steps are indicated below.

1. Detection of the facial zone present in the image and removal of the background [Fig. 4(a)]. Standard Viola Zones algorithm based on Haar wavelets and AdaBoost classifier bank [43,44] is used to segment the facial zone from the background.
2. The segmented facial image is normalized to $128 \times 128$ image [Fig. 4(b)].
3. The 14 fiducial points are automatically located using standard facial feature extraction algorithms widely utilized in literature [31,32]. The algorithms used in our feature extraction module can be broadly categorized into two groups, namely, the (i) generic, and (ii) feature template based methods. Generic methods are based on edges, lines and curves. The fiducial points situated at the eyebrows, lip ends are extracted in this manner. Feature template based methods, on the other hand, are used to detect facial features such as inner and outer canthii of eyes, feature points placed at the middle and upper eyelids, tip of the nose and the centre of the nose bridge.
   Fig. 4(c) shows a typical facial image with the fourteen extracted fiducial points. The sequence of automatically locating fourteen fiducial points are as follows [45].

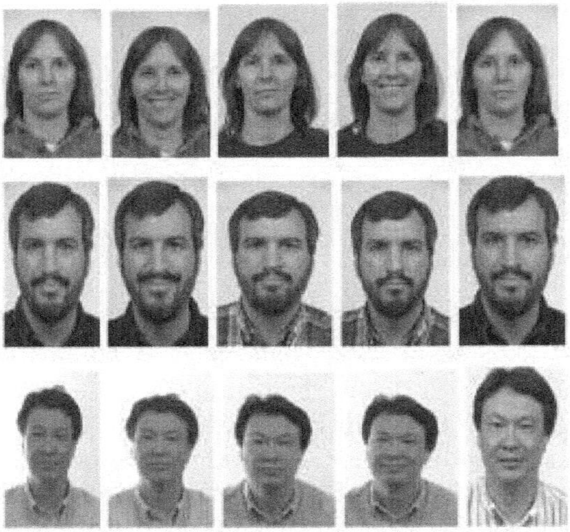

(a)

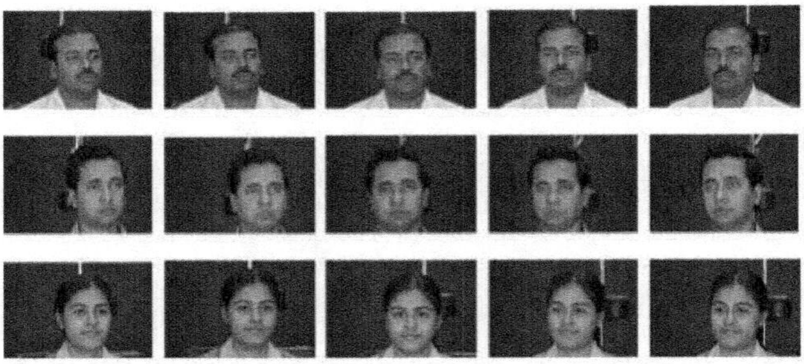

(b)

**Fig. 3.** Sample images from databases (a) FERET, and (b) CDAC

   (a) Extract left and right eyebrows, and designate fiducial points at their two ends.
   (b) Extract inner and outer canthus of left and right eyes.
   (c) Extract middle points of the upper and lower eyelids of left and right eyes.
   (d) Extract the center of the nose bridge and nose tip.
   (e) Identify two endpoints of the lip as landmark points.
4. Convolve the 40 Gabor wavelets [five scales and eight orientations, by eqns. (1)-(4) at each fiducial point, to generate a feature vector of dimension $N = 560$ for each face. All the faces are now represented as points in a phase space of dimension 560.

Using the 40 Gabor jets around each fiducial point (or node) we generate a face bunch graph representing the set of face images. Here each node is endowed with a stack of intensity patches (textural information) extracted by convolving Gabor

(a) (b) (c)

**Fig. 4.** Preprocessing and feature extraction from a sample facial image. (a) Detection of facial zone, and (b) segmented facial image marked with (c) fiducial points.

wavelets at the corresponding fiducial point. This stack of values is collected from a large number of faces ($5m$), so that during matching the graph can accommodate variations in pose, illumination and expression across the gallery images as well as test images.

### 4.3 Dataset CDAC

Here we analyze the results from clustering and feature selection, using the CDAC database.

**Clustering.** The patterns representing facial images are clustered using the well-known c-means algorithm [40]. Four data subsets, each having 40 images from the CDAC database (containing faces of eight randomly selected persons at five different poses), have been considered for carrying out clustering experiments. Different number of clusters $c$ are chosen. The generated partitions are validated in terms of Davies Bouldin cluster validity index (DB) [41]. The optimal clustering, for $c = c_0$, minimizes

$$DB = \frac{1}{c} \sum_{k=1}^{c} \max_{l \neq k} \frac{S(U_k) + S(U_l)}{d(U_k, U_l)}, \tag{9}$$

for $1 \leq k, l \leq c$, where $S(U_k)$ is the within-cluster distance of cluster $U_k$ and $d(U_k, U_l)$ refers to the between-cluster separation amongst cluster pair $U_k$, $U_l$. Note that a minimum value of $DB$ corresponds to the most compact and well separated partitioning with $c_0$ clusters.

In all the four cases $DB$ was found to be minimum for $c_0 = 8$, which corresponded to the eight persons in the phase space. The respective confusion matrices are provided in Fig. 5(a). Next we analyzed the confusion matrix for the first subset of 40 faces using $c = 8$. Fig. 5(b) depicts the actual faces in each of the eight clusters. It is observed that two out of five members of category 1 have been wrongly identified as belonging to partition 6. Similarly two members of group 7 and one member of partition 8 have been misplaced into categories 5 and 6 respectively. This makes the overall accuracy of the clustering to be 87.5%. A closer investigation reveals that the misplaced faces in the same partition also bear a strong resemblance to each other in human perception. The

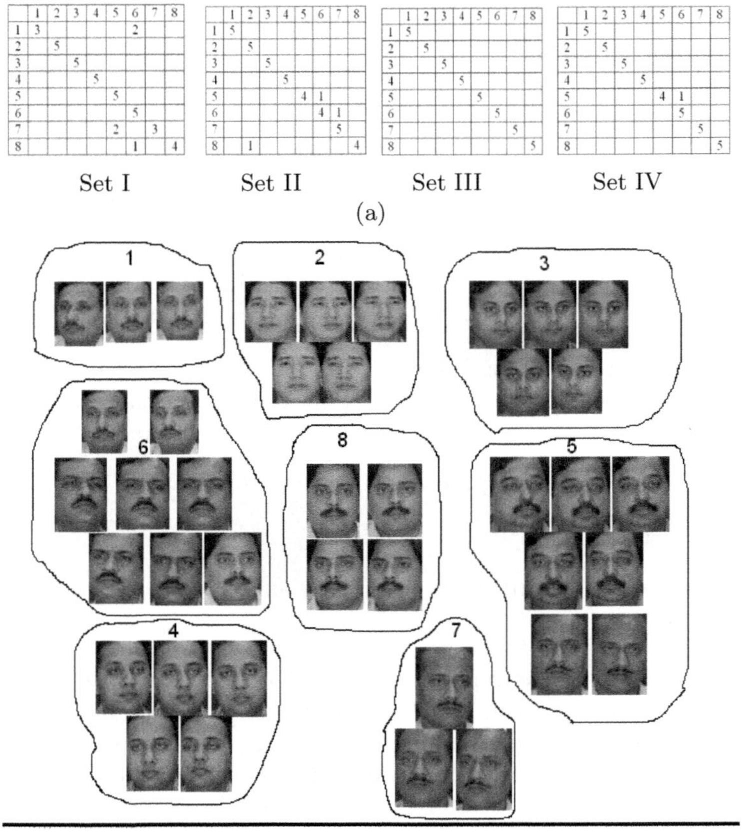


(b)

**Fig. 5.** Sample CDAC image set in original feature space with (a) confusion matrices for four subsets, and (b) cluster structure for Set I

**Table 1.** Average variation of cluster validity index over original and reduced domains, for different cluster numbers, using dataset *CDAC*

| Validity index | Feature space | Cluster # $c =$ | | | | | | |
|---|---|---|---|---|---|---|---|---|
| | | 4 | 5 | 6 | 7 | 8 | 9 | 10 |
| *DB* | Original | 1.06 | 0.98 | 1.04 | 0.86 | **0.54** | 0.62 | 0.66 |
| | Reduced | 0.89 | 0.84 | 0.74 | 0.60 | **0.50** | 0.67 | 0.60 |
| *ARI* | Original | 0.09 | 0.22 | 0.32 | 0.49 | **0.69** | 0.46 | 0.37 |
| | Reduced | 0.09 | 0.25 | 0.36 | 0.53 | **0.75** | 0.42 | 0.44 |

first row of Table 1 indicates the average $DB$ values, over the four data subsets, corresponding to $c = 4, \ldots, 10$. The minimum value, marked in bold, occurs for eight clusters.

Next the clustering result was compared against external criteria using the Adjusted Rank Index ($ARI$) [46]. We assume that each face cluster ideally consists of different poses (here five) of the same person. Given a set of $n$ $(= 5m)$ facial images $S = \{O_1, \ldots, O_n\}$, let $V_1 = \{v_1^1, \ldots, v_1^r\}$ and $V_2 = \{v_2^1, \ldots, v_2^t\}$ represent two different partitions of the objects in $S$ such that $\bigcup_{i=1}^{r} v_1^i = S = \bigcup_{j=1}^{t} v_2^j$ and $v_1^i \bigcap v_1^{i'} = \emptyset = v_2^j \bigcap v_2^{j'}$ for $1 \leq i \neq i' \leq r$ and $1 \leq j \neq j' \leq t$. Suppose $V_1$ is our external criterion and $V_2$ is the clustering result. Consider a confusion matrix of Fig. 5(a). Here the rows correspond to the classes to which the faces belong, whereas the columns indicate the clusters to which the algorithm assigns them. Let $V_1$ and $V_2$ be picked at random, such that the total number of objects in the classes and clusters is fixed. Let $n_{ij}$ be the number of objects that are in both class $v_1^i$ and cluster $v_2^j$, with $n_i$ and $n_j$ being the number of objects in class $v_1^i$ and cluster $v_2^j$ (individually) respectively. Considering the generalized hypergeometric model of randomness, the Adjusted Rank Index (ARI) is expressed as [46]

$$ARI = \frac{\sum_{i,j} {}^{n_{ij}}C_2 - \frac{[\sum_i {}^{n_i}C_2 \sum_j {}^{n_j}C_2]}{{}^{n}C_2}}{\frac{1}{2}[\sum_i {}^{n_i}C_2 + \sum_j {}^{n_j}C_2] - \frac{[\sum_i {}^{n_i}C_2 \sum_j {}^{n_j}C_2]}{{}^{n}C_2}}. \tag{10}$$

Here a higher value of $ARI$ is indicative of a better clustering. The third row of Table 1 shows the average $ARI$ values, for $c = 4, \ldots, 10$, over the four data subsets. The highest value of $ARI = 0.69$ occurred for eight clusters.

Finally, we explored the feature set extracted from the left eyebrow (fiducial point) of all the members of each partition. Sample results for partition 5 are depicted in Fig. 6. While patterns (a) to (e) correspond to actual samples from this group, patterns (f) and (g) refer to the two misclassified faces from partition 7. Presence of three major peaks are detected in all the seven cases, thereby resulting in confusion. Although the remaining part of the feature maps exhibit variations in patterns (f)-(g), with respect to the corresponding part of patterns (a)-(e), it is somehow dominated by the stronger peaks. This is due to misinformation carried by the redundant features. Hence the relevance of dimensionality reduction becomes evident. Similar behavior is observed at the other fiducial points as well.

**Soft feature selection.** In this section we outline the results of our investigation in feature selection using the evolutionary rough algorithm of Section 3. Starting with 560 features, the reduced feature vectors turned out to be 255, 248, 248 and 261 components in the four runs on the image subsets. It is observed from the second row of Table 1 that the values of $DB$ [eqn. (9)] are, in general, always lower in the reduced feature space. This is indicative of better (more compact) partitioning after elimination of redundant features. The least value, once again, corresponds to $c_0 = 8$.

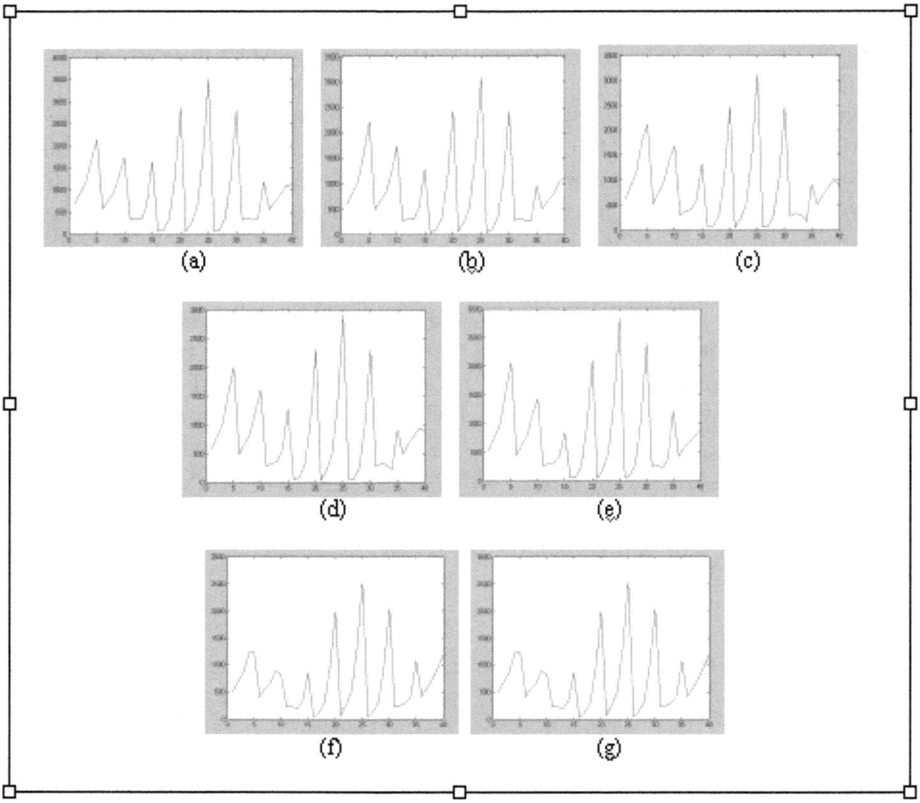

**Fig. 6.** Gabor components corresponding to left eyebrow feature of seven members in category 5. Patterns (a)-(e) belong to actual members of this group. Patterns (f)-(g) belong to misclassified members from category 7.

After feature selection the value of $ARI$ [eqn. (10)] increases to 0.75 for $c_0 = 8$. This increment of $ARI$ in the reduced feature space confirms that the clustering result is now in better agreement against the chosen external criterion (*i.e.,* all facial images in one cluster should belong to a single person). This also validates the selected features to be more informative, as compared to those that are eliminated. It should, therefore, be possible to build an accurate recognition system on top of this.

Consider the first image subset (Set I) of Fig. 5(a) and the corresponding cluster structure in Fig. 5(b). The 40 Gabor components extracted from the left eyebrow for partition 5 are depicted in Fig. 6. After feature selection, the 17 Gabor components in the reduced space are presented in Fig. 7. The patterns in Figs. 6 [parts (a)-(e)] are extracted from the left eyebrows of correctly mapped members of cluster 5, whereas parts (f) and (g) (of the same figure) belong to the two misplaced cluster members from group 7. The 40-feature space makes the two sets look indistinguishable. However, after dimensionality reduction, the patterns

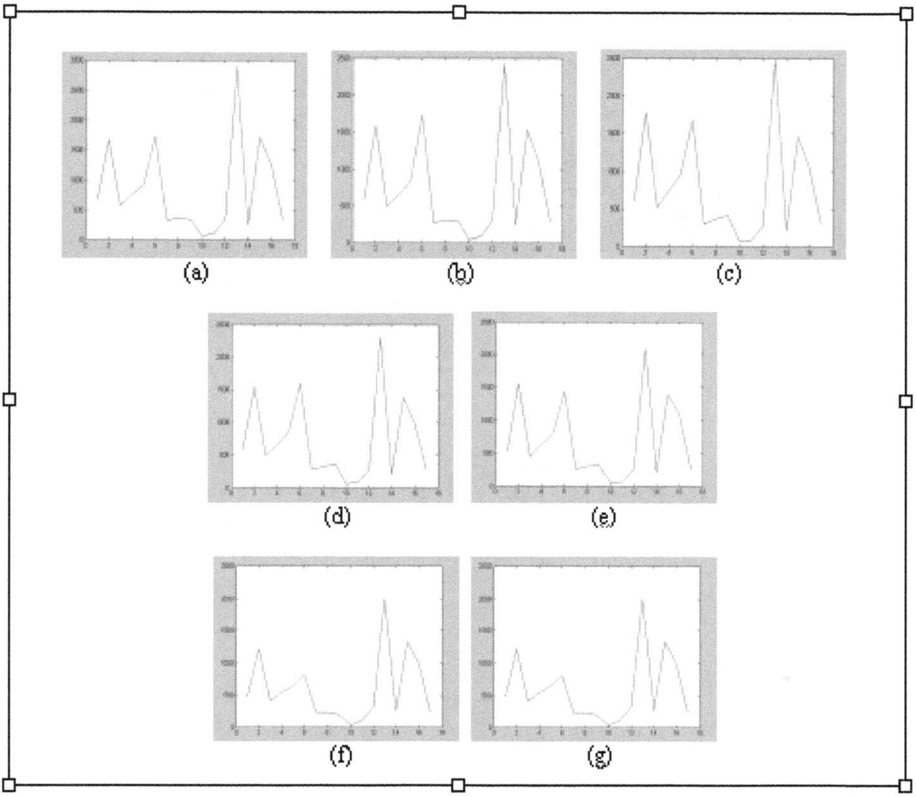

**Fig. 7.** Gabor components, corresponding to left eyebrow, in the reduced domain. Patterns (a)-(e) belong to actual members of cluster 5, while patterns (f) and (g) belong to misclassified members from category 7 and are distinguishable,

(a) to (e) of Fig. 7 demonstrate a unique profile attributable to the biometric feature at the left eyebrow of the face corresponding to partition 5. The profile extracted from the face belonging to cluster 7 [Fig. 6(f)-(g)], after dimensionality reduction, in Fig. 7(f)-(g) become distinguishable from those of Fig. 7 (a)-(e). This, therefore, increases the accuracy of the partitioning from 87.5% to 92.5%, upon applying the evolutionary-rough feature selection algorithm [29].

### 4.4 Dataset FERET

In the second experiment, our dataset constituted six randomly selected faces in five poses from the FERET database. The minimum $DB$ index corresponded to a partition of the dataset into six clusters, as shown in Fig. 8(a). The corresponding confusion matrix is shown in Fig. 8(b). The overall accuracy of clustering was 90%. Eliminating redundant features, using the evolutionary-rough algorithm, the accuracy of the partitioning increased from 90% to 93.33%.

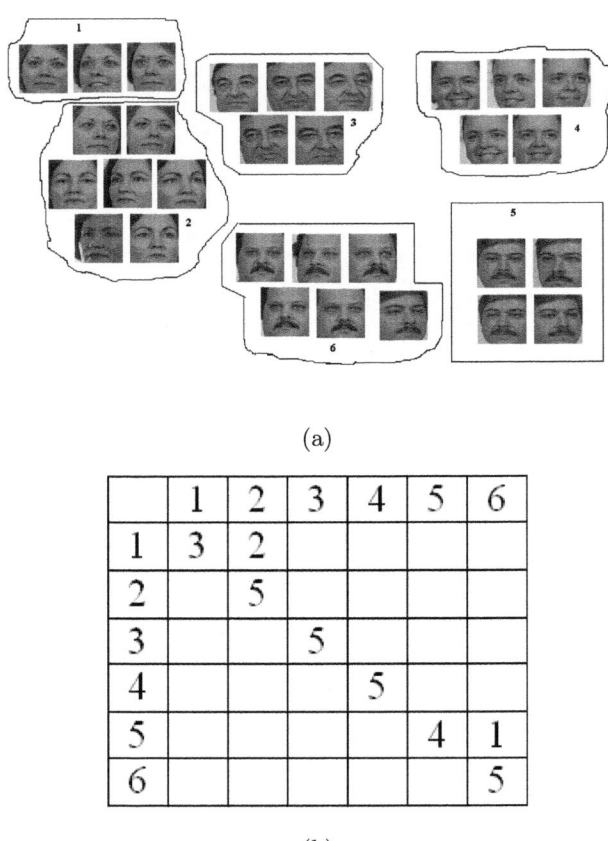

(a)

| | 1 | 2 | 3 | 4 | 5 | 6 |
|---|---|---|---|---|---|---|
| 1 | 3 | 2 | | | | |
| 2 | | 5 | | | | |
| 3 | | | 5 | | | |
| 4 | | | | 5 | | |
| 5 | | | | | 4 | 1 |
| 6 | | | | | | 5 |

(b)

**Fig. 8.** Sample FERET image set in original feature space, with (a) cluster structure, and (b) confusion matrix

**Table 2.** Improvement in the accuracy of partitioning with soft feature selection on the larger datasets

| Data set | Experimental run | Reduced dimension | Accuracy of partitioning (%) | |
|---|---|---|---|---|
| | | | original space | reduced space |
| CDAC | Set I | 297 | 61 | 81 |
| | Set II | 300 | 67 | 78 |
| | Set III | 299 | 65 | 82 |
| | Set IV | 305 | 66 | 80 |
| FERET | Set I | 302 | 63 | 78 |
| | Set II | 308 | 62 | 76 |

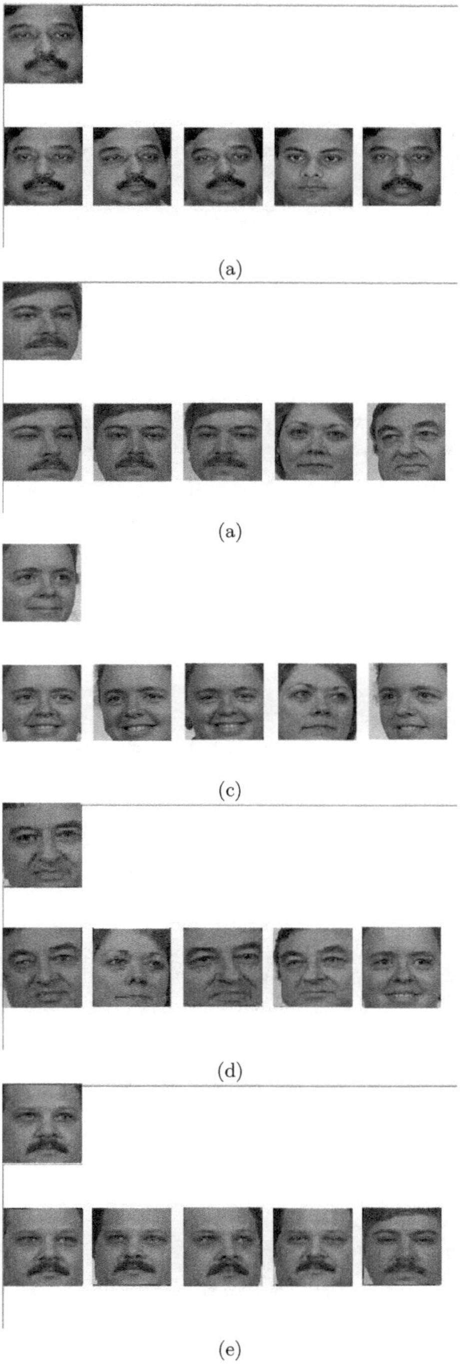


**Fig. 9.** Sample query image and eight best matches retrieved from (a) CDAC, and (b)-(e) FERET databases

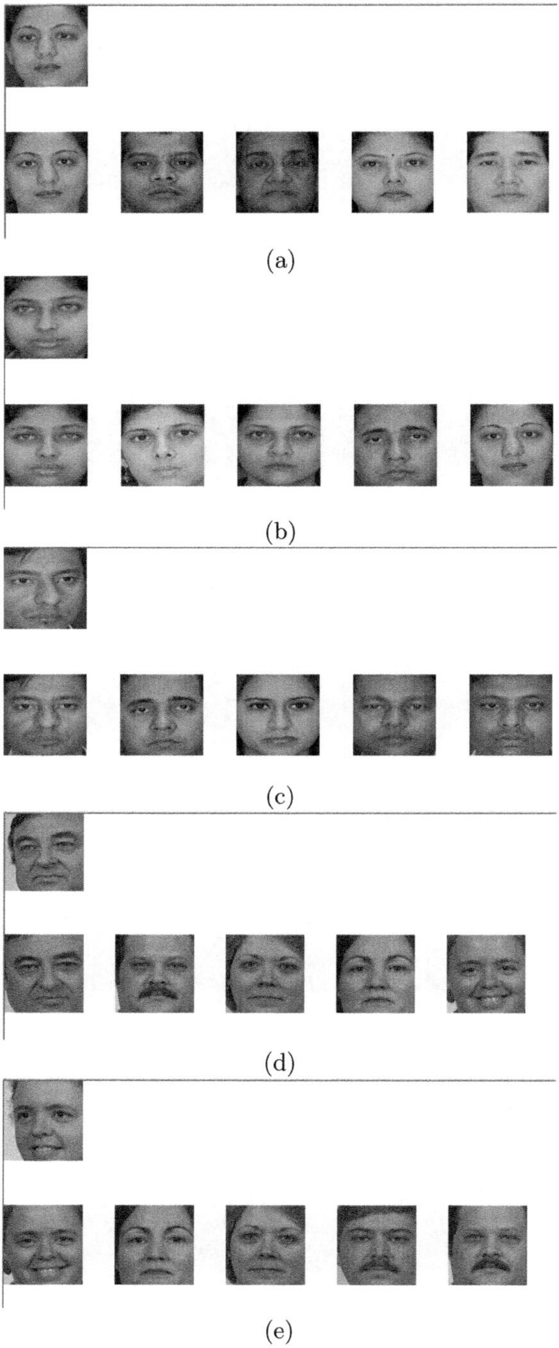

**Fig. 10.** Sample query image and five best matches retrieved from (a)-(c) CDAC database and (d)-(e) FERET database, with only one pose per person

### 4.5 Larger Data

Next the size of the data (for both CDAC and FERET) was increased to 40, with five poses each. This resulted in 200 face images. Four such facial image sets were prepared from the CDAC database. A pair of such randomly selected image sets was prepared from the FERET database. As before, the evolutionary-rough algorithm was applied to choose the reduced set of features out of the original 560 attributes. Such dimensionality reduction again lead to an improvement in the accuracy of partitioning, as demonstrated in Table 2. This reveals that the good discernibility information, existing among a smaller set of attributes, can be effectively selected by our system for the face recognition purpose.

### 4.6 Retrieval

Next we used the enhanced CDAC and FERET databases for testing the retrieval capability of the system in the reduced feature space. Fig. 9 depicts a sample query image and its corresponding five most similar retrievals. The retrieved images demonstrate that in all cases the first three correct facial views were recognized.

Finally we validated the effectiveness of this EBGM-inspired system in recognizing a person from a database consisting of only one image per individual. A set of 40 randomly selected images from the CDAC database, consisting of a single pose per person, was considered for this experiment. It was observed that in all cases the system correctly recognized the right person (in a different pose w.r.t. the query image). The retrieval results are depicted in Fig. 10, for the reduced set of features, corresponding to the five best matches.

## 5 Conclusion

In this article we have described an application of the evolutionary-rough feature selection algorithm for face recognition. The hybridization of MOGA with RST helped in the elimination of redundant and confusing features, while improving the accuracy of the recognition system in the reduced space. Elastic Bunch Graph Matching was used as the basic face recognition engine, with Gabor jets serving as feature components at each node. Clustering validity indices helped in evaluating the goodness of the partitioning. Successful retrieval of faces was also performed. The model served as an interesting study in face biometry.

Automatic extraction of features was done in the pre-processing stage. Since the facial biometric features typically consist of a large number of superfluous attributes, hence initial redundancy reduction served to retain only those features that played a significant role in discerning between objects. It aided faster convergence in the reduced search space.

Selection of the more frequently occurring attributes has revealed significant results. It was observed that all the feature components did not possess the same degree of discernible properties; rather, a ranking was possible among them – as a function of their discernible capabilities. This finding is in agreement with anthropometric results. Comparison has been made between the recognition accuracy in the original feature space and that with the reduced feature set. Significant improvement in performance was observed in the reduced domain. Simulations were conducted on the FERET and CDAC facial image databases. These allowed us to experiment with the Caucasian faces as well as those from the different Indian ethnic groups.

## References


1. Kelly, M.D.: Visual identification of people by computer. Technical Report AI-130, Stanford AI Project, Stanford, CA (1970)
2. Kanade, T.: Computer recognition of human faces. Interdisciplinary Systems Research 47 (1977)
3. Zhao, W., Chellappa, R., Phillips, P.J., Rosenfeld, A.: Face recognition: A literature survey. ACM Computing Surveys 35, 399–458 (2003)
4. Bruce, V., Young, A.: In the Eye of the Beholder. Oxford University Press, Oxford (1998)
5. Ellis, H.D.: Introduction to aspects of face processing: Ten questions in need of answers. In: Ellis, H., Jeeves, M., Newcombe, F., Young, A. (eds.) Aspects of Face Processing, Nijhoff, Dordrecht, pp. 3–13 (1986)
6. Shepherd, J.W.: Social factors in face recognition. In: Davies, G., Ellis, H., Shepherd, J. (eds.) Perceiving and Remembering Faces, pp. 55–78. Academic Press, London (1981)
7. Sirovich, L., Kirby, M.: Low-dimensional procedure for the characterization of human face. Optical Science American 4, 519–524 (1987)
8. Kirby, M., Sirovich, L.: Application of the Karhunen-Loeve procedure for the characterization of human faces. IEEE Transactions on Pattern Analysis and Machine Intelligence 12, 103–108 (1990)
9. Wiskott, L., Fellous, J.M., Kruger, N., Malsburg, C.V.D.: Face recognition in elastic bunch graph matching. In: Jain, L.C., Halici, U., Hayashi, I., Lee, S.B., Tsutsui, S. (eds.) Intelligent Biometric Techniques in Fingerprint and Face Recognition, pp. 357–396. CRC Press, Boca Raton (1999)
10. Sun, Y., Yin, L.: A genetic algorithm based feature selection approach for 3D face recognition. In: Proceedings of the Biometric Consortium Conference, vol. 35 (2005)
11. Carreira-Perpinan, M.A.: Continuous Latent Variable Models for Dimensionality Reduction and Sequential Data Reconstruction. PhD thesis, Dept. of Computer Science, Univ. of Sheffield, Sheffield, U.K (2001)
12. Fodor, J.K.: A survey of dimension reduction techniques. Technical Report UCRL-ID-148494, Lawrence Livermore National Laboratory, Centre for Applied Scientific Computing (2002)
13. Jain, A.K., Duin, R.P.W., Mao, J.: Statistical pattern recognition: A review. IEEE Transactions on Pattern Analysis and Machine Intelligence 22, 4–37 (2000)
14. Webb, A.R.: Statistical Pattern Recognition. Wiley, Cambridge (2002)
15. Mitra, P., Murthy, C.A., Pal, S.K.: Unsupervised feature selection using feature similarity. IEEE Transactions on Pattern Analysis and Machine Intelligence 24, 301–312 (2002)

16. Krishnapuram, R., Hartemink, A.J., Carin, L., Figueiredo, M.A.T.: A Bayesian approach to joint feature selection and classifier design. IEEE Transactions on Pattern Analysis and Machine Intelligence 26, 1105–1111 (2004)
17. Law, M.H.C., Figueiredo, M.A.T., Jain, A.K.: Simultaneous feature selection and clustering using mixture models. IEEE Transactions on Pattern Analysis and Machine Intelligence 26, 1154–1166 (2004)
18. Kohavi, R.: Wrappers for feature subset selection. Artificial Intelligence 97, 273–324 (1995)
19. Miller, A.: Subset Selection in Regression. CRC Press, Washington (1990)
20. Wei, H.L., Billings, S.A.: Feature subset selection and ranking for data diensionality reduction. IEEE Transactions on Pattern Analysis and Machine Intelligence 29, 162–166 (2007)
21. Jolliffe, I.T.: Discarding variables in a principal component analysis - I: Artificial data. Applied Statistics 21, 199–215 (1972)
22. Zadeh, L.A.: Fuzzy logic, neural networks, and soft computing. Communications of the ACM 37, 77–84 (1994)
23. Pal, S.K., Mitra, S.: Neuro-fuzzy Pattern Recognition: Methods in Soft Computing. John Wiley, New York (1999)
24. Pawlak, Z.: Rough Sets, Theoretical Aspects of Reasoning about Data. Kluwer Academic, Dordrecht (1991)
25. Skowron, A., Rauszer, C.: The discernibility matrices and functions in information systems. In: Slowiński, R. (ed.) Intelligent Decision Support, Handbook of Applications and Advances of the Rough Sets Theory, pp. 331–362. Kluwer Academic, Dordrecht (1992)
26. Goldberg, D.E.: Genetic Algorithms in Search. In: Optimization and Machine Learning. Addison-Wesley, Reading (1989)
27. Bjorvand, A.T.: 'Rough Enough' – A system supporting the rough sets approach. In: Proceedings of the Sixth Scandinavian Conference on Artificial Intelligence, Helsinki, Finland, pp. 290–291 (1997)
28. Deb, K.: Multi-Objective Optimization using Evolutionary Algorithms. John Wiley, London (2001)
29. Banerjee, M., Mitra, S., Banka, H.: Evolutionary rough feature selection in gene expression data. IEEE Transactions on Systems, Man and Cybernetics - Part C 37, 622–632 (2007)
30. Wilkinson, C.: Forensic Facial Reconstruction. Cambridge University Press, Cambridge (2004)
31. Nguyen, D., Halupka, D., Aarabi, P., Sheikholeslami, A.: Real time face detection and lip feature extraction using field-programmable gate arrays. IEEE Transactions on Systems, Man, and Cybernetics, Part B 36, 902–912 (2006)
32. Vukadinovic, D., Pantic, M.: Fully automatic facial feature point detection using Gabor feature based boosted classifiers. In: Proceedings of IEEE ICCV Workshop on Statistical and Computational Theories of Vision, Waikoloa, Hawaii, pp. 1692–1698 (2005)
33. Wiskott, L., Fellous, J.M., Kruger, N., Malsburg, C.V.D.: Face recognition by elastic bunch graph matching. IEEE Transactions on Pattern Analysis and Mechine Intelligence 19, 775–779 (1997)
34. Okada, K., Steffens, J., Maurer, T., Hong, H., Elagin, E., Neven, H., Malsburg, C.V.D.: The Bochum/USC face recognition system and how it fared in the FERET phase III test. In: Wechsler, H., Phillips, P.J., Bruce, V., Soulie, F.F., Huang, T.S. (eds.) Face Recognition: From Theory to Applications. NATO ASI Series, vol. 163, pp. 186–205. Springer, Berlin (1998)

35. Daugman, J.G.: Complete discrete 2D Gabor transform by neural networks for image analysis and compression. IEEE Transactions on Acoustics, Speech and Signal Processing 36, 1169–1179 (1988)
36. Bruce, V.: Recognizing Faces. Lawrence Erlbaum Associates, London (1988)
37. Bruce, V., Hancock, P.J.B., Burton, A.M.: Human face perception and identification. In: Wechsler, H., Phillips, P.J., Bruce, V., Soulie, P.F., Huang, T.S. (eds.) Face Recognition: From Theory to Applications. Lawrence Erlbaum Associates, London (1988)
38. Yin, R.K.: Looking at upside-down faces. Journal of Experimental Physics 81, 141–151 (1969)
39. Bichsel, M., Pentland, A.P.: Human face recognition and the face image set's topology. CVGIP: Image Understanding 59, 254–261 (1994)
40. Tou, J.T., Gonzales, R.C.: Pattern Recognition Principles. Addison-Wesley, Reading (1974)
41. Jain, A.K., Dubes, R.C.: Algorithms for Clustering Data. Prentice Hall, New Jersey (1988)
42. Bac, L.H., Tuan, N.A.: Using rough set in feature selection and reduction in face recognition problem. In: Ho, T.-B., Cheung, D., Liu, H. (eds.) PAKDD 2005. LNCS (LNAI), vol. 3518, pp. 226–233. Springer, Heidelberg (2005)
43. Viola, P., Zones, M.: Rapid object detection using a boosted cascade of simple features. In: Proceedings of IEEE Computer Society Conference on Computer Vision and Pattern Recognition, Kauai, Hawaii, pp. 12–14 (2001)
44. Viola, P., Zones, M.: Robust real time object detection. In: Proceedings of IEEE ICCV Workshop on Statistical and Computational Theories of Vision, Vancouver, Canada, p. 747 (2001)
45. Zhang, X., Mersereau, R.M.: Lip feature extraction towards an automatic speechreading system. In: Proceedings of International Conference on Image Processing, Vancouver, Canada, vol. 3, pp. 226–229 (2000)
46. Hubert, L., Arabie, P.: Comparing partitions. Journal of Classification 2, 193–218 (1985)

# Spatial Reasoning Based on Rough Mereology: A Notion of a Robot Formation and Path Planning Problem for Formations of Mobile Autonomous Robots*

Paweł Ośmiałowski and Lech Polkowski

Polish–Japanese Institute of Information Technology
Koszykowa 86, 02008 Warsaw, Poland
polkow@pjwstk.edu.pl, newchief@king.net.pl

**Abstract.** We address in this work problems of path planning for autonomous robots; we extend this topic by introducing a new definition of a robot formation and we give a parallel treatment of planning and navigation problems for robot formations. In our investigations into problems of multi–robot planning and navigation, we apply rough mereological theory of spatial reasoning to problems of formations of many robots in a rigorous way and we address the planning and navigation problems for formations of many robots.

In approaching those problems, we employ rough mereology – a theory for approximate reasoning based on the notion of a part to a degree. Using the primitive predicate of a rough inclusion, we construct counterparts of classical predicates of elementary geometry as introduced by Alfred Tarski, which serve us in building a description of robot environment.

The software system Player/Stage is employed as the environment in which predicates of rough mereological geometry are implemented as SQL functions and as means of simulation and visualization of robot trajectories to chosen goals.

**Keywords:** mobile autonomous robotics, spatial reasoning, rough mereology, formations of robots, Player/Stage software system.

## 1 Introduction

Robotics of autonomous mobile robots presents the most intricate field for applications of techniques of artificial intelligence, decision making and cognitive methods. Among the basic problems in this area are: localization, navigation and planning problems and we are concerned with the third one, though it is obvious that the three problems are not separable. Research on planning in mobile robotics has brought on many ideas, algorithms and heuristics, see, e.g., [7], like bug algorithms, potential functions and fields, probabilistic planners, graph based methods (roadmaps) etc., etc.

* This work does extend the scope of the Special Session on Rough Mereology at RSKT 2008, Chengdu, Sichuan, China, May 2008.

J.F. Peters et al. (Eds.): Transactions on Rough Sets XII, LNCS 6190, pp. 143–169, 2010.

From among those methods, we choose to adopt the method of potential field, see sect.6. In classical setting, the potential field is built as the sum of two components: repulsive, induced by obstacles, and attractive, induced by goals. The field force is defined as the gradient of the repulsive, respectively, attractive, potential, see [7]. In either case, the potential is defined with the use of a metric, in analogy to classical physical examples of a potential field like Coulomb or gravitational fields. Our approach is different: the potential field is constructed by means of a chosen rough inclusion – the primitive predicate of rough mereology, see sect.6. A robot is driven to the goal by following areas of increasing density of the field as shown in sect.6. The problem for a single robot is presented fully in [27] where mereological potential fields were constructed and applied in planning of paths.

### 1.1 On Problems of Work with Many Robots

Problems of cooperative mobile robotics are demanding as they require an accounting for group behavior of many autonomous mobile robots.

There is the increasing need for making use of such teams in practical problems of performing complex tasks inaccessible for a single robot (like pushing large objects, rescue operations, assembling); there is also a theoretical interest in research on aspects of their behavior: cooperative mechanisms, leadership, conflict resolution, consensus making, many of which belong as well in biology and environmental studies, see, e.g., [2], [4], [6], [18], [39] and also a discussion in [5]. These motifs have propelled research in direction of multi–robot planning.

Cooperative behavior is perceived by many authors, see, e.g., [5] and references therein, as a specialization of collective behavior having the tint of achieving jointly some goal. The goal may mean an economic advantage, reaching a specified position, learning jointly a feature or a category of objects, etc., etc.

Main directions of research in this area of schemes for cooperative mobile robotics, include, as distinguished in the literature, see, e.g., [5], [12], a study on group architectures, conflicts of resources, motivations for cooperation, learning of a cooperative behavior, spatiotemporal aspects: path planning, moving to formations, pattern generation.

In this work, which extends our earlier results [35], [26], we are concerned with the last aspect, i.e., moving to formations and path planning in order to make robots into a given formation. Alongside, we are concerned with problems of repairing formations and navigating formations in static environments.

### 1.2 Path Planning Methods

Path planning methods, according to Latombe [17], [7] can be divided into *centralized*, in which case planning considers all robots in a team, or *decoupled*, when path is planned for each robot independently. Another division of path planning methods consists in local vs. global approach; in the local method, planning is provided for some neighbourhood of a robot, whereas in the global approach, the whole environment is taken into consideration.

Path planning consists in finding controls which secure the desired path of a robot toward a goal with, e.g., obstacle avoidance. As with a single robot, the path planning problem arises for teams of robots. In particular, centralized methods for single robots are extended to teams of robots see, e.g., [43] where such planning is applied with help of relational structures called super–graphs on which admissible paths are searched for. This approach however assumes that the situation is static, i.e., no changes in the environment happen during plan execution.

The assumption of environment stability is not valid in real world situations and the reactive approach [1] to planning considers simple, sensor–actuator coupling schemes expressible as low–level behaviors [48]; in these schemes, potential field methods, vector field histograms, dynamic window approach are used [47].

Some reactive methods use variants of search algorithms like A*, e.g., D* [4].

In this work we will study the problem of path planning in order to make robots in a team into a formation.

We apply as a theoretical framework for our approach, a qualitative theory of spatial reasoning as provided by Rough Mereology, see [30], [34], [35] [26]. In this framework, we give a definition of a formation by a team of robots, by means of a rough mereological beetweenness relation among them, see [34], [26].

This definition, which in our case is translation and rotation invariant, allows for considering formally various means of robot arranging into a pattern.

In the same framework, we study the problem of path planning for moving into a given formation. We propose some procedures to this end.

We model our robots on Roomba [1] robots by iRobot$^{(R)}$ , i.e., we assume our robots to be planar disk–shaped objects. We use Player/Stage system, see [25], [26], see [29], as a tool for simulation and visualization.

## 2 On Formations of Autonomous Mobile Robots

A study of the concept of a robot formation was initially based on a perception of animal behavior like herding, swarming, flocking or schooling. In this respect, a few principles emerged, see, e.g., [2]: keeping all animals within a certain distance from one another (e.g., to ensure mutual visibility), moving away when the distance becomes too close (to avoid congestion, collision, or resource conflict), adapting own movement to movement of neighbours (e.g., velocity of motion), orienting oneself on a leader.

From those observations a geometric approach to formations has been derived: a formally simplest approach [2] uses *referencing* techniques; reference is made either to the team center or to the team leader, or to a specified neighbour in a coordinate system given by the position of the team center or the leader along with the orientation given by the nearest navigation point; positions are determined, e.g., with the help of GPS or dead reckoning.

Another method for forming a geometric formation relies on a direct usage of a metric, say $\rho$, see, e.g., [6], [42]: given a threshold $\delta$, and a parameter $\varepsilon$, for

---
[1] Roomba is the trademark of iRobot Inc.

each robot $r$ in a team, its farthest neighbour $r_1$ and the nearest neighbour $r_2$, if $\rho(r, r_1) > \delta$ then $r$ moves toward $r_1$, if $\rho(r, r_1) < \delta - \varepsilon$ then $r$ moves away from $r_1$, if $\delta - \varepsilon < \rho(r, r_1) < \delta$ then $r$ moves away from $r_2$. By this method, robots are arranged on a circle.

Some methods rely on the potential field technique [18]; in this approach, the potential of the field is defined dependent on the distance among robots in the team in order to keep distances among them as prescribed. In addition, also the technique of a virtual leader is involved to keep robots in a team at a prescribed distance from their current leaders; in some approaches the relation the leader - the follower is expressed by means of control laws in a given coordinate system [39], [9] with execution of movement controlled by an omnidirectional camera.

It seems desirable to propose an approach which in principle would be metric independent and which would take into account only relative positions of robots one to another. In this work we propose a definition of a formation which is based on spatial predicates defined within rough mereological theory of spatial reasoning, see, e.g., [30], [34], [35], [26].

## 3 Qualitative Spatial Reasoning

In this Section, we recall elements of spatial theory induced in the rough mereological framework which have already been presented elsewhere [35], [26].

A formulation of spatial theory called Qualitative Spatial Reasoning have emerged on the basis of mereological theory based on the notion of a Connection [8]. This last theory in turn was conceived as a modification of Calculus of Individuals due to Leonard and Goodman [19], an elaboration on an idea by A. N. Whitehead of a spatial calculus based on the idea of an extension [49]. The idea of an extension is the dual to the notion of a part, on which the first (1916) mereological theory of concepts due to Stanislas Leśniewski [20]–[24] is based. Qualitative Spatial Reasoning as invoked above, can be considered as a variant of reasoning based on mereology.

Qualitative Spatial Reasoning is a basic ingredient in a variety of problems in mobile robotics, see, e.g., [14]. Spatial reasoning which deals with objects like solids, regions etc., by necessity refers to and relies on mereological theories of concepts based on the opposition part–whole [10]. Mereological ideas have been early applied toward axiomatization of geometry of solids, see [16], [44].

Mereological theories rely either on the notion of a part [20]–[24] or on the notion of objects being connected [8], [10]. Our approach to spatial reasoning is developed within the paradigm of rough mereology, see sect. 4. Rough mereology is based on the predicate of being a part to a degree and thus it is a natural extension of mereology based on part relation, as proposed by S. Leśniewski.

## 4 Rough Mereology

We recall here some already presented rudiments of rough mereology, see, e.g., [31], [33], [34], [36].

Rough mereology begins with the notion of a *rough inclusion*, cf., [36], which is a parameterized relation $\mu_r$ such that for any pair of objects $u, v$ the formula $\mu(u, v, r)$ means that $u$ is a *part of* $v$ *to a degree of* $r$ where $r \in [0, 1]$.

The following is the list of basic postulates for rough inclusions; *el* is the element (ingredient) relation of a mereology system based on a part relation, see [20]–[24]; informally, it is a partial ordering on the given class of objects defined from the strict order set by the part relation: for objects $u, v$, one has $u$ is el $v$ if and only if $u{=}v$ or $u$ is part $v$.

RM1. $\mu(u, v, 1) \Longleftrightarrow u$ is *el* $v$ (a part in degree 1 is equivalent to an element).

RM2. $\mu(u, v, 1) \Longrightarrow$ for all $w, r$ $(\mu(w, u, r) \Longrightarrow \mu(w, v, r))$ (monotonicity of $\mu$).

RM3. $\mu(u, v, r) \wedge s \leq r \Longrightarrow \mu(u, v, s)$ (assuring the meaning "a part to degree *at least* $r$").

In our applications to spatial reasoning, objects will be regions in Euclidean spaces, notably rectangles, in particular squares, discs, in 2–dimensional space, and the rough inclusion applied will predominantly be the one defined by the equation,

$$\mu^0(u, v, r) \text{ if and only if } \frac{|u \cap v|}{|u|} \geq r, \tag{1}$$

where $|u|$ is the area of the region $u$.

On the basis of a given rough inclusion $\mu$, we can introduce predicates of a certain geometry of regions in low–dimensional spaces. Points in such geometries are recovered usually be means of the technique of ultrafilters of regions proposed by Alfred Tarski [44].

## 5 Mereogeometry: A Geometry of Regions

We are interested in introducing into the mereological world defined by $\mu^0$ a geometry in whose terms it will be possible to express spatial relations among objects; a usage for this geometry will be found in navigation and control tasks of multiagent mobile robotics.

### 5.1 A Notion of a Quasi–distance

We first introduce a notion of a quasi–distance $\kappa$ in our rough mereological universe by letting,

$$\kappa(u, v) = min\{argmax_r \mu^0(u, v, r), argmax_s \mu^0(v, u, s)\}. \tag{2}$$

Observe that mereological distance differs essentially from the standard distance: the closer are objects, the greater is the value of $\kappa$: $\kappa(u, v) = 1$ means $u = v$, whereas $\kappa(u, v) = 0$ means disjointness in the sense of $\mu^0$ of $u$ and $v$ regardless of the Euclidean distance between them.

### 5.2 A Notion of Betweenness

We now introduce the notion of betweenness $T(z, u, v)$, see [45],

$$T(z, u, v) \iff \text{for all w } \kappa(z, w) \ \in [\kappa(u, w), \kappa(v, w)]. \tag{3}$$

Here, [, ] means the non–oriented interval. We check that the predicate $T$ satisfies the axioms of Tarski [45] for *betweenness.*

**Proposition 1.** *1. $z$ is $T(u, u) \Longrightarrow z = u$ (identity).*
*2. $v$ is $T(u, w)$ and $z$ is $T(v, w) \Longrightarrow v$ is $T(u, z)$ (transitivity).*
*3. $v$ is $T(u, z)$ and $v$ is $T(u, w)$ and $u \neq v \Longrightarrow z$ is $T(u, w)$or $w$ is $T(u, z)$ (connectivity).*

*Proof.* By means of $\kappa$, the properties of betweenness in our context are translated into properties of betweenness in the real line which hold by the Tarski theorem [45], Theorem 1.

### 5.3 The Notion of Nearness

We apply $\kappa$ to define in our context the predicate $N$ of *nearness* proposed in van Benthem [3],

$$N(z, u, v) \iff (\kappa(z, u) = r, \kappa(u, v) = s \Longrightarrow s < r). \tag{4}$$

Here, nearness means that $w$ is closer to $u$ than $v$ is to $u$.

Then the following hold, i.e., $N$ does satisfy all axioms for nearness in [3],

**Proposition 2.** *1. $z$ is $N(u, v)$ and $v$ is $N(u, w) \Longrightarrow z$ is $N(u, w)$ (transitivity).*
*2. $z$ is $N(u, v)$ and $u$ is $N(v, z) \Longrightarrow u$ is $N(z, v)$ (triangle inequality).*
*3. non($z$ is $N(u, z)$) (irreflexivity).*
*4. $z = u$ or $z$ is $N(z, u)$ (selfishness).*
*5. $z$ is $N(u, v) \Longrightarrow z$ is $N(u, w)$ or $w$ is $N(u, v)$ (connectedness).*

For the proof see [35].

### 5.4 A Variant of Betweenness Due to vanBenthem

We make an essential use of the other betweenness predicate $T_B$ proposed by van Benthem [3], on the basis of the nearness predicate,

$$T_B(z, u, v) \iff [\text{for all } w \ (z \text{ is } w \text{ or } N(z, u, w) \text{ or } N(z, v, w))]. \tag{5}$$

**Proposition 3.** *The functor $T_B$ of betweenness in the sense of van Benthem does satisfy the Tarski axioms (TB1)–(TB2) whereas (TB3) is refuted by simple examples.*

For the proof see [35].

*Example 1.* We consider a context in which objects are rectangles positioned regularly, i.e., having edges parallel to axes in $R^2$. The measure $\mu$ is $\mu^0$ of (1). In this setting, given two disjoint rectangles $C$, $D$, the only object between $C$ and $D$ in the sense of the predicate $T_B$ is the *extent* $ext(C, D)$ of $C, D$, , i.e., the minimal rectangle containing the union $C \cup D$.

As linear stretching or contracting along an axis does not change the area relations, it is sufficient to consider two unit squares $A, B$ of which $A$ has (0,0) as one of vertices whereas $B$ has (a,b) with $a, b > 1$ as the lower left vertex (both squares are regularly positioned). Then the distance $\kappa$ between the extent $ext(A, B)$ and either of $A, B$ is $\frac{1}{(a+)(b+1)}$. For a rectangle $R : [0, x] \times [0, y]$ with $x \in (a, a+1), y \in (b, b+1)$, we have that $\kappa(R, A) = \frac{(x-a)(y-b)}{xy} = \kappa(R, B)$. For $\phi(x, y) = \frac{(x-a)(y-b)}{xy}$, we find that $\frac{\partial \phi}{\partial x} = \frac{a}{x^2} \cdot (1 - \frac{b}{y}) > 0$, and, similarly, $\frac{\partial \phi}{\partial y} > 0$, i.e., $\phi$ is increasing in $x, y$ reaching the maximum when $R$ becomes the extent of $A, B$. An analogous reasoning takes care of the case when $R$ has some (c,d) with $c, d > 0$ as the lower left vertex.

### 5.5 On the Notion of a Line

A line segment may be defined via the auxiliary notion of a pattern; we introduce this notion as a functor $Pt$.

We let

$$Pt(u, v, z) \Longleftrightarrow z \text{ is } T_B(u, v) \text{ or } u \text{ is } T_B(z, v) \text{ or } v \text{ is } T_B(u, z).$$

We will say that a finite sequence $u_1, u_2, ..., u_n$ of objects *belong in a line segment* whenever $Pt(u_i, u_{i+1}, u_{i+2})$ for $i = 1, , ..., n-2$; formally, we introduce the functor *Line* of finite arity defined via

$$Line(u_1, u_2, ..., u_n) \Longleftrightarrow \text{for all } i < n - 1. Pt(u_i, u_{i+1}, u_{i+2}).$$

*Example 2.* With reference to Example 1, rectangles $C, D$ and their extent $ext(C, D)$ form a line segment.

## 6 Mereological Potential Fields

As mentioned in sect.2, the technique of potential fields, see [11], [13] for seminal ideas, and, cf., [17], [7], well–known from planning in case of a single robot, has been extended to the case of robot teams. An example of this approach is given in [18] where robots in a team are organized around a set of beacons called *leaders* and are subjected to repulsive and attractive forces induced by potential fields generated for pairs of robots and pairs of the form robot – leader in such a way as to prevent too close distance among robots and to keep them along leaders.

In our case, we apply the idea already exposed, see [27], [35], [26], of building a potential field from the rough inclusion $\mu^0$.

Our path planner accepts target point coordinates and provides a list of *way–points* from a given robot position to the goal. It takes as an input a map of static obstacles that a robot should avoid while approaching the target point. A robot and a target should both lay within the area delimited by surrounding static obstacles that form borders of the robot environment. There can be other static obstacles within the area, all marked on the provided map. After the path is proposed, a robot is lead through the path until it reaches given target. If a robot cannot move towards the target position for some longer time (e.g., it keeps on hitting an other robot reaching its target or some unknown non–static obstacle), a new path is proposed.

We tested our planner device by running simulations in which we have had a model of Roomba robot travelling inside an artificially created environment. Real Roomba robots are disc–shaped and therefore easy to model, but they do not provide many useful sensor devices (except bumpers which we were using to implement lower-level reaction to hitting unexpected obstacles). Also, odometry of Roomba robots is unreliable [46] hence we assume that simulated robots are equipped with a global positioning system.

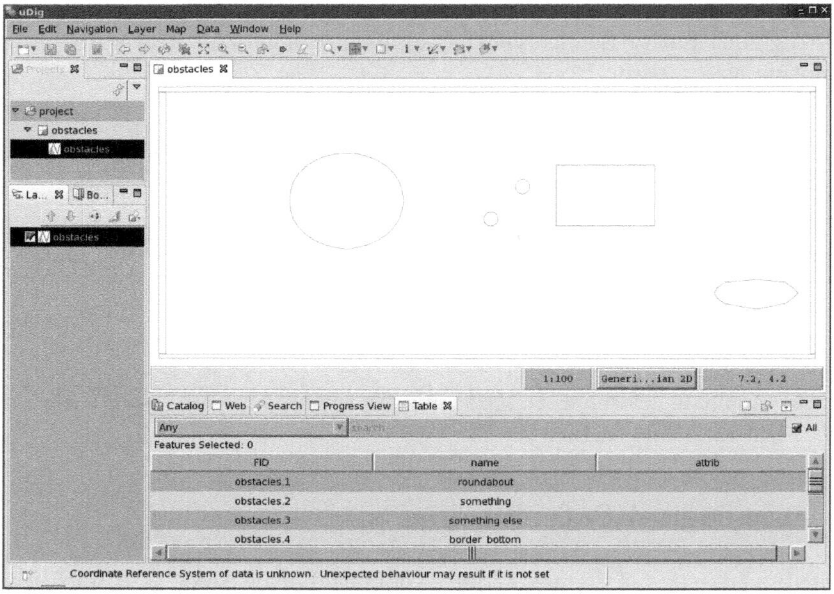


**Fig. 1.** Map of our artificial world edited by the uDig application (created and maintained by Refractions Research). The map consists of number of layers whose can be edited individually; on the figure we can see how solid obstacles are situated within *obstacles* layer.

Right after the target position is given, our planner builds the mereological potential field filled with squared areas each of the same size. The field is delimited by environment's borders. Only space free of obstacles is filled. To compute a value of the potential field at a given place, we are taking mereological feature of one object being a part of another to a degree where our objects are squared areas that fill the potential field. Near the goal any two squared areas are parts of each other to a higher degree and this value goes low as the distance to the goal increases. It can happen that for bigger workspace, areas too far from the goal are not filled as the potential is limited to values from 0 to 1, where value 0 means that two squares are not part of each other (maximal mereological distance between two areas) while 1 means that two areas are part of each other to a maximal degree (minimal mereological distance between two areas). As a result our potential field is dense with squared areas close to the target and it gets sparse far from it.

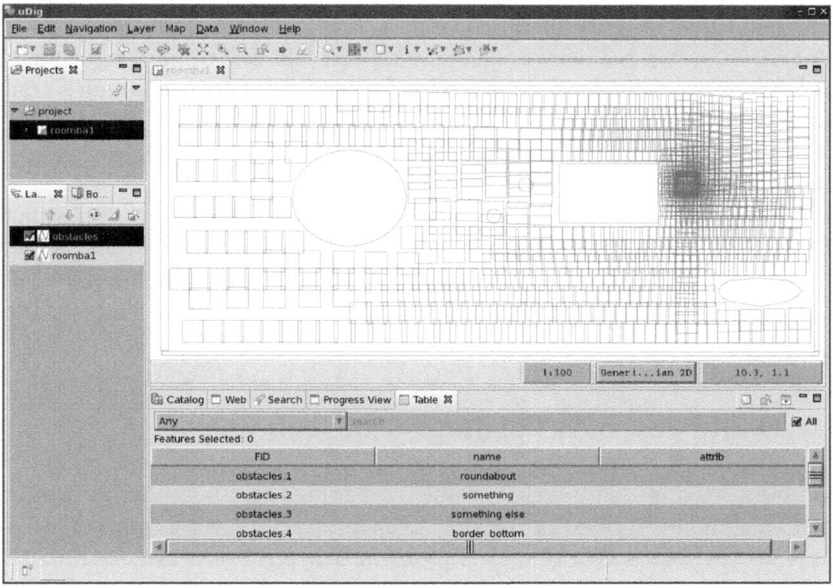


**Fig. 2.** Obstacles layer together with potential field layer (potential field generated for given goal is stored as another map layer). Observe increasing density towards the goal.

We recall [26] the algorithm for building the potential field with squared areas.

## ALGORITHM SQUARE FILL

Structure: a queue Q

1. Add to the queue Q, x and y coordinates of a given goal together with 0 as current distance from current squared area to the next neighboring area (so they will be part of each other to the maximal degree). Also put clockwise as current direction of exploration. These are initial values.

2. Spin in the main loop until there are no more elements in the queue Q:

2.1. Extract x, y, current distance and current direction of exploration from the beginning of queue Q.

2.2. Check if there is any other squared area already present in potential field to which the distance from current x and y coordinates is equal or shorter than current distance. If so, skip taken element and run new main loop turn.

2.3. Form new squared area with current x and y as the coordinates of the centroid of this new area. Check if there are any common part with any static obstacle within this new squared area. If so, skip taken element and run new main loop turn.

2.4. Add new squared area to the potential field.

2.5. Increase current distance by 0.01.

2.6. Add eight neighbour areas to the queue Q (for each area add these data: x and y coordinates, current distance and direction of exploration opposite to current); if direction is clockwise neighbours are: left, left-up, up, right-up, right, right-down, down, left-down; if direction is anti-clockwise neighbours are: left-down, down, right-down, right, right-up, up, left-up, left.

2.7. Run new main loop turn.

Mereological potential fields play a vital role in our approach to formations: when moving robots in a formation, to a chosen goal(s), each robot acquires a path to a goal planned with the aid of the mereological potential field, and it does pursue this path, subject to controllers guiding the formation execution or change.

## 7 A Definition of a Formation of Robots

We address in this section the problem of organizing a set of robots into a team with some prescribed structure. As mentioned in sect.2, many basic ideas about ways in which to organize robots into a structure are borrowed from behavioral biology [2], [18], and they rest on adaptation of classical ideas from single robot world like metric distance, potential fields induced by robots, metric and angular orientation on a leader. In those approaches, the behavioral aspect takes precedence before any attempt at formally imposing some structure on the team. We here, to the contrary, propose a theory of structures, based on the predicates of rough mereological geometry, of which foremost is the predicate $T_B$ of betweenness.

We restrict ourselves to the case when our robots are Roomba robots [46]. A Roomba robot is a disc–shaped robot and due to this we model it as the square circumscribing the robot with edges parallel to coordinate axes of the reference system. This allows for the extent of two given robots to be always oriented as a regular rectangle, i.e., with edges parallel to coordinate axes. In particular, this feature allows for translational and rotational invariance of extents, more generally under affine transformation of the plane.

For robots $A, B, C$, we say that

**Definition 1.** *We say that a robot $B$ is between robots $A$ and $C$, in symbols (between $B$ $A$ $C$), in case the rectangle $ext(B)$ is contained in the extent of rectangles $ext(A), ext(C)$, i.e., $\mu_0(ext(B), ext(ext(A), ext(C)), 1)$.*

This allows as well for a generalization to the notion of *partial betweenness* which models in a more precise manner spatial relations among $A, B, C$ (we say in this case that robot $B$ is between robots $A$ and $C$ to a degree of at least $r$): in symbols,

$$\text{(between–deg } r\ B\ A\ C\ \text{)} \tag{6}$$

if and only if

$$\mu^0(ext(B), ext[ext(A), ext(C)], r). \tag{7}$$

We now give the central definition in this work: the definition of a formation. By a formation, we mean a set of robots along with a structure imposed on it as a set of spatial relations among robots.

**Definition 2.** *For a team of robots, $T(r_1, r_2, ..., r_n) = \{r_1, r_2, ..., r_n\}$, an ideal formation $IF$ on $T(r_1, r_2, ..., r_n)$ is a betweenness relation (between...) on the set $T(r_1, r_2, ..., r_n)$ of robots.*

In practice, ideal formations will be given as a list of expressions of the form,

$$\text{(between } r_0\ r_1\ r_2\text{)}, \tag{8}$$

indicating that the object $r_0$ is between $r_1, r_2$, for all such triples, along with a list of expressions of the form,

$$\text{(not–between } r_0\ r_1\ r_2\text{)}, \tag{9}$$

indicating triples which are not in the given betweenness relation.

To account for dynamic nature of the real world, in which due to sensory perception inadequacies, dynamic nature of the environment etc., we allow for some deviations from ideal formations by allowing that the robot which is between two neighbours can be between them to a degree in the sense of (6).

This leads to the notion of a real formation,

**Definition 3.** *For a team of robots, $T(r_1, r_2, ..., r_n) = \{r_1, r_2, ..., r_n\}$, a real formation $RF$ on $T(r_1, r_2, ..., r_n)$ is a betweenness to degree relation (between–deg ....) on the set $T(r_1, r_2, ..., r_n)$ of robots.*

In practice, real formations will be given as a list of expressions of the form,

$$\text{(between–deg } \delta\ r_0\ r_1\ r_2\text{)}, \tag{10}$$

indicating that the object $r_0$ is to degree of $\delta$ in the extent of $r_1, r_2$, for all triples in the relation (between–deg ....), along with a list of expressions of the form,

$$\text{(not–between } r_0\ r_1\ r_2\text{)}, \tag{11}$$

indicating triples which are not in the given betweenness relation.

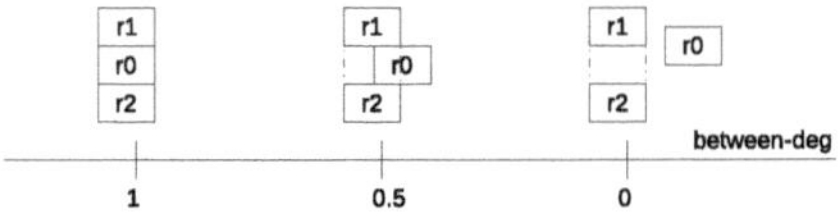


**Fig. 3.** Object $r_0$ is in extent of $r_1$ and $r_2$ to degree $\delta$

In Fig. 1, we sketch some cases of instances of relations (between–deg $\delta\, r_0\, r_1\, r_2$ ).

# 8 On Complexity of Formation Description

Description of formations, as proposed in Def. 1, 2 of sect. 7, can be a list of relation instances of large cardinality, cf., Examples 3 and 4, below. The problem can be posed of finding a minimal set of instances wholly describing a given formation.

To address this problem, we construct an *information system* $form-Inf$ see, e.g., [28], i.e. a triple $(U, A, f)$ where $U$ is a set of objects, $A$ is a set of attributes and $f$ is a value assignment, i.e. a mapping $f : A \times U \rightarrow V$, where $V$ is a set of possible values of attributes in $A$ on objects in $U$. For a formation $IF$, in order to define the information system $form - Inf\ (IF)$, we let,

$U = T(r_1, ..., r_n)$, a team of robots;

$A = \{[r_k, r_l, r_m] : r_k, r_l, r_m$ pairwise distinct robots$\}$;

for a given formation $IF$ of robots $r_1, ..., r_n$, the value assignment $f$ is defined as follows,

$$f([r_k, r_l, r_m], r_i) = \begin{cases} 1 \text{ in case } r_i = r_l \text{ and (between } r_l\ r_k\ r_l) \\ \frac{1}{2} \text{ in case } r_i = r_l \text{ } or \text{ } r_i = r_m \text{ and (between } r_l\ r_k\ r_m) \\ 0 \text{ in case } r_i\ \neq r_l\, r_k\, r_m \end{cases} \tag{12}$$

The system $form - Inf(IF)$ describes the formation $IF$. We now recall the notion of a reduct, see, e.g., [28]. For an information system $(U, A, f)$, a reduct is a set $B \subseteq A$ of attributes which determines functionally the system, i.e., the subsystem $(U, B, f)$ has the property that: there exists a mapping $H$ such that for each object $u \in U$: $\{a(u) : a \in A\} = H(\{a(u) : a \in B\})$, meaning that values of all attributes on $u$ are functionally determined by values on $u$ of attributes in $B$.

Clearly, reducts of the system $form-Inf(IF)$ provide a complete description of the formation $IF$. Now, we can state the negative in a sense result.

**Proposition 4.** *The problem of finding a minimum size description of a given formation is NP–hard.*

*Proof.* As shown in Skowron and Rauszer [40], the problem of finding a minimum size reduct of a given information system is NP–hard.

## 9 Implementation in Player/Stage Software System

Player/Stage is an Open-Source software system designed for many UNIX-com--patible platforms, widely used in robotics laboratories [25], [12], [29]. Main two parts are Player – message passing server (with bunch of drivers for many robotics devices, extendable by plug–ins) and Stage – a plug–in for Player's bunch of drivers which simulates existence of real robotics devices that operate in the simulated 2D world. Player/Stage offers client–server architecture. Many clients can connect to one Player server, where clients are programs (robot controllers) written by a roboticist who can use Player client-side API. Player itself uses drivers to communicate with devices,and in this activity it does not make distinction between real and simulated hardware. It gives roboticist means for testing programmed robot controller in both real and simulated world.

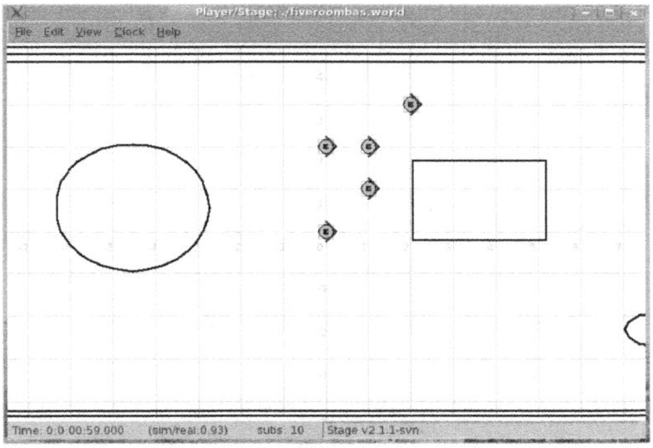


**Fig. 4.** Stage simulator in use – five Roomba robots inside simulated world

Among all Player drivers that communicate with devices (real or simulated), there are drivers not intended for controlling hardware, instead those drivers offer many facilities for sensor data manipulation, for example, camera image compression, retro–reflective detection of cylindrical markers in laser scans, path planning. One of the new features added to Player version 2.1 is the PostGIS driver: it connects to PostgreSQL database in order to obtain and/or update stored vector map layers.

PostGIS itself is an extension to the PostgreSQL object–relational database system which allows GIS (Geographics Information Systems) objects to be stored in the database [38]. It also offers new SQL functions for spatial reasoning. Maps which are to be stored in SQL database can be created and edited by graphical tools like uDig or by C/C++ programs written using GEOS library of GIS functions. PostGIS, uDig and GEOS library are projects maintained by Refractions Research.

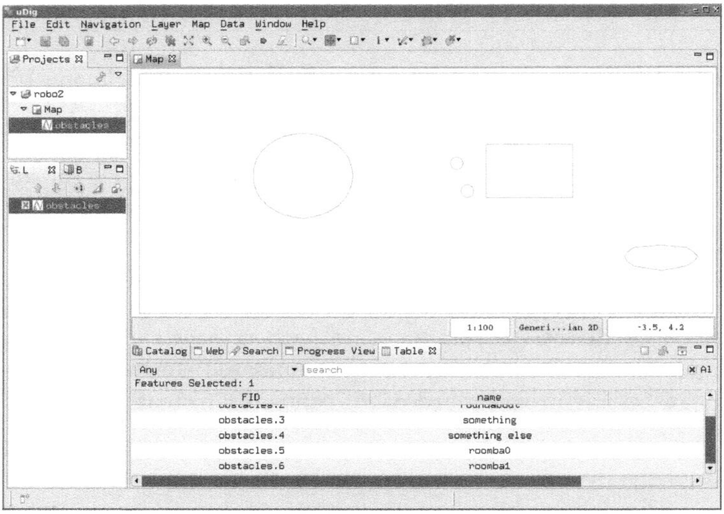


**Fig. 5.** uDig application in use - modification of *obstacles* layer

A map can have many named layers, and for each layer a table in SQL database is created. We can assume that the layer named *obstacles* consists of objects which a robot cannot walk through. Other layers can be created in which we can divide robot's workspace into areas with an assigned attribute which for example tells whether a given area is occupied by an obstacle or not. During our experimentations, we have created a plug–in for Players bunch of drivers which constantly tracks changes of position of every robot and updates *obstacles* layer so robots are marked as obstacles. As a result, the map stored in SQL database is kept always up to date. This feature is also useful in multi–agent environments: at any time a robot controller can send a query to SQL database server regarding every other robot position.

### 9.1 SQL Queries Representing Rough Mereogeometric Predicates

A roboticist can write a robot controller using Player client-side API which obtains information about current situation through the *vectormap* interface. Additionally, to write such a program, PostgreSQL client–side API can be used in order to open direct connection to the database server on which our mereogeometry SQL functions are stored together with map database. These functions can be called using this connection and results are sent back to the calling program. This gives robot controller program ability to perform spatial reasoning based on rough mereology.

Using PostGIS SQL extensions we have created our mereogeometry SQL functions [15]. Rough mereological distance is defined as such:

```
CREATE FUNCTION meredist(object1 geometry, object2 geometry)
RETURNS DOUBLE PRECISION AS
$$
    SELECT min(degrees.degree) FROM
            ((SELECT
                    ST_Area(ST_Intersection(extent($1), extent($2)))
                    / ST_Area(extent($1))
                    AS degree)
            UNION (SELECT
                    ST_Area(ST_Intersection(extent($1), extent($2)))
                    / ST_Area(extent($2))
                    AS degree))
            AS degrees;

$$ LANGUAGE SQL STABLE;
```

Having mereological distance function we can derive nearness predicate:

```
CREATE FUNCTION merenear(obj geometry, o1 geometry, o2 geometry)
RETURNS BOOLEAN AS
$$
    SELECT meredist($1, $2) > meredist($3, $2)
$$ LANGUAGE SQL STABLE;
```

The equi-distance can be derived as such:

```
CREATE FUNCTION mereequ(obj geometry, o1 geometry, o2 geometry)
RETURNS BOOLEAN AS
$$
    SELECT (NOT merenear($1, $2, $3))
            AND (NOT merenear($1, $3, $2));

$$ LANGUAGE SQL STABLE;
```

Our implementation of the betweenness predicate makes use of a function that produces an object which is an extent of given two objects:

```
CREATE FUNCTION mereextent(object1 geometry, object2 geometry)
RETURNS geometry AS
$$
    SELECT GeomFromWKB(AsBinary(extent(objects.geom))) FROM
            ((SELECT $1 AS geom)
            UNION (SELECT $2 AS geom))
            AS objects;

$$ LANGUAGE SQL STABLE;
```

The betweenness predicate is defined as follows:

```
CREATE FUNCTION merebetb(obj geometry, o1 geometry, o2 geometry)
RETURNS BOOLEAN AS
$$
    SELECT
            meredist($1, $2) = 1
            OR meredist($1, $3) = 1
            OR
                    (meredist($1, $2) > 0
                    AND meredist($1, $3) > 0
                    AND meredist(mereextent($2, $3),
                            mereextent(mereextent($1, $2), $3)) = 1);

$$ LANGUAGE SQL STABLE;
```

Using the betweenness predicate we can check if three objects form a pattern:

```
CREATE FUNCTION merepattern
    (object1 geometry, object2 geometry, object3 geometry)
RETURNS BOOLEAN AS
$$
    SELECT merebetb($3, $2, $1)
            OR merebetb($1, $3, $2)
            OR merebetb($2, $1, $3);

$$ LANGUAGE SQL STABLE;
```

Also having pattern predicate we can check if four objects form a line:

```
CREATE FUNCTION mereisline4
    (obj1 geometry, obj2 geometry, obj3 geometry, obj4 geometry)
RETURNS BOOLEAN AS
$$
    SELECT merepattern($1, $2, $3) AND merepattern($2, $3, $4);

$$ LANGUAGE SQL STABLE;
```

To figure out if a set of objects form a line an aggregate can be used:

```
CREATE FUNCTION mereisline_state
    (state_array geometry[4], input_data geometry)
RETURNS geometry[4] AS
$$
    SELECT ARRAY[$1[2], $1[3], $2, result.object]
            FROM (SELECT CASE
                    WHEN $1[4] IS NOT NULL
                            THEN $1[4]
                    WHEN $1[3] IS NULL
                            THEN NULL
```

```
                WHEN ($1[2] IS NULL) AND (meredist($1[3], $2) > 0)
                     THEN NULL
                WHEN ($1[2] IS NULL) AND (meredist($1[3], $2) = 0)
                     THEN $2
                WHEN ($1[1] IS NULL) AND merepattern($1[2], $1[3], $2)
                     THEN NULL
                WHEN ($1[1] IS NULL) AND (NOT merepattern($1[2], $1[3], $2))
                     THEN $2
                WHEN merepattern($1[1], $1[2], $1[3]) AND merepattern($1[2], $1[3], $2)
                     THEN NULL
                ELSE $2
           END AS object)
           AS result;
$$ LANGUAGE SQL STABLE;
CREATE FUNCTION mereisline_final
   (state_array geometry[4])
RETURNS BOOLEAN AS
$$
   SELECT ($1[4] IS NULL)
          AND ($1[3] IS NOT NULL)
          AND ($1[2] IS NOT NULL);
$$ LANGUAGE SQL STABLE;
CREATE AGGREGATE mereisline
(
   SFUNC = mereisline_state,
   BASETYPE = geometry,
   STYPE = geometry[],
   FINALFUNC = mereisline_final,
   INITCOND = '{}'
);
```

For our convenience we have derived betweenness predicate in more general form:

```
CREATE FUNCTION merebet
   (object geometry, object1 geometry, object2 geometry)
RETURNS BOOLEAN AS
$$
   SELECT (
          ST_Area(ST_Intersection(extent($1), mereextent($2, $3)))
          / ST_Area(extent($1))
          ) = 1.0;

$$ LANGUAGE SQL STABLE;
```

## 9.2 A Driver for Player Server to Maintain Formations

We have created a plug–in driver (written in C++ programming language) for Player server that keeps on tracking all robots in a team in order to make sure their positions form desired formation. If formation is malformed, our driver tries to repair it by moving robots to their proper positions within the formation. Also our driver is responsible for processing incoming orders: position commands which are dispatched to formation leader (selected member of a team) and geometry queries which are replied with information about current formation extent size and its global position. As such, our driver can be considered as a finite state machine which by default is constantly switching between two states: *process orders* and *formation integrity check*. If formation integrity check fails it switches to *repair formation* state.

Formation integrity check is done according to a given description. As pointed earlier, description of formation is a list of s-expressions (LISP–style symbolic expressions). To parse those descriptions efficiently we have used *sfsexp* programming library written by Matthew Sottile [41]. Each relation between robots in given description is checked and if related robots positions do not fulfil requirements, error value is incremented. Also while traversing through a description, overall error value is computed in order to figure out what could be the maximum error value for the given description. Finally, error value derived from each robot position is divided by computed overall error value which gives the

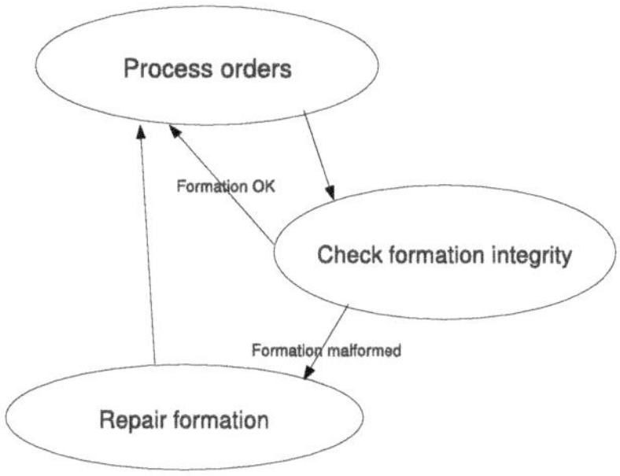


**Fig. 6.** States of our formation keeping driver for Player server

normalised formation fitness value between 0 (all requirements were fulfilled) and 1 (none of requirements were fulfilled). If the fitness value is below some threshold (typically 0.2), then we can conclude that robots are in their desired positions.

Typical formation description may look like below.

*Example 3.*
```
(cross
  (set
        (max-dist 0.25 roomba0 (between roomba0 roomba1 roomba2))
        (max-dist 0.25 roomba0 (between roomba0 roomba3 roomba4))
        (not-between roomba1 roomba3 roomba4)
        (not-between roomba2 roomba3 roomba4)
        (not-between roomba3 roomba1 roomba2)
        (not-between roomba4 roomba1 roomba2)
  )
)
```

This is a description of a formation of five Roomba robots arranged in a cross shape. The *max–dist* relation is used to bound formation in space by keeping all robots close one to another. Example below describes diamond shape formed by team of eight Roomba robots.

*Example 4.*
```
(diamond
  (set
        (max-dist 0.11 roomba1 (between roomba1 roomba0 roomba2))
        (max-dist 0.11 roomba3 (between roomba3 roomba1 roomba4))
        (max-dist 0.11 roomba5 (between roomba5 roomba4 roomba6))
        (max-dist 0.11 roomba7 (between roomba7 roomba0 roomba6))
        (between roomba1 roomba0 roomba2)
        (between roomba1 roomba0 roomba3)
        (between roomba1 roomba2 roomba7)
        (between roomba1 roomba3 roomba7)
        (between roomba3 roomba2 roomba4)
        (between roomba3 roomba2 roomba5)
        (between roomba3 roomba1 roomba5)
        (between roomba3 roomba1 roomba4)
        (between roomba5 roomba4 roomba6)
        (between roomba5 roomba4 roomba7)
        (between roomba5 roomba3 roomba7)
        (between roomba5 roomba3 roomba6)
        (between roomba7 roomba0 roomba6)
        (between roomba7 roomba0 roomba5)
        (between roomba7 roomba1 roomba5)
        (between roomba7 roomba1 roomba6)
        (not-between roomba1 roomba0 roomba4)
        (not-between roomba1 roomba2 roomba6)
        (not-between roomba1 roomba2 roomba3)
        (not-between roomba3 roomba0 roomba4)
        (not-between roomba3 roomba2 roomba6)
        (not-between roomba3 roomba1 roomba2)
        (not-between roomba5 roomba6 roomba7)
```

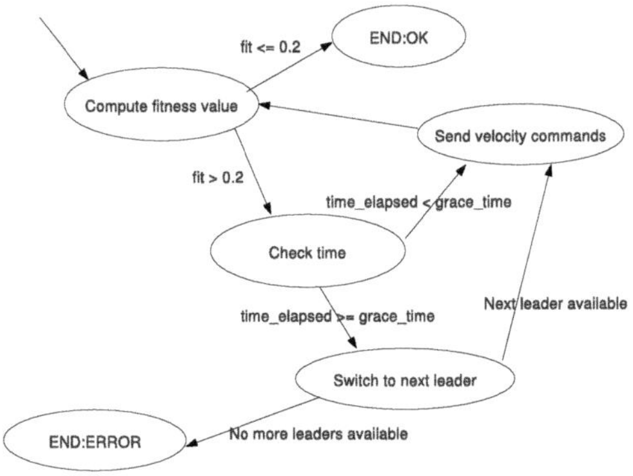


**Fig. 7.** Pure behavioral method of repairing the formation

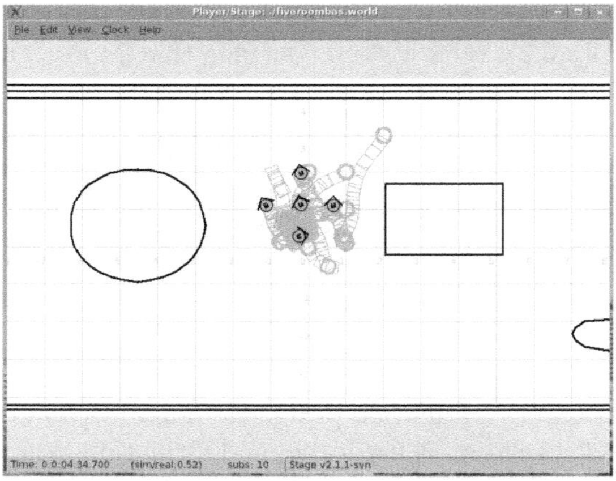

**Fig. 8.** Trails of robots that moved to their positions using pure behavioral method of repairing the formation; due to change of the leader robot which occurred during repairing process we can observe that initially formation was unsuccessfully built in a different place

```
        (not-between roomba5 roomba2 roomba6)
        (not-between roomba5 roomba0 roomba4)
        (not-between roomba7 roomba5 roomba6)
        (not-between roomba7 roomba2 roomba6)
        (not-between roomba7 roomba0 roomba4)
        (not-between roomba0 roomba1 roomba5)
        (not-between roomba0 roomba3 roomba7)
        (not-between roomba2 roomba1 roomba5)
        (not-between roomba2 roomba3 roomba7)
        (not-between roomba4 roomba1 roomba5)
        (not-between roomba4 roomba3 roomba7)
        (not-between roomba6 roomba1 roomba5)
        (not-between roomba6 roomba3 roomba7)
    )
)
```

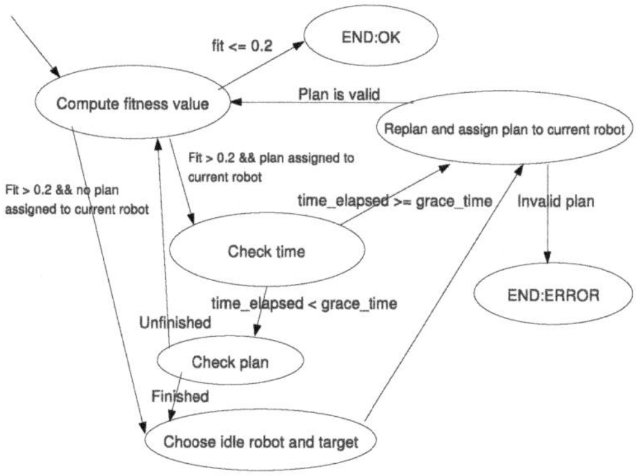


**Fig. 9.** *One robot at a time* method for repairing the formation

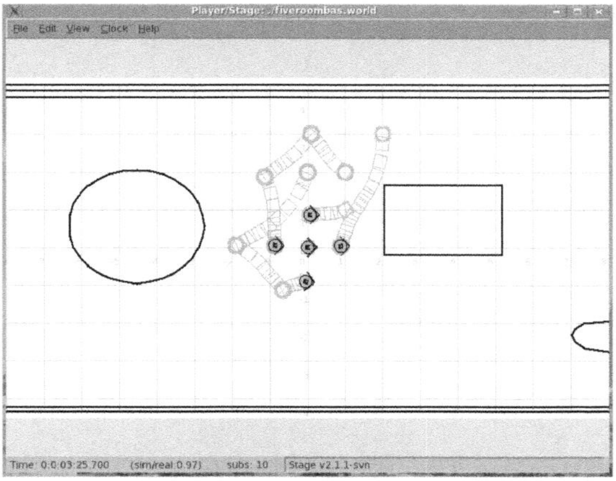


**Fig. 10.** Trails of robots moved to their positions on planned paths (*one robot at a time* method)

If formation is malformed our driver can try to repair it. A run–time parameter of the driver indicates which one of three methods should be used to move robots into their desired positions within the formation.

**Three methods of formation repairing.** We propose three methods for restoring a team to its prescribed formation shape. The first method is behavioral and does not use any planning. The second one is decoupled as planning is made for each robot separately, and global as all robots are taken into consideration at the same time. The third method is decoupled and global, and in addition is behavioral, as all robots move simultaneously.

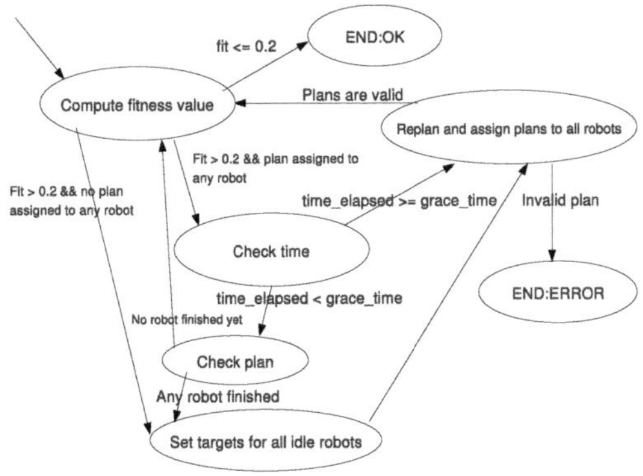


**Fig. 11.** *All robots at a time* method of repairing the formation

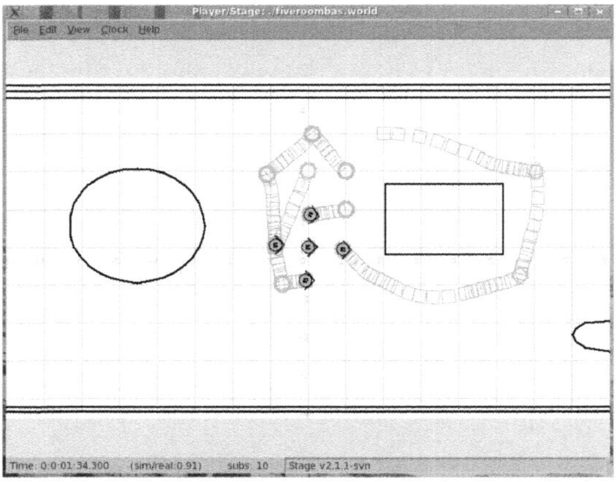

**Fig. 12.** Trails of robots moved to their positions on planned paths (*all robots at a time* method)

**Method 1: Pure behavioral** Each robot (except a selected leader) moves to the goal position. Whenever collision is detected (on the robot bumper device), robot goes back for a while then turns left or right for a while and from this new situation, it tries again to go towards goal position. Due to the nature of this method, formation repair process is time–consuming (reactions to possible collisions take additional time) and may be even impossible. Formation is repaired relatively to one selected member of a team called a leader (therefore this selected member sticks in place while all other robot moves to their positions). If formation is not repaired after some grace time, a next member of a team is selected to be the new leader (therefore this new selected member sticks in place while all other robot moves which changes whole situation). If there are no members left to be new leaders, this method signals that the formation shape is impossible to be restored.

**Method 2: One robot at a time** The procedure is repeated for each robot in a team: a path is planned by using any available planner (e.g., *wavefront* planner shipped with Player, or *mereonavigator* planner created during our experimentations [26]); then a robot moves to the goal position. This is the most reliable method, however it is time too consuming for bigger teams.

**Method 3: All robots at a time** Paths are planned and executed for all robots simultaneously. Whenever collision occurs during plan execution, lower level behavior causes involved robots to go back for a while, turn left or right for a while and new paths for those robots are planned. This is the fastest method and despite the fact that it is not collision aware, it is reliable enough.

## 10 Navigation by Obstacles with Robot Formations: Formation Changing and Repairing

The final stage of planning is in checking its soundness by navigating robots in an environment with obstacles. We show results of navigating with a team of robots in the initial formation of cross–shape in a crowded environment, see Fig. 13. In order to bypass a narrow avenue between an obstacle and the border of the environment, the formation changes to a line, and after bypassing it can use repairing to restore to the initial formation (in case it is required), or to any other required by local constraints formation, see Figs. 14–18.

The initial cross–shaped formation is shown in Fig. 13 along with obstacles in the environment.

Reaching the target requires passing by a narrow passage between the border and the rightmost obstacle. To carry out this task, robots in the formation are

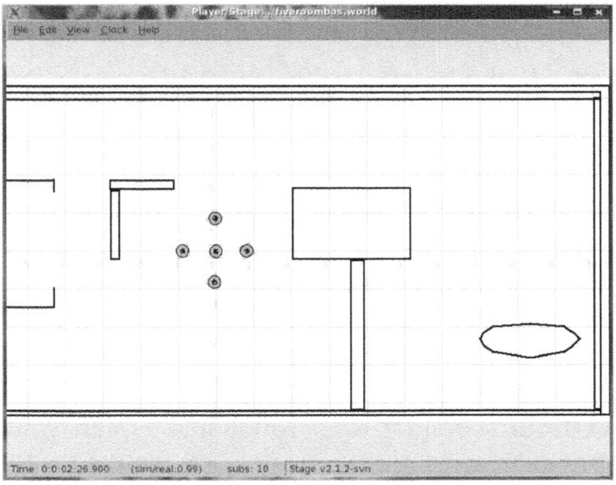


**Fig. 13.** Initial formation of robots and the obstacle map

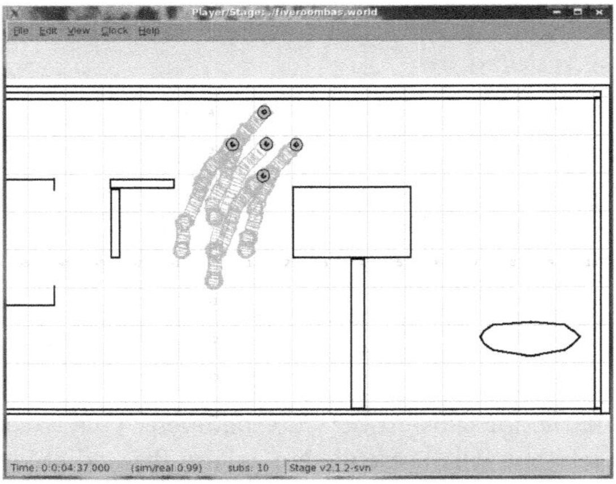

**Fig. 14.** Trails of robots moved to their positions on the line formation

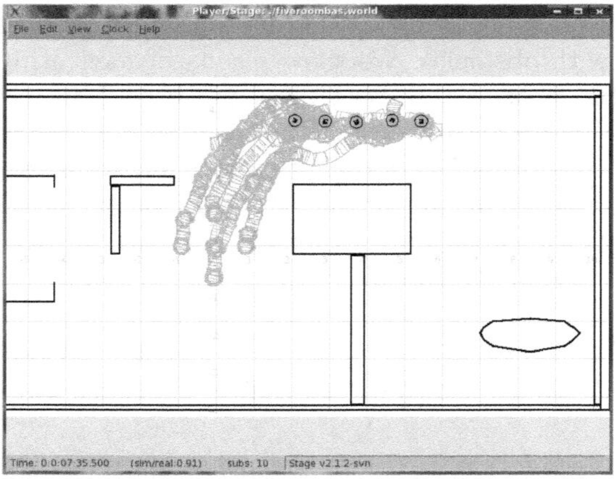

**Fig. 15.** Trails of robots moved to their positions on the line formation

bound to change the initial formation. They try the line formation, see Figs. 14–15.

However, making the line formation at the entrance to narrow passage is coupled with some difficulties: when the strategy *all robots at a time* is applied, robots at the lower part of the formation perceive robots at the upper part as obstacles and wait until the latter move into passage, which blocks whole routine as it assumes that from the start each robot has a plan valid until it reaches the goal or until it collides with another robot. To avoid such blocking of activity the behavior *wander* was added, see clouded area in Figs. 15, 16, which permitted robots to wander randomly until they find that they are able to plan their paths into the line. It can be observed that this wandering consumes some extra time.

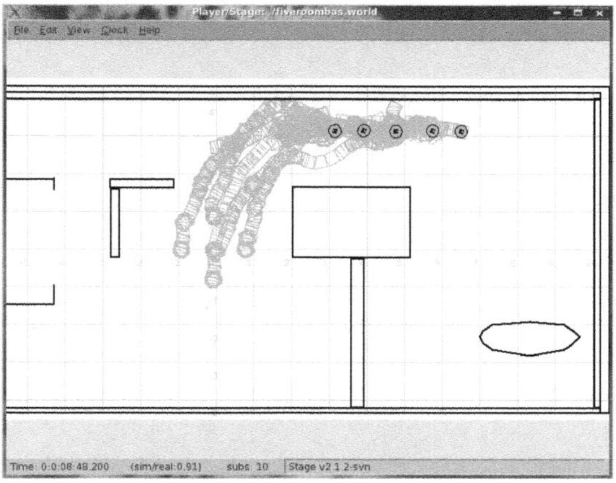

**Fig. 16.** Trails of robots moving in the line formation through the passage

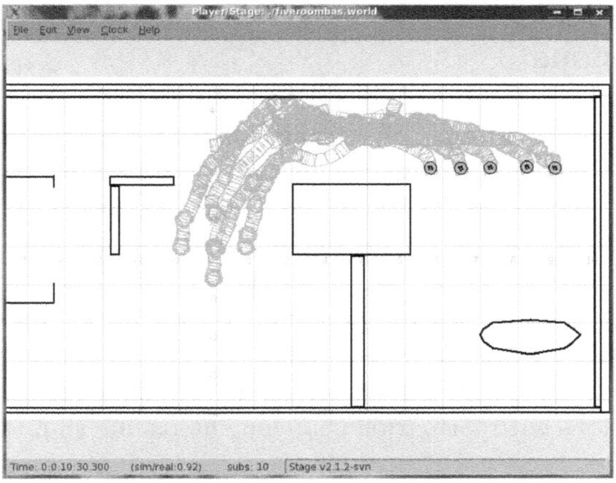

**Fig. 17.** Trails of robots moving in the line formation through and after the passage

When the strategy *one robot at a time* is applied it is important to carefully select the order in which robots are moved: the robots that have a clear pass to their target positions not obstructed by other robots should go first, so some additional priority queue structure can be added.

Surprisingly, *pure behavioral* strategy showed good performance in managing with these difficulties, however, as we have expected, when this strategy is applied, it is time–consuming to reshape the formation.

After the line was formed and robots passed through the passage, see Figs. 17–18, the line formation could be restored to the initial cross–shape formation, if necessary, with the help of a strategy for repairing formation of section 9.2.

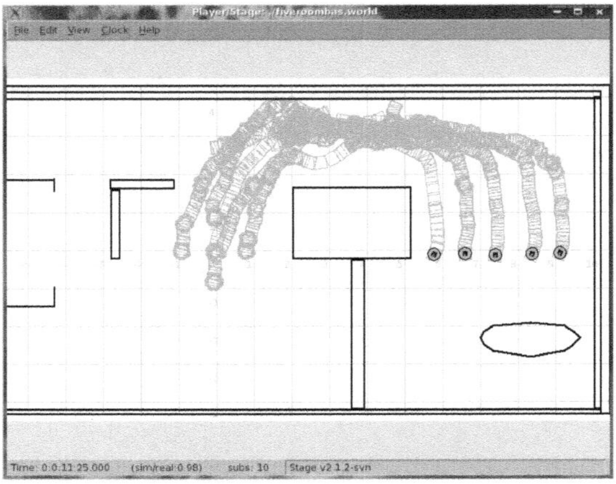

Fig. 18. Trails of robots in the line formation in the free workspace after passing through the passage

## 11 Conclusions

We have proposed a definition of a formation and we have presented a Player driver for making formations according to our definition. Our definition of a formation is based on a set of rough mereological predicates which altogether define a geometry of the space. The definition of a formation makes formations independent of a metric in the space and it is invariant under affine transformations. We have examined three methods of formation restoring, based on a reactive (behavioral) model as well as on decoupled way of planning.

Finally, we have performed simulations in Player/Stage system of planning paths for formations with formation change. The results show the validity of the approach.

Further research will be directed at improving the effectiveness of execution by studying divisions into sub–formations and merging sub–formations into formations.

## References

1. Arkin, R.C.: Behavior–Based Robotics. MIT Press, Cambridge (1998)
2. Balch, T., Arkin, R.C.: Behavior–based formation control for multirobot teams. IEEE Transactions on Robotics and Automation 14(6), 926–939 (1998)
3. van Benthem, J.: The Logic of Time. Reidel, Dordrecht (1983)
4. Brumitt, B., Stentz, A., Hebert, M.: CMU UGV Group: Autonomous driving with concurrent goals and multiple vehicles: Mission planning and architecture. Autonomous Robots 11, 103–115 (2001)
5. Uny Cao, Y., Fukunaga, A.S., Kahng, A.B.: Cooperative mobile robotics: Antecedents and directions. Autonomous Robots 4, 7–27 (1997)

6. Chen, Q., Luh, J.Y.S.: Coordination and control of a group of small mobile robots. In: Proceedings of IEEE Intern. Conference on Robotics and Automation, pp. 2315–2320 (1998)
7. Choset, H., Lynch, K.M., Hutchinson, S., Kantor, G., Burgard, W., Kavraki, L., Thrun, S.: Principles of Robot Motion. In: Theory, Algorithms, and Implementations. MIT Press, Cambridge (2005)
8. Clarke, B.L.: A calculus of individuals based on connection. Notre Dame Journal of Formal Logic 22(2), 204–218 (1981)
9. Das, A., Fierro, R., Kumar, V., Ostrovski, J.P., Spletzer, J., Taylor, C.J.: A vision-based formation control framework. IEEE Transactions on Robotics and Automation 18(5), 813–825 (2002)
10. Gotts, N.M., Gooday, J.M., Cohn, A.G.: A connection based approach to common-sense topological description and reasoning. The Monist 79(1), 51–75 (1996)
11. Khatib, O.: Real–time obstacle avoidance for manipulators and mobile robots. In: Proceedings IEEE Intern. Conf. on Robotics and Automation, St. Louis MO, pp. 500–505 (1986); Also see: International Journal of Robotic Research 5, 90–98 (1986)
12. Kramer, J., Scheutz, M.: Development environments for autonomous mobile robots: A survey. Autonomous Robots 22, 101–132 (2007)
13. Krogh, B.: A generalized potential field approach to obstacle avoidance control, SME-I Technical paper MS84–484, Society of Manufacturing Engineers, Dearborn MI (1984)
14. Kuipers, B.I., Byun, Y.T.: A qualitative approach to robot exploration and map learning. In: Proceedings of the IEEE Workshop on Spatial Reasoning and Multi-Sensor Fusion, pp. 390–404. Morgan Kaufmann, San Mateo (1987)
15. Ladanyi, H.: SQL Unleashed. Sams Publishing, USA (1997)
16. De Laguna, T.: Point, line, surface as sets of solids, J. Philosophy 19, 449–461 (1922)
17. Latombe, J.: Robot Motion Planning. Kluwer, Boston (1991)
18. Ehrich Leonard, N., Fiorelli, E.: Virtual leaders, artificial potentials and coordinated control of groups. In: Proceedings of the 40th IEEE Conference on Decision and Control, Orlando Fla, pp. 2968–2973 (2001)
19. Leonard, H., Goodman, N.: The calculus of individuals and its uses. The Journal of Symbolic Logic 5, 45–55 (1940)
20. Leśniewski, S.: O Podstawach Ogolnej Teorii Mnogosci (On Foundations of General Theory of Sets. The Polish Scientific Circle in Moscow, Moscow (1916) (in Polish)
21. Leśniewski, S.: Grundzüge eines neuen Systems der Grundlagen der Mathematik. Fundamenta Mathematicae 24, 242–251 (1926)
22. Leśniewski, S.: Über die Grundlegen der Ontologie. C.R. Soc. Sci. Lettr. Varsovie III, 111–132 (1930)
23. Leśniewski, S.: On the Foundations of Mathematics. Przegląd Filozoficzny 30, 164–206 (1927) (in Polish); 31, 261–291 (1928); 32, 60–101 (1929); 33, 77–105 (1930); 34, 142–170 (1931)
24. Leśniewski, S.: On the foundations of mathematics. Topoi 2, 7–52 (1982)
25. Ośmiałowski, P.: Player and Stage at PJIIT Robotics Laboratory. Journal of Automation, Mobile Robotics and Intelligent Systems 2, 21–28 (2007)
26. Ośmiałowski, P.: On path planning for mobile robots: Introducing the mereological potential field method in the framework of mereological spatial reasoning. Journal of Automation, Mobile Robotics and Intelligent Systems 3(2), 24–33 (2009)

27. Ośmiałowski, P., Polkowski, L.: Spatial reasoning based on rough mereology: path planning problem for autonomous mobile robots. In: Transactions on Rough Sets. LNCS (in print)
28. Pawlak, Z.: Rough Sets: Theoretical Aspects of Reasoning about Data. Kluwer, Dordrecht (1991)
29. Player/Stage, `http://playerstage.sourceforge.net`
30. Polkowski, L.: On connection synthesis via rough mereology. Fundamenta Informaticae 46, 83–96 (2001)
31. Polkowski, L.: A rough set paradigm for unifying rough set theory and fuzzy set theory (a plenary lecture). In: RSFDGrC 2003. LNCS (LNAI), vol. 2639, pp. 70–78. Springer, Heidelberg (2003)
32. Polkowski, L.: Toward rough set foundations. Mereological approach (a plenary lecture). In: Tsumoto, S., Słowiński, R., Komorowski, J., Grzymała-Busse, J.W. (eds.) RSCTC 2004. LNCS (LNAI), vol. 3066, pp. 8–25. Springer, Heidelberg (2004)
33. Polkowski, L.: A unified approach to granulation of knowledge and granular computing based on rough mereology: A survey. In: Pedrycz, W., Skowron, A., Kreinovich, V. (eds.) Handbook of Granular Computing, pp. 375–400. John Wiley and Sons, Chichester (2008)
34. Polkowski, L., Ośmiałowski, P.: Spatial reasoning with applications to mobile robotics. In: Jing, X.-J. (ed.) Mobile Robots Motion Planning. New Challenges, pp. 43–55. I-Tech Education and Publishing KG, Vienna (2008)
35. Polkowski, L., Ośmiałowski, P.: A framework for multiagent mobile robotics: Spatial reasoning based on rough mereology in Player/stage system. In: Chan, C.-C., Grzymala-Busse, J.W., Ziarko, W.P. (eds.) RSCTC 2008. LNCS (LNAI), vol. 5306, pp. 142–149. Springer, Heidelberg (2008)
36. Polkowski, L., Skowron, A.: Rough mereology: a new paradigm for approximate reasoning. International Journal of Approximate Reasoning 15(4), 333–365 (1997)
37. Polkowski, L., Skowron, A.: Rough mereology in information systems with applications to qualitative spatial reasoning. Fundamenta Informaticae 43, 291–320 (2000)
38. Ramsey, P.: PostGIS Manual, In: Postgis. pdf file downloaded from Refractions Research home page
39. Shao, J., Xie, G., Yu, J., Wang, L.: Leader–following formation control of multiple mobile robots. In: Proceedings of the 2005 IEEE Intern. Symposium on Intelligent Control, Limassol, Cyprus, pp. 808–813 (2005)
40. Skowron, A., Rauszer, C.: The discernibility matrices and functions in decision systems. In: Słowiński, R. (ed.) Intelligent Decision Support. Handbook of Applications and Advances of the Rough Sets Theory, pp. 311–362. Kluwer, Dordrecht (1992)
41. sfsexp, `http://sexpr.sourceforge.net`
42. Sugihara, K., Suzuki, I.: Distributed motion coordination of multiple mobile robots. In: Proceedings 5th IEEE Intern. Symposium on Intelligent Control, Philadelphia PA, pp. 138–143 (1990)
43. Švestka, P., Overmars, M.H.: Coordinated path planning for multiple robots. Robotics and Autonomous Systems 23, 125–152 (1998)
44. Tarski, A.: Les fondements de la géométrie des corps. In: Supplement to Annales de la Sociéte Polonaise de Mathématique, Cracow, pp. 29–33 (1929)
45. Tarski, A.: What is elementary geometry? In: The Axiomatic Method with Special Reference to Geometry and Physics, pp. 16–29. North-Holland, Amsterdam (1959)

46. Tribelhorn, B., Dodds, Z.: Evaluating the Roomba: A low-cost, ubiquitous platform for robotics research and education. In: 2007 IEEE International Conference on Robotics and Automation, ICRA 2007, Roma, Italy, April 10-14, pp. 1393–1399 (2007)
47. Thrun, S., Burgard, W., Fox, D.: Probabilistic Robotics. MIT Press, Cambridge (2005)
48. Urdiales, C., Perez, E.J., Vasquez-Salceda, J., Sànchez-Marrè, M., Sandoval, F.: A purely reactive navigation scheme for dynamic environments using Case–Based Reasoning. Autonomous Robots 21, 65–78 (2006)
49. Whitehead, A.N.: Process and Reality. In: An Essay in Cosmology. Macmillan, New York (1929); (corr. ed.: Griffin, D. R., Sherbourne, D. W. (eds.) (1978))

# Perceptually Near Pawlak Partitions*

Sheela Ramanna

University of Winnipeg,
Dept. Applied Computer Science,
515 Portage Ave., Winnipeg, Manitoba, R3B 2E9, Canada
s.ramanna@uwinnipeg.ca

**Abstract.** The problem considered in this paper is how to compare perceptually indiscernible partitions of disjoint, non-empty sets such as pairs of digital images viewed as sets of points. Such partitions are called perceptual Pawlak partitions, named after Z. Pawlak, who introduced a attribute-based equivalence relation in 1981 (the well-known indiscernibility relation from rough set theory). The solution to the problem stems from an approach to pairwise comparison reminiscent of the G. Fechner's 1860 approach to comparing perceptions in psychophysics experiments. For Fechner, one perception of an object is indistinguishable from the perception of a different object, if there is no perceptible difference in the particular sensed feature value of the objects, *e.g.*, perceptions resulting from lifting small objects where the object feature is weight. In comparing visual perceptions, partitions of images determined by a particular form of indiscernibility relation $\sim_B$ are used. The $L_1$ (Manhattan distance) norm form of what is known as a perceptual indiscernibility relation defined within the context of a perceptual system is used in this article to define what are known as perceptually indiscernible Pawlak partitions (PIPs). An application of PIPs and near sets is given in this article in terms of a new form of content-based image retrieval (CBIR). This article investigates the efficacy of perceptual CBIR using Hausdorff and Mahalanobis distance measures to determine the degree of correspondence between pairs of perceptual Pawlak partitions of digital images. The contribution of this article is the introduction of an approach to comparing perceptually indiscernible image partitions.

**Keywords:** Digital image, Hausdorff measure, image retrieval, indiscernibility relation, Mahalanobis measure, near sets, perception, perceptually near, representative space.

* Many thanks to James F. Peters, Amir H. Meghdadi, Som Naimpally, Christopher Henry and Piotr Wasilewski for the suggestions and insights concerning topics in this paper. We especially want to thank Amir H. Meghdadi for the use of his implementation of near set theory in a comprehensive image comparison toolset. This research has been supported by the Natural Science & Engineering Research Council of Canada (NSERC) grant 185986.

J.F. Peters et al. (Eds.): Transactions on Rough Sets XII, LNCS 6190, pp. 170–191, 2010.

# 1 Introduction

This article considers a solution to the problem of how to compare perceptually indiscernible Pawlak partitions of digital images. Of interest here is a solution to a particular form of what is known as the image correspondence problem, *i.e.*, how to establish a correspondence between regions of pairs of images using image comparison strategies. The particular form of partitions of sets considered here are named after Z. Pawlak because the partitions are defined by an equivalence relation inspired by the original indiscernibility relation introduced by Z. Pawlak in 1981 [1] and elaborated in [2,3,4,5]. The indiscernibility relation is a basic building block in defining rough sets [2].

The proposed solution to the image correspondence problem stems from an approach to pairwise comparison of images that is similar to G. Fechner's 1860 approach to comparing perceptions in psychophysics experiments [6]. For Fechner, a pair of perceptions are indistinguishable if there is no perceptible difference in a particular feature of the perceived objects, *e.g.*, perception resulting from lifting small objects where the feature is weight. In solving the image correspondence problem, each image is viewed as set of points. Partitions of images are defined by a perceptual indiscernibility relation introduced by J.F. Peters and S. Ramanna in 2008 and published in 2009 [7] and elaborated in [8,9,10], where the descriptions of regions of images are compared. Let $x, y$ denote a pair of subimages in an image $X$ and let $\mathcal{B}$ denote a set of real-valued functions that represent subimage features. In the simplest form of a perceptual indiscernibility relation $\sim_{\mathcal{B}}$, put $\sim_{\mathcal{B}} = \{(x, y) \in X \times X \mid \forall \phi \in \mathcal{B}, \phi(x) = \phi(y)\}$. In comparing a pair of images $X, Y$, a partition of each image is defined by $\sim_{\mathcal{B}}$. Then pairs of digital images are near sets to the extent that partitions of the images resemble each other, *i.e.*, image resemblance is present in the case where pairs of classes $A \subset X, B \subset Y$ contain subimages $x \in A, y \in B$ where $x \sim_{\mathcal{B}} y$. In this paper, the degree of image resemblance is determined by specialized forms of the Hausdorff [11,12,13] or Mahalanobis [14,15] distance measures.

Rough set-based approach to image analysis dates back to the early 1990s. The application of rough sets in image analysis was launched in a seminal paper published by A. Mrózek and L. Plonka [16]. The early work on the use of rough sets in image analysis can be found in [17,18,19,20,21,22,23]. A review of rough sets and near sets in medical imaging can be found in [24]. More recently, D. Sen and S.K. Pal [25] introduce an entropy based, average image ambiguity measure to quantify greyness and spatial ambiguities in images. This measure has been used for standard image processing tasks such as enhancement, segmentation and edge detection. Forms of rough entropy have also been used in a clustering approach to image segmentation [26,10]. The OBIR system [27] is an example of a rough set based object retrieval system for specific single objects in digital images. OBIR includes an object similarity ratio which is a ratio of number of pixels which have the same value in the lower approximation array of the two images(query image and database image) to a size of lower approximation array. This similarity measure is region-based and uses only two regions, namely, object and the background of grey scale images. Papers related to the foundations and applications of fuzzy sets, rough sets and near sets approaches in image analysis can be found in S.K. Pal and J.F. Peters [10].

This article includes an application of the proposed approach to solving the image correspondence problem in terms of a new form of content-based image retrieval (CBIR). In a CBIR system, image retrieval from large image databases is based on some similarity measure of the actual contents of images rather than metadata such as captions or keywords. Image content can include colors, shapes, textures, or any other information that can be derived from an image. Most CBIR systems have one thing in common: images are represented by numeric values that signify properties of the images such as features or descriptors that facilitates meaningful retrieval [28]. In general, there are two approaches (i) the discrete approach is inspired by textual information retrieval and uses text retrieval metrics. This approach requires all features to be mapped to binary features; the presence of a certain image feature is treated like the presence of a word in a text document, (ii) the continuous approach is similar to nearest neighbor classification. Each image is represented by a feature vector and features are compared and subsequently ranked using various distance measures. Image patches (*i.e.*, sub-images of images) or features derived from image patches offer very promising approaches to CBIR [28]. Different types of local features can then be extracted for each subimage and used in the retrieval process. An extensive survey of CBIR systems can be found in [29].

Although features such as color, shape, texture, or combination of them are considered to be helpful in human judgment of image similarity, human perception is not fully considered at the algorithmic stage of feature extraction. In [30], a cognitive discriminative biplot, is used to capture perceptual similarity. This method is structural where point diagrams capture the objects, their relationships and data variations. The evaluation of the biplot is conducted using a standard nearest neighbour algorithm. Other systems that include some form of user feedback are relevance feedback with Bayesian methods [31] and active learning procedures with support vector machine [32]. Our approach to perceptual CBIR can be viewed as an *image patch* approach where local feature values are first extracted from subimages contained in perceptually indiscernible Pawlak partitions.

This paper has the following organization. The basic notation used in this article as well as an introduction to object description, perceptual systems and perceptual Pawlak partitions (PIPs) are introduced in Sect. 2. Then, in Sect. 3, a brief introduction to near sets is given. Probe functions for image features investigated in this article are given in Sect. 4. The Hausdorff and Mahalanobis distance measures are briefly explained in Sect. 5. Then an extensive set of experiments to illustrate the new approach to CBIR are given in Sect. 6.

## 2 Basic Notions

This section briefly presents the basic notions underlying perceptually near Pawlak partitions used in this work.

**Table 1.** Notation

| Symbol | Interpretation |
|---|---|
| $O, X, Y$ | Set of perceptual objects, $X, Y \subseteq O, x \in X, y \in Y$, |
| $\mathbb{F}, \mathcal{B}$ | Sets of probe functions, $\mathcal{B} \subseteq \mathbb{F}, \phi_i \in \mathcal{B}$, |
| $\phi_i(x)$ | $\phi_i : X \rightarrow \Re, i^{th}$ probe function representing feature of $x$, |
| $\boldsymbol{\phi}_{\mathcal{B}}(x)$ | $(\phi_1(x), \phi_2(x), \ldots, \phi_i(x), \ldots, \phi_k(x))$,description of $x$ of length $k$, |
| $\parallel \cdot \parallel_1$ | $= \sum_{i=1}^{k} \mid \cdot \mid, L_1$ norm, |
| $\sim_{\mathcal{B}}$ | $\{(x, y) \in O \times O : \parallel \boldsymbol{\phi}_{\mathcal{B}}(x) - \boldsymbol{\phi}_{\mathcal{B}}(y) \parallel_1 = 0\}$, perceptual indiscernibility, |
| $x_{/\sim_{\mathcal{B}}}$ | $= \{y \mid y \in X, x \sim_{\mathcal{B}} y\}$, class containing $x$, |
| $X_{/\sim_{\mathcal{B}}}$ | $= \bigcup_{x \in X} x_{/\sim_{\mathcal{B}}}$, partition of $X$ defined by $\sim_{\mathcal{B}}$, |
| $X \bowtie_{\mathcal{B}} Y$ | $X$ resembles (is near) $Y \iff \exists x \in X, y \in Y, x \sim_{\mathcal{B}} y$, |
| $\langle O, \bowtie_{\mathcal{B}} \rangle$ | perceptual representative space [9]. |

## 2.1 Description and Perceptual Systems

An object description in rough set theory is defined by vectors of attribute values in an information table. However, in near set theory, description of an object $x$ is defined by means of a vector of real-valued function values $\phi(x)$, named *probe functions*, that gives some measurable (or perceivable features) of an object in the physical world.

$$\phi_{\mathcal{B}}(x) = (\phi_1(x), \phi_2(x), ..., \phi_i(x), ..., \phi_l(x)), \tag{1}$$

where $\phi_i : X \rightarrow \Re$ (reals). The object description representation in (1), was first introduced in [33] based on the intuition that object descriptions are analogous to recorded measurements from sensors and hence the name probe function. Near set theory is defined in the context of a perceptual system [34]. A perceptual system is defined as a set of perceptual objects $O$ along with a set of probe functions $\mathbb{F} = \{\phi_1, \phi_2, ..., \phi_L\}$ and is denoted by $\langle O, \mathbb{F} \rangle$. Nearness relation is then defined between sets of perceptual objects from the perceptual system relative to the probe functions defined in the perceptual system. Sets $X, Y \subseteq O$ are *weakly near* to each other if, and only if there are $x \in X$ and $y \in Y$ and there is $\mathcal{B} \subseteq \mathbb{F}$ such that $x$ and $y$ are indiscernible to each other (*i.e*, $\forall \phi_i \in \mathcal{B} \quad \phi_i(x) = \phi_i(y)$).

**Definition 1. Perceptual System**
A perceptual system $\langle O, \mathbb{F} \rangle$ consists of a sample space $O$ containing a finite, non-empty set of sensed sample objects and a countable, non-empty set $\mathbb{F}$ containing probe functions representing object features.

The perception of physical objects and their description within a perceptual system facilitates pattern recognition and the discovery of sets of similar objects.

## 2.2 Perceptual Pawlak Partitions

It is possible to extend the original idea of an indiscernibility relation between objects [1,3,35] to what is known as perceptual indiscernibility (introduced in [8]) that befits the perception of objects in the physical world, especially in science and engineering applications where there is an interest in describing, classifying and measuring device signals and various forms of digital images.

**Definition 2. Perceptual Indiscernibility Relation** [8]
Let $\langle O, \mathbb{F}\rangle$ be a perceptual system. Let $\mathcal{B} \subseteq \mathbb{F}$ and let $\boldsymbol{\phi}_{\mathcal{B}}(x)$ denote a description of an object $x \in O$ of the form $(\phi_1(x), \phi_2(x), \ldots, \phi_i(x), \ldots, \phi_k(x))$. A perceptual indiscernibility relation $\sim_{\mathcal{B}}$ is defined relative to $\mathcal{B}$, *i.e.*,

$$\sim_{\mathcal{B}} = \{(x, y) \in O \times O \mid \| \boldsymbol{\phi}_{\mathcal{B}}(x) - \boldsymbol{\phi}_{\mathcal{B}}(y) \|_1 = 0\},$$

where $\| \cdot \|_1 = \sum_{i=1}^{k} | \cdot | (L_1 \text{ norm})$.

In this section, classes are defined in the context of perceptual partitions of digital images.

**Definition 3. Perceptual Pawlak Partitions**
Let $\langle O, \mathbb{F}\rangle$ be a perceptual system. Let $X \subseteq O, \mathcal{B} \subseteq \mathbb{F}$. A perceptual Pawlak partition (PIP) of $X$ defined by $\sim_{\mathcal{B}}$ is denoted by $X_{/\sim_{\mathcal{B}}}$, *i.e.*,

$$X_{/\sim_{\mathcal{B}}} = \bigcup_{x \in X} x_{/\sim_{\mathcal{B}}}.$$

A perceptual Pawlak partition is a cover of $X$ [36], *i.e.*, a separation of the elements of $X$ into non-overlapping classes where $x_{/\sim_{\mathcal{B}}}$ denotes an equivalence class containing $x$.

**Example 1 Sample Perceptual Pawlak Partition**
We now illustrate the notion of perceptual partitions in terms of two classical images of Lena and Barbara. The images are shown in Figures 1.1 and 1.4 and their respective partitions based on a single feature are shown in Figures 1.2 and 1.5. The different coloured regions in Figures 1.2 and 1.5 represent different equivalence classes based on $\mathcal{B} = \{\overline{gs}\}$ which represents an average greyscale feature of the pixel and where $\phi_{\{\overline{gs}\}}(x) = \phi_{\{\overline{gs}\}}(y)$ in each region. A single equivalence class in Figures 1.3 and 1.6 is represented with ▫ boxes that identify subimages scattered throughout the partitions.

**Example 2 Sample Perceptual Edge Classes in Pawlak Partition**
Fig. 2 presents two examples of perceptual Pawlak partitions. The ▫ boxes (subimages) in Fig. 2.3 and Fig. 2.6 represent edge orientation classes. Each class contains subimages with the matching edge orientations.

The different coloured regions in Figures 2.2 and 2.5 represent different equivalence classes based on $\mathcal{B} = \{e_o\}$. It is worth noting that the size (and the nature) of the equivalence classes are much smaller for the $e_o$ feature when compared with $\overline{gs}$ feature.

1.1: Lena

1.2: Partition based on Grey Scale

1.3: Equivalence Class in Fig. 1.2

1.4: Barbara

1.5: Partition based on Grey Scale

1.6: Equivalence Class in Fig. 1.5

**Fig. 1.** Sample Perceptual Partitions and Classes based on Grey Scale Feature

2.1: Lena

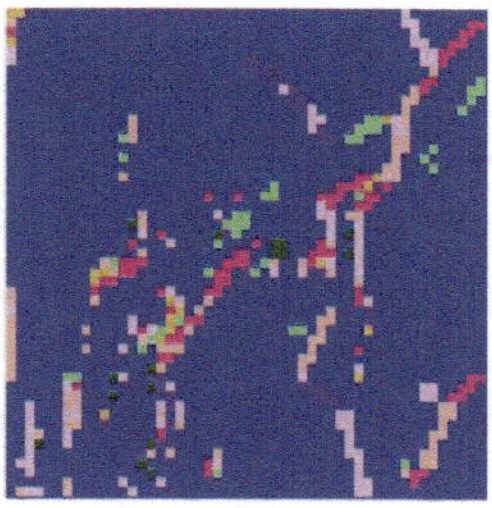

2.2: Partitions based on Edge Orientation Feature

2.3: Exemplary Equivalence Class

2.4: Barbara

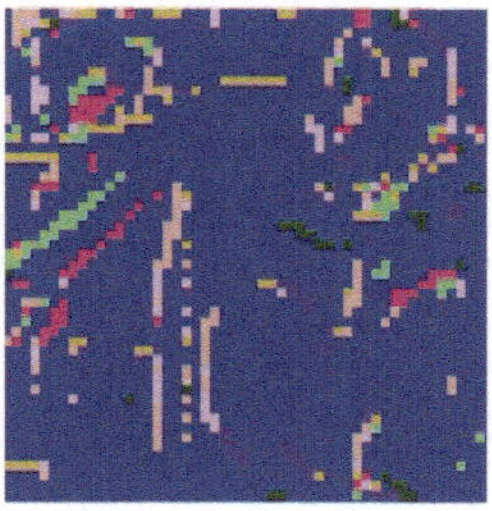

2.5: Partitions based on Edge Orientation Feature

2.6: Exemplary Equivalence Class

**Fig. 2.** Sample Perceptual Partitions and Classes Based on Edge Orientation

# 3 Near Sets and Perceptually Near Pawlak Partitions

The basic idea in the near set approach to object recognition is to compare object descriptions. Sets of objects $X, Y$ are considered near each other if the sets contain objects with at least partial matching descriptions.
–Near sets. General theory about nearness of objects,
–J.F. Peters, 2007.

## 3.1 Near Sets

Near sets are disjoint sets that resemble each other [37]. Resemblance between disjoint sets occurs whenever there are observable similarities between the objects in the sets. Similarity is determined by comparing lists of object feature values. Each list of feature values defines an object's description. Comparison of object descriptions provides a basis for determining the extent that disjoint sets resemble each other. Objects that are perceived as similar based on their descriptions are grouped together. These groups of similar objects can provide information and reveal patterns about objects of interest in the disjoint sets.

Near set theory provides methods that can be used to extract resemblance information from objects contained in disjoint sets, i.e., it provides a formal basis for the observation, comparison, and classification of objects. The discovery of near sets begins with choosing the appropriate method to describe observed objects. This is accomplished by the selection of probe functions representing observable object features. A basic model for a probe function was introduced by M. Pavel [38] in the text of image registration and image classification. In near set theory, a probe function is a mapping from an object to a real number representing an observable feature value [39]. For example, when comparing objects in digital images, the texture feature (observed object) can be described by a probe function representing contrast, and the output of the probe function is a number representing the degree of contrast between a pixel and its neighbour.

Probe functions provide a basis for describing and discerning affinities between objects as well as between groups of similar objects [7]. Objects that have, in some degree, affinities are considered near each other. Similarly, groups of objects (i.e. sets) that have, in some degree, affinities are also considered near each other.

**Definition 4. Near Sets**
Let $\langle O, \mathbb{F} \rangle$ be a perceptual system. Let $X, Y \subset O$ be disjoint sets in $O$ and $\mathcal{B} \subseteq \mathbb{F}$. $X \bowtie_{\mathcal{B}} Y$ ($X$ and $Y$ are near each other) if, and only if there are $x \in X, y \in Y$, where $x \sim_{\mathcal{B}} y$, *i.e.*, $x$ and $y$ have matching descriptions.

This leads to what are known as perceptual representative spaces introduced in [9].

**Definition 5. Perceptual Representative Space** [9]
A *perceptual representative space* is denoted by $\langle O, \sim_{\mathcal{B}} \rangle$ where $O$ is a non-empty set of *perceptual objects*, $\mathcal{B}$ a countable set of *probe functions*, and $\sim_{\mathcal{B}}$ is a perceptual indiscernibility relation (see Table 2).

Perceptual representative spaces have close affinity with proximity spaces [40,41,42,43,44,45], especially in terms of recent work on nearness in digital images [46]. Originally, the notion of nearness between sets (proximity relation between sets) was derived from a spatial meaning of distance by V. Efremovič [44,45]. Later, the notion of proximity between sets became more general and closer to the notion of nearness underlying near sets, *i.e.*, proximity not limited to a spatial interpretation was introduced by S.A. Naimpally in 1970 [40] (see, *e.g.*, [41,42]). This later form of proximity relation permits either a quantitative (spatial) or qualitative (non-spatial) interpretation.

Perceptual representative spaces are patterned after the original notion of a representative space introduced by J.H. Poincaré during the 1890s [47]. Poincaré's form of representative space was used to specify a physical continuum associated with a source of sensation, *e.g.*, visual, tactile, motor. A perceptual representative space $\mathcal{P} = \langle O, \sim_{\mathcal{B}} \rangle$ is a generalization of Poincaré's representative space inasmuch as $\mathcal{P}$ represents sets of similar perceptual objects in a covering of $O$ determined by a relation such as $\sim_{\mathcal{B}}$.

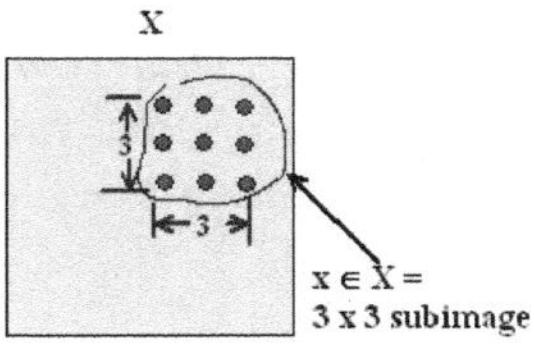


**Fig. 3.** Sample $3 \times 3$ subimage

For example, a digital image $X$ can be viewed as a set of points represented by image pixels (picture elements). In what follows, an element $x \in X$ is understood to be a $p \times p$ subimage, where $p$ denotes the number of pixels on a subimage edge. For example, a $3 \times 3$ subimage is shown in Fig. 3. Pairs of digital images containing pixels with matching descriptions are near sets. Now consider the partition pairs of disjoint sets.

**Proposition 1.** *Let $\langle O, \sim_{\mathcal{B}} \rangle$ denote a perceptual representative space and let $X, Y \subset O$ denote disjoint sets in $O$ and $\mathcal{B}$ a set of probe functions representing object features. Then $X \underline{\bowtie}_{\mathcal{B}} Y$ if, and only if $X_{/\sim_{\mathcal{B}}} \underline{\bowtie}_{\mathcal{B}} Y_{/\sim_{\mathcal{B}}}$.*

*Proof.* Let $\langle O, \sim_{\mathcal{B}} \rangle$ denote a perceptual representative space, where $O$ is a non-empty set of perceptual objects, $\mathcal{B}$ a set of probe functions representing object features. And let $X, Y \subset O$ denote a pair of disjoint sets. A perceptual indiscernibility relation $\sim_{\mathcal{B}}$ determines, for example, a partition of $X$, *i.e.* separation of elements of $X$ into disjoint subsets that cover $X$. Consider quotient sets $X_{/\sim_{\mathcal{B}}}, Y_{/\sim_{\mathcal{B}}}$ determined by $\sim_{\mathcal{B}}$ (See Table 1).

$\Rightarrow$ Assume $X \bowtie_{\mathcal{B}} Y$. Then, from Def. 4, there are $x \in X, y \in Y$ such that $x \sim_{\mathcal{B}} y$. Hence, $x \in x_{/\sim_{\mathcal{B}}} \bowtie_{\mathcal{B}} y \in y_{/\sim_{\mathcal{B}}}$, and $x_{/\sim_{\mathcal{B}}} \in X_{/\sim_{\mathcal{B}}}, y_{/\sim_{\mathcal{B}}} \in Y_{/\sim_{\mathcal{B}}}$. Then, from Def. 4, $X_{/\sim_{\mathcal{B}}} \bowtie_{\mathcal{B}} Y_{/\sim_{\mathcal{B}}}$.
$\Leftarrow$ Assume $X_{/\sim_{\mathcal{B}}} \bowtie_{\mathcal{B}} Y_{/\sim_{\mathcal{B}}}$. The proof that therefore $X \bowtie_{\mathcal{B}} Y$ again follows from Def. 4 and the approach in the proof is symmetric with the proof of $\Rightarrow$. $\square$

**Corollary 1.** *Let $\langle O, \sim_{\mathcal{B}} \rangle$ denote a perceptual representative space, where $O$ is an arbitrary, non-empty set of perceptual objects, $\mathcal{B}$ a set of probe functions representing features of elements of $O$, and $\sim_{\mathcal{B}}$ is a perceptual indiscernibility relation. Also, let $X, Y \subset O$ denote disjoint sets in $O$. Then $X \bowtie_{\mathcal{B}} Y$ if, and only if there are $x_{/\sim_{\mathcal{B}}} \in X_{/\sim_{\mathcal{B}}}$ and $y_{/\sim_{\mathcal{B}}} \in Y_{/\sim_{\mathcal{B}}}$ such that $x_{/\sim_{\mathcal{B}}} \bowtie_{\mathcal{B}} y_{/\sim_{\mathcal{B}}}$.*

It is now possible to specialize Cor. 1 by considering $O$ to be a non-empty set of sample digital images (sets of points). The partition of a digital image $X$ (set of points) entails the selection of a subimage size, usually containing $p \times p$ pixels. In picture terms, $p$ is the number of points on a subimage edge (see, *e.g.*, $3 \times 3$ subimage in Fig. 3). Let $x \in X$ denote a subimage in $X$ and let $x_{/\sim_{\mathcal{B}}}$ denote a class containing $x$, where every other subimage in $x_{/\sim_{\mathcal{B}}}$ has description similar to the description of $x$. In other words, a class in a partition of $X$ consists of one more more subimages with similar descriptions. In image processing terms, such a partition determines an image segmentation consisting of non-overlapping regions that are identified with classes.

**Corollary 2.** *Let $\langle O, \sim_{\mathcal{B}} \rangle$ denote a perceptual representative space, where $O$ is a non-empty set of digital images, $\mathcal{B}$ a set of probe functions representing subimage features, and $\sim_{\mathcal{B}}$ is a perceptual indiscernibility relation. Also, let $X, Y \subset O$ denote digital images in $O$. Then $X \bowtie_{\mathcal{B}} Y$ if, and only if $X_{/\sim_{\mathcal{B}}} \bowtie_{\mathcal{B}} Y_{/\sim_{\mathcal{B}}}$.*

**Example 3 Near Digital Images**
Let $\langle O, \sim_{\mathcal{B}} \rangle$ denote a perceptual representative space, where $O$ is a non-empty set of digital images, $\mathcal{B}$ a set of probe functions representing subimage features, and $\sim_{\mathcal{B}}$ is a perceptual indiscernibility relation. Consider the images $Im1, Im2 \subset O$ in Fig. 4.1 and Fig. 4.2, respectively. Let $\overline{gs}$ denote a function that returns the average greyscale value for the pixels in a subimage and assume $\mathcal{B} = \{\overline{gs}\}$. Let

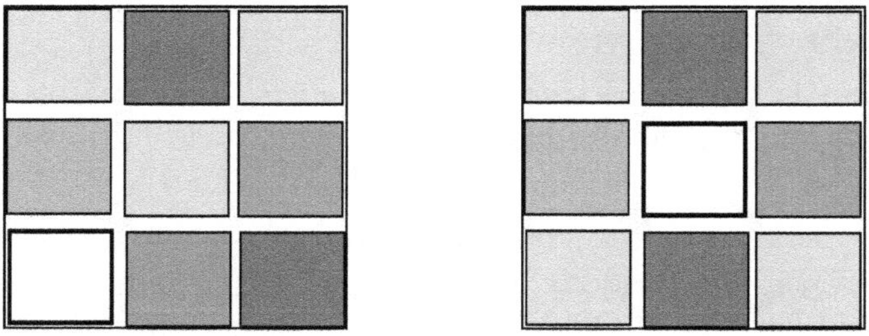

4.1: Image im1 partition    4.2: Image im2 partition

**Fig. 4.** Sample Perceptually Near Image Partitions

$x_{/\sim_{\{\overline{gs}\}}} \in X_{/\sim_{\{\overline{gs}\}}}, y_{/\sim_{\{\overline{gs}\}}} \in Y_{/\sim_{\{\overline{gs}\}}}$ denote classes represented by a grey-shaded box ▢ in the partition of the images in Fig. 4.1 and Fig. 4.2. In effect, the classes with grey-shaded boxes ▢ represent collections of subimages that have matching descriptions. However, unlike the partition in Fig. 1, the individual subimages in Fig. 4 are not visible.

Since these images contain classes (represented by shaded boxes) with matching shades of grey determined by $\sim_{\mathcal{B}}$, such classes are examples of near sets. That is, since there are classes in the segmentations in Fig 4 that are near sets, we know from Def. 4 that $Im1_{/\sim_{\{\overline{gs}\}}} \underline{\bowtie}_{\mathcal{B}} Im2_{/\sim_{\{\overline{gs}\}}}$. Then, from Cor. 2, $Im1 \underline{\bowtie}_{\mathcal{B}} Im2$.

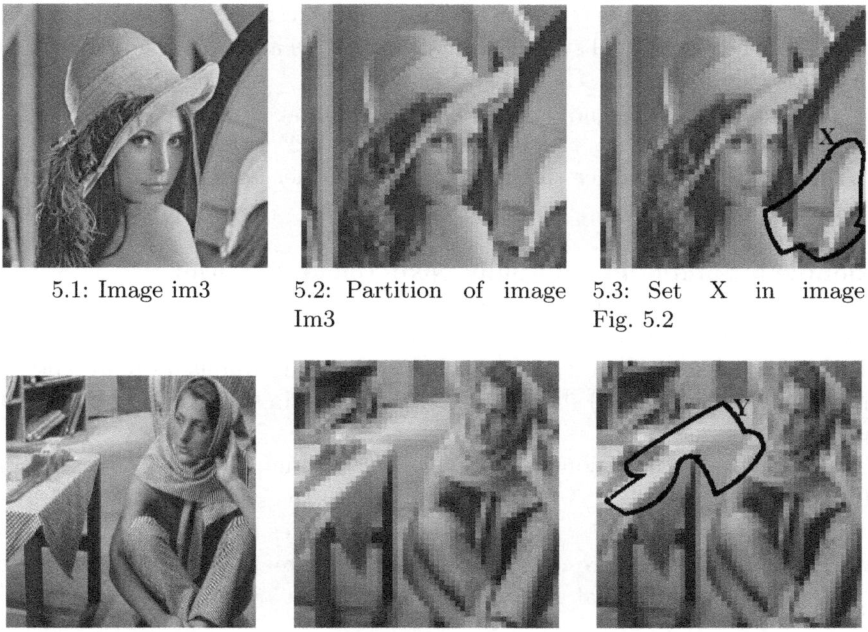


5.1: Image im3

5.2: Partition of image Im3

5.3: Set X in image Fig. 5.2

5.4: Image im4

5.5: Partition of image Im4

5.6: Set Y in image Fig. 5.5

**Fig. 5.** Sample Perceptually Near Partitions

### 3.2 Perceptually Near Pawlak Partitions

The partition of an image defined by $\sim_{\mathcal{B}}$ results in the separation of the parts of the image into equivalence classes, *i.e.*, results in an image segmentation. A $\sim_{\mathcal{B}}$-based segmentation is called a *perceptually indiscernible partition* (PIP). Notice that, by definition, a class in the partition of an image is set of subimages with matching descriptions.

In the partition in Fig. 5.2, for example, a single class is represented by the ■ (dark grey) shaded boxes indicating subimages scattered throughout the partition (the individual pixels in a subimage have been assigned the average grey level intensity for the subimage). Depending on the features chosen, PIPs will be more or less perceptually near each other. The notion of perceptually near partitions was introduced in [48,49] and elaborated in [8,34,37,50,51].

**Definition 6. Perceptually Near Pawlak Partitions**
Let $\langle O, \mathbb{F} \rangle$ be a perceptual system, where $O$ is a non-empty set of picture elements (points) and $X, Y \subset O$, *i.e.*, $X, Y$ are digital images, and $\mathcal{B} \subseteq \mathbb{F}$. $X_{/\sim_{\mathcal{B}}} \underline{\bowtie}_{\mathcal{B}} Y_{/\sim_{\mathcal{B}}}$ if, and only if there are $x_{/\sim_{\mathcal{B}}} \in X_{/\sim_{\mathcal{B}}}, y_{/\sim_{\mathcal{B}}} \in Y_{/\sim_{\mathcal{B}}}$ such that $x_{/\sim_{\mathcal{B}}} \underline{\bowtie}_{\mathcal{B}}\, y_{/\sim_{\mathcal{B}}}$.

**Example 4 Subimages That Resemble Each Other**
Let $\langle O, \mathbb{F} \rangle$ be a perceptual system. Assume that $O$ is a set of images (each image is a set of points). Let $I1, I2 \subset O$ and let $I1_{/\sim_{\mathcal{B}}}, I2_{/\sim_{\mathcal{B}}}$ be partitions defined by $\sim_{\mathcal{B}}$. $I1_{/\sim_{\mathcal{B}}} \underline{\bowtie}_{\mathcal{B}} I2_{/\sim_{\mathcal{B}}}$ if, and only if there are subimages $X \subseteq I1_{/\sim_{\mathcal{B}}}, Y \subseteq I2_{/\sim_{\mathcal{B}}}$, where $X \bowtie_{\mathcal{B}} Y$. That is, there are subimages $X, Y$ that resemble each other ($X, Y$ are near sets). Notice that $I1$ and $I2$ can either be the same image or two different images.

**Example 5 Sample Perceptually Near Image Partitions**
Consider Images $Im3, Im4$ in Fig. 5.1, 5.4, respectively. Each shaded box in Figures 5.2, 5.5 have a uniform greylevel representing the greylevel of the pixels inside the shaded area. That is, since both $X, Y$ contain subimages represented by, for example, ■ (light grey) shaded areas ($X, Y$ contain subimages with matching descriptions, namely, ■).

From Def. 6, the partitions $Im3_{/\sim_{\mathcal{B}}}, Im4_{/\sim_{\mathcal{B}}}$ in Figures 5.2, 5.5 are perceptually near each other ($Im3_{/\sim_{\mathcal{B}}} \underline{\bowtie}_{\mathcal{B}} Im4_{/\sim_{\mathcal{B}}}$), since the subimages $X \subset Im1_{/\sim_{\mathcal{B}}}$ in Fig. 5.3 and $Y \subset Im2_{/\sim_{\mathcal{B}}}$ in Fig. 5.6 are examples of perceptually near sets, *i.e.* from Def. 4 and the presence of subimages with matching grey levels in both $X$ and $Y$, we conclude that $X \bowtie_{\mathcal{B}} Y$.

# 4 Probe Functions for Image Features

This section briefly introduces probe functions for image features used in the CIBR experiments reported in this article.

## 4.1 Edge Features

Edge features include edge orientation ($e_o$) and edge intensity ($e_i$). Edge features are extracted using a wavelet-based multiscale edge detection method from [52]. This method involves calculation of the gradient of a smoothed image using wavelets, and defines edge pixels as those that have locally maximal gradient magnitudes in the direction of the gradient.

### 4.2 Region Features

Region features include both texture and colour from [53]. Hue and Saturation colour characteristics taken together are called Chromaticity. The texture features carry information about the relative position of pixels in an image rather than just the histogram of the intensities. Let G be the co-occurrence matrix whose element $g_{ij}$ is the number of times that pixel pairs with intensities $z_i$ and $z_j$ occur in image $f$ specified by operator $Q$ with $L$ possible intensity levels where $1 \leq i,\ j \leq L$. $Q$ defines position of two pixels relative to each other. $p_{ij}$ is the probability that a pair of points satisfying Q with intensities $z_i$ and $z_j$ defined as follows:

$$p_{ij} = \frac{g_{ij}}{n}, \text{where n is the total number of pixel pairs}$$

$m_r$ and $m_c$ are means computed along rows and columns respectively and $\sigma_r$ and $\sigma_c$ are standard deviations computed along rows and columns respectively of the normalized $G$. The following probe function for texture features are considered in this paper:

$$\phi_C = \sum_{i=1}^{K}\sum_{j=1}^{K} \frac{(i-m_r)(j-m_c)p_{ij}}{\sigma_r\dot{\sigma}_c}, \text{Correlation}, \tag{2}$$

$$\phi_{Ct} = \sum_{i=1}^{K}\sum_{j=1}^{K} (i-j)^2\dot{p}_{ij}, \text{Contrast}, \tag{3}$$

$$\phi_E = \sum_{i=1}^{K}\sum_{j=1}^{K} p_{ij}^2, \text{Energy}, \tag{4}$$

$$\phi_H = \sum_{i=1}^{K}\sum_{j=1}^{K} \frac{p_{ij}}{1+|i-j|}, \text{Homogeneity}, \tag{5}$$

$$\phi_h = \sum_{i=1}^{K}\sum_{j=1}^{K} \dot{p}_{ij}\dot{\log}_2\dot{p}_{ij}, \text{Entropy}. \tag{6}$$

## 5 Nearness Measures

This section briefly presents the two nearness measures used to quantify the degree of resemblance between perceptually indiscernible image partitions. It should also be observed that $O$ in a perceptually representative space is viewed as a metric space with a metric defined in obvious ways by means of, for example, the $L_1$ norm from Def. 2.

### 5.1 Hausdorff Nearness Measure ($H_dNM$)

Hausdorff distance [11,12,13,36] is defined between two finite point sets in a metric space. Assume $d(x, y)$ is a distance defined between points $x$ and $y$ in a metric space. Let $X$ and $Y$ be sets of points in the space. The Hausdorff distance $\rho_H(X, Y)$ between sets $X$ and $Y$ [36] is defined in (7). Then to such a space, we apply the Hausdorff or Mahalanobis distance between sets in measuring the resemblance between sets.

$$\rho_H(X, Y) = \max\{d_H(X, Y), d_H(Y, X)\}, \tag{7}$$

where

$$d_H(X, Y) = \max_{x \in X}\{\min_{y \in Y}\{d(x, y)\}\}, \tag{8}$$

$$d_H(Y, X) = \max_{y \in Y}\{\min_{x \in X}\{d(x, y)\}\}. \tag{9}$$

$d_H(X, Y)$ and $d_H(Y, X)$ are directed Hausdorff distances from $X$ to $Y$ and from $Y$ to $X$, respectively. In comparing images $X, Y$, a Hausdorff nearness measure $H_dNM(X, Y) = \frac{1}{1+\rho_H(X,Y)}$. The Hausdorff nearness measure $H_dNM$ is typically used to find a part of a test image that matches a given query or template image.

### 5.2 Generalized Mahalanobis Nearness Measure (gMNM)

The generalized Mahalanobis nearness measure (gMNM) was first introduced in [54]. The Mahalanobis distance [15] is a form of distance between two points in the feature space with respect to the variance of the distribution of points. The original Mahalanobis distance is usually defined between two sample multivariate vectors $\boldsymbol{x}$ and $\boldsymbol{y}$ as follows [55].

$$D_M(\boldsymbol{x}, \boldsymbol{y}) = (\boldsymbol{x} - \boldsymbol{y})^T \Sigma^{-1}(\boldsymbol{x} - \boldsymbol{y}), \tag{10}$$

where the vectors are assumed to have a normal multivariate distribution with the covariance matrix $\Sigma$. This formula is usually used to measure the distance $D_M(\boldsymbol{x}, \boldsymbol{m})$ between a vector $\boldsymbol{x}$ and the mean of the distribution $\boldsymbol{m}$. Following the same approach, the Mahalanobis distance can be used to define a distance measure between two separate distributions. Let us assume $\chi_1 = (\Sigma_1, \boldsymbol{m_1})$ and $\chi_2 = (\Sigma_2, \boldsymbol{m_2})$ are two normal multivariate distributions with means $\boldsymbol{m_1}$,$\boldsymbol{m_1}$ and covariance matrices $\Sigma_1, \Sigma_2$. Moreover, assume that $P(\omega_1)$ and $P(\omega_2)$ represent prior probabilities of the given distributions. A generalized Mahalanobis distance between the two distributions is defined as follows [56,55].

$$gMD = \sqrt{(\boldsymbol{m_1} - \boldsymbol{m_2})^T \Sigma_W^{-1}(\boldsymbol{m_1} - \boldsymbol{m_2})}, \tag{11}$$

where $\Sigma_W^{-1}$ refers to the within-class covariance matrix defined as in equation 12

$$\Sigma_W = \sum_{i=1,2}\left(P(\omega_i)\sum_{x \in \chi_i}\frac{(x - m_i)(x - m_i)^T}{n_i}\right). \tag{12}$$

Therefore, a generalized Mahalanobis distance-based nearness measure (gMNM) between two images is defined in this paper as follows. Let $X$ and $Y$ denote sets of perceptual objects (images). Let $\bar{\Phi}_X$ and $\bar{\Phi}_Y$ represent the mean feature vector for all the perceptual objects $x \in X$ and $y \in Y$, respectively. Also, let $\Sigma_X$ and $\Sigma_Y$ be the covariance matrices of the multivariate distributions of $\Phi_X$ and $\Phi_Y$ (feature values), respectively. Then

$$gMD(X,Y) = \sqrt{(\bar{\Phi}_X - \bar{\Phi}_Y)^T \Sigma_{X,Y}^{-1} (\bar{\Phi}_X - \bar{\Phi}_Y)}, \tag{13}$$

$$gMNM(X,Y) = \frac{1}{1 + gMD(X,Y)}, \tag{14}$$

where

$$\Sigma_{X,Y} = \frac{1}{2} \left( \Sigma_X + \Sigma_Y \right). \tag{15}$$

## 6 Sample CBIR Results with Image Nearness Measures

We illustrate the application of the image nearness measures in an image retrieval context with images drawn from the SIMPLIcity database [57] and using the Meghdadi toolset [58]. Table 2 includes pair-wise comparison of the query image numbered *420* with sample test images drawn from different categories such as dinosaurs, buildings and so on (see section 7 for a complete list). The experiments include a number of different image features for various subimage sizes ($p$). The notation for image features used in experiments is as follows:

$$\begin{aligned} \mathcal{T} &= \text{texture features denoted by CCtEH from Sec. 4.2}, \\ h &= \text{entropy}, \\ e_i &= \text{edge intensity}, \\ e_o &= \text{edge orientation}, \\ e_{io} &= \text{edge intensity and edge orientation taken together}, \\ \mathcal{C} &= \text{chromaticity feature with hue and saturation taken together}. \end{aligned}$$

For conciseness (column compression), symbols represented as features are concatenated in Table 2. For example, $\mathcal{T}e_{io}h$ denotes the fact that the texture features (correlation, contrast, energy, homogeneity, entropy) and edge features (edge intensity and edge orientation). The measurements reflect various subimage sizes from pixel matrices ranging in size from $20 \times 20$ to $5 \times 5$. Our choice of an $8 \times 8$ matrix was influenced by the retrieval quality of the nearness measurements and also the computation time. Image retrieval typically involves comparison of a large number of images and the issue of computation time is important. It can also be observed that adding entropy and chromaticity features does not add to the discriminatory power of nearness measures.

**Table 2.** Nearness Measurements

| p | Feature | Images | $gMNM$ | $H_dNM$ | p | Feature | Images | $gMNM$ | $H_dNM$ |
|---|---|---|---|---|---|---|---|---|---|
| **20** | $\mathcal{T}e_o$ | **420,423** | 0.61 | 0.80 | 8 | grey$e_{io}$ | 420,304 | 0.10 | 0.75 |
| 5 | $\mathcal{T}e_o$ | 420,423 | 0.82 | 0.90 | 8 | grey | 420,304 | 0.10 | 0.86 |
| **5** | $\mathcal{T}e_{io}$ | **420,423** | 0.81 | 0.87 | 10 | $\mathcal{T}e_{io}$h | 420,474 | 0.89 | 0.81 |
| 10 | $\mathcal{T}e_{io}$ | 420,474 | 0.97 | 0.83 | 5 | $\mathcal{T}e_{io}$h | **420,423** | 0.75 | 0.86 |
| 5 | $\mathcal{T}$ | 420,474 | 0.99 | 0.96 | 5 | $\mathcal{T}e_{io}$h$\mathcal{C}$ | **420,423** | 0.4 | 0.76 |
| 5 | $\mathcal{T}e_{io}$ | 420,474 | 0.95 | 0.75 | 5 | $\mathcal{T}e_{io}$h$\mathcal{C}$ | 420,460 | 0.90 | 0.75 |
| 5 | $\mathcal{T}e_o$ | 420,432 | 0.90 | 0.85 | 5 | $\mathcal{T}e_{io}$h | 420,460 | 0.90 | 0.75 |
| 5 | $\mathcal{T}e_{io}$ | 420,432 | 0.90 | 0.85 | 5 | $\mathcal{T}e_{io}$ | 420,460 | 0.90 | 0.76 |
| 5 | $\mathcal{T}e_{io}$h | 420,432 | 0.79 | 0.83 | 5 | $\mathcal{T}e_o$ | 420,460 | 0.91 | 0.79 |
| 5 | $\mathcal{T}e_{io}$h$\mathcal{C}$ | 420,432 | 0.60 | 0.715 | 5 | $\mathcal{T}e_{io}$h | 420,456 | 0.24 | 0.74 |
| 5 | $\mathcal{T}e_o$ | 420,436 | 0.85 | 0.87 | 5 | $\mathcal{T}e_{io}$ | 420,456 | 0.24 | 0.74 |
| 5 | $\mathcal{T}e_{io}$ | 420,436 | 0.85 | 0.83 | 5 | $\mathcal{T}e_o$ | 420,456 | 0.24 | 0.74 |
| 5 | $\mathcal{T}e_{io}$h | 420,436 | 0.78 | 0.80 | 5 | $\mathcal{T}e_{io}$h$\mathcal{C}$ | 420,436 | 0.36 | 0.68 |
| 8 | $\mathcal{T}e_{io}$ | 420,478 | 0.77 | 0.83 | 8 | $e_{io}$ | 420,478 | 0.79 | 0.98 |
| 8 | $\mathcal{T}e_{io}$ | 420,484 | 0.94 | 0.86 | 8 | $e_{io}$ | 420,484 | 0.95 | 0.98 |
| 8 | $\mathcal{T}e_{io}$ | 420,424 | 0.73 | 0.81 | 8 | $e_{io}$ | 420,424 | 0.74 | 0.96 |
| 8 | $\mathcal{T}e_{io}$ | 420,473 | 0.94 | 0.82 | 8 | $e_{io}$ | 420,473 | 0.99 | 0.98 |
| 8 | $\mathcal{T}e_{io}$ | 420,490 | 0.87 | 0.79 | 8 | $e_{io}$ | 420,490 | 0.95 | 0.99 |
| 8 | $\mathcal{T}e_{io}$ | 420,514 | 0.128 | 0.72 | 8 | $e_{io}$ | 420,514 | 0.42 | 0.99 |
| 8 | $\mathcal{T}e_{io}$ | 420,703 | 0.07 | 0.76 | 8 | $e_{io}$ | 420,703 | 0.34 | 0.97 |
| 8 | $\mathcal{T}e_{io}$ | 420,600 | 0.32 | 0.72 | 8 | $e_{io}$ | 420,600 | 0.48 | 0.97 |
| 8 | $\mathcal{T}e_{io}$ | 420,200 | 0.29 | 0.77 | 8 | $e_{io}$ | 420,200 | 0.37 | 0.97 |
| 8 | $\mathcal{T}e_{io}$ | 420,304 | 0.34 | 0.74 | 8 | $e_{io}$ | 420,304 | 0.48 | 0.97 |
| 8 | $\mathcal{T}e_{io}$ | 420,499 | 0.93 | 0.85 | 8 | $e_{io}$ | 420,499 | 0.95 | 0.98 |
| 8 | $\mathcal{T}e_{io}$ | 420,408 | 0.61 | 0.78 | 8 | $e_{io}$ | 420,408 | 0.73 | 0.97 |
| 8 | $\mathcal{T}e_{io}$ | 420,410 | 0.71 | 0.75 | 8 | $e_{io}$ | 420,410 | 0.77 | 0.96 |
| **8** | te$_{io}$ | **420,423** | 0.84 | 0.85 | 8 | $e_{io}$ | **420,423** | 0.93 | 0.98 |
| 8 | $\mathcal{T}e_{io}$ | 420,432 | 0.88 | 0.81 | 8 | $e_{io}$ | 420,432 | 0.95 | 0.9 |
| 8 | $\mathcal{T}e_{io}$ | 420,436 | 0.79 | 0.78 | 8 | $e_{io}$ | 420,436 | 0.94 | 0.97 |
| 8 | $\mathcal{T}e_{io}$ | 420,456 | 0.93 | 0.84 | 8 | $e_{io}$ | 420,456 | 0.98 | 0.98 |
| 8 | $\mathcal{T}e_{io}$ | 420,460 | 0.96 | 0.86 | 8 | $e_{io}$ | 420,460 | 0.99 | 0.98 |
| 8 | $\mathcal{T}e_{io}$ | 420,474 | 0.95 | 0.85 | 8 | $e_{io}$ | 420,474 | 0.95 | 0.98 |

### 6.1 Analysis of Hausdorff Nearness Measure: Image Retrieval Experiments

In this section, we discuss the significance of the Hausdorff nearness measure values drawn from Table 2.

Fig. 6 shows 20 images (1 query and 19 test) ordered by their $H_dNM$ values using only the $e_{io}$ edge features. It can be observed that i) the difference between the measure values of the nearest and furthest images is very small (0.99 vs 0.96) and ii) images 6.3, 6.19 and 6.20 are also out of sequence. Fig. 7 shows the same 20 images ordered by their $H_dNM$ values using both edge features and texture features $\mathcal{T}e_{io}$. It is noteworthy that all images belonging to the same category as the query image are now retrieved in the proper sequence. Also the difference between the measure values of the nearest and furthest images shows some improvement(0.86 vs 0.72).

### 6.2 Analysis of Generalized Mahalanobis Nearness Measure: Image Retrieval Experiments

In this section, we discuss the significance of the generalized Mahalanobis nearness measure values drawn from Table 2. Fig. 8 shows 20 images (1 query and 19

6.1: Query, M=1; 6.2: M=0.99; 6.3: M=0.99; 6.4: M=0.98; 6.5: M=0.98; 6.6: M=0.98; 6.7: M=0.98; 6.8: M=0.98; 6.9: M=0.98; 6.10: M=0.98; 6.11: M=0.98; 6.12: M=0.98; 6.13: M=0.97; 6.14: M=0.97; 6.15: M=0.97; 6.16: M=0.97; 6.17: M=0.97; 6.18: M=0.97; 6.19: M=0.96; 6.20: M=0.96

**Fig. 6.** Images ordered by $H_d NM$ measure values with Edge Features

7.1: Query, M=1; 7.2: M=0.86; 7.3: M=0.86; 7.4: M=0.85; 7.5: M=0.85; 7.6: M=0.85; 7.7: M=0.84; 7.8: M=0.83; 7.9: M=0.82; 7.10: M=0.81; 7.11: M=0.81; 7.12: M=0.79; 7.13: M=0.78; 7.14: M=0.78; 7.15: M=0.77; 7.16: M=0.76; 7.17: M=0.75; 7.18: M=0.74; 7.19: M=0.72; 7.20: M=0.72

**Fig. 7.** Images ordered by $H_d NM$ measure values with Edge and Texture Features

test images) ordered by their $gMNM$ values using only the edge features $e_{io}$. It can be observed that i) the difference between the measure values of the nearest and furthest images is fairly large (0.99 vs 0.34) compared with $H_dNM$ measure ii) all images belonging to the same category as the query image are retrieved in the proper sequence. Fig. 9 shows the same 20 images ordered by their $gNM$ values using both edge features and texture features $\mathcal{T}e_{io}$. It can be observed that the difference between the measure values of the nearest and furthest images (0.96 vs. 0.07) is considerable, when additional features are added and that the texture features contribute significantly to the quality of the image retrieval.

### 6.3 Remarks on quality of retrieval

The observations in sections 6.2 and 6.1 reveal that the generalized Mahalanobis nearness measure is a better measure in terms of its discriminatory power and its stability. The Hausdorff measure calculates distance between two sets of feature-valued vectors extracted from subimages in two images using the $L_1$ norm. In addition, the edge intensity and edge orientation features are not sufficient to get a good match between the query image and the test images, even though the measure is modified to compensate for the effect of outliers (see eqns. 8 and 9). The Mahalanobis measure calculates distance between feature vectors with respect to the covariance of the multivariate distribution of mean feature values extracted from subimages in two images. It can be seen that with just the edge intensity and edge orientation features, the Mahalanobis nearness measure is able to find a good match for the query image. This is because this measure takes into account not only in-class co-variance, but also prior probabilities.

## 7 Performance Measure

Several performance evaluation measures have been proposed based on the well-known precision(P) and the recall(R) to evaluate the performance of CBIR methods [59]. We use the following definitions for P and R:

$$P = \frac{\text{Number of relevant images retrieved}}{\text{Total number of images retrieved}}, \tag{16}$$

$$R = \frac{\text{Number of relevant images retrieved}}{\text{Total number of \quad relevant images}}. \tag{17}$$

We have used a total of 578 images drawn from the SIMPLIcity database in the six categories (see Fig.6). Table 3 shows the average measure values for images in each category.

However, the individual pair-wise measure values vary within each category. In order to establish relevancy of the test images, a threshold $th$ for the nearness value which acts as a cut-off needs to be determined. Notice that the total number

8.1: Query | 8.2: M=0.99 | 8.3: M=0.99 | 8.4: M=0.98 | 8.5: M=0.95
8.6: M=0.95 | 8.7: M=0.95 | 8.8: M=0.95 | 8.9: M=0.95 | 8.10: M=0.94
8.11: M=0.93 | 8.12: M=0.79 | 8.13: M=0.77 | 8.14: M=0.74 | 8.15: IM=0.73
8.16: M=0.48 | 8.17: M=0.48 | 8.18: M=0.42 | 8.19: M=0.37 | 8.20: M=0.34

**Fig. 8.** Ordered by gMNM-measure values with Edge Features

9.1: Query | 9.2: M=0.96 | 9.3: M=0.95 | 9.4: M=0.94 | 9.5: M=0.94
9.6: M=0.93 | 9.7: M=0.93 | 9.8: M=0.88 | 9.9: M=0.87 | 9.10: M=0.84
9.11: M=0.79 | 9.12: M=0.77 | 9.13: M=0.73 | 9.14: M=0.71 | 9.15: M=0.61
9.16: M=0.34 | 9.17: M=0.32 | 9.18: M=0.29 | 9.19: M=0.13 | 9.20: M=0.07

**Fig. 9.** Images ordered by gMNM-measure values with Edge and Texture Features

**Table 3.** Nearness Measurements

| Category | Number of Images | *Average gMNM* | *Average$H_dNM$* |
|---|---|---|---|
| Building | 100 | 0.27 | 0.77 |
| Bus | 94 | 0.24 | 0.75 |
| Dinosaurs | 100 | 0.84 | 0.82 |
| Elephant | 100 | 0.19 | 0.74 |
| Flower | 84 | 0.31 | 0.73 |
| Horse | 100 | 0.12 | 0.74 |

of relevant images should not be > 100 (size of the query image category set). Using this approach, we obtain the following results for $P$ and $R$ with $th = 0.6$:

$$P_{Te_{io}} = \frac{100}{578} = 0.173 \text{ using Hausdorff } H_dNM, \tag{18}$$

$$R_{Te_{io}} = \frac{100}{100} = 1.0 \text{ using Hausdorff } H_dNM, \tag{19}$$

$$P_{Te_{io}} = \frac{99}{99} = 1.0 \text{ using Mahalanobis gMNM}, \tag{20}$$

$$R_{Te_{io}} = \frac{99}{100} = 0.99 \text{ using Mahalanobis gMNM}. \tag{21}$$

It is noteworthy that on a large sample of images, Mahalanobis $gMNM$ distance measure is more precise than the Hausdorff $H_dNM$ distance measure for the same threshold value(see, *eqn.* (18)).

## 8 Conclusion

This paper introduces the notion of perceptually near Pawlak partitions. Such partitions are defined with a perceptual indiscernibility relation that is an extension of the original indiscernibility relation introduced by Z. Pawlak in 1981. It can be observed that each perceptually indiscernible Pawlak partition (PIP) of the picture elements (viewed as points in this work) of a digital image results in the separation of an image into non-overlapping regions. By definition, a PIP of a digital image is a particular form of image segmentation. In this article, near set theory is used to discover correspondences between images and to measure the degree of nearness between pairs of images. This leads to a new form of content-based image retrieval (CBIR). The sample experiments reported in this article suggest that the proposed approach to CBIR is quite promising. Future work will include near set-based object recognition and more advanced forms of CBIR.

## References

1. Pawlak, Z.: Classification of objects by means of attributes. Polish Academy of Sciences 429 (1981)
2. Pawlak, Z.: Rough sets. International J. Comp. Inform. Science 11, 341–356 (1982)

3. Pawlak, Z., Skowron, A.: Rudiments of rough sets. Information Sciences 177, 3–27 (2007)
4. Pawlak, Z., Skowron, A.: Rough sets: Some extensions. Information Sciences 177, 28–40 (2007)
5. Pawlak, Z., Skowron, A.: Rough sets and boolean reasoning. Information Sciences 177, 41–73 (2007)
6. Fechner, G.: Elemente der Psychophysik. In: Elements of Psychophysics trans. by Adler, H.E. (ed.). Holt, Rinehart & Winston, London (1860/1966)
7. Peters, J., Ramanna, S.: Affinities between perceptual granules: Foundations and perspectives. In: Bargiela, A., Pedrycz, W. (eds.) Human-Centric Information Processing Through Granular Modelling. SCI, vol. 182, pp. 49–66. Springer, Berlin (2009)
8. Peters, J.: Tolerance near sets and image correspondence. Int. J. of Bio-Inspired Computation 4(1), 239–445 (2009)
9. Peters, J.: Corrigenda and addenda: Tolerance near sets and image correspondence. Int. J. Bio-Inspired Computation 2(5), 1–8 (2010) (in Press)
10. Pal, S., Peters, J.: Rough Fuzzy Image Analysis: Foundations and Methodologies. CRC Press, Boca Raton (2010)
11. Hausdorff, F.: Grundzüge der mengenlehre. Verlag Von Veit & Comp., Leipzig (1914)
12. Hausdorff, F.: Set theory. Chelsea Publishing Company, New York (1962)
13. Rogers, C.: Hausdorff Measures. Cambridge U. Press, Cambridge (1970)
14. Mahalanobis, P.: On tests and measures of group divergence i. theoretical formulae. J. and Proc. Asiat. Soc. of Bengal 26, 541–588 (1930)
15. Mahalanobis, P.: On the generalized distance in statistics. Proc. Nat. Institute of Science (Calcutta) 2, 49–55 (1936)
16. Mrózek, A., Plonka, L.: Rough sets in image analysis. Foundations of Computing and Decision Sciences F18(3-4), 268–273 (1993)
17. Pal, S., Mitra, P.: Multispectral image segmentation using rough set initialized em algorithm. IEEE Transactions on Geoscience and Remote Sensing 11, 2495–2501 (2002)
18. Peters, J., Borkowski, M.: k-means indiscernibility over pixels. In: Tsumoto, S., Słowiński, R., Komorowski, J., Grzymała-Busse, J.W. (eds.) RSCTC 2004. LNCS (LNAI), vol. 3066, pp. 580–585. Springer, Heidelberg (2004)
19. Pal, S., Shankar, B.U., Mitra, P.: Granular computing, rough entropy and object extraction. Pattern Recognition Letters 26(16), 401–416 (2005)
20. Borkowski, M., Peters, J.: Matching 2d image segments with genetic algorithms and approximation spaces. In: Peters, J.F., Skowron, A. (eds.) Transactions on Rough Sets V. LNCS, vol. 4100, pp. 63–101. Springer, Heidelberg (2006)
21. Borkowski, M.: 2D to 3D Conversion with Direct Geometrical Search and Approximation Spaces. PhD thesis, Dept. Elec. Comp. Engg. (2007), `http://wren.ee.umanitoba.ca/`
22. Maji, P., Pal, S.: Maximum class separability for rough-fuzzy c-means based brain mr image segmentation. In: Peters, J.F., Skowron, A., Rybiński, H. (eds.) Transactions on Rough Sets IX. LNCS, vol. 5390, pp. 114–134. Springer, Heidelberg (2008)
23. Mushrif, M., Ray, A.: Color image segmentation: Rough-set theoretic approach. Pattern Recognition Letters 29(4), 483–493 (2008)
24. Hassanien, A., Abraham, A., Peters, J., Schaefer, G., Henry, C.: Rough sets and near sets in medical imaging: A review. IEEE Trans. Info. Tech. in Biomedicine 13(6), 955–968 (2009), doi:10.1109/TITB.2009.2017017

25. Sen, D., Pal, S.: Generalized rough sets, entropy, and image ambiguity measures. IEEE Transactions on Systems, Man, and Cybernetics–PART B 39(1), 117–128 (2009)
26. Malyszko, D., Stepaniuk, J.: Standard and fuzzy rough entropy clustering algorithms in image segmentation. In: Chan, C.-C., Grzymala-Busse, J.W., Ziarko, W.P. (eds.) RSCTC 2008. LNCS (LNAI), vol. 5306, pp. 409–418. Springer, Heidelberg (2008)
27. Sharawy, G.A., Ghali, N.I., Ghoneim, W.A.: Object-based image retrieval system based on rough set theory. IJCSNS International Journal of Computer Science and Network Security 9(1), 160–166 (2009)
28. Deselaers, T.: Image Retrieval, Object Recognition, and Discriminative Models. Ph.d. thesis, RWTH Aachen University (2008)
29. Deselaers, T., Keysers, D., Ney, H.: Features for image retrieval: An experimental comparison. Information Retrieval 11(1), 77–107 (2008d)
30. Christos, T., Nikolaos, A.L., George, E., Spiros, F.: On the perceptual organization of image databases using cognitive discriminative biplots. EURASIP Journal on Advances in Signal Processing, doi:10.1155/2007/68165
31. Su, Z., Zhang, H., Li, S., Ma, S.: Relevance feedback in content-based image retrieval: Bayesian framework, feature subspaces, and progressive learning. IEEE Transactions on Image Processing 12(8), 924–937 (2003)
32. Matthieu, C., Philippe, H.G., Sylvie, P.-F.: Stochastic exploration and active learning for image retrieval. Image and Vision Computing 25(1), 14–23 (2007)
33. Peters, J.F.: Classification of objects by means of features. In: Proc. IEEE Symposium Series on Foundations of Computational Intelligence (IEEE SCCI 2007), Honolulu, Hawaii, pp. 1–8 (2007)
34. Peters, J., Wasilewski, P.: Foundations of near sets. Information Sciences. An International Journal 179, 3091–3109 (2009), doi:10.1016/j.ins.2009.04.018
35. Pawlak, Z.: Rough sets. International Journal of Computer and Information Sciences 11, 341–356 (1982)
36. Engelking, R.: General topology. Sigma series in pure mathematics. Heldermann Verlag, Berlin (1989)
37. Henry, C., Peters, J.: Near sets, Wikipedia (2009), `http://en.wikipedia.org/wiki/Near_sets`
38. Pavel, M.: Fundamentals of Pattern Recognition, 2nd edn. Marcel Dekker, Inc., New York (1993)
39. Peters, J.: Near sets. General theory about nearness of objects. Applied Mathematical Sciences 1(53), 2029–2609 (2007)
40. Naimpally, S., Warrack, B.: Proximity Spaces. Cambridge University Press, Cambridge (1970); Cambridge Tract in Mathematics No. 59
41. DiMaio, G., Naimpally, S.: D-proximity spaces. Czech. Math. J. 41(116), 232–248 (1991)
42. DiMaio, G., Naimpally, S.: Proximity approach to semi-metric and developable spaces. Pacific J. Math. 44, 93–105 (1973)
43. Efremovič, V.: Infinitesimal spaces. Dokl. Akad. Nauk SSSR 76, 341–343 (1951)
44. Efremovič, V.: The geometry of proximity. Mat. Sb. 31, 189–200 (1952)
45. Efremovič, V., Švarc, A.: A new definition of uniform spaces. Metrization of proximity spaces. Dokl. Akad. Nauk SSSR 89, 393–396 (1953)
46. Pták, P., Kropatsch, W.: Nearness in digital images and proximity spaces. In: Nyström, I., Sanniti di Baja, G., Borgefors, G. (eds.) DGCI 2000. LNCS, vol. 1953, pp. 69–77. Springer, Heidelberg (2000)

47. Poincaré, J.H.: L'espace et la géomètrie. Revue de m'etaphysique et de morale 3, 631–646 (1895)
48. Peters, J.: Near sets. Special theory about nearness of objects. Fundamenta Informaticae 75(1-4), 407–433 (2007)
49. Henry, C., Peters, J.: Image pattern recognition using approximation spaces and near sets. In: An, A., Stefanowski, J., Ramanna, S., Butz, C.J., Pedrycz, W., Wang, G. (eds.) RSFDGrC 2007. LNCS (LNAI), vol. 4482, pp. 475–482. Springer, Heidelberg (2007)
50. Henry, C., Peters, J.: Perception-based image analysis. Int. J. of Bio-Inspired Computation 2(2) (2009) (in Press)
51. Henry, C., Peters, J.: Near set evaluation and recognition (near) system. Technical report, Computationa Intelligence Laboratory, University of Manitoba, UM CI (2009), Laboratory Technical Report No. TR-2009-015
52. Mallat, S., Zhong, S.: Characterization of signals from multiscale edges. IEEE Transactions on Pattern Analysis and Machine Intelligence 14(7), 710–732 (1992)
53. Gonzalez, R.C., Woods, R.E.: Digital Image Processing. Prentice-Hall, Upper Saddle Rv. (2002), ISBN 0-20-118075-8
54. Meghdadi, A., Peters, J.: Content-based image retrieval using a tolerance near set approach to image similarity. Image and Vision Computing (2009) (under review)
55. Duda, R., Hart, P., Stork, D.: Pattern Classification, 2nd edn. John Wiley & Sons, Chichester (2001)
56. Arai, R., Watanabe, S.: A quantitative method for comparing multi-agent-based simulations in feature space. In: David, N., Sichman, J.S. (eds.) MAPS 2008. LNCS, vol. 5269, pp. 154–166. Springer, Heidelberg (2009)
57. Wang, J.Z.: Simplicity-content-based image search engine. Content Based Image Retrieval Project (1995-2001)
58. Meghdadi, A., Peters, J., Ramanna, S.: Tolerance classes in measuring image resemblance. In: Velásquez, J.D., Ríos, S.A., Howlett, R.J., Jain, L.C. (eds.) KES 2009. LNCS, vol. 5712, pp. 127–134. Springer, Heidelberg (2009)
59. Muller, H., Muller, W., Squire, D., Marchand-Maillet, S., Pun, T.: Performance evaluation in content-based image retrieval: Overview and proposals. Pattern Recognition Letters 22(5), 593–601 (2001)

# A Novel Split and Merge Technique for Hypertext Classification

Suman Saha, C.A. Murthy, and Sankar K. Pal

Center for Soft Computing Research, Indian Statistical Institute
{ssaha_r,murthy,sankar}@isical.ac.in

**Abstract.** As web grows at an increasing speed, hypertext classification is becoming a necessity. While the literature on text categorization is quite mature, the issue of utilizing hypertext structure and hyperlinks has been relatively unexplored. In this paper, we introduce a novel split and merge technique for classification of hypertext documents. The splitting process is performed at the feature level by representing the hypertext features in a tensor space model. We exploit the local-structure and neighborhood recommendation encapsulated in the this representation model. The merging process is performed on multiple classifications obtained from split representation. A meta level decision system is formed by obtaining predictions of base level classifiers trained on different components of the tensor and actual category of the hypertext document. These individual predictions for each component of the tensor are subsequently combined to a final prediction using rough set based ensemble classifiers. Experimental results of classification obtained by using our method is marginally better than other existing hypertext classification techniques.

**Keywords:** Hypertext classification, tensor space model, rough ensemble classifier.

## 1 Introduction

As the web is expanding, where most web pages are connected with hyperlinks, the role of automatic categorization of hypertext is becoming more and more important. The challange of retrieval engine is, it need to search and retrieve toolarge number of web pages. By categorizing documents a priori, the search space can be reduced dramatically and the quality of ad-hoc retrieval improved. Besides, web users often prefer navigating through search directories as in portal sites.

Vector Space Model (VSM), the footstone of many web mining and information retrieval techniques [1], is used to represent the text documents and define the similarity among them. Bag of word (BOW) [2] is the earliest approach used to represent document as a bag of words under the VSM. In the BOW representation, a document is encoded as a feature vector, with each element in the vector indicating the presence or absence of a word in the document by TFIDF (Term Frequency Inverse Document Frequency) indexing. A document

J.F. Peters et al. (Eds.): Transactions on Rough Sets XII, LNCS 6190, pp. 192–210, 2010.

vector has no memory about the structure of the hypertext. Information about the HTML markup structure and hyperlink connectivity is ignored in VSM representation. For web page categorization, a frequently used approach is to use hyperlink information, which improves categorization accuracy [3]. Often hyperlink structure is used to support the predictions of a learned classifier, so that documents that are pointed to by the same page will be more likely to have the same classification. There exist many articles using different kinds of features (URL, anchortext, meta-tags, neighborhood, etc.....), but finally they are represented in a single vector, thereby losing the information about the structural component of hypertext where the word appeared. As an example, the fact that a word appearing in the title or URL is more important than the same word appearing in the text content, is ignored. Details of hypertext features have been given in section 2. Based on the assumption that each source of information provides a different viewpoint, a combination has the potential to have better knowledge than any single method. Methods utilizing different sources of information are combined to achieve further improvement, especially when the information considered is orthogonal. In web classification, combining link and content information is quite popular. A common way to combine multiple information is to treat information from different sources as different (usually disjoint) feature sets, on which multiple classifiers are trained. After that, these classifiers are combined together to generate the final decision.

In this article we have proposed a novel split and merge classification of hypertext documents. The splitting process relies on different types of features, which are extracted from a hypertext document and its neighbors. The split features are represented in a tensor space model, which consists of a sixth order tensor for each hypertext document that is a different vector for each of the different types of features. In this representation the features extracted from URL or Title or any other part are assigned to different tensor components (vector). Note that, unlike a matrix, components of a tensor differ in size, and feature space of each tensor component may be different from others. This representation model does not ignore the information about internal markup structure and link structure of the hypertext documents. In each tensor component a base level classification has been performed. The different types of classifications have been combined using rough set based ensemble classifier [4]. Our experiments demonstrate that splitting the feature set based on structure improves the performance of a learning classifier. By combining different classifiers it is possible to improve the performance even further.

In order to realize the specified objectives, the features of hypertext documents are discussed in section 2. Sections 3 and 4 present tensor space model and rough set based ensemble classifier respectively. Section 5 covers the proposed methodology. Finally, the experimental results are reported in section 6.

## 2 Hypertext Features

A hypertext document consists of different types of features which are found to be useful for representing a web page [5]. Written in HTML, web pages contain additional information other than text content, such as HTML tags, hyperlinks and anchor text (Fig 1). These features can be divided into two broad classes: on-page features, which are directly located on the page to be represented, and features of neighbors, which are found on the pages related in some way with the page to be represented.

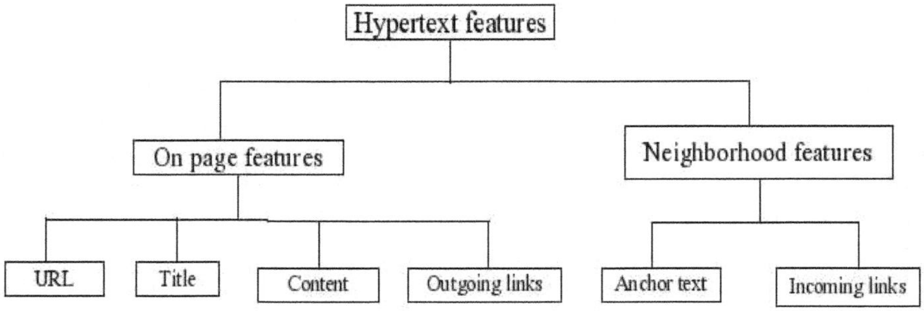


**Fig. 1.** Different type of features of hypertext document

Most commonly used on-page features are URL of the web page, outgoing links of web page, HTML tags, title-headers and text body content of the web page.

1) Features of URL: Uniform resource locators (URLs), which mark the address of a resource on the world wide web, provides valuable information about the document and can be used to predict the category of the resource [6]. A URL is first divided to yield a baseline segmentation into its components as given by the URI protocol (e.g., scheme :// host / path elements / document . extension), and further segmented wherever one or more non-alphanumeric characters appear (e.g., faculty-info − > faculty info).

2) Anchor text: Anchor text usually provides relevant descriptive or contextual information about the content of the link's destination. Thus it can be used to predict the category of the target page. Anchor text can provide a good source of information about a target page because it represents how people linking to the page actually describe it. Several studies have tried to use either the anchor text or the text near it to predict a target page's content [7].

3) Text content: The text on a page is the most relevant component for categorization. However, due to a variety of uncontrolled noise in web pages, a bag-of-words representation for all terms may not result in top performance. Researchers have tried various methods to make better use of the textual features. Popular methods are feature selection, vector of features, N-gram representation, which includes not only single terms, but also up to 5 consecutive words

[2]. The advantage of using n-gram representation is that it is able to capture the concepts expressed by a sequence of terms (phrases), which are unlikely to be characterized using single terms. However, an n-gram approach has a significant drawback; it usually generates a space with much higher dimensionality than the bag-of-words representation does. Therefore, it is usually performed in combination with feature selection [2].
4) Title and headers: Title and headers can be the most significant features found in a hypertext document, because they generally summarize the content of the page. Researchers have shown that incorporating features of title and headers improve the categorization results [8].
5) In-links: Link structure of the web offers some important information for analyzing the relevance and quality of web pages. Intuitively, the author of a web page A, who places a link to web page B, believes that B is relevant to A. The term in-links refers to the hyperlinks pointing to a page. Usually, the larger the number of in-links, the higher a page will be rated. The rationale is similar to citation analysis, in which an often-cited article is considered better than the one never cited. The assumption is made that if two pages are linked to each other, they are likely to be on the same topic. One study actually found that the likelihood of linked pages having similar textual content was high, if one considered random pairs of pages on the web [9]. Researchers have developed several link-analysis algorithms over the past few years. The most popular link-based web analysis algorithms include PageRank [10] and HITS [11].
6) Out-links: Category of the already classified neighboring pages can be used to determine the categories of unvisited web pages. In general, features of neighbors provide an alternative view of a web page, which supplement the view from on-page features. Therefore, collectively considering both can help in reducing the categorization error. Underlying mechanism of collective inference has been investigated by the researchers and has been argued that the benefit does not only come from a larger feature space, but also from modelling dependencies among neighbors and utilizing known class labels [8]. Such explanations may also apply to why web page classification benefits from utilizing features of neighbors.

## 3 Split Representation of Hypertexts in Tensor Space Model

Tensors provide a natural and concise mathematical framework for formulating and solving problems in high dimensional space analysis [12].Tensor algebra and multilinear analysis have been applied successfully in many domains such as; face recognition, machine vision, document analysis, feature decomposition, text mining etc. [13,14,15,16,17,18,19].

An n-order tensor in m-dimensional space is a mathematical object that has $n$ indices and each ranges from 1 to $m$, i.e., each index of a tensor ranges over the number of dimensions of space. Tensors are generalizations of scalars (0-order, which have no indices), vectors (1-order, which have a single index), and matrices (2-order, which have two indices) to an arbitrary number of indices.

Document indexing and representation has been a fundamental problem in information retrieval for many years. Most of the previous works are based on the Vector Space Model (VSM). The documents are represented as vectors, and each word corresponds to a dimension. In this section, we introduce a new Tensor Space Model (TSM) for document representation. In Tensor Space Model, a document is represented as a tensor (Fig 2), where domain of the tensor is the product of different vector spaces. Each vector space is associated with a particular type of features of the hypertext documents. The vector spaces considered here are corresponding to 1) features of URL, 2) features of anchor text, 3) features of title and headers, 4) features of text content, 5) features of outgoing links and 6) features of incoming links, the features are word in our case.

In this paper, we propose a novel Tensor Space Model (TSM) for hypertext representation. The proposed TSM is based on different types of features extracted from the HTML document and their neighbors. It offers a potent mathematical framework for analyzing the internal markup structure and link structure of HTML documents along with text content. The proposed TSM for hypertext consists of a $6^{th}$ order tensor, for each order the dimension is the number of terms of the corresponding types extracted from the hypertexts.

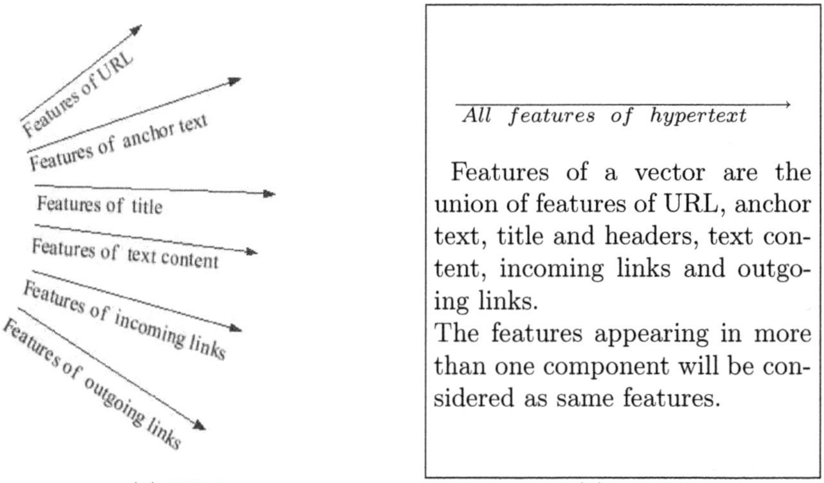


(a) TSM (b) VSM

**Fig. 2.** Hypertext representation using (a) tensor space model and (b) vector space model

### 3.1 Mathematical Formulation of TSM

Let $\mathcal{H}$ be any hypertext document. Let, $S^{\mathcal{H}u} = \{e_1^{\mathcal{H}u}, e_2^{\mathcal{H}u}, \ldots, e_{n_1}^{\mathcal{H}u}\}$ be the set corresponding to features of URL, $S^{\mathcal{H}a} = \{e_1^{\mathcal{H}a}, e_2^{\mathcal{H}a}, \ldots, e_{n_2}^{\mathcal{H}a}\}$ be the set corresponding to features of anchor text, $S^{\mathcal{H}h} = \{e_1^{\mathcal{H}h}, e_2^{\mathcal{H}h}, \ldots, e_{n_3}^{\mathcal{H}h}\}$ be the set corresponding to features of title and headers, $S^{\mathcal{H}c} = \{e_1^{\mathcal{H}c}, e_2^{\mathcal{H}c}, \ldots, e_{n_4}^{\mathcal{H}c}\}$ be the set corresponding to features of text content, $S^{\mathcal{H}out} = \{e_1^{\mathcal{H}out}, e_2^{\mathcal{H}out}, \ldots, e_{n_5}^{\mathcal{H}out}\}$ be the set corresponding to features of outgoing links and $S^{\mathcal{H}in} = \{e_1^{\mathcal{H}in}, e_2^{\mathcal{H}in}, \ldots, e_{n_6}^{\mathcal{H}in}\}$ be the set corresponding to features of incoming links.

Let $S^{\mathcal{H}}$ be the set representing all features of $\mathcal{H}$ in a vector space, $\mathcal{V}$. Then, $S^{\mathcal{H}} = S^{\mathcal{H}u} \cup S^{\mathcal{H}a} \cup S^{\mathcal{H}h} \cup S^{\mathcal{H}c} \cup S^{\mathcal{H}out} \cup S^{\mathcal{H}in}$. Let $S_1^{\mathcal{H}}$ be the set of features which are present in more than one component. So, $S_1^{\mathcal{H}} = \cup_{(x,y \in F) \& x \neq y} S^{\mathcal{H}x} \cap S^{\mathcal{H}y}$, where, $F = \{u, a, h, c, out, in\}$. Note that, features present in more than two components is already considered in the above expression. Let $s$ be an element of $S_1^{\mathcal{H}}$. That is $s$ has occurred in more than one component of the hypertext documents. For each appearance of $s$ in different components, $s$ may have different significance regarding the categorization of the hypertext documents. Now the multiple appearance of $s$ is ignored in $S^{\mathcal{H}}$, as it is a set of union of the sets corresponding to the components of hypertext.

In the vector space model for hypertext representation, vectors are constructed on $S^{\mathcal{H}}$, that is, occurrence of $s \in S_1^{\mathcal{H}}$ in different components is ignored. In some advanced vector space models elements of different components are tagged [8], that is $S_1^{\mathcal{H}} = \phi$. Let $|.|$ denote cardinality of a set. Number of features of different components may have a large variance value. For example, $|S^{\mathcal{H}u}| << |S^{\mathcal{H}c}|$. In this representation, importance of the elements corresponding to the components with low cardinality, is ignored during magnitude normalization.

In tensor space model the features corresponding to different components of hypertext are represented as different components of a tensor. Let $\mathcal{T}$ be the tensor space corresponding to hypertext documents. Each member $T$ of $\mathcal{T}$ is of the form $T = T_{xi}$ where, $x \in F$ and $1 \leq i \leq |S^{\mathcal{H}x}|$, i.e. the value of $T$ at $(x, i)$ is $e_i^{\mathcal{H}x}$. Note that $i$ depends on $x$, so it is not just a matrix.

**Similarity measures on TSM.** Cosine similarity is a measure of similarity between two vectors of n dimensions by finding the angle between them, often used to compare documents in text mining. Given two vectors of attributes, $A$ and $B$, the cosine similarity, $Sim(A, B) = (|A.B|)/(|A|.|B|)$ where the word vectors $A$, $B$ , represented after removing stop words and stemming. For text matching, the attribute vectors $A$ and $B$ are usually the tf-idf vectors of the documents. The resulting similarity will yield the value of 0 meaning, the vectors are independent, and 0 meaning, the vectors are same, with in-between values indicating intermediate similarities or dissimilarities.

Let $\mathcal{T}$ be the tensor space corresponding to hypertext documents. Each member $T$ of $\mathcal{T}$ is of the form $T = T_{rs}$ where $r$ ranges on the types of features considered and $s$ ranges on number of terms extracted of particular types. The tensor similarity between two tensors $T_i$ and $T_j$ of $\mathcal{T}$ is defined as $sim(T_i, T_j) = \sum_r sim(T_ir, T_jr)$, where $sim(T_ir, T_jr)$ is the similarity between $r^{th}$ component of $T_i$ and $T_j$. Now, for each $r$, the $r^{th}$ components of a tensor $T_i$ is a vector. So, $sim(T_ir, T_jr)$ is basically the similarity between two vectors. Note that, here, cosine similarity is considered as vector similarity measure.

**Computational complexity on TSM.** Let $n$ be the total number of features of hypertext documents. Let $n_1, n_2, \ldots, n_r$ be the number of features associated with the $1^{st}, 2^{nd}, \ldots, r^{th}$ components of the tensor respectively. From the definition of TSM we obtain $\sum_{i=1}^{r} n_i = n$. Let $m$ be the number of documents. The complexity of an algorithm, $\mathcal{A}$ constructed on VSM can be expressed as $f(m, n, \alpha)$, where $\alpha$ is corresponding to specific parameters of $\mathcal{A}$. The expression of complexity $f(m, n, \alpha)$ is written as: $O(m^i n^j \alpha^k)$. The complexity of the same algorithm, $\mathcal{A}$ constructed on TSM can be written as: $O(m^i n_t^j \alpha^k)$, where $n_t = max_{s=1}^{r}\{n_1, n_2, \ldots, n_r\}$. Since, $n_t < n$, we can write $(n_t)^j \leq n^j$. Hence, $O(m^i n_t^j \alpha^k) \leq O(m^i n^j \alpha^k)$. Thus the following theorem holds.

**Theorem:** Computational complexity of an algorithm performing on tensor space model using tensor similarity measure as distance is at most the computational complexity of the same algorithm performing on vector space model using vector similarity measure as distance.

## 4 Merging Classifications Using Rough Ensemble Classifier

The rough ensemble classifier (REC) is designed to extract decision rules from trained classifier ensembles that perform classification tasks [4]. REC utilizes trained ensembles to generate a number of instances consisting of prediction of individual classifiers as conditional attribute values and actual classes as decision attribute values. Then a decision table is constructed using all the instances with one instance in each row. Once the decision table is constructed, rough set attribute reduction is performed to determine core and minimal reducts. The classifiers corresponding to a minimal reduct are then taken to form classifier ensemble for REC classification system. From the minimal reduct, the decision rules are computed by finding mapping between decision attribute and conditional attributes. These decision rules obtained by rough set technique are then used to perform classification tasks (Fig.3). Following theorems exists in this regard.

- **Theorem 1:** Rough set based combination is an optimal classifier combination technique [4].
- **Theorem 2:**The performance of the rough set based ensemble classifier is at least same as every one of its constituent single classifiers [4].

## 5 Split and Merge Classification of Hypertexts

Here we propose a split and merge classification of hypertexts. A hypertext document is split on the basis of the different types of features existing in it. Tensor space model has been used to represent the hypertext using the information of text content, internal mark-up structure and link structure. Classification of hypertext documents, represented as tensor, can be obtained in two ways: (1) by integrating classifier's parameters of different tensor components and (2) by integrating classifiers output obtained on different tensor components. In the first

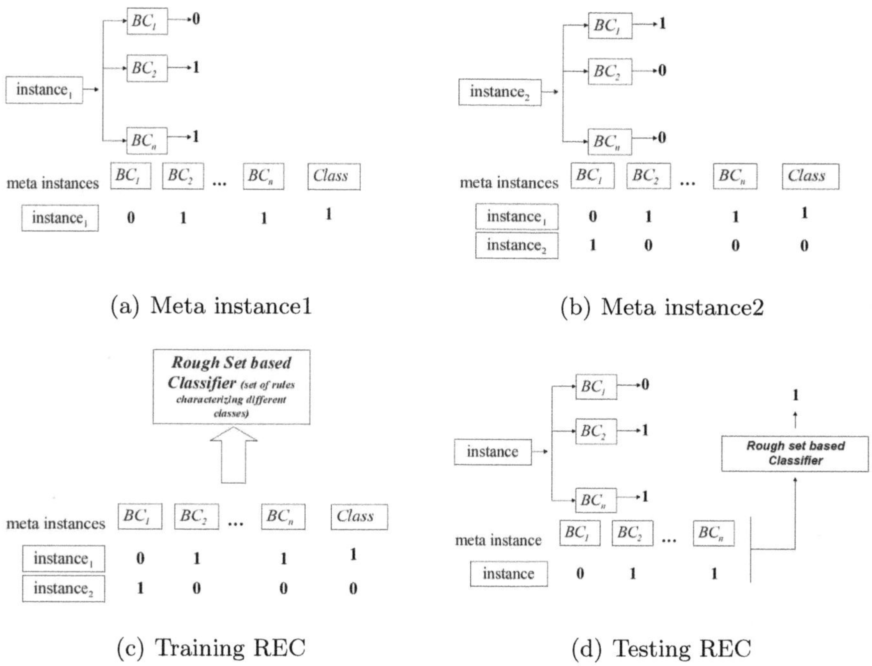


(a) Meta instance1

(b) Meta instance2

(c) Training REC

(d) Testing REC

**Fig. 3.** Example describing different steps of REC. Subfigures (a) and (b) show the classification of instances by different base classifiers (denoted as BC). Outputs of the base classifiers along with the actual class have been considered to construct meta data. Sub figure (c) show the training of REC. Subfigure (d) show the output of REC.

way, a K-NN classification has been performed using tensor similarity measure. In the second way ensemble classification has been performed (Fig. 4). For ensemble classification, base level classification has been carried out on individual tensor components and combined classification has been obtained using rough set based ensemble classifier.

### 5.1 Preprocessing for Split Representation

Hypertext documents are tokenized with syntactic rules and canonical forms. First we select a set of relevant features from a HTML document. For each type of feature an individual tensor component is constructed. A tensor component is a vector, which represents the terms of particular type corresponding to the component. Note that the tensor space model captures the structural representation of hypertext document.

1) Preprocessing text content:

- The text is stemmed using Porter's stemming algorithm and stop words are removed.

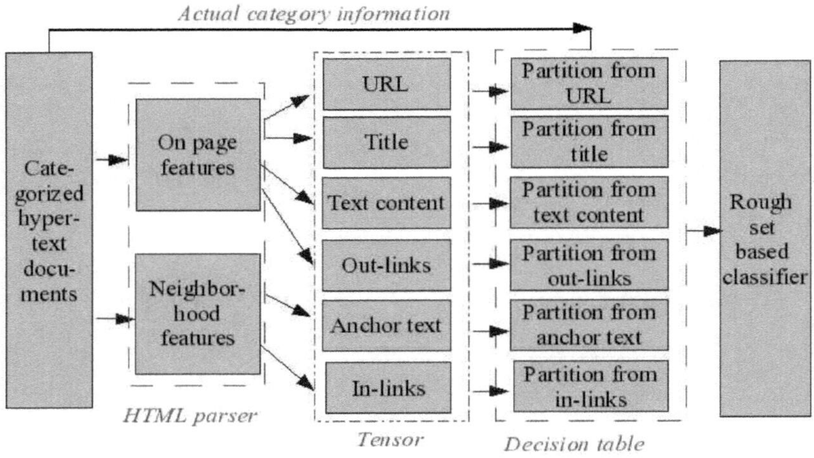


**Fig. 4.** Block diagram of proposed method

– Unique words present in the text are represented as a tensor component. This tensor component corresponds to the text content of the hypertext documents.

2) Preprocessing URL:

– A URL is first divided to yield a baseline segmentation into its components as given by the URI protocol (e.g., scheme :// host / path elements / document . extension), and further segmented wherever one or more non-alphanumeric characters appear.
– These segmented substrings are treated as words. All these words found in a URL will be represented as a tensor component corresponding to features of URLs.

3) Preprocessing anchor text:

– Anchor text is a small text content. The text is stemmed using Porter's stemming algorithm and stop words are removed.
– This is computed in the same way as text content, except substituting each document by a virtual document consisting of all the anchor text inside that document.
– Unique words present in this virtual document are represented as a tensor component corresponding to features of anchor text.

4) Preprocessing of title and headers:

– Title and headers are text contents. The text is stemmed using Porter's stemming algorithm and stop words are removed.

- Unique words present in these text are represented as a tensor component corresponding to features of title and headers.

5) Preprocessing in-links:

- All in-links are first divided to yield a baseline segmentation into its components as given by the URI protocol and further segmented wherever one or more non-alphanumeric characters appear.
- The tokens obtained by segmentation of the in-links are stored in a tensor component corresponding to features of in-links.

6) Preprocessing out-links:

- All out-links are first divided to yield a baseline segmentation into its components as given by the URI protocol and further segmented wherever one or more non-alphanumeric characters appear.
- The tokens obtained by segmentations of the out-links are stored in a tensor component corresponding to features of out-links.

The above methodologies are applied on split hypertext documents. The merging will take place during classification, which is described below.

### 5.2 Merging of Classifications

We now describe how to merge the classifications obtained from each one of the components of the tensor using naive bayes as base level classifiers. To generate the initial classifications for rough ensemble classifier, we assume a base level classifier and train it on each tensor component. These trained classifiers provide different classifications on the tensor space. Outputs of the base classifiers and the actual class information are used to construct meta level decision table. Output of a base level classifier contributes to the existence of an attribute, values of this attribute can be any class label that is determined by the base classifier corresponding to the tensor component. This meta data represented in the form of decision table is the input of rough set based ensemble classifier and its output is the merged classification. So the number of attributes is the same as number of tensor components. Rough set based attribute reduction techniques eliminate superfluous attributes and create a minimal sufficient subset of attributes for a decision table. Such minimal sufficient subset of attributes is called a reduct. Once the reduct is computed we remove redundant classifiers from the ensemble and construct new reduced decision table. Rough set based decision rules extracted from this reduced decision table are applied to obtain final classification. These decision rules perform merging of the base level decisions into a final decision.

## 6 Experimental Results

We performed a large number of experiments to test the output of proposed methods. We now describe the data corpuses, methodologies and results.

### 6.1 Data Collection

We used four data sets, Looksmart, Dmoz, webkb and Yahoo for our experiments. We crawled the Looksmart and Dmoz web directories. These directories are well known for maintaining a categorized hypertext documents. The web directories are multi-level tree-structured hierarchy. The top level of the tree, which is the first level below the root of the tree, contains 13 categories in Looksmart (Table 2) and 16 categories for Dmoz (Table 1). Each of these categories contains sub-categories that are placed in the second level below the root. We use the top-level categories to label the web pages in our experiments.

**Table 1.** Class distribution and features of the dmoz data in links and pages

(a)

| Class | #Pages | %Pages | #Links | %Links |
|---|---|---|---|---|
| Arts | 1855 | 6.27 | 4292 | 8.25 |
| Business | 1672 | 5.65 | 3665 | 7.04 |
| Computers | 2017 | 6.82 | 3946 | 7.58 |
| Games | 1500 | 5.07 | 2124 | 4.08 |
| Health | 1343 | 4.54 | 3210 | 6.17 |
| Home | 1786 | 6.04 | 2895 | 5.56 |
| Sports | 2537 | 8.58 | 3374 | 6.48 |
| Kids and Teens | 2290 | 7.74 | 2978 | 5.72 |
| News | 2626 | 8.88 | 3702 | 7.11 |
| Recreation | 2631 | 8.89 | 2996 | 5.76 |
| Reference | 1032 | 3.49 | 3389 | 6.51 |
| Regional | 1492 | 5.04 | 5441 | 10.46 |
| Science | 2387 | 8.07 | 2977 | 5.72 |
| Shopping | 1596 | 5.39 | 2020 | 3.88 |
| World | 1529 | 5.17 | 2093 | 4.02 |
| Society | 1271 | 4.29 | 2896 | 5.56 |
| Total | 29564 | 100 | 51998 | 100 |

(b)

| Components | # features |
|---|---|
| URL | 27935 |
| Anchor | 25111 |
| Title | 36965 |
| Text | 104126 |
| In-link | 23903 |
| Out-link | 21878 |
| Total | 239918 |
| Union | 188519 |
| $S_1$* | 51399 |

*$S_1$ includes the features appearing in at least two components.

The webkb data set was collected from the WebKB project. The pages in the WebKB dataset are classified into one of the categories Student, Course, Department, Faculty, Project, Staff and Other (Table 3). Here there are 8077 documents in 7 categories. The largest category (Other) consists of 3025 pages; while the smallest category (Staff) consists of only 135 pages.

Another data set consists of 40000 web pages crawled from the Yahoo topic directory (http://dir.yahoo.com). This is a large hypertext corpous, manually classified by the human experts. The extracted subset includes 33253 pages, which are distributed among 14 top level categories. The largest category (Science) consists of 4627 pages; while the smallest category (Regional) consists of

**Table 2.** Class distribution and features of the looksmart data

(a)

| Class | #Pages | %Pages | #Links | %Links |
|---|---|---|---|---|
| Auto | 677 | 5.38 | 1859 | 7.12 |
| Education | 1211 | 9.64 | 3463 | 13.26 |
| Health | 1087 | 8.65 | 2655 | 10.17 |
| Money | 631 | 5.02 | 1193 | 4.57 |
| Recreation | 131 | 1.04 | 654 | 2.50 |
| Style | 976 | 7.76 | 1353 | 5.18 |
| Travel | 595 | 4.73 | 1622 | 6.21 |
| Cities | 1245 | 9.91 | 2396 | 9.17 |
| Food | 1203 | 9.57 | 2371 | 9.08 |
| HomeLiving | 1676 | 13.34 | 2796 | 10.71 |
| Music | 1236 | 9.83 | 2971 | 11.38 |
| Sports | 742 | 5.90 | 1483 | 5.68 |
| Tech Games | 1152 | 9.17 | 1285 | 4.92 |
| Total | 12562 | 100 | 26101 | 100 |

(b)

| Components | # features |
|---|---|
| URL | 17469 |
| Anchor | 17766 |
| Title | 11463 |
| Text | 41153 |
| In-link | 16599 |
| Out-link | 13272 |
| Total | 117722 |
| Union | 86822 |
| $S_1$* | 30900 |

*$S_1$ includes the features appearing in at least two components.

**Table 3.** Class distribution and features of the webkb data

(a)

| Class | #Pages | %Pages | #Links | %Links |
|---|---|---|---|---|
| Student | 1639 | 20.29 | 2544 | 19.07 |
| Faculty | 1121 | 13.87 | 2147 | 16.09 |
| Course | 926 | 11.46 | 1229 | 9.21 |
| Project | 701 | 8.67 | 1083 | 8.11 |
| Department | 530 | 6.56 | 1194 | 8.95 |
| Other | 3025 | 37.45 | 4730 | 35.45 |
| Staff | 135 | 1.67 | 413 | 3.09 |
| Total | 8077 | 100 | 13340 | 100 |

*$S_1$ includes the features appearing in at least two components.

(b)

| Components | # features |
|---|---|
| URL | 12898 |
| Anchor | 10515 |
| Title | 16193 |
| Text | 23582 |
| In-link | 14529 |
| Out-link | 14094 |
| Total | 91811 |
| Union | 72071 |
| $S_1$* | 19740 |

only 782 pages. Detailed information about number of pages and number of links in the each category of the Yahoo data set is given in the Table 4.

We processed the data sets to remove images and scripts followed by stop-words removal and stemming. Link graph has been constructed for each of the datasets for extracting neighborhood features. URLs have been segmented for extracting URL features. Finally features extracted from all the components of hypertext have been represented using both the models (i.e., tensor space model and vector space model) in our experiments. We have considered vector space model for the purpose of comparison.

**Table 4.** Class distribution and features of the yahoo data

(a)

| Class | #Pages | %Pages | #Links | %Links |
|---|---|---|---|---|
| Arts | 2731 | 8.21 | 4269 | 7.70 |
| Business | 4627 | 13.91 | 6092 | 11.00 |
| Computers | 3205 | 9.63 | 6444 | 11.63 |
| Education | 2976 | 8.94 | 5357 | 9.67 |
| Entertainment | 1592 | 4.78 | 2184 | 3.94 |
| Government | 782 | 2.35 | 1703 | 3.07 |
| Health | 2542 | 7.64 | 3999 | 7.22 |
| NewsMedia | 3716 | 11.17 | 6580 | 11.88 |
| Recreation | 1482 | 4.45 | 2965 | 5.35 |
| Reference | 1183 | 3.55 | 3165 | 5.71 |
| Regional | 1020 | 3.06 | 2219 | 4.00 |
| Science | 3350 | 10.07 | 4486 | 8.10 |
| SocialScience | 2859 | 8.59 | 3493 | 6.30 |
| SocietyCulture | 1188 | 3.57 | 2424 | 4.37 |
| Total | 33253 | 100 | 55380 | 100 |

(b)

| Components | # features |
|---|---|
| URL | 34045 |
| Anchor | 31863 |
| Title | 43428 |
| Text | 127459 |
| In-link | 44720 |
| Out-link | 40163 |
| Total | 321678 |
| Union | 256118 |
| $S_1$* | 65560 |

*$S_1$ includes the features appearing in at least two components.

### 6.2 Evaluation Measure

We have employed the standard measures to evaluate the performance of hypertext classification (i.e. precision, recall and $F_1$ measures). Precision ( $P$ ) is the proportion of actual positive class members returned by the system among all positive class members. Recall ( $R$ ) is the proportion of predicted positive members among all actual positive class members in the data. $F_1$ is the harmonic average of precision and recall as shown below:

$$F_1 = \frac{2PR}{P+R}$$

To evaluate the average performance across multiple categories, there are two conventional methods: micro-average-$F_1$ and macro-average-$F_1$. Micro-average-$F_1$ is the global calculation of $F_1$ measure regardless of categories. Macro-average-$F_1$ is the average on $F_1$ scores of all categories. Micro-average gives equal weight to every document, while macro-average gives equal weight to every category, regardless of its frequency. In our experiments, precision, recall micro-average-$F_1$ and macro-average-$F_1$ will be used to evaluate the classification performance.

### 6.3 Classification Results on TSM

Decisions of many vector space classifiers are based on a notion of distance, e.g., when computing the nearest neighbors in k-NN classification. For evaluation of the tensor space model for hypertext representation, we have constructed two k-NN classifiers. In the first case, k-NN classification on vector space representation for hypertext document is considered and vector similarity measure is used to compute nearest neighbor. In the second case, k-NN classification on tensor space model for hypertext representation is considered and tensor similarity measure is used to compute nearest neighbour. The performances of these two classifiers have been observed on four different datasets, Yahoo, webKB, Looksmart and Dmoz. The classification results of comparisons are shown in tables 5(a), 5(b), 5(c) and 5(d). The results has been shown in terms of Precession, recall, micro-average-$F_1$ and macro-average-$F_1$. It can be observed from the tables that classification results are better when tensor space model for hypertext representation is considered compared to classification results when vector space model for representation is considered.

**Table 5.** Results of k-NN classification on VSM and TSM

(a)

| Data set | VSM | TSM | Better? |
|---|---|---|---|
| Dmoz | 92.94 | 95.07 | √ |
| Looksmart | 90.13 | 93.55 | √ |
| WebKB | 91.85 | 94.67 | √ |
| Yahoo | 88.24 | 89.12 | √ |

(b)

| Data set | VSM | TSM | Better? |
|---|---|---|---|
| Dmoz | 83.27 | 88.89 | √ |
| Looksmart | 87.36 | 90.26 | √ |
| WebKB | 85.80 | 84.91 | √ |
| Yahoo | 82.36 | 86.42 | √ |

(c)

| Data set | VSM | TSM | Better? |
|---|---|---|---|
| Dmoz | 87.83 | 91.87 | √ |
| Looksmart | 88.72 | 91.87 | √ |
| WebKB | 88.72 | 89.52 | √ |
| Yahoo | 85.19 | 87.74 | √ |

(d)

| Data set | VSM | TSM | Better? |
|---|---|---|---|
| Dmoz | 85.69 | 89.32 | √ |
| Looksmart | 83.18 | 87.63 | √ |
| WebKB | 84.28 | 87.23 | √ |
| Yahoo | 84.50 | 86.34 | √ |

### 6.4 Classification Results on Individual Components and Combined Results

In this subsection we have provided the results of experiments regarding classifications of hypertext documents. Classifications of hypertext have been performed on different components of tensor space model corresponding to different types of feature sets using naive bayes classifier. Here naive bayes classifier is used for its simplicity in implementation. We have also provided the results of classification with tensor space model using k-NN classifier where tensor similarity measure is distance. In addition, the combined results of classification are provided using rough set based ensemble classifier. The cases considered are given below.

*A*) Classification based on URL features (2, 5.1) using naive bayes classifier.

*B*) Classification based on Anchor text features (2, 5.1) using naive bayes classifier.

*C*) Classification based on features of Title and headers (2, 5.1) using naive bayes classifier.

**Table 6.** Classification results on individual components and their rough set based combination in terms of precision

| Data set | A | B | C | D | E | F | G | H |
|---|---|---|---|---|---|---|---|---|
| Dmoz | 62.01 | 70.30 | 76.61 | 81.96 | 66.37 | 68.42 | 95.07 | 95.32 |
| Looksmart | 67.50 | 70.11 | 74.51 | 82.64 | 62.29 | 61.28 | 93.55 | 94.02 |
| WebKB | 64.90 | 72.02 | 69.92 | 86.16 | 67.30 | 61.71 | 94.67 | 95.72 |
| Yahoo | 61.64 | 68.98 | 67.93 | 79.43 | 59.93 | 61.38 | 89.12 | 90.24 |

**Table 7.** Classification results on individual components and their rough set based combination in terms of recall

| Data set | A | B | C | D | E | F | G | H |
|---|---|---|---|---|---|---|---|---|
| Dmoz | 57.83 | 71.56 | 68.95 | 79.59 | 56.58 | 57.85 | 88.89 | 90.15 |
| Looksmart | 60.04 | 73.32 | 69.57 | 77.63 | 60.84 | 64.04 | 90.26 | 91.67 |
| WebKB | 56.97 | 65.48 | 67.75 | 79.27 | 54.72 | 56.53 | 84.91 | 85.81 |
| Yahoo | 57.90 | 67.51 | 66.86 | 78.85 | 54.41 | 54.39 | 86.42 | 87.14 |

**Table 8.** Classification results on individual components and their rough set based combination in terms of micro average $F_1$

| Data set | A | B | C | D | E | F | G | H |
|---|---|---|---|---|---|---|---|---|
| Dmoz | 59.85 | 70.92 | 72.57 | 80.76 | 61.08 | 62.69 | 91.87 | 92.66 |
| Looksmart | 63.55 | 71.68 | 71.96 | 80.06 | 61.56 | 62.63 | 91.87 | 92.83 |
| WebKB | 60.68 | 68.60 | 68.82 | 82.57 | 60.36 | 59.01 | 89.52 | 90.49 |
| Yahoo | 59.71 | 68.24 | 67.39 | 79.14 | 57.04 | 57.67 | 87.74 | 88.66 |

**Table 9.** Classification results on individual components and their rough set based combination in terms of macro average $F_1$

| Data set | A | B | C | D | E | F | G | H |
|---|---|---|---|---|---|---|---|---|
| Dmoz | 59.79 | 70.72 | 72.71 | 79.95 | 58.14 | 59.36 | 89.32 | 90.04 |
| Looksmart | 62.86 | 71.89 | 67.11 | 78.28 | 61.14 | 61.26 | 87.63 | 88.74 |
| WebKB | 59.37 | 69.15 | 65.68 | 82.46 | 58.81 | 57.71 | 87.23 | 91.85 |
| Yahoo | 58.53 | 66.91 | 65.84 | 77.50 | 54.58 | 58.73 | 86.34 | 88.17 |

*D*) Classification based on features of Text content (2, 5.1) using naive bayes classifier.

*E*) Classification based on features of In-coming links (2, 5.1) using naive bayes classifier.

*F*) Classification based on features of Out-going links (2, 5.1) using naive bayes classifier.

*G*) Classification based on Tensor similarity measure (6.3) using using k-NN classifier.

*H*) Classification based on split marge classification (5.2).

Results on precision, recall, micro-$F_1$ and macro-$F_1$ of A, B, C, D, E, F, G and H have been reported in tables 6, 7, 8, 9 respectively. It can be observed that classification results are poor for link based features than the text based features, and combined results corresponding to proposed methods are far better.

### 6.5 Comparisons with Some Recent Hypertext Classification Techniques

We have compared the performance of the proposed methods with existing classification techniques. A brief review of existing hypertext classification techniques is given below and these methods are considered for comparisons.

$A_1$) The article "Enhanced hypertext categorization using hyperlinks"[8], is the first hypertext classification system that combines textual and linkage features into a general statistical model to infer the of interlinked documents. Relaxation labelling technique is used for better classification by exploiting link information in a small neighborhood around documents.

$B_1$) The article "Improving A Page Classifier with Anchor Extraction and Link Analysis"[20], describes a technique that improves a simple web page classifier's performance on pages from a new, unseen web site, by exploiting link structure within a site as well as page structure within hub pages. On real-world test cases, this technique significantly and substantially improves the accuracy of a bag-of-words classifier, reducing error rate by about half, on average.

$C_1$) The article "Fast webpage classification using URL features"[6], explores the use of URLs for web page categorization via a two-phase pipeline of word segmentation and classification. This technique quantify its performance against document-based methods, which require the retrieval of the source documents.

$D_1$) The article, "Link-Local Features for Hypertext Classification"[21], demonstrates that the need to focus on relevant parts of predecessor pages, namely on the region in the neighborhood of the origin of an incoming link. Authors have investigated different ways for extracting such features, and compared several different techniques for using them in a text classifier.

$E_1$) The article "Graph based Text Classification: Learn from Your Neighbors" [22], presents a new method for graph-based classification, with particular emphasis on hyperlinked text documents but broader applicability. This approach is based on iterative relaxation labelling and can be combined with either Bayesian or SVM classifiers on the feature spaces of the given data items. The graph neighborhood is taken into consideration to exploit locality patterns. $F_1$) In the article, "Web Page Classification with Heterogeneous Data Fusion" [23], the contextual and structural information, of web pages has been represented into a common format of kernel matrix, via a kernel function. A generalized similarity measure between a pair of web pages is proposed. The experimental results on a collection of the ODP database validate the advantages of

**Table 10.** Comparison of the eight hypertext classification methods in terms of precision

| Data set | $A_1$ | $B_1$ | $C_1$ | $D_1$ | $E_1$ | $F_1$ | $G_1$ | $H_1$ |
|---|---|---|---|---|---|---|---|---|
| Dmoz | 86.4 | 87.79 | 92.11 | 87.62 | 93.17 | 93.98 | 95.07 | 95.32 |
| Looksmart | 91.82 | 86.15 | 89.81 | 87.27 | 87.78 | 88.9 | 93.55 | 94.02 |
| WebKB | 87 | 89.9 | 86.43 | 90.5 | 93.18 | 87.39 | 94.67 | 95.72 |
| Yahoo | 82.34 | 83.78 | 83.51 | 84.68 | 86.12 | 87.72 | 89.12 | 90.24 |

**Table 11.** Comparison of the eight hypertext classification methods in terms of recall

| Data set | $A_1$ | $B_1$ | $C_1$ | $D_1$ | $E_1$ | $F_1$ | $G_1$ | $H_1$ |
|---|---|---|---|---|---|---|---|---|
| Dmoz | 82.62 | 86.78 | 85.18 | 84.21 | 83.95 | 84.71 | 88.89 | 90.15 |
| Looksmart | 83.44 | 88.56 | 84.62 | 82.11 | 87.26 | 91.29 | 90.26 | 91.67 |
| WebKB | 80 | 81.38 | 83.61 | 84.33 | 83.26 | 83.46 | 84.91 | 85.81 |
| Yahoo | 80.5 | 82.54 | 83.61 | 85.11 | 81.56 | 83.13 | 86.42 | 87.14 |

**Table 12.** Comparison of the eight hypertext classification methods in terms of micro average $F_1$

| Data set | $A_1$ | $B_1$ | $C_1$ | $D_1$ | $E_1$ | $F_1$ | $G_1$ | $H_1$ |
|---|---|---|---|---|---|---|---|---|
| Dmoz | 84.46 | 87.28 | 88.50 | 85.88 | 88.32 | 89.10 | 91.87 | 92.66 |
| Looksmart | 87.42 | 87.33 | 87.13 | 84.61 | 87.51 | 90.07 | 91.87 | 92.83 |
| WebKB | 83.35 | 85.42 | 84.99 | 87.30 | 87.94 | 85.37 | 89.52 | 90.49 |
| Yahoo | 81.40 | 83.15 | 83.55 | 84.89 | 83.77 | 85.36 | 87.74 | 88.66 |

**Table 13.** Comparison of the eight hypertext classification methods in terms of macro average $F_1$

| Data set | $A_1$ | $B_1$ | $C_1$ | $D_1$ | $E_1$ | $F_1$ | $G_1$ | $H_1$ |
|---|---|---|---|---|---|---|---|---|
| Dmoz | 81.26 | 85.78 | 87.82 | 82.43 | 83.28 | 87.99 | 89.32 | 90.04 |
| Looksmart | 86.09 | 85.45 | 83.25 | 83.87 | 87.69 | 88.89 | 87.63 | 88.74 |
| WebKB | 81 | 84.32 | 81.45 | 86.46 | 85.16 | 84.81 | 87.23 | 91.85 |
| Yahoo | 80.7 | 83.17 | 81.77 | 83.45 | 81.69 | 84.93 | 86.34 | 88.17 |

the proposed method over traditional methods based on any single data source and the uniformly weighted combination of them.

$G_1$) Here a k-NN classifier on tensor space model is considered where tensor similarity measure used is the distance between hypertext documents 6.3.

$H_1$) Proposed split and merge classification where rough ensemble classifier on tensor space model is considered 5.2.

Results in terms of precision, recall, micro-$F_1$ and macro-$F_1$ of $A_1$, $B_1$, $C_1$, $D_1$, $E_1$, $F_1$, $G_1$ and $H_1$ have been reported in Tables 10, 11,12 and 13 respectively. It can be observed that performance of the proposed methods (i.e., $G_1$ and $H_1$) are better than others in terms of precision, recall, micro-$F_1$ and macro-$F_1$. Among $G_1$ and $H_1$, $H_1$ is found to be the better than $G_1$.

# 7 Conclusion

We proposed a split and merge classification of hypertext documents. In the split process the hypertext documents is represented in tensor space model. Tensor space model consists of a sixth order tensor for each hypertext document, that is a different vector for each of the different types of features. In this representation, the features extracted from URL or Title are assigned in different tensor components. In the merging process, base level classification has been performed on individual tensor components and combined classification has been obtained by using rough set based ensemble classifier. Two step improvement on the existing classification results of hypertext has been shown. In the first step we achieve better classification results by merging similarity distances. In the second step further improvement of the results has been obtained by merging the output of base level classifiers using rough ensemble classifier. Improvement took place because of the initial split process.

# Acknowledgment

The authors would like to thank the Department of Science and Technology, Government of India, for funding the Center for Soft Computing Research: A National Facility. This paper was done when one of the authors, S. K. Pal, was a J.C. Bose Fellow of the Government of India.

# References

1. Wong, S.K.M., Raghavan, V.V.: Vector space model of information retrieval: a reevaluation. In: Proceedings of the 7th Annual International ACM SIGIR Conference on Research and Development in Information Retrieval, Swinton, UK, pp. 167–185. British Computer Society (1984)
2. McCallum, A., Nigam, K.: A comparison of event models for naive bayes text classification. In: AAAI 1998 Workshop on Learning for Text Categorization (1998)
3. Yang, Y., Slattery, S., Ghani, R.: A study of approaches to hypertext categorization. Journal of Intelligent Information Systems 18(2-3), 219–241 (2002)
4. Saha, S., Murthy, C.A., Pal, S.K.: Rough set based ensemble classifier for web page classification. Fundamentae Informetica 76(1-2), 171–187 (2007)
5. Furnkranz, J.: Web mining. In: The Data Mining and Knowledge Discovery Handbook, pp. 899–920. Springer, Heidelberg (2005)
6. Kan, M.Y., Thi, H.O.N.: Fast webpage classification using url features. In: CIKM 2005: Proceedings of the 14th ACM International Conference on Information and Knowledge Management, pp. 325–326. ACM, New York (2005)

7. Eiron, N., McCurley, K.S.: Analysis of anchor text for web search. In: SIGIR 2003: Proceedings of the 26th Annual International ACM SIGIR Conference on Research and Development in Informaion Retrieval, pp. 459–460. ACM, New York (2003)
8. Chakrabarti, S., Dom, B., Indyk, P.: Enhanced hypertext categorization using hyperlinks. In: SIGMOD 1998: Proceedings of the 1998 ACM SIGMOD International Conference on Management of Data, pp. 307–318. ACM, New York (1998)
9. Chakrabarti, S., Roy, S., Soundalgekar, M.V.: Fast and accurate text classification via multiple linear discriminant projections. The International Journal on Very Large Data Bases 12(2), 170–185 (2003)
10. Brin, S., Page, L.: The anatomy of a large-scale hypertextual Web search engine. Computer Networks and ISDN Systems 30(1-7), 107–117 (1998)
11. Zong, X., Shen, Y., Liao, X.: Improvement of hits for topic-specific web crawler. In: Huang, D.-S., Zhang, X.-P., Huang, G.-B. (eds.) ICIC 2005. LNCS, vol. 3644, pp. 524–532. Springer, Heidelberg (2005)
12. Borisenko, A.I., Tarapov, I.E.: Vector and Tensor Analysis with Applications. Dover Publications, Mineola (1979)
13. Resnik, P.: Signal processing based on multilinear algebra. PhD thesis, Katholieke, University of Leuven, Belgium (1997)
14. Vasilescu, M.A.O., Terzopoulos, D.: Multilinear analysis of image ensembles: Tensorfaces. In: Heyden, A., Sparr, G., Nielsen, M., Johansen, P. (eds.) ECCV 2002. LNCS, vol. 2350, pp. 447–460. Springer, Heidelberg (2002)
15. Kolda, T.G., Bader, B.W., Kenny, J.P.: Higher-order web link analysis using multilinear algebra. In: International Conference on Data Mining. IEEE Press, Los Alamitos (2005)
16. Liu, N., Zhang, B., Yan, J., Chen, Z., Liu, W., Bai, F., Chien, L.: Text representation: From vector to tensor. In: International Conference on Data Mining. IEEE Computer Society, Los Alamitos (2005)
17. Cai, D., He, X., Han, J.: Tensor space model for document analysis. In: Proceedings of ACM SIGIR 2006 Conference, pp. 625–626. ACM, New York (2006)
18. Cai, D., He, X., Han, J.: Beyond streams and graphs: Dynamic tensor analysis. In: International Conference on Knowledge Discovery and Data Mining (SIGKDD 2006), pp. 374–383. ACM, New York (2006)
19. Plakias, S., Stamatatos, E.: Tensor space models for authorship identification. In: Darzentas, J., Vouros, G.A., Vosinakis, S., Arnellos, A. (eds.) SETN 2008. LNCS (LNAI), vol. 5138, pp. 239–249. Springer, Heidelberg (2008)
20. Cohen, W.: Improving a page classifier with anchor extraction and link analysis (2002)
21. Utard, H., Furnkranz, J.: Link-local features for hypertext classification. In: Ackermann, M., Berendt, B., Grobelnik, M., Hotho, A., Mladenič, D., Semeraro, G., Spiliopoulou, M., Stumme, G., Svátek, V., van Someren, M. (eds.) EWMF 2005 and KDO 2005. LNCS (LNAI), vol. 4289, pp. 51–64. Springer, Heidelberg (2006)
22. Angelova, R., Weikum, G.: Graph-based text classification: learn from your neighbors. In: SIGIR 2006: Proceedings of the 29th Annual International ACM SIGIR Conference on Research and Development in Information Retrieval, pp. 485–492. ACM, New York (2006)
23. Xu, Z., King, I., Lyu, M.R.: Web page classification with heterogeneous data fusion. In: WWW 2007: Proceedings of the 16th International Conference on World Wide Web, pp. 1171–1172. ACM, New York (2007)

# A Non-boolean Lattice Derived by Double Indiscernibility

Yukio-Pegio Gunji and Taichi Haruna

Department of Earth and Planetary Sciences, Faculty of Science, Kobe University
yukio@kobe-u.ac.jp

**Abstract.** The central notion of a rough set is the indiscernibility that is based on an equivalence relation. Because an equivalence relation shows strong bondage in an equivalence class, it forms a Galois connection and the difference between the upper and lower approximations is lost. Here, we introduce two different equivalence relations, one for the upper approximation and one for the lower approximation, and construct a composite approximation operator consisting of different equivalence relations. We show that a collection of fixed points with respect to the operator is a lattice and there exists a representation theorem for that construction.

## 1 Introduction

This paper addresses the difference between topological space and rough set theory [1,2] in terms of lattice theory [3,4]. Rough set theory provides a method for data analysis based on the notion of indiscernibility, which is defined by an equivalence relation [5,6]. Because equivalence classes can be used analogously in open sets, similar concepts in topological space can be defined. The upper and lower approximations in rough set theory correspond to the concept of closure and an internal set, respectively [7,8].

On one hand, the closure and internal set are defined under the constraints of topological space (i.e., closed with respect to finite intersection and any union). On the other hand, the upper and lower approximations can be defined without such a constraint. The essential difference between operations in a topological space and approximations in a rough set is the relationship between an element and a set (open set or equivalence class) containing the element. Any elements in an equivalence class have the same equivalence class, a property that is different from that in topological space. The property of a strong equivalence class leads to a fixed point with respect to an approximation that is not found in a closure operator in topological space.

The logical structure of a rough set has been studied in terms of modal logic [7,8,9] and lattice theory [10,11,12,13,14]. A topological space equipped with closure leads to modal logic. A rough set equipped with approximation similarly leads to modal logic, but the approximation operator is redefined by the modal style binary relation differently than is an original approximation in a rough

J.F. Peters et al. (Eds.): Transactions on Rough Sets XII, LNCS 6190, pp. 211–225, 2010.

set [15,16]. A lattice is an ordered set that is closed with respect to join and meet, and happens to be useful in computer science [4,12]. Recently, a lattice was constructed based on the approximation operator of a rough set, while the equivalence relation was generalized with a binary relation [11,12,14,16] (Note that [12] refers to a lattice derived from a specific operator that is defined by an equivalence relation). Indiscernibility is lost due to the generalization and one cannot examine how equivalence relations play a role in lattice structure.

Here, we study approximations and rough sets that are defined by equivalence relations. In particular, we introduce many different equivalence relations and define a pseudo-closure operator based on the upper and lower approximations of different equivalence relations. A lattice is defined by a collection of fixed points of a pseudo-closure operator. If one equivalence relation is included by the other, we show that a derived lattice is a Boolean set lattice; otherwise, a derived lattice is not restricted to a Boolean lattice. Finally, we show that any lattice can be expressed as a collection of fixed points of a pseudo-closure operator. The existence of many different indiscernibilities plays a role in the diversity of lattices.

## 2 Lattice Derived by a Single Indiscernibility

First we review the necessary tools defined in a rough set theory [1,2,5].

**Definition.1 Rough Set.** Given a universal set $U$, let $R \subseteq U \times U$ be an equivalence relation of $U$. For $X \subseteq U$, we define the $R$-upper and $R$-lower approximations of $X$, denoted by $R^*(X)$ and $R_*(X)$, respectively, as follows:

$$R^*(X) = \{x \in U | [x]_R \cap X \neq \emptyset\} \quad (1)$$
$$R_*(X) = \{x \in U | [x]_R \subseteq X\} \quad (2)$$

where $[x]_R$ is an equivalence class of $R$ such that $[x]_R = \{y \in U | xRy\}$.

The difference between the upper and lower approximations produces the boundary of an object $X$, which depends on a given equivalence relation. The lower and upper approximations mimic the internal set and closure operator in a topological space, respectively. If $U$ is a topological space, $X \subseteq U$ and an open set containing an element $x$ is expressed as $O_x$, the closure operator is expressed as $C(X) = \{x \in U| \ O_x \cap X \neq \emptyset\}$ and the internal set is expressed as $X^{int} = \{x \in U| \ O_x \subseteq X\}$. Although in an open set $y \in U_x$ does not imply $U_x = U_y$, $y \in [x]_R$, it implies $[x]_R = [y]_R$ in an equivalence class. This difference is the central notion of indiscernibility.

The Basic properties of the approximations follow. They are used in defining a pseudo-closure operator and a lattice derived by approximations.

**Theorem 2. Basic Properties of Rough Set([2]).** If $R \subseteq U \times U$ is an equivalence relation of $U$, for a subset $X \subset U$ the following statements hold:

$$\text{(i)}\ R_*(X) \subseteq X \subseteq R^*(X) \tag{3}$$
$$\text{(ii)}\ R^*(\emptyset) = R_*(\emptyset) = \emptyset, \tag{4}$$
$$R^*(U) = R_*(U) = U \tag{5}$$
$$\text{(iii)}\ R_*(X \cap Y) = R_*(X) \cap R_*(Y), \tag{6}$$
$$R^*(X \cup Y) = R^*(X) \cup R^*(Y) \tag{7}$$
$$\text{(iv)}\ X \subseteq Y \Rightarrow R_*(X) \subseteq R_*(Y), R^*(X) \subseteq R^*(Y) \tag{8}$$
$$\text{(v)}\ R_*(X \cup Y) \supseteq R_*(X) \cup R_*(Y), \tag{9}$$
$$R^*(X \cap Y) \subseteq R^*(X) \cap R^*(Y) \tag{10}$$
$$\text{(vi)}\ R_*(U - X) = U - R^*(X), \tag{11}$$
$$R^*(U - X) = U - R_*(X) \tag{12}$$
$$\text{(vii)}\ R_*(R_*(X)) = R^*(R_*(X)) = R_*(X) \tag{13}$$
$$\text{(viii)}\ R^*(R^*(X)) = R_*(R^*(X)) = R^*(X) \tag{14}$$

These statements mimic the properties of closure and internal set in a topological space. Statements like $R^*(R_*(X)) = R_*(X)$ and $R_*(R^*(X)) = R^*(X)$ are not found in a topological space. They hold because $y \in [x]_R$ implies $[x]_R = [y]_R$ In fact, it is easy to see that $R_*(X) \subseteq R^*(R_*(X))$. Conversely, in supposing $x \in R^*(R_*(X))$, $[x]_R \cap R_*(X) \neq \emptyset \Leftrightarrow \exists y \in [x]_R,\ y \in R_*(X) \Leftrightarrow \exists y \in [x]_R, [y]_R \subseteq X$. Because $y \in [x]_R$ implies that $[x]_R = [y]_R$, $[x]_R \subseteq X$ and $x \in R_*(X)$. Finally, we obtain $R_*(X) = R^*(R_*(X))$. This kind of operation plays a central role in generating a non-Boolean lattice, given plural equivalence relations.

**Theorem 3. Galois Connection.** Given a universal set $U$ and an equivalence relation $R \subseteq U \times U$, for a subset $X$, $Y \subseteq U$, the following Galois connection holds:

$$R^*(X) \subseteq Y \Leftrightarrow X \subseteq R_*(Y). \tag{15}$$

*Proof.*
It is proved in [10].

**Proposition 4. Properties of Galois Connection.** Given a universal set $U$ and an equivalence relation $R \subseteq U \times U$, for a subset $X \subseteq U$,

$$R^*(R_*(X)) \subseteq X, \tag{16}$$
$$X \subseteq R_*(R^*(X)) \tag{17}$$

holds.

*Proof.*
It is proved in [17] that $R_*R^* : \wp(U) \rightarrow \wp(U)$ is a lattice-theoretical closure operator and $R^*R_* : \wp(U) \rightarrow \wp(U)$ is a lattice-theoretical interior operator.

**Theorem 5. Duality of Fixed Points.** Given a universal set $U$ and an equivalence class $R \subseteq U \times U$, for a subset $X \subseteq U$,

$$R^*(X) = X \Leftrightarrow R_*(X) = X. \tag{18}$$

*Proof.*
It is proved in [10].

Because of the duality of a fixed point, a collection of fixed points of $R_*(X) = X$ forms a set lattice that is Boolean.

**Theorem 6. Lattice Derived from a Single Equivalence Class.** Given a universal set $U$ and an equivalence class $R \subseteq U \times U$, a partially ordered set ordered by inclusion, such as $\langle P; \subseteq \rangle$, with $P = \{X \subseteq U | R_*(X) = X\}$, is a set lattice. Similarly, $\langle Q; \subseteq \rangle$, with $Q = \{X \subseteq U | R^*(X) = X\}$, is also a set lattice.

*Proof.*
It is proved in [11] that if $(f, g)$ is a Galois connection on a complete Boolean lattice such that $f$ is extensive and a self-conjugate, then the set of fixed points of $f$ forms a complete lattice. This is now the case.

As shown in Fig.1, the lattice defined in theorem 6 is a $\mathbf{2}^n$-Boolean lattice, where $n$ is the number of equivalence classes. Each equivalence class corresponds to an atom of the Boolean lattice, and any union of equivalence classes exists in the lattice. In the next section, we introduce the pseudo-closure operator and examine the gap between the upper and lower approximations.

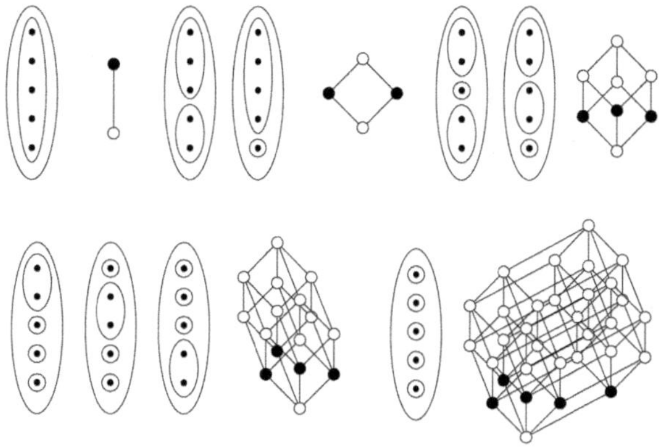

**Fig. 1.** The diagram of nested loops represents an equivalence relation in a set consisting of five elements. The inner loops represent equivalence classes. Right handed diagram of nested loop is a Hasse diagram of a lattice defined by $\langle P = \{X \subseteq U | R_*(X) = X\}; \subseteq \rangle$ from the left handed equivalence relation. The number of equivalence classes, $n$, forms a $\mathbf{2}^n$-Boolean lattice that is a set lattice. An equivalence class corresponds to an atom represented by a black circle in each Hasse diagram.

## 3 Lattice Derived by Ordered Indiscernibility

As mentioned above, the difference between the upper and lower approximations produces the boundary of an object, $X$, that depends on an equivalence relation. Indiscernibility plays a central part in forming a thick boundary. We are interested in how such a thick boundary contributes to a lattice structure. To estimate the role of a thick boundary, we introduce the composition of the upper and lower approximations. In this section we introduce two equivalence relations, where one relation includes the other. The case of a single equivalence relation included by itself also applies. Ordered indiscernibility is also studied under the name of dependency in [18].

**Definition 7. Order of Equivalence Relations.** Given a universal set $U$ and equivalence relations $R$ and $S \subseteq U \times U$, we define an order of equivalence relations $R \leq S$ if, for any $x, y \in U$, $xRy \Rightarrow xSy$.

**Proposition 8. Properties of Equivalence Relations related to Order.** Given a universal set $U$ and equivalence relations $R, S \subseteq U \times U$, suppose $R \leq S$. Then the following statements hold:

$$\text{(i) } S_*(X) \subseteq R_*(X),\ R^*(X) \subseteq S^*(X) \tag{19}$$

$$\text{(ii) } S^*(X) \subseteq R_*(S^*(X)),\ R^*(S_*(X)) \subseteq S_*(X) \tag{20}$$

*Proof.*

(i) We would prove $x \in S_*(X) \Rightarrow x \in R_*(X)$. The statement is equivalent to:$[x]_S \subseteq X \Rightarrow [x]_R \subseteq X$. Then under the assumption $[x]_S \subseteq X$, we would prove $[x]_R \subseteq X$. Supposing $y \in [x]_R$, $xRy$. From $R \leq S$, we obtain $xSy$ and then $y \in [x]_S$. Because of the assumption that $[x]_S \subseteq X$ and $y \in [x]_S$, $y \in X$. Finally, we obtain $[x]_R \subseteq X$. Similarly, we would prove $R^*(X) \subseteq S^*(X)$, that is $x \in R^*(X) \Rightarrow x \in S^*(X)$ from the assumption such that $[x]_R \cap X \neq \emptyset$, $\exists y \in [x]_R$, $y \in X$. Then, we obtain $xRy$ and from $R \leq S$, $xSy$. This means that $y \in [x]_S$. Thus, there exists $y \in X$ such that $y \in [x]_S$. Finally we obtain $[x]_S \cap X \neq \emptyset$.

(ii) From (i) and Theorem 2-(vii), $S^*(X) = S_*(S^*(X)) \subseteq R_*(S^*(X))$. Similarly, $R^*(S_*(X)) \subseteq S^*(S_*(X)) = S_*(X)$.

**Theorem 9. Lattices Ordered by Equivalence Relations.** Given a universal set $U$, an order of equivalence relations $R, S \subseteq U \times U$ is defined by $R \leq S$. Then $\langle P; \subseteq\rangle$ with $P = \{X \subseteq U | R_*(S^*(X)) = X\}$ is a set lattice. Similarly, $\langle Q; \subseteq\rangle$ with $Q = \{X \subseteq U | R^*(S_*(X)) = X\}$ is also a set lattice. The notation $\langle P; R\rangle$ implies a partially ordered set $P$ that is ordered by relation $R$.

*Proof.*

From theorem 2-(i), $R_*(S^*(X)) \subseteq S^*(X)$, and from proposition 8-(ii), $S^*(X) \subseteq R_*(S^*(X))$. Thus, $S^*(X) = R_*(S^*(X))$. Since $R_*(S^*(X)) = X$ in $P$, we obtain $S^*(X) = X$ in $P$. This means that a lattice $\langle P; \subseteq\rangle$ with $P = \{X \subseteq U | R_*(S^*(X)) = X\}$ is equivalent to $\langle \{X \subseteq U | S^*(X) = X\}; \subseteq\rangle$, which is a set lattice from Theorem 6. Similarly, it is easy to see that $\langle Q; \subseteq \rangle$ with $Q = \{X \subseteq U | R^*(S_*(X)) = X\}$ is equivalent to $\langle \{X \subseteq U | S_*(X) = X\}; \subseteq \rangle$, which is a set lattice.

From theorem 9, if two equivalence relations are ordered, the composition of the two approximations is reduced to one approximation. Thus, the effect of a thick boundary of an object is lost and the derived lattice consists of all unions of equivalence classes.

**Corollary 10. Lattices driven by Single Equivalence Relations.** Given a universal set $U$, $R \subseteq U \times U$ is defined as an equivalence relation. Then, $\langle P; \subseteq \rangle$ with $P = \{X \subseteq U | R_*(R^*(X)) = X\}$ is a set lattice. Similarly, $\langle Q; \subseteq \rangle$ with $Q = \{X \subseteq U | R^*(R_*(X)) = X\}$ is a set lattice.

*Proof.*
It is simple to prove from theorem 9, since $R \leq R$.

Corollary 10 shows that the composition of the lower and upper approximations is reduced to a single approximation. Even if the recognition of objects depends on the approximation based on an equivalence relation, the structure of a lattice is invariant. This is a Boolean lattice. How does the diversity of a lattice structure arise? In the next section we introduce two equivalence relations that are not ordered and show that the composition of the upper and lower approximations cannot be reduced to one.

# 4 Lattice Driven by Plural Indiscernibility

In this section we introduce two non-ordered equivalence relations and the operator that is composed of the upper and lower approximations, where the upper approximation is based on one relation and the lower one is based on the other relation. First we examine this operator in terms of the closure operator. In fact, since this operator does not satisfy all the conditions of the closure operator, we call this the pseudo-closure operator.

**Theorem 11. Pseudo-Closure Derived from Plural Equivalence Relations.** Given a universal set $U$, $R$ and $S \subseteq U \times U$ are defined as different equivalence relations. The operations $T$ and $K$ are defined by $T = S_*R^*$, $K = R^*S_*$. Then, for $X, Y \subseteq U$,

$$\text{(i) } X \subseteq Y \Rightarrow T(X) \subseteq T(Y),\ K(X) \subseteq K(Y) \tag{21}$$

$$\text{(ii) } T(T(X)) = T(X),\ K(K(X)) = K(X) \tag{22}$$

*Proof.*

(i) Supposing $x \in T(X)$, from Theorem 2-(vii) $x \in S_*(R^*(X)) \Leftrightarrow [x]_S \subseteq R^*(X) \Leftrightarrow \forall y \in [x]_S,\ [y]_R \cap X \neq \emptyset \Leftrightarrow \forall y \in [x]_S,\ \exists z \in [y]_R,\ z \in X$. From the assumption $z \in X \subseteq Y$, $\forall y \in [x]_S$, $\exists z \in [y]_R$, $z \in Y \Leftrightarrow \forall y \in [x]_S$, $[y]_R \cap Y \neq \emptyset \Leftrightarrow [x]_S \subseteq R^*(Y) \Leftrightarrow x \in S_*(R^*(Y)) = T(Y)$.Similarly, supposing $x \in K(X)$, $x \in R^*(S_*(X)) \Leftrightarrow [x]_R \cap S_*(X) \neq \emptyset \Leftrightarrow \exists y \in [x]_R$, $[y]_S \subseteq X$. From the assumption $\exists y \in [x]_R$, $[y]_S \subseteq Y \Leftrightarrow y \in [x]_R$, $y \in S_*(Y) \Leftrightarrow [x]_R \cap S_*(Y) \neq \emptyset \Leftrightarrow x \in R^*(S_*(Y)) = K(Y)$.

(ii) Supposing $x \in T(T(X))$, then

$$\begin{aligned} & x \in S_*(R^*(S_*(R^*(X)))) \\ \Leftrightarrow\ & [x]_S \subseteq R^*(S_*(R^*(X))) \Leftrightarrow \forall y \in [x]_S, [y]_R \cap S_*(R^*(X)) \neq \emptyset \\ \Leftrightarrow\ & \forall y \in [x]_S, \exists z \in [y]_R, [z]_S \subseteq R^*(X) \\ \Leftrightarrow\ & \forall y \in [x]_S, \exists z \in [y]_R, \forall w \in [z]_S, [w]_R \cap X \neq \emptyset. \end{aligned}$$

From $z \in [y]_R$, $[y]_R = [z]_R$. Because it is trivially true that $z \in [z]_S$, $z$ can be applied to the statement, $w \in [z]_S$, $[w]_R \cap X \neq \emptyset$. Thus we obtain $[z]_R \cap X \neq \emptyset$ and $[y]_R \cap X \neq \emptyset$. Finally,

$$\begin{aligned} & x \in S_*(R^*(S_*(R^*(X)))) \\ \Leftrightarrow\ & \forall y \in [x]_S, \exists z \in [y]_R, \forall [w] \in [y]_S, [w]_R \cap X \neq \emptyset \Rightarrow \forall y \in [x]_S, [y]_R \cap X \neq \emptyset \\ \Leftrightarrow\ & [x]_S \subseteq R^*(X) \Leftrightarrow x \in S_*(R^*(X)) \Leftrightarrow x \in T(X). \end{aligned}$$

Conversely, supposing $x \in T(X)$, $x \in S_*(R^*(X)) \Leftrightarrow [x]_S \subseteq R^*(X) \Leftrightarrow \forall y \in [x]_S$, $[y]_R \cap X \neq \emptyset$. Because $y \in [y]_R$, there exists an element in both $[x]_S$ and $[y]_R$. This implies $\exists z \in [y]_R$, $z \in [x]_S$, and $[x]_S = [z]_S$. Thus, in the statement $\forall y \in [x]_S$, $[y]_R \cap X \neq \emptyset$, we can replace $[x]_S$ by $[z]_S$ and $y$ by $w$. Therefore, the statement $\forall w \in [z]_S$, $[w]_R \cap X \neq \emptyset$ holds under the condition $\exists z \in [y]_R$. It follows that

$$\begin{aligned} & x \in S_*(R^*(X)) \\ \Leftrightarrow\ & \forall y \in [x]_S, [y]_R \cap X \neq \emptyset \Rightarrow \forall y \in [x]_S, \exists z \in [y]_R, \forall w \in [z]_S, [w]_R \cap X \neq \emptyset \\ \Leftrightarrow\ & \forall y \in [x]_S, \exists z \in [y]_R, \forall w \in [z]_S, w \in R^*(X) \\ \Leftrightarrow\ & x \in S_*(R^*(S_*(R^*(X)))) \Leftrightarrow x \in T(T(X)). \end{aligned}$$

We introduce a lattice as a collection of fixed points with respect to the pseudo-closure operators. Because join and meet for a lattice are not defined by union and intersection, information with respect to combinations of equivalence classes is lost in a derived lattice. Thus, we can see not only Boolean but non-Boolean lattices.

**Theorem 12. Lattice driven by Pseudo-Closure.** Given a universal set $U$, $R$ and $S \subseteq U \times U$ are defined as different equivalence relations, and $T = S_* R^*$ and $K = R^* S_*$. A partially ordered set $\langle \mathfrak{L}_T; \subseteq \rangle$, with $\mathfrak{L}_T = \{X \subseteq U | T(X) = X\}$, is a lattice. Similarly, $\langle \mathfrak{L}_K; \subseteq \rangle$, with $\mathfrak{L}_K = \{X \subseteq U | K(X) = X\}$, is a lattice. Meet and join for a lattice are defined by the following:

For $X, Y \in \mathfrak{L}_T$

$$X \wedge Y = T(X \cap Y), \tag{23}$$
$$X \vee Y = T(X \cup Y). \tag{24}$$

Similarly, for $X, Y \in \mathfrak{L}_K$

$$X \wedge Y = K(X \cap Y), \tag{25}$$
$$X \vee Y = K(X \cup Y). \tag{26}$$

Indeed,$\langle \mathfrak{L}_T; \subseteq \rangle$and$\langle \mathfrak{L}_K; \subseteq \rangle$are complete lattices.

*Proof.*
We would prove that $\langle \mathfrak{L}_T; \subseteq \rangle$ with $\mathfrak{L}_T = \{X \subseteq U | T(X) = X\}$ is a lattice. The statement on the operation $K$ can be proven in a similar manner.

(i) First we would check that meet and join are well-defined. Because $X \cap Y \subseteq X$ and $X \cap Y \subseteq Y$, applying $T = S_* R^*$ to these inequalities leads to $T(X \cap Y) \subseteq T(X)$ and $T(X \cap Y) \subseteq T(Y)$, per theorem 2-(iv). Thus, we obtain $X \wedge Y = T(X \cap Y) \subseteq T(X) = X$ and $X \wedge Y = T(X \cap Y) \subseteq T(Y) = Y$, implying that $X \wedge Y$ is a lower bound of $\{X, Y\}$. Supposing that $Z \in \mathfrak{L}_T$ is a lower bound of $\{X, Y\}$, we obtain $Z \subseteq X$ and $Z \subseteq Y$, and then $Z \subseteq X \cap Y$. It follows that $T(Z) \subseteq T(X \cap Y)$ and $Z = T(Z) \subseteq T(X \cap Y) = X \wedge Y$. This implies that $X \wedge Y$ is the greatest lower bound. Similarly, for $X \subseteq X \cup Y$ and $Y \subseteq X \cup Y$, we obtain $X = T(X) \subseteq T(X \cup Y) = X \vee Y$ and $Y = T(Y) \subseteq T(X \cup Y) = X \vee Y$, implying that $X \vee Y$ is an upper bound of $\{X, Y\}$. Supposing that $Z \in \mathfrak{L}_T$ is an upper bound, $X \subseteq Z$, $Y \subseteq Z$ and $X \cup Y \subseteq Z$, and $X \vee Y = T(X \cup Y) \subseteq T(Z) = Z$. This implies that $X \vee Y$ is the least upper bound.

(ii) We would prove that a partially ordered set $\langle \mathfrak{L}_T; \subseteq \rangle$ is closed with respect to join and meet. In order to show $X \wedge Y \in \mathfrak{L}_T$, we have to show $T(X \cap Y) \in \mathfrak{L}_T$. In fact, $T(X \wedge Y) = T(T(X \cap Y)) = T(X \cap Y) = X \wedge Y$. Thus, $X \wedge Y \in \mathfrak{L}_T$. Similarly, $T(X \vee Y) = T(T(X \cup Y)) = T(X \cup Y) = X \vee Y$ and $X \vee Y \in \mathfrak{L}_T$.

(iii) For any subset $M \subseteq \mathfrak{L}_T$, $T(\wedge M) = TT(\cap M)$. For any $X_i \in M, TT(X_i) = T(X_i)$ and $TT(\cap M) = T(\cap M) = \wedge M$. Thus we obtain $\wedge M \in \mathfrak{L}_T$. Similarly we obtain $\vee M \in \mathfrak{L}_T$. It can be verified straightforwardly that $\langle \mathfrak{L}_K; \subseteq \rangle$ is also a complete lattice.

next proposition shows that an element of a lattice in theorem 12 is expressed as a union of equivalence classes based on a single equivalence relation.

**Lemma 13. Elements of a lattice.** Given a universal set $U$, $R$ and $S \subseteq U \times U$ are defined as different equivalence relations. The operations $T$ and $K$ are defined by $T = S_* R^*$, $K = R^* S_*$. Then, for $X \subseteq U$,

$$\text{(i) } T(X) = X \Rightarrow S_*(X) = X \tag{27}$$
$$\text{(ii) } K(X) = X \Rightarrow R^*(X) = X \tag{28}$$

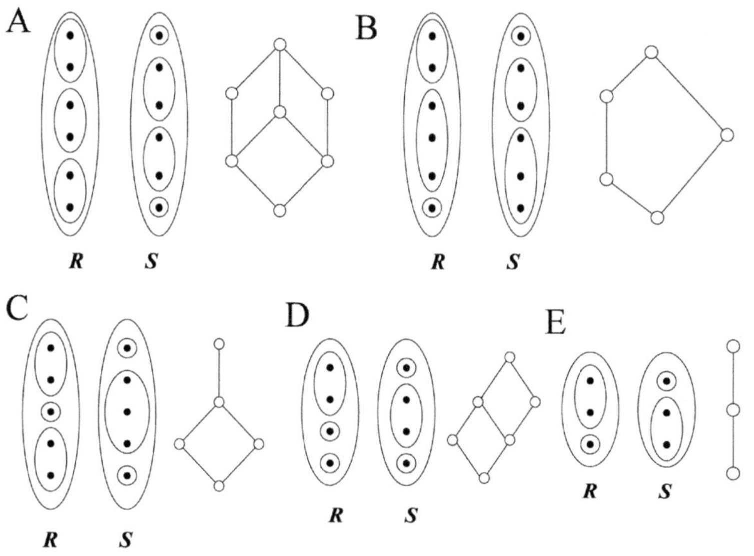


**Fig. 2.** The diagram of nested loops represents an equivalence relation, $R$ or $S$, in a universal set. A lattice derived from a pair of equivalence relations, $R$ and $S$ is expressed by a Hasse diagram. Each lattice is defined by $\langle \mathfrak{L}_T; \subseteq \rangle$ with $\mathfrak{L}_T = \{X \subseteq U | T(X) = X\}$. Each elements of a lattice is expressed as a union of equivalence classes of $S$.

*Proof.*

(i) Supposing $S_*R^*(X) = X$, we obtain $S_*S_*R^*(X) = S_*(X)$. From the left-handed form of theorem 2-(vii), $S_*S_*R^*(X) = S_*R^*(X)$. Then, $S_*R^*(X) = S_*(X)$ and, finally, $S_*(X) = X$.
(ii) In addition to (i), it is simple to obtain $R^*(X) = X$ by using theorem 2-(viii).

**Theorem 14. Galois connections in a derived lattice.** Given a universal set $U$, $R$ and $S \subseteq U \times U$ are defined as different equivalence relations. Then, for $X, Y \in \mathfrak{L}_T = \{X \subseteq U | T(X) = X\}$, the following Galois connection holds:

$$R^*(X) \subseteq Y \Leftrightarrow X \subseteq S_*(Y). \tag{29}$$

*Proof.*
In supposing $X \subseteq S_*(Y)$, we obtain $S_*(R^*(X)) \subseteq S_*(Y)$. From lemma 13, $S_*(R^*(X)) = R^*(X)$ and $S_*(Y) = Y$. We then obtain $R^*(X) \subseteq Y$. Conversely, in supposing $R^*(X) \subseteq Y$ we obtain $S_*(R^*(X)) \subseteq S_*(Y)$ from theorem 2-(iv). Then, $X \subseteq S_*(Y)$.

Fig. 2 shows some examples of a lattice defined in theorem 12. As mentioned in lemma 13, if a lattice is defined as a collection of fixed points with respect to the operator $T$, elements of a lattice are expressed as a union of equivalence classes of $S$. Here, we enumerate all elements of the lattices in Fig. 2A, where

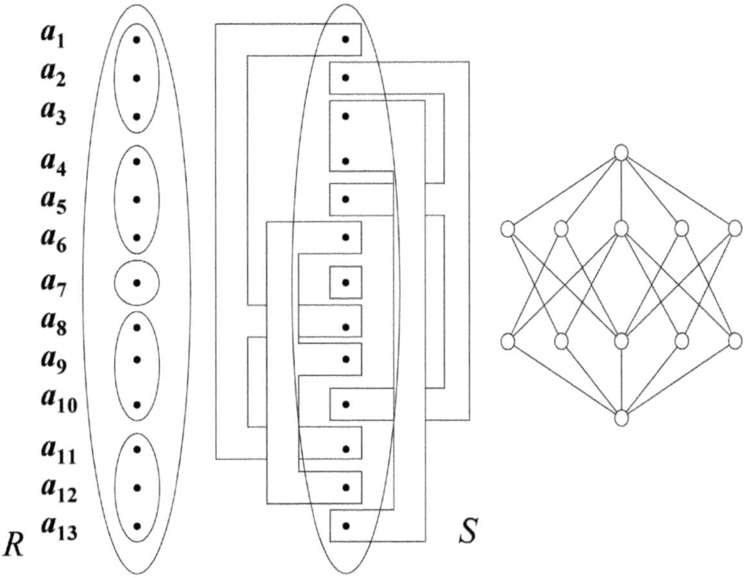


**Fig. 3.** Hasse diagram of an orthocomplemented lattice (right) defined by a collection of $T(X) = X$. A universal set is $\{a_1, a_2, \ldots, a_{13}\}$. Equivalence classes of $R$ are represented by loops, and those of $S$ are represented by polygons.

all elements of a universal set are denoted by $a, b, c, d, e, f$ in descending order. Equivalence classes of $R$ are $\{a, b\}, \{c, d\}$ and $\{e, f\}$, and equivalence classes of $S$ are $\{a\}, \{b, c\}, \{d, e\}$ and $\{f\}$ if one collects all fixed points with respect to $T$, that is $\{\{a\}, \{f\}, \{a, b, c\}, \{a, f\}, \{d, e, f\}, X\}$.

A more complex orthocomplemented lattice is also constructed by a collection of fixed points with respect to $T$, as shown in Fig. 3. Atoms of a lattice are equivalence classes of $S$. It is easy to verify that a subset $\{a_1, a_8, a_{11}\}$ which itself is an equivalence class of $S$, $\{a_1, a_8, a_{11}\}$, forms a fixed point by $S_*R^*(\{a_1, a_8, a_{11}\}) = S_*(\{a_1, a_2, a_3, a_8, a_9, a_{10}, a_{11}, a_{12}, a_{13}\}) = \{a_1, a_8, a_{11}\}$.

It is easy to see that there exists a lattice isomorphism between $\mathfrak{L}_T$ and $\mathfrak{L}_K$ To verify this property, we first show a Galois connection between $\langle \mathfrak{L}_T; \subseteq \rangle$ and $\langle \mathfrak{L}_K; \subseteq \rangle$.

**Theorem 15. Galois connection between $\mathfrak{L}_T$ and $\mathfrak{L}_K$.** Given a universal set $U$, $R$ and $S \subseteq U \times U$ are defined as different equivalence relations. The pair $(R^*, S_*)$ is a Galois connection between $\langle \mathfrak{L}_T; \subseteq \rangle$ and $\langle \mathfrak{L}_K; \subseteq \rangle$, where $T = S_*R^*$ and $K = R^*S_*$.

*Proof.*
For $X \in \mathfrak{L}_T$ and $Y \in \mathfrak{L}_K$, supposing $X \subseteq S_*(Y)$, we obtain $R^*(X) \subseteq R^*(S_*(Y)) = KY = Y$. Conversely, supposing $R^*(X) \subseteq Y$, we obtain $S_*(R^*(X)) \subseteq S_*(Y)$. Because $S_*(R^*(X)) = TX = X$, $X \subseteq S_*(Y)$. Thus, $R^*(X) \subseteq Y \Leftrightarrow X \subseteq S_*(Y)$.

**Theorem 16. Lattice Isomorphism among $\mathfrak{L}_T, \mathfrak{L}_K, \mathfrak{L}_N$ and $\mathfrak{L}_M$.** Given a universal set $U$, $R$ and $S \subseteq U \times U$ are defined as equivalence relations, and the operators are defined by $T = S_*R^*$, $K = R^*S_*$, $M = S^*R_*$ and $N = R_*S^*$. A partially ordered set $\langle \mathfrak{L}_M; \subseteq \rangle$ with $\mathfrak{L}_M = \{X \subseteq U | M(X) = X\}$ and $\langle \mathfrak{L}_N; \subseteq \rangle$ with $\mathfrak{L}_N = \{X \subseteq U | N(X) = X\}$ are lattices, and all lattices are isomorphic such that

$$\langle \mathfrak{L}_T; \subseteq \rangle \cong \langle \mathfrak{L}_K; \subseteq \rangle \cong \langle \mathfrak{L}_M; \subseteq \rangle \cong \langle \mathfrak{L}_N; \subseteq \rangle \tag{30}$$

*Proof.*
First we prove $\langle \mathfrak{L}_T; \subseteq \rangle \cong \langle \mathfrak{L}_K; \subseteq \rangle$.

(i) A map $\varphi : \mathfrak{L}_T \to \mathfrak{L}_K$ is defined as a lattice homomorphism, where, for $X \in \mathfrak{L}_T$, $\varphi(X) = R^*(X)$. First, we would show that $\varphi : \mathfrak{L}_T \to \mathfrak{L}_K$ is well defined. For $X \in \mathfrak{L}_T$ , $S_*R^*(X) = X$. Thus, $\varphi(X) = R^*(X) = R^*(S_*R^*(X)) = K(R^*(X)) = K(\varphi(X))$. This implies that $\varphi(X) \in \mathfrak{L}_K$. Indeed, $\varphi(X \vee Y) = \varphi(T(X \cup Y)) = R^*S_*R^*(X \cup Y) = R^*S_*(R^*(X) \cup R^*(Y)) = K(\varphi(X) \cup \varphi(Y)) = \varphi(X) \vee \varphi(Y)$. Also, $\varphi(X \wedge Y) = R^*(X \wedge Y) = R^*(T(X \cap Y)) = R^*(S_*R^*(X \cap Y)) = R^*S_*(R^*(X \cap Y))$. Since $\varphi(X \wedge Y) \in \mathfrak{L}_K$ , $\varphi(X \wedge Y) = R^*(X \cap Y)$ is a fixed point with respect to $R^*S_*$, $R^*S_*(R^*(X \cap Y)) = R^*(X \cap Y)$. Thus, $R^*(X \cap Y) = R^*(S_*R^*(X) \cap S_*R^*(Y))$, since $X$ and $Y \in \mathfrak{L}_T$ are fixed points with respect to $S_*R^*$. Therefore, from theorem 2-(iii), $R^*(S_*R^*(X) \cap S_*R^*(Y)) = R^*S_*(R^*(X) \cap R^*(Y)) = K(R^*(X)?R^*(Y)) = K(\varphi(X) \cap \varphi(Y)) = \varphi(X) \wedge \varphi(Y)$.

(ii) First we would show that $\varphi : \mathfrak{L}_T \to \mathfrak{L}_K$ is an injection. Supposing $\varphi(X) = \varphi(Y)$ for $X, Y \in \mathfrak{L}_T$, we obtain $R^*(X) = R^*(Y)$. Thus, applying $S_*$ to both sides, $S_*R^*(X) = S_*R^*(Y)$. Because $X, Y \in \mathfrak{L}_T$ , we obtain $S_*R^*(X) = X$, $S_*R^*(Y) = Y$. Thus, we obtain $X = Y$. This implies that $\varphi(X) = \varphi(Y) \Rightarrow X = Y$, and that $\varphi$ is an injection. Second, we would prove that $\varphi : \mathfrak{L}_T \to \mathfrak{L}_K$ is a surjection. Since $Y \in \mathfrak{L}_K$ is a fixed point with respect to $R^*S_*$, for any $Y \in \mathfrak{L}_K$ , $Y = R^*S_*(Y) = \varphi(S_*(Y))$. Also, from $S_*R^*(S_*(Y)) = S_*(R^*S_*(Y)) = S_*(Y)$, $S_*(Y)$ is a fixed point with respect to $S_*R^*$ and $S_*(Y) \in \mathfrak{L}_T$. This implies that $\varphi$ is a surjection.

(iii) It is easy to show that $\langle \mathfrak{L}_M; \subseteq \rangle$ and $\langle \mathfrak{L}_N; \subseteq \rangle$ are lattices by theorem 9. A map $\psi : \mathfrak{L}_T \to \mathfrak{L}_M$ is defined for $X \in \mathfrak{L}_T$, $\psi(X) = U - X$. In particular, elements in $\langle \mathfrak{L}_M; \subseteq \rangle$ are ordered opposite those in $\langle \mathfrak{L}_T; \subseteq \rangle$.Because $\psi(X) = U - X = U - S_*R^*(X) = S^*(U - R^*(X)) = S^*(R_*(U - X)) \in \mathfrak{L}_M$ , the map is well defined. If $\psi^{-1}(Y) = U - Y$ for $Y \in \mathfrak{L}_M$ then $\psi^{-1}(Y) = U - Y = U - S^*R_*(Y) = S_*(U - R_*(Y)) = S_*(R^*(U - Y)) \in \mathfrak{L}_T$ . Thus, $\psi^{1}$ is also well defined. For any $X \in \mathfrak{L}_T, \psi^{-1}(\psi(X)) = U - S^*(R_*(U - X)) = S_*(U - R_*(U - X)) = S_*(R^*(U - (U - X))) = S_*(R^*(X)) = X$. Thus, $\psi : \mathfrak{L}_T \to \mathfrak{L}_M$ is a bijection. If $X \subseteq Y$, then $U - Y \subseteq U - X$. Therefore, we obtain that $S^*(R_*(U - Y))) \subseteq S^*(R_*(U - X)))$. It can be easily shown that $\psi(X \wedge Y) = \psi(X) \vee \psi(Y)$, where $X \vee Y = M(X \cup Y)$ and $X \wedge Y = M(X \cap Y)$ in $\langle \mathfrak{L}_M, \subseteq \rangle$. First, $\psi(X \wedge Y) = U - X \wedge Y = U - T(X \cap Y) = U - S_*R^*(X \cap Y) = S^*(U - R^*(X \cap Y)) = S^*(R_*(U - (X \cap Y)) = M((U - X) \cup (U - Y)) = M(\psi(X) \cup \psi(Y)) = \psi(X) \vee \psi(Y)$.

Next, $\psi(X \vee Y) = \psi(T(X \cup Y)) = U - S_* R^*(X \cup Y) = S^*(U - R^*(X \cup Y)) = S^*(R_*(U - (X \cup Y)) = S^*(R_*((U - X) \cap (U - Y))) = M(\psi(X) \cap \psi(Y)) = \psi(X) \wedge \psi(Y)$. Isomorphism is verified straightforwardly.

(iv) Since $R^*$ and $R_*$ are mutually dual, $\mathfrak{L}_M$ is dually order isomorphic to $\mathfrak{L}_T$. $\mathfrak{L}_N$ is dually order isomorphic to $\mathfrak{L}_K$ by the same argument.

Given two equivalence relations, there can be four kinds of operations consisting of upper and lower approximations, such as $S_* R^*$, $S^* R_*$, $R_* S^*$ and $R^* S_*$. Theorem 16 shows that a lattice obtained as a collection of fixed points of those operators is unique up to isomorphism. Fig. 4 shows an example of four lattices as a collection of fixed points with respect to $S_* R^*$, $S^* R_*$, $R_* S^*$ and $R^* S_*$. Although some of them are opposite, they are the same structure.

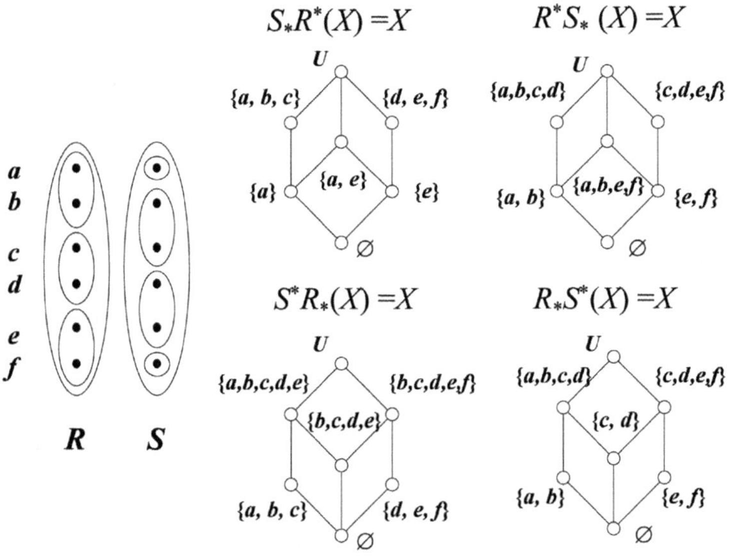


**Fig. 4.** Given two equivalence relations $R$ and $S$ on a universal set $U$, a collection of fixed points with respect to four kinds of operators forms a unique lattice up to isomorphism

## 5 Representation Theorem for Complete Lattices in Terms of Double Indiscernibility

In the above section we observe that two kinds of equivalence relations form a complete lattice. Conversely, we can verify that any lattice can be represented by a collection of fixed points with respect to an operator $T$. First we define a universal set and equivalence relations derived from a given lattice.

**Definition 17. Universal set derived from a lattice.** Let $\langle L; \leq \rangle$ be a lattice. A universal set $U_L \subseteq L \times L$, derived from $L$ is defined by $U_L = \{\langle x, y\rangle \in L \times L | \langle x, y\rangle \notin \leq\}$. Two equivalence classes derived from $L$, denoted by $R$ and $S \subseteq U_L \times U_L$, are defined by $\langle x, y\rangle R\langle z, y\rangle$ and $\langle x, y\rangle S\langle x, w\rangle$.

**Lemma 18.** Let $\langle L; \leq \rangle$ be a lattice. Given $x$ in $L$, the lower and upper sets of $x$, $X_x^l$ and $X_x^u$, are defined by $X_x^l = \{\langle y, z\rangle \in U_L | y \leq x\}$ and $X_x^u = \{\langle y, z\rangle \in U_L | x \leq z\}$, respectively. Then

$$\text{(i) } R^*(X_x^l) = U - X_x^u \tag{31}$$

$$\text{(ii) } S_*(U - X_x^u) = X_x^l \tag{32}$$

*Proof.*

(i) Supposing $\langle y, z\rangle \in R^*(X_x^l)$, $[\langle y, z\rangle]R \cap X_x^l \neq \emptyset \Leftrightarrow \exists\langle w, z\rangle \in X_x^l$, $\langle w, z\rangle \in [\langle y, z\rangle]_R$

$$\langle y, z\rangle \notin R_x^*(X_x^l) \Leftrightarrow \forall\langle w, z\rangle \in X_x^l, \langle w, z\rangle \notin [\langle y, z\rangle]_R.$$

From $\langle w, z\rangle \notin [\langle y, z\rangle]_R$, $w \leq z$. Because $\forall\langle w, z\rangle \in X_x^l$ implies that any $w \leq x$, $z$ is larger than the maximal $w$ that is $x$. Then $x \leq z$ and

$$\langle y, z\rangle \notin R_x^*(X_x^l) \Leftrightarrow \langle y, z\rangle \in \{\langle y, z\rangle \in U_L | x \leq z\} = X_x^u.$$

Finally, we obtain $R^*(X_x^l) = U - X_x^u$.

(ii) Supposing $\langle y, z\rangle \in S_*(U - X_x^u)$,

$$\begin{aligned}\langle y, z\rangle \in S_*(U - X_x^u) &\Leftrightarrow [\langle y, z\rangle]_S \subseteq U - X_x^u \\ &\Leftrightarrow \forall\langle y, z\rangle \in [\langle y, z\rangle]_S \Rightarrow \langle y, w\rangle \notin X_x^u \\ &\Leftrightarrow \forall\langle y, w\rangle \in X_x^u \Rightarrow \langle y, w\rangle \notin [\langle y, z\rangle]_S.\end{aligned}$$

From $\langle y, w\rangle \notin [\langle y, z\rangle]_S$, we obtain $y \leq w$. Because it holds for any $\langle y, w\rangle \in X_x^u$ (i.e. for $\langle y, w\rangle$ with $x \leq w$, $y$ is smaller than the minimal $w$ that is $x$). Then, $y \leq x$ and

$$\langle y, z\rangle \in S_*(U - X_x^u) \Leftrightarrow \langle y, z\rangle \in \{\langle y, z\rangle \in U_L | y \leq x\} = X_x^l.$$

Finally we obtain the following representation theorem.

**Theorem 19. Representation Theorem.** Let $L$ be a lattice. Then the map $\eta : \langle L; \leq \rangle \to \langle \mathfrak{L}_T; \subseteq \rangle$ defined by

$$\eta(x) = X_x^l. \tag{33}$$

for $x \in L$, is an isomorphism of $L$ onto $\mathfrak{L}_T$, where $T = S_*R^*$, and $S$ and $R \subseteq U_L \times U_L$, are equivalence relations on $U_L$, with the inverse of $\eta$ given by $\eta^{-1}(X) = \vee\{y \in L | \langle y, z\rangle \in X\}$ for $X \in \mathfrak{L}_T$.

*Proof.*
The map $\eta$ is well defined, since $T(\eta(x)) = T(X_x^l) = X_x^l$ from lemma 18. Isomorphism is verified by $\eta^{-1}(\eta(x)) = \eta^{-1}(X_x^l) = \vee\{y \in L | \langle y, z\rangle \in X_x^l\} = \vee\{y \in L | y \leq x\} = x$. It can be verified straightforwardly that $\eta$ is a lattice homomorphism.

Therefore, we can represent any lattice as a collection of fixed points with respect to an operator $T$ consisting of two equivalence relations.

## 6 Conclusion

Here we concentrate on the concept of indiscernibility based on equivalence relations, because indiscernibility is a central notion of a rough set but not of a topological space. Because indiscernibility approximation operators form a Galois connection that shows a strong bondage between two perspectives, a collection of a fixed point with respect to approximation operator forms a trivial set lattice. There is no diversity in terms of lattice structure.

A diversity of lattice structure results from a discrepancy between two equivalence relations. In a double equivalence relations an operator consists of an upper approximation based on one equivalence relation and a lower approximation based on the other. A Galois connection where $R^*(X) \subseteq Y \Leftrightarrow X \subseteq S_*(Y)$ does not hold because $R$ is different from $S$. If we collect fixed points with respect to $S_*R^*$, $R^*(X) \subseteq Y \Leftrightarrow X \subseteq S_*(Y)$ holds in that collection (i.e., in a derived lattice). Collecting fixed points results in a loss of information and join and meet in a lattice cannot be defined by union and intersection, respectively. A wide variety of lattices can be accessed as is verified by the existence of the representation theorem.

## References


1. Pawlak, Z.: Information systems-theoretical foundations. Information Systems 6, 205–218 (1981)
2. Pawlak, Z.: Rough Sets. Intern. J. Comp. Inform. Sci. 11, 341–356 (1982)
3. Birkhoff, G.: Lattice Theory, Coll. Publ., XXV. American Mathematical Society, Providence (1967)
4. Davey, B.A., Priestley, H.A.: Introduction to Lattices and Order, 2nd edn. Cambridge University Press, Cambrdige (2002)
5. Polkowski, L.: Rough Sets. In: Mathematical Foundations. Physical-Verlag/Springer, Heidelberg (2002)
6. Doherty, P., Lyukaszewicz, W., Skowron, A., Szalas, A.: Knowledge Representation Techniques. In: A rough Set Approach. Springer, Berlin (2006)
7. Orlowska, E.: Logic approach to information systems. Fundamenta Informaticae 8, 359–378 (1985)
8. Vakarelov, D.: Modal logic for knowledge representation systems. In: Meyer, A.R., Taitslin, M.A. (eds.) Logic at Botik 1989. LNCS, vol. 363, pp. 257–277. Springer, Heidelberg (1989)
9. Orlowska, E.: Modal logics in the theory of information systems. Z. Math. Logik u. Grund. d. Math. 30, 213–222 (1984)

10. Järvinen, J.: Pawlakfs information systems in terms of Galois connections and functional dependencies. Fundamenta Informaticae 75, 315–330 (2007)
11. Järvinen, J., Kondo, M., Kortelainen, J.: Modal-like operators in Boolean lattices, Galois connections and fixed points. Fundamenta Informaticae 76, 129–145 (2007)
12. Järvinen, J.: Lattice theory for rough sets. In: Peters, J.F., Skowron, A., Düntsch, I., Grzymała-Busse, J.W., Orłowska, E., Polkowski, L. (eds.) Transactions on Rough Sets VI. LNCS, vol. 4374, pp. 400–498. Springer, Heidelberg (2007)
13. Yao, Y.Y.: A comparative study of formal concept analysis and rough set theory in data analysis. In: Tsumoto, S., Słowiński, R., Komorowski, J., Grzymała-Busse, J.W. (eds.) RSCTC 2004. LNCS (LNAI), vol. 3066, pp. 59–68. Springer, Heidelberg (2004)
14. Yao, Y.Y.: Concept lattices in rough set theory. In: Fuzzy Information, Processing NAFIPS f04, vol. 2, pp. 796–801 (2004)
15. Gediga, G., Düntsch, I.: Modela-style operators in qualitative data analysis. In: Proceedings of the 2002 IEEE International Conference on Data Mining, pp. 155–162 (2002)
16. Yao, Y.Y., Wong, S.K.M., Lin, T.Y.: A review of rough set models. In: Lin, T.Y., Cercone, N. (eds.) Rough Sets and Data Mining: Analysis for Imprecise Data, pp. 47–75. Kluwer Academic Pub., Boston (1997)
17. Ore, O.: Galois connexions. Transactions of American Mathematical Society 55, 493–513 (1944)
18. Pawlak, Z.: Rough Sets. In: Theoretical Aspects of Reasoning about Data. Kluwer Academic Pub., Boston (1991)

# Rough Set Approximations in Formal Concept Analysis

Daisuke Yamaguchi[1], Atsuo Murata[1], Guo-Dong Li[2], and Masatake Nagai[3]

[1] Graduate School of Natural Science and Technology, Okayama University, 3-1-1 Tsushimanaka, Okayama, 700–8530 Japan
daicom0204@yahoo.co.jp

[2] Graduate School of System Design, Tokyo Metroporitan University, 6-6 Asahigaoka, Hino, 191–0065 Japan
guodong_li2004@yahoo.co.jp

[3] Faculty of Engineering, Kanagawa University, 3-27-1 Rokkakubashi, Kanagawa-ku, Yokohama, 221–8686 Japan
masatake4263@oregano.ocn.ne.jp

**Abstract.** Conventional set approximations are based on a set of attributes; however, these approximations cannot relate an object to the corresponding attribute. In this study, a new model for set approximation based on individual attributes is proposed for interval-valued data. Defining an indiscernibility relation is omitted since each attribute value itself has a set of values. Two types of approximations, single- and multi-attribute approximations, are presented. A multi-attribute approximation has two solutions: a maximum and a minimum solution. A maximum solution is a set of objects that satisfy the condition of approximation for at least one attribute. A minimum solution is a set of objects that satisfy the condition for all attributes. The proposed set approximation is helpful in finding the features of objects relating to condition attributes when interval-valued data are given. The proposed model contributes to feature extraction in interval-valued information systems.

**Keywords:** rough sets, set approximations, grey numbers, grey system theory, interval data, indeterministic information systems.

## 1 Introduction

In recent years, database systems have become more complex as both large amounts and various types of data, such as nominal values, real values, set values and missing values, are stored in them. Extracting valuable information from such database systems is an important paradigm in data mining. Lipski [13] was the first to explore such database systems. Pawlak [19, 20] introduced the rough set, which is a powerful tool for extracting information from databases (i.e., information systems). One of the most important concepts of rough sets is set approximation, which is a model to determine whether objects certainly or possibly belong to a class. However, the relationship between each object and attribute is not determined, because such an approximation is obtained from a

J.F. Peters et al. (Eds.): Transactions on Rough Sets XII, LNCS 6190, pp. 226–235, 2010.

set of attributes by means of an indiscernibility relation. In addition, the original approach focuses mainly on nominal values and does not consider other types of data.

To solve this problem, the rough set approach is extended to deal with a wider range of values. For example, Kryszkiewicz [7,8] extended the rough set approach to deal with both nominal and missing values; however, set values and interval values were not considered. Leung and Li [9] considered the concept of a maximal consistent block for set approximation, which includes nominal and missing values. In addition, this set approximation is also given in terms of a set of attributes. Nakamura [16] discussed the procedure to extract decision rules from set-valued information systems with missing values, in which set approximation was not considered. Sakai and Okuma [21] suggested algorithms on attribute dependency, attribute reduction and rule generation. These algorithms were defined for set values and missing values; however, interval values were once again not considered. Guan and Wang [5] suggested a maximum tolerance relation and a relative attribute reduction model for set values, in which set approximation is based on a set of attributes. These set values are integers since each value is an alternative label for a nominal value. Leung et al. [10] proposed a decision rule induction model for interval-valued information systems, which considered only real values, although set approximation was not discussed.

In the above-mentioned discussion, conventional set approximations are based on multiple condition attributes as a whole; however, the feature of an individual condition attribute cannot be determined from these. In this study, a new model called the attribute-oriented set approximation, is proposed for set approximation for interval-valued data based on individual attributes. In this model, defining an indiscernibility relation is omitted so that each attribute value has a set of values. The remainder of this paper is organized as follows. Interval-valued information systems are explained briefly in Section 2, together with two types of interpretation of interval data. The attribute-oriented set approximation is presented in Section 3. In Section 4, the proposed model is applied to numerical examples to illustrate the model in detail. In addition, the proposed model is compared to a conventional model. Our conclusions are presented in Section 5.

## 2 Interval-Valued Information Systems

Let $IS = (\mathbb{U}, A \cup \{d\}, V, f)$ be an information system, where

- $\mathbb{U}$ is a set of objects,
- $A$ is a set of condition attributes,
- $d$ is a decision attribute,
- $V$ is a set of values called the domain, $V = \{V_a | \forall a \in A\}$, $V_a = \mathbb{R}$ for all $a$, and $\mathbb{R}$ is the set of real numbers in this paper, and
- $f$ is an information function given as $f : \mathbb{U} \times A \longrightarrow V$.

Based on these definitions, $f(x, a)$ is a value assigned to $x$ on $a$. In general, $a(x)$ is used to represent $f(x, a)$. In the case of interval values, a value range is assigned to $x$ such that $a(x) = [\underline{a}(x), \overline{a}(x)]$, where $\underline{a}(x) < \overline{a}(x)$.

From the discussion by Orłowska [18] on the semantic interpretation of given data, interval values can be interpreted as follows:

**Type I.** The result of measuring lengths includes an error due to the selected scale. In error analysis [6,15,22], $x = x_{\text{best}} + \delta x$ is used to represent the result, where $x$ is the measured value, $x_{\text{best}}$ is the best estimate and $\delta x$ is the error in the measurement of $x$. This form can be represented by an interval where $x = [x_{\text{best}} - \delta x, x_{\text{best}} + \delta x]$. The best estimate exists in this interval, but is indeterministic. This type of interval includes uncertainty.

**Type II.** Recently, a new statistical approach called symbolic data analysis (SDA) [4] has been suggested, which deals with a unit of data called symbolic data. For example, a large number of observations are summarized as an interval. Let $X = \{167, 168, 165, 180, 167\}$ be a set of heights measured for five persons, which is represented by the interval $X = [165, 180]$. Billard and Diday [2] suggested that it is difficult to deal with large amounts of data with conventional statistical analysis, and that such data should first be summarized in order to extract information effectively. This type of interval can be interpreted as a summarization of many observations.

Leung [10] considered only Type II interval values. If the given data include Type I interval values, $IS$ is called an incomplete information system [13] or indeterministic information system. Grey system theory [3,14,17,23] is applicable to Type I interval data and such information systems are called grey information systems [11,12,24,25]. In grey system theory, grey numbers represent uncertain values, and in this study, are represented by an information function. When two values $\underline{a}(x)$ and $\overline{a}(x)$ ($\underline{a}(x) = \inf V_a, \overline{a}(x) = \sup V_a$) are given in $x$ on $a$, $\otimes a(x)$ is called as follows:

- If and only if $\underline{a}(x) \to -\infty$ and $\overline{a}(x) \to +\infty$, $\otimes a(x)$ is a black number denoting a missing (null) value.
- If and only if $\underline{a}(x) = \overline{a}(x)$, $\otimes a(x)$ is a white number denoting a deterministic value.
- Else, $\otimes a(x) = [\underline{a}(x), \overline{a}(x)]$ is a grey number denoting an indeterministic value.

For interval-valued information systems, a tolerance relation [5] on $a$ can be defined as follows:

$$T_a = \{(x, y) | a(x) \cap a(y) \neq \emptyset\}, \tag{1}$$

where $a \in A$. A tolerance relation on $B$ can be defined as follows:

$$T_B = \{(x, y) | a(x) \cap a(y) \neq \emptyset, \forall a \in B\}, \tag{2}$$

where $B \subseteq A$. Furthermore, let $T_B(x) = \{y \in \mathbb{U} | y T_B x\}$ be a tolerant class on $B$. These are the essential definitions of a rough set model for interval-valued information systems.

# 3 Attribute-Oriented Set Approximation

Conventional set approximation is based on equivalence or tolerance classes. A subset $B$ of $A$ usually contains multiple attributes as a unit of approximation. In contrast, attribute-oriented set approximation is defined on each attribute of $A$ so that each attribute value has a set of real values. Single- and multi-attribute approximations are defined in this section. Defining the indiscernibility relation is omitted.

**Definition 1.** Let $x$ be an object of $\mathbb{U}$, $a$ be an attribute of $A$ and $a(x)$ be a value for $x$ on $a$, where $a(x) = [\underline{a}(x), \overline{a}(x)]$. Let $o$ be an object denoting the objective of the approximation, $a(o)$ be a value for $o$ on $a$, $U(a(o))$ be an upper approximation, $L(a(o))$ be a lower approximation, $B(a(o))$ be a boundary region and $N(a(o))$ be a negative region. These approximations are defined as follows:

$$U(a(o)) = \{x \in \mathbb{U} | a(x) \cap a(o) \neq \emptyset\}, \tag{3}$$
$$L(a(o)) = \{x \in \mathbb{U} | a(x) \subseteq a(o)\}, \tag{4}$$
$$B(a(o)) = U(a(o)) - L(a(o)), \tag{5}$$
$$N(a(o)) = \mathbb{U} - U(a(o)), \tag{6}$$

where $B(a(o)) \subseteq L(a(o)) \subseteq U(a(o)) \subseteq \mathbb{U}$.

**Definition 2.** Let $A = \{a_1, a_2, \ldots, a_n\}$ be a set of condition attributes, and $O = \{a_1(o), a_2(o), \ldots, a_n(o)\}$ be a set of values denoting the objective of the approximation. A lower approximation $L(O)$, a boundary region $B(O)$ and a negative region $N(O)$ are defined as follows:

$$L(O) = [\underline{L}(O), \overline{L}(O)], \tag{7}$$
$$B(O) = [\underline{B}(O), \overline{B}(O)], \tag{8}$$
$$N(O) = [\underline{N}(O), \overline{N}(O)], \tag{9}$$
$$\underline{L}(O) = \bigcap_{i=1}^{n} L(a_i(o)), \tag{10}$$
$$\overline{L}(O) = \bigcup_{i=1}^{n} L(a_i(o)), \tag{11}$$
$$\underline{B}(O) = \bigcap_{i=1}^{n} B(a_i(o)), \tag{12}$$
$$\overline{B}(O) = \bigcup_{i=1}^{n} B(a_i(o)), \tag{13}$$
$$\underline{N}(O) = \bigcap_{i=1}^{n} N(a_i(o)), \tag{14}$$
$$\overline{N}(O) = \bigcup_{i=1}^{n} N(a_i(o)), \tag{15}$$

where $\underline{L}(O) \subseteq \overline{L}(O)$, $\underline{B}(O) \subseteq \overline{B}(O)$ and $\underline{N}(O) \subseteq \overline{N}(O)$.

Definition 1 is a single-attribute set approximation on $a$ and Definition 2 is a multi-attribute set approximation on $A$. In conventional upper and lower approximations, each result has only one solution. In contrast, the proposed model may have two solutions, a minimum solution $\underline{\circ}$ and a maximum solution $\overline{\circ}$, where $\circ = \{L, B, N\}$. A minimum solution is a set of objects that satisfy the condition of approximation for all attributes of $A$. A maximum solution is a set of objects that satisfy the condition for at least one attribute of $A$. Let $L(a(o))$ denote a set of objects that are indistinguishable from the objective $o$ on $a$, $B(a(o))$ denote a set of objects that are similar to objective $o$ on $a$ and $N(a(o))$ denote a set of objects that are distinguishable to objective $o$ on $a$. It is known that $L(a(o)) \subseteq U(a(o))$ and $U(a(o)) \subseteq \mathbb{U}$. An upper approximation $U(a(o))$ is required to compute $B(a(o))$ and $N(a(o))$ so that more significant objects can be extracted. The proposed model is therefore known as the attribute-oriented set approximation.

# 4 Example

## 4.1 Subject

Consumers cannot decide which automobile is suitable for their family or company. In this example, the proposed model is applied to evaluate the difference in specifications among automobile manufacturers. The Auto-MPG data set available from the UCI Repository [1] is used here; Table 1 contains ten automobile manufacturers as objects, seven condition attributes and one decision attribute. In the original data set, each automobile model has a single specification value for each condition attribute. This data set can be summarized as an interval-valued information system by applying the following steps. First, each automobile model is classified as belonging to one of the ten manufacturers, with the result that a manufacturer may have multiple automobile models. Then for each manufacturer, specification values for each condition attribute are obtained by selecting the maximum and minimum specification values from the relevant attributes of the models belonging to the manufacturer. These values form the upper and lower endpoints of an interval that becomes the attribute value for the particular manufacturer. This is an approach to discover information from large information systems in symbolic data analysis [2]. Each attribute value in Table 1 becomes a Type II interval.

Table 2 focuses on the attributes, Cylinders ($a_2$) and Model Year ($a_7$), to describe the following example simply.

## 4.2 Applying the Conventional Model

The conventional approach is applied to Table 2. Let $\mathbb{U}/A$ ={{Audi}, {BMW}, {Cadillac}, {Ford, Toyota}, {Honda, Volkswagen}, {Mazda}, {Subaru}, {Volvo}} be the indiscernibility relations and $X$ ={Toyota} be a set for approximation. The lower and upper approximations are obtained as $R_*(X) = \emptyset$ and $R^*(X)$ ={Ford, Toyota}, respectively.

**Table 1.** An interval-valued AUTO-MPG data set

| $x$ No. | Manufacturer | $a_1$ M.P.G. | $a_2$ Cylinders | $a_3$ Displacement | $a_4$ Horsepower |
|---|---|---|---|---|---|
| 1 | Audi | [20,36.4] | [4,5] | [97,131] | [67,103] |
| 2 | BMW | [21.5,26] | [4,4] | [121,121] | [110,113] |
| 3 | Cadillac | [16.5,23] | [8,8] | [350,350] | [125,180] |
| 4 | Ford | [15,36.1] | [4,6] | [98,250] | [65,98] |
| 5 | Honda | [24,44.6] | [4,4] | [81,120] | [53,97] |
| 6 | Mazda | [18,46.6] | [3,4] | [70,120] | [52,110] |
| 7 | Subaru | [26,33.8] | [4,4] | [97,108] | [67,93] |
| 8 | Toyota | [19,39.1] | [4,6] | [71,168] | [52,122] |
| 9 | Volkswagen | [25,43.1] | [4,4] | [79,105] | [48,78] |
| 10 | Volvo | [17,30.7] | [4,6] | [121,163] | [76,125] |

| $x$ No. | Manufacturer | $a_5$ Weight | $a_6$ Acceleration | $a_7$ Model-year | $d$ Country |
|---|---|---|---|---|---|
| 1 | Audi | [2.19,2.95] | [14,19.9] | [70,80] | Germany |
| 2 | BMW | [2.23,2.60] | [12.5,12.8] | [70,77] | Germany |
| 3 | Cadillac | [3.90,4.38] | [12.1,17.4] | [76,79] | USA |
| 4 | Ford | [1.80,3.57] | [13.6,21] | [70,81] | USA |
| 5 | Honda | [1.76,2.49] | [13.8,18.5] | [74,82] | Japan |
| 6 | Mazda | [1.97,2.72] | [12.5,19.4] | [72,82] | Japan |
| 7 | Subaru | [1.99,2.39] | [15.5,18] | [74,81] | Japan |
| 8 | Toyota | [1.65,2.93] | [12.6,21] | [70,81] | Japan |
| 9 | Volkswagen | [1.83,2.22] | [12.2,21.5] | [74,82] | Germany |
| 10 | Volvo | [2.87,3.16] | [13.6,19.6] | [72,81] | Sweden |

The original data set is available from the UCI repository [1].

### 4.3 Applying the Proposed Model

Assuming that $o$ = Toyota, $O = \{a_2(o), a_7(o)\}$, $a_2(o) = [4,6]$ and $a_7(o) = [70,81]$, the proposed approximation is performed.

Single-attribute upper approximations are obtained as follows:

- $U(a_2(o)) = \{$Audi, BMW, Ford, Honda, Mazda, Subaru, Volkswagen, Volvo$\}$.
- $U(a_7(o)) = \{$Audi, BMW, Cadillac, Ford, Honda, Mazda, Subaru, Volkswagen, Volvo$\}$.

Single-attribute lower approximations are obtained as follows:

- $L(a_2(o)) = \{$Audi, BMW, Ford, Honda, Mazda, Subaru, Volkswagen, Volvo$\}$.
- $L(a_7(o)) = \{$Audi, BMW, Cadillac, Ford, Subaru, Volvo$\}$.

Boundary regions are given as follows:

- $B(a_2(o)) = U(a_2(o)) - L(a_2(o)) = \emptyset$.
- $B(a_7(o)) = U(a_7(o)) - L(a_7(o)) = \{$Honda, Matsuda, Volkswagen$\}$.

**Table 2.** The rearranged interval-valued AUTO-MPG data set

| $x$ No. Manufacturer | $a_2$ Cylinders | $a_7$ Model Year | $d$ Country |
|---|---|---|---|
| 1  Audi | [4,5] | [70,80] | Germany |
| 2  BMW | [4,4] | [70,77] | Germany |
| 3  Cadillac | [8,8] | [76,79] | USA |
| 4  Ford | [4,6] | [70,81] | USA |
| 5  Honda | [4,4] | [74,82] | Japan |
| 6  Mazda | [3,4] | [72,82] | Japan |
| 7  Subaru | [4,4] | [74,81] | Japan |
| 8  Toyota | [4,6] | [70,81] | Japan |
| 9  Volkswagen | [4,4] | [74,82] | Germany |
| 10 Volvo | [4,6] | [72,81] | Sweden |

**Table 3.** An approximation table

| $o$ =Toyota | $\underline{o}$ | $\overline{o}$ | $\overline{o} - \underline{o}$ |
|---|---|---|---|
| $L$ | Audi, BMW, Ford, Subaru, Volvo | Audi, BMW, Ford, Honda, Mazda, Subaru, Volkswagen, Volvo | Honda, Matsuda, Volkswagen |
| $B$ | ∅ | Honda, Mazda, Volkswagen | Honda, Mazda, Volkswagen |
| $N$ | ∅ | Cadillac | Cadillac |

Negative regions are given as follows:

- $N(a_2(o)) = \mathbb{U} - U(a_2(o)) = \{\text{Cadillac}\}$.
- $N(a_7(o)) = \mathbb{U} - U(a_7(o)) = \emptyset$.

Multi-attribute lower approximations are obtained as follows:

- $\underline{L}(O) = L(a_2(o)) \cap L(a_7(o)) = \{\text{Audi, BMW, Ford, Subaru, Volvo}\}$.
- $\overline{L}(O) = L(a_2(o)) \cup L(a_7(o)) = \{\text{Audi, BMW, Ford, Honda, Mazda, Subaru, Volkswagen, Volvo}\}$.

Multi-attribute boundary regions are given as follows:

- $\underline{B}(O) = B(a_2(o)) \cap B(a_7(o)) = \emptyset$.
- $\overline{B}(O) = B(a_2(o)) \cup B(a_7(o)) = \{\text{Honda, Mazda, Volkswagen}\}$.

Multi-attribute negative regions are given as follows:

- $\underline{N}(O) = N(a_2(o)) \cap N(a_7(o)) = \emptyset$.
- $\overline{N}(O) = N(a_2(o)) \cup N(a_7(o)) = \{\text{Cadillac}\}$.

Table 3 gives a summary of the results.

## 5 Discussion

According to $R^*(X)$, the conventional model asserts that Ford is not a Japanese manufacturer, but has the same specification as Toyota. However, the conventional approach cannot find the relationship between the manufacturer and the attributes. In contrast, the proposed model can find the following information.

Objects of $U(a(o))$ are similar to the objective in the specification on $a$, because each attribute value overlaps with $a(o)$ . However, these upper approximations are used to compute $B$ and $N$ below.

Objects of $L(a(o))$ have an interval value completely included in the interval of $a(o)$. The practical meaning is that there are no differences between these objects and the objective on $a$. In this example, many manufacturers belong to $L(a_2(o))$ and $L(a_7(o))$. The detailed information is extracted by the multi-attribute approximation.

Objects of $B(a(o))$ are similar to the objective in the specification on $a$. These attribute values overlap but are not included in the interval of $a(o)$. According to $B(a_2(o))$, no manufacturer is similar to Toyota on $a_2$. However, according to $B(a_7(o))$, Honda, Matsuda and Volkswagen are similar to Toyota on $a_7$.

Objects of $N(a(o))$ are distinguishable to the objective on $a$. According to $N(a_2(o))$, Cadillac has different features to Toyota on $a_2$. For example, only Cadillac does not belong to $U(a_2(o))$, which suggests that the number of cylinders in Cadillac automobiles is less than four or more than six. In fact, according to Table 2, Cadillac is equipped with eight cylinders, more than those of the other manufacturers. This seems to be one of the features of Cadillac. According to $N(a_7(o))$, none of the manufacturers have features on $a_7$.

Objects of $\underline{L}(O)$ are manufacturers, all attribute values of which are completely included in the interval of $a(o)$. Therefore, Audi, BMW, Ford, Subaru and Volvo have no differences in specification when compared to Toyota. Objects of $\overline{L}(O)$ are manufacturers whose attribute values are included in the interval of $a(o)$ for at least one attribute, although $\underline{L}(O) \subseteq \overline{L}(O)$. To extract more significant manufacturers, the subtraction between the two is performed. In this example, Honda, Mazda and Volkswagen have no differences compared to Toyota on $a_2$ or $a_7$ since $\overline{L}(O) - \underline{L}(O) = \{$Honda, Mazda, Volkswagen$\}$.

$\underline{B}(O)$ is related to Pawlak's boundary region; it is obtained by the subtracting the upper and lower approximations for all condition attributes. However, in this example, none of the manufacturers are similar to Toyota either in $a_2$ or $a_7$. Manufacturers of $\overline{B}(O)$ are similar to the objective for at least one condition attribute, although $\underline{B}(O) \subseteq \overline{B}(O)$. The subtraction of the two is also performed to extract more significant manufacturers. In this example, Honda, Mazda and Volkswagen are similar to Toyota on $a_2$ or $a_7$ since $\overline{B}(O) - \underline{B}(O) = \{$Honda, Mazda, Volkswagen$\}$, where $\underline{B}(O) = \emptyset$.

Objects of $\underline{N}(O)$ are distinguishable to the objective in all condition attributes. In the example, none of the manufacturers are distinguishable from Toyota in the two attributes. Objects of $\overline{N}(O)$ are distinguishable to the objective in at least one condition attribute. In this example, Cadillac can be distinguished from Toyota based on the number of cylinders since $\overline{N}(O) - \underline{N}(O) =$ {Cadillac}, although $\underline{N}(O) = \emptyset$.

The example can be concluded by the proposed model:

- Audi, BMW, Ford, Subaru and Volvo cannot be distinguished from Toyota in the two attributes.
- Cadillac is different from Toyota in terms of cylinders.
- Honda, Mazda and Volkswagen are similar to Toyota in terms of model year.

Therefore, the proposed model can relate the objects to the given condition attributes and is available to extract information.

The proposed approximation is based on the set-operator, union and intersection. In this paper, the proposed model focuses on interval-valued information systems, but it is also applicable to set-valued information systems [5].

Multi-attribute approximations are given as interval sets; the interval-set algebra proposed by Yao [26] might be effective if further analysis is required. In [26], application of the interval-set algebra is not discussed. Applying the interval-set algebra to the proposed model and discussing its effectiveness remain as future works.

## 6 Conclusions

In this study, a new set approximation model based on individual attributes has been proposed without defining an indiscernibility relation. Conventional set approximation is based on a set of condition attributes, and the relationship between individual objects and attributes cannot be determined. The proposed set approximation is helpful in finding the features of objects in terms of condition attributes, when interval-valued data are given. This paper has presented an example to find which automobiles are similar to Toyota in specifications. The conventional model finds that Toyota has the same specification as Ford. In contrast, the proposed model finds additional information, such as: (1) Audi, BMW, Ford, Subaru and Volvo cannot be distinguished from Toyota, (2) Cadillac is different to Toyota in terms of cylinders, and (3) Honda, Mazda and Volkswagen are similar to Toyota in terms of model year. The proposed model contributes to feature extraction in interval-valued information systems, and can be extended further with the interval-set algebra [26].

## References

1. Asuncion, A., Newman, D.J.: UCI Machine Learning Repository. University of California, Department of Information and Computer Science, Irvine, CA (2007), `http://www.ics.uci.edu/~mlearn/MLRepository.html`
2. Billard, L., Diday, E.: Symbolic Data Analysis. John Wiley & Sons, Chichester (2006)
3. Deng, J.-L.: Grey System. China Ocean Press, Beijing (1988)
4. Diday, E., Esposito, F.: An introduction to symbolic data analysis and the SODAS software. Intelligent Data Analysis 7, 583–601 (2003)

5. Guan, Y.-Y., Wang, H.-K.: Set-valued information systems. Information Sciences 176, 2507–2525 (2006)
6. Hansen, E., Walster, G.W.: Global Optimization Using Interval Analysis, 2nd edn. Marcel Dekker, New York (2004)
7. Kryszkiewicz, M.: Rough set approach to incomplete information systems. Information Sciences 112, 39–49 (1998)
8. Kryszkiewicz, M.: Rules in incomplete information systems. Information Sciences 113, 271–292 (1999)
9. Leung, Y., Li, D.: Maximal consistent block technique for rule acquisition in incomplete information systems. Information Sciences 153, 85–106 (2003)
10. Leung, Y., Fischer, M.M., Wu, W.-Z., Mi, J.-S.: A rough set approach for the discovery of classification rules in interval-valued information systems. International Journal of Approximate Reasoning (2007), doi:10.1016/j.ijar.2007.05.001
11. Li, G.-D., Yamaguchi, D., Nagai, M.: A grey-based decision making approach to suppliers selection problem. Mathematical and Computer Modeling 46, 573–581 (2007)
12. Li, G.-D., Yamaguchi, D., Nagai, M.: A grey-based rough decision-making approach to supplier selection. International Journal of Advanced Manufacturing Technology (2007), doi:10.1007/s00170-006-0910-y
13. Lipski Jr., W.: On semantic issues connected with incomplete information databases. ACM Transactions on Database Systems 4, 262–296 (1979)
14. Liu, S.-F., Lin, Y.: Grey Information. Springer, Heidelberg (2006)
15. Moore, R.E.: Reliability in Computing. Academic Press, London (1988)
16. Nakamura, A.: A rough logic based on incomplete information and its application. International Journal of Approximate Reasoning 15, 367–378 (1996)
17. Nagai, M., Yamaguchi, D.: Elements on Grey System Theory and its Applications. Kyoritsu-Shuppan (2004) (in Japanese)
18. Orłowska, E., Pawlak, Z.: Representation of nondeterministic information. Theoretical Computer Science 29, 27–39 (1984)
19. Pawlak, Z.: Rough sets. International Journal of Computer and Information Sciences 11, 341–356 (1982)
20. Pawlak, Z.: Rough sets Theoretical Aspects of Reasoning about Data. Kluwer Academic Publishers, Dordrecht (1991)
21. Sakai, H., Okuma, A.: Basic algorithms and tools for rough non-deterministic information analysis. In: Peters, J.F., Skowron, A., Grzymała-Busse, J.W., Kostek, B.z., Świniarski, R.W., Szczuka, M.S. (eds.) Transactions on Rough Sets I. LNCS, vol. 3100, pp. 209–231. Springer, Heidelberg (2004)
22. Taylor, J.R.: An Introduction to Error Analysis. University Science Books (1982)
23. Wen, K.-L.: Grey Systems: Modeling and Prediction. Yang's Scientific Research Institute (2004)
24. Yamaguchi, D., Li, G.-D., Nagai, M.: On the combination of rough set theory and grey theory based on grey lattice operations. In: Greco, S., Hata, Y., Hirano, S., Inuiguchi, M., Miyamoto, S., Nguyen, H.S., Słowiński, R. (eds.) RSCTC 2006. LNCS (LNAI), vol. 4259, pp. 507–516. Springer, Heidelberg (2006)
25. Yamaguchi, D., Li, G.-D., Nagai, M.: A grey-rough set approach for interval data reduction of attributes. In: Kryszkiewicz, M., Peters, J.F., Rybiński, H., Skowron, A. (eds.) RSEISP 2007. LNCS (LNAI), vol. 4585, pp. 400–410. Springer, Heidelberg (2007)
26. Yao, Y.-Y.: Two views of the theory of rough sets in finite universes. International Journal of Approximate Reasoning 15, 291–317 (1996)

# On the Relation between Jumping Emerging Patterns and Rough Set Theory with Application to Data Classification

Paweł Terlecki

Institute of Computer Science, Warsaw University of Technology
Nowowiejska 15/19, 00-665 Warsaw, Poland
pterleck@ii.pw.edu.pl

**Abstract.** Contrast patterns are an essential element of classification methods based on data mining. Among many propositions, jumping emerging patterns (JEPs) have gained significant recognition due to their simplicity and strong discrimination capabilities. This thesis considers JEPs in terms of discovery and classification. The focus is put on their correspondence to the rough set theory. Transformations between transactional data and decision tables allow us to demonstrate relations of JEPs and global/local reducts. As a part of this discussion, we introduce the concept of a jumping emerging pattern with negation (JEPN). Our observations lead to two novel JEP mining methods based on local reducts: global condensation and local projection. Both attempt to decrease dimensionality of subproblems prior to reduct computation. We show that JEP mining can be reduced to the reduct set problem. The latter is addressed with a new approach, called RedApriori, that follows an Apriori candidate generation scheme and employs pruning based on the notion of attribute set dependence. In addition, we discuss different ways of storing pattern collections and propose a CC-Trie, a tree structure that ensures compactness of information and fast pattern lookups.

A classic mining method for highly-supported JEPs employs a structure called a CP-Tree. We show how attribute set dependence can be employed in this approach to extend the pruning capabilities. Moreover, the problem of finding top-$k$ most supported minimal JEPs is proposed. We discuss a solution that gradually raises minimal support while a CP-Tree is being mined. Small training sets are a challenge in classification. To improve accuracy, we propose AdaAccept, an adaptive classification meta-scheme that analyzes testing instances in turns. It employs an internal classifier with reject option that modifies itself only with accepted instances. Furthermore, we consider a concretization of this scheme in the field of emerging patterns, AdaptiveJEP-Classifier. Two adaptation methods, support adjustment and border recomputation, are put forward. The work has both theoretical and experimental character. The proposed methods and optimizations are evaluated and compared against solutions known in the literature.

J.F. Peters et al. (Eds.): Transactions on Rough Sets XII, LNCS 6190, pp. 236–338, 2010.

## 1 Introduction

Identification of interesting patterns is a fundamental stage of knowledge discovery ([1]). Historically, the first direction of data mining was to look for intrinsic relationships within a considered entity set. Popular forms of knowledge, like association or exception rules ([2,3]), express such relationships in terms of features possessed by entities. An ultimate goal is often to generalize results obtained for specific entities to the broader domain they originate from.

Contrast patterns go beyond this classic formulation and describe differences against a certain background entity class - a context for further interpretation and utilization. Among many propositions ([4,5,6]), the field of emerging patterns (EPs, [7,8,9,10]) has gained a lot of attention and continues to evolve dynamically. According to the original definition, EPs are patterns with a high support ratio between a *background* set and an examined *target* set. This property makes EPs naturally useful in the classification process ([8,11]). Specifically, they are intensively utilized in the area of bioinformatics ([12,13,14,15]). Patterns usually refer to sets of genes whose co-activity is much more frequently observed in one class of organisms than in the others.

Perception and analysis of contrasts are an important part of our cognition and have been the subject of many scientific discipline.s. One interesting contemporary perspective is provided by rough set theory ([16,17,18,19]). It considers a universum of entities (objects) characterized by means of features (attributes). Indiscernibility of objects with regard to a certain attribute set induces a partition of the universum. Blocks of this partition allow us to approximate certain object sets. The most fundamental concept of a rough set approximates a conventional set by means of objects that are definitely included in the set and the ones it may contain. These concepts are useful in classification problems. In particular, in [20], the authors propose to use generated decision rules to characterize a decision class.

Rough set theory provides a powerful framework for incomplete, vague and uncertain data ([21,22,17,18,23,24,19]). It has a rich heritage of conferences, journals and independent publications with numerous contributions across such fields like the probability theory ([25,26]), Dempster-Shafer theory of evidence ([27]), neural networks ([28]) and granular computing ([29,30]). Its applications include rule induction and classification ([31,32,33,34]), feature selection and data reduction ([35,36,37]), clustering ([38,39]) and discretization ([40,41]). Furthermore, as far as decision problems are concerned, a derivative concept, information maps ([42,43]), provides a convenient way to represent domain knowledge about objects, their parts and their changes.

The primary goal of this work is to investigate relationships between the field of emerging patterns and rough set theory. We focus on jumping emerging patterns (JEPs, [7]), defined as EPs supported in a target set and completely absent from a background one. This highly discriminative notion is contrasted with rough set decision reducts, i.e. minimal attribute sets that differentiate objects into classes as well as all available attributes do. Reducts play an important role in data reduction ([35,36,37]) and classification ([31,32,33,34]). Furthermore, they have been extended to several derivative notions ([44,45,46,47,48,37]) and combined with a variety of knowledge discovery techniques ([28,35,49,29]).

The problem of JEP discovery has been tackled with several approaches including operations on set intervals ([50,51]), closed frequent patterns ([52]), binary decision diagrams ([53]) and search space enumeration ([54]). Our discussion goes in three different directions: formulating JEP finding in terms of reduct computation ([55,37,56,57]), investigating pruning conditions in space enumeration techniques and discovering only a certain number of most supported minimal JEPs. The latter approach is dictated by practical reasons and its version for frequent pattern mining has been examined in the literature ([58,59]).

The field of emerging patterns puts a lot of attention on interesting types of patterns and their use in data classification. Constrained EPs ([9]), Strong JEPs, Noise-tolerant EPs and Generalized Noise-tolerant EPs ([54]) are examples of handling noisy data, while chi-EPs ([10,60]) and delta-discriminative EPs ([52]) attempt to capture non-trivial statistical properties. A part of our consideration looks at capabilities of emerging patterns with negated items. This general direction has been investigated in association rule mining resulting in propositions, like negative association rules ([61]), confined association rules ([62]), unexpected rules ([63]) or exception rules ([64,3]).

Training data are an essential input in supervised learning. In some cases, time-consuming or costly expertise may result in relatively small sets of pre-classified entities and inaccurate classification hypothesis. Model adaptation is one possible way to address this problem. In fact, such methods enhance the semi-supervised learning paradigm, where a model is trained with both classified and unclassified entities ([65,66]). Seeking improvement for EP-based classification, we follow the path of online learning and adjust a model with confidently classified entities. This requires appropriate strategies on estimating decision confidence and the way adaptation is performed.

Emerging patterns were originally defined as certain itemsets in transactional data. However, the concept is constantly evolving and becoming more general. Recent interesting propositions include Generalized Emerging Patterns ([67]) for hierarchical data and Contrast Graphs ([68,69]) in graph databases ([70]).

### 1.1 Focus

This article focuses on

1. Conceptual relations between jumping emerging patterns (JEPs) and rough set reducts. JEP discovery can be expressed by means of reduct computation.
2. Attribute set dependence with pruning opportunities for reducts and JEP finding.
3. For small training sets, adaptive classification may improve overall accuracy.

### 1.2 Contributions

We make the following major contributions:

1. Relations between global/local reducts in a decision table and JEPs in a transaction database representing the same data. In particular, the fact that global reducts in the positive region correspond to JEPs and a bidirectional correspondence between local reducts and minimal JEPs.
2. JEPRedLoc, a JEP mining algorithm for input data given as a decision table. JEPs are mined by finding local reducts.
3. The concept of a JEP with negation and its relation to regular JEPs and local reducts in binary decision tables based on classified transactional data.
4. JEPGlobalCond, a JEP mining approach for transaction databases originating from decision tables. In the preprocessing phase, *global condensation*, it employs graph coloring to group items, so that the data can be transformed to a decision table of lower dimensionality. JEPs are found by computing local reducts.
5. JEPLocalProj, a JEP mining approach for sparse transactional data. JEPs supported by a given transaction are found by computing local reducts for a respective object in a certain, *locally projected*, decision table of usually low dimensionality
6. RedApriori, an algorithm for reduct computation that follows the Apriori scheme. We propose two pruning conditions based on attribute set dependence: by examining a previous generation and by computing coverage count. A CC-Trie is put forward to concisely store and efficiently look up attribute sets.
7. A pruning approach, that eliminates patterns with dispensable items, for SJEP mining methods based on a CP-Tree.
8. The problem of finding top-$k$ most supported minimal JEPs. Our algorithm uses a CP-Tree, checks minimality of patterns upfront and gradually increases minimum support as minimal JEPs are discovered.
9. AdaAccept, an adaptive classification meta-scheme that works with small training sets. It is concretized to AdaptiveJEP-Classifier with adaptation condition based on distance and ambiguity and two adaptation methods: support adjustment and border recomputation.

# 2 Theoretical Foundations

## 2.1 Elements of Rough Set Theory

Rough set theory provides a robust knowledge discovery framework (see Section 3.4). This work utilizes concepts and a formal apparatus related to data reduction. We describe approaches for both unclassified and classified data, and the material of this section is divided accordingly. Definitions and conventions are based on [19,71].

**Information Systems.** In inductive reasoning, general conclusions on a certain domain are formulated based on partial knowledge. Input data are represented by a set of observations (objects) characterized by a set of attributes.

**Definition 2.1.** *An* information system *is a pair* $(\mathcal{U}, \mathcal{A})$*, where*

> $\mathcal{U}$ *is a non-empty, finite set of objects, also referred to as* universum,
> $\mathcal{A}$ *is a non-empty finite set of attributes.*

*Each attribute* $a \in \mathcal{A}$ *is a function of the form* $a : \mathcal{U} \mapsto V_a$*, where* $V_a$ *is the value domain for* $a$*.*

Attributes describe objects of an information system and, therefore, make them comparable. Indiscernibility of objects can be expressed by an equivalence relation based on attribute equality. This relation contains all pairs of objects that cannot be discerned with a given set of attributes. The rest of the definitions is formulated for a given information system $\mathcal{IS} = (\mathcal{U}, \mathcal{A})$.

**Definition 2.2.** *For* $B \subseteq \mathcal{A}$*, an* indiscernibility relation *is defined as follows* $IND_{\mathcal{IS}}(B) = \{(u, v) \in \mathcal{U} \times \mathcal{U} : \forall_{a \in B}\ a(u) = a(v)\}$.
*The indiscernibility relation defines a partition of the universum* $\mathcal{U}$ *denoted by* $\mathcal{U}/IND_{\mathcal{IS}}(B)$*. A block of this partition that contains an object* $u \in \mathcal{U}$ *is denoted by* $B_{\mathcal{IS}}(u)$*.*

An attribute set $B \subseteq \mathcal{C}$ represents a certain amount of knowledge on the universum. Since each block of the corresponding partition $\mathcal{U}/IND_{\mathcal{IS}}(B)$ contains mutually indiscernible objects, one may use these blocks to approximate sets of objects.

**Definition 2.3.** *For* $B \subseteq \mathcal{C}$*, a* $B$-lower approximation *of a set* $X \subseteq \mathcal{U}$ *is defined as follows:*
$\underline{B_{\mathcal{IS}}}(X) = \{u \in \mathcal{U} \mid B_{\mathcal{IS}}(u) \subseteq X\}$.

If information carried by an attribute set is understood as its ability to discern objects, an indiscernibility relation becomes a convenient tool to examine the informative role of attributes and attribute sets. In particular, one may want to verify if the absence of an attribute in an attribute set preserves discrimination capabilities of the latter.

**Definition 2.4.** *An attribute* $a \in B$ *is* dispensable *in* $B$ *iff* $IND_{\mathcal{IS}}(B - \{a\}) = IND_{\mathcal{IS}}(B)$*, otherwise* $a$ *is* indispensable.

Consequently, one may want to reduce attribute sets by eliminating dispensable attributes.

**Definition 2.5.** *An attribute set* $B \subseteq \mathcal{A}$ *is*

> independent *iff all its members are indispensable, otherwise, it is* dependent,
> *a* super reduct *iff* $IND_{\mathcal{IS}}(B) = IND_{\mathcal{IS}}(\mathcal{A})$,
> *a* reduct *iff* $B$ *is an independent super reduct.*

*The sets of all super reducts and reducts are denoted by* $SRED_{\mathcal{IS}}(\mathcal{U})$ *and* $RED_{\mathcal{IS}}(\mathcal{U})$*, respectively.*

All available information on object discrimination in $\mathcal{IS}$ is represented by $\mathcal{A}$. The same amount of information is carried by super reducts. Among them, only reducts are minimal with respect to inclusion. The remaining ones contain dispensable attributes.

Super reducts and reducts considered in this work are usually described in the literature as *global*, since all possible object pairs are taken into account.

An indiscernibility relation contains all pairs of indiscernible objects, but does not allow us to directly examine impact of particular attributes on discrimination. For this purpose, one usually operates on a *discernibility matrix* ([55]), whose each element corresponds to a pair of objects and is equal to a set of all attributes that are individually capable of discerning these objects. In particular, empty elements correspond to pairs of identical objects with respect to $\mathcal{A}$. In our discussion only non-empty elements of the matrix are substantial.

**Definition 2.6.** *A* discernibility set $DS_{\mathcal{IS}}$ *is defined by*
$DS_{\mathcal{IS}} = \{\{a \in \mathcal{A} : a(u) \neq a(v)\} : u, v \in \mathcal{U}\}$.

If a given attribute set intersects all non-empty elements of a discernibility matrix, it is a super reduct ([55]). Trivially, a similar fact is true for a discernibility set.

The discrimination power of an attribute set can be measured by the number of elements of a discernibility set this set intersects. In fact, this approach was used to formulate fitness functions of evolutionary algorithms for reduct computation ([72]), where attribute sets discerning more object pairs were preferred.

**Definition 2.7.** *For* $B \subseteq \mathcal{A}$, coverage count *is defined as follows:*
$covcount_{\mathcal{IS}}(B) = |\{X \in DS_{\mathcal{IS}} : X \cap B \neq \emptyset\}|$.

As long as we are interested in discerning all object pairs, one may want to reduce a discernibility set, so that only non-empty minimal elements remain. If one intersects all these elements, the eliminated ones are also guaranteed to be intersected. In accordance with the convention for boolean functions, this reduction is performed by means of the absorption laws ([55]).

**Definition 2.8.** *A* reduced discernibility set *is defined as follows:*
$RDS_{\mathcal{IS}} = \{B \in DS : B \neq \emptyset \wedge \forall_{S \in DS}(S = \emptyset \vee S \not\subset B)\}$

Both discernibility and reduced discernibility set are considered in other works of the field, such as [73].

Whenever an information system is known from the context, respective subscripts in corresponding symbols are omitted, e.g. $B(X)$ instead of $B_{\mathcal{IS}}(X)$, $IND(X)$ instead of $IND_{\mathcal{IS}}(X)$ or $DS$ instead of $DS_{\mathcal{IS}}$.

*Example 2.1.* Let us consider the information system $(\mathcal{U}, \mathcal{A}) = (\{u_1, u_2, u_3, u_4, u_5, u_6, u_7, u_8\}, \{a_1, a_2, a_3, a_4, d\})$ given in Table 2.1. The indiscernibility relation for all the attributes of the system is equal to:

$$IND(\mathcal{A}) = \{(u_1, u_1), (u_2, u_2), (u_3, u_3), (u_4, u_4), (u_5, u_5)\\ (u_6, u_6), (u_7, u_7), (u_8, u_8), (u_3, u_4), (u_4, u_3)\}.$$

**Table 2.1.** A sample information system $(\{u_1, u_2, u_3, u_4, u_5\}, \{a_1, a_2, a_3, a_4, d\})$

| | $a_1$ | $a_2$ | $a_3$ | $a_4$ | $d$ |
|---|---|---|---|---|---|
| $u_1$ | 0 | 0 | 2 | 4 | 0 |
| $u_2$ | 0 | 1 | 0 | 1 | 0 |
| $u_3$ | 0 | 1 | 1 | 3 | 0 |
| $u_4$ | 0 | 1 | 1 | 3 | 1 |
| $u_5$ | 1 | 1 | 0 | 1 | 1 |
| $u_6$ | 1 | 0 | 1 | 0 | 1 |
| $u_7$ | 0 | 2 | 0 | 5 | 2 |
| $u_8$ | 1 | 0 | 1 | 2 | 2 |

Relations for subsets of $\mathcal{A}$ can potentially contain more object pairs. For example, we have $IND(\{a_1, a_2\}) = IND(\mathcal{A}) \cup \{(u_2, u_7), (u_7, u_2), (u_3, u_4), (u_4, u_3), (u_6, u_8), (u_8, u_6)\}$.

Let us consider the attribute sets: $\{a_4, d\}$ and $\{a_3, a_4, d\}$. Because we have $IND(\mathcal{A}) = IND(\{a_4, d\}) = IND(\{a_3, a_4, d\})$, both sets are super reducts. Since $\{a_4, d\} \subseteq \{a_3, a_4, d\}$, the set $\{a_3, a_4, d\}$ is dependent and the attribute $a_3$ is dispensable. Meanwhile, $IND(\{a_4\}) \neq IND(\{a_4, d\})$ and $IND(\{d\}) \neq IND(\{a_4, d\})$, thus, $\{a_4, d\}$ is independent and is a reduct. In fact, we have $RED(\mathcal{U}) = \{\{a_1, a_3, d\}, \{a_2, a_3, d\}, \{a_4, d\}\}$.

The discernibility set and reduced discernibility set for this system are as follows

$$\begin{aligned} DS =&\{\{a_1, a_2, a_3, a_4, d\}, \{a_1, a_2, a_3, a_4\}, \{a_1, a_2, a_4, d\}, \{a_1, a_2, a_4\} \\ &\{a_1, a_3, a_4, d\}, \{a_1, d\}, \{a_1, a_3, a_4\}, \{a_2, a_3, a_4, d\}, \{a_2, a_3, a_4\} \\ &\{a_2, a_4, d\}, \{a_3, a_4, d\}, \{a_3, a_4\}, \{a_4, d\}, \{d\}\}, \\ RDS =&\{\{a_1, a_2, a_4\}, \{a_3, a_4\}, \{d\}\}. \end{aligned}$$

Also, we have: $covcount(\{a_1, a_2\}) = 10$ and $covcount(\mathcal{A}) = 14$.

**Decision Tables.** In decision problems each observation carries information on conditions and corresponding decisions. Such data can be represented as an information system, whose attributes are appropriately grouped in order to indicate their different interpretation. We consider the most common case of a single decision attribute. In fact, one may always combine several decision attributes into a single one or include each of them at a time. A decision attribute interpreted as a class label allows us to treat the observations as classified.

**Definition 2.9.** *A* decision table *is a triple* $(\mathcal{U}, \mathcal{C}, d)$, *where*

$(\mathcal{U}, \mathcal{C} \cup \{d\})$ *is an information system,*
$d$ *is a distinguished attribute.*

*A set of all attributes is denoted by* $\mathcal{A} = \mathcal{C} \cup \{d\}$, *elements of* $\mathcal{C}$ *are called* condition attributes *and* $d$ *is called a* decision attribute.

Let us consider a decision table $\mathcal{DT} = (\mathcal{U}, \mathcal{C}, d)$. The concepts of an indiscernibility relation and a lower approximation for $\mathcal{DT}$ are understood as defined for the information system $(\mathcal{U}, \mathcal{A})$. The corresponding symbols are subscripted with $DT$.

The decision attribute allows us to partition the universum into blocks determined by possible decisions.

**Definition 2.10.** *For $k \in \mathcal{V}_d$, a* decision class $k$ *is defined by* $U_k = \{u \in \mathcal{U} : d(u) = k\}$.

If information on the objects is available through an attribute set $B \subseteq \mathcal{C}$, the decision classes can be approximated by means of the blocks of the partition $\mathcal{U}/IND(B)$. The lower approximations of the classes indicate objects that can be consistently classified. Note that, in general, this property does not hold. Some objects may be indiscernible with respect to $B$ but belong to different decision classes.

**Definition 2.11.** *For $B \subseteq \mathcal{C}$, a* $B$-positive region *with respect to a decision attribute $d$ is defined as follows*

$$POS_{\mathcal{DT}}(B, d) = \bigcup_{X \in \mathcal{U}/IND_{\mathcal{DT}}(\{d\})} \underline{B_{\mathcal{DT}}}(X).$$

*A decision table $\mathcal{DT}$ is* consistent *or* deterministic *if $POS_{\mathcal{DT}}(\mathcal{C}) = \mathcal{U}$. Otherwise, we call it* inconsistent *or* non-deterministic.

In case of decision tables, one is often interested in differentiating objects across classes, whereas objects from the same class can remain indiscernible. This interpretation is followed in the definitions of dispensability of attributes and types of attribute sets. In particular, an attribute is dispensable iff its absence does not introduce new pairs of indiscernible objects that belong to different classes.

**Definition 2.12.** *An attribute $a \in B$ is* dispensable *in $B$ iff $IND_{\mathcal{DT}}(B - \{a\}) \subseteq IND_{\mathcal{DT}}(B) \cup IND_{\mathcal{DT}}(\{d\})$, otherwise $a$ is* indispensable.

**Definition 2.13.** *An attribute set $B \subseteq \mathcal{C}$ is*

independent *iff all its members are indispensable, otherwise, it is* dependent,

*a* global super reduct *iff $IND_{\mathcal{DT}}(B) \subseteq IND_{\mathcal{DT}}(\mathcal{A}) \cup IND_{\mathcal{DT}}(\{d\})$,*

*a* global reduct *iff $B$ is an independent global super reduct.*

*The sets of all global super reducts and global reducts are denoted by $SRED_{\mathcal{DT}}(\mathcal{U}, d)$, $RED_{\mathcal{DT}}(\mathcal{U}, d)$, respectively.*

One may also define super reducts by means of a positive region, which fact we give here as a lemma.

**Lemma 2.1 ([55]).** *For $B \subseteq \mathcal{C}$, we have* $POS_{\mathcal{DT}}(B, d) = POS_{\mathcal{DT}}(\mathcal{C}, d) \Longleftrightarrow B \in SRED_{\mathcal{DT}}(\mathcal{U}, d)$.

In order to distinguish super reducts and reducts for a decision table from analogous notions in an information system, one usually calls the former *decision* or *with respect to a decision* ([74]).

Furthermore, in the context of a single object $u \in \mathcal{U}$, one may consider attribute sets that differentiate $u$ from all objects from other classes as well as $\mathcal{C}$.

**Definition 2.14.** *For an object $u \in \mathcal{U}$, an attribute set $B \subseteq \mathcal{C}$ is*

*a* local super reduct *iff* $\forall_{c \in V_d} (\mathcal{C}(u) \cap U_c = \emptyset \Longrightarrow B(u) \cap U_c = \emptyset)$,

*a* local reduct *iff $B$ is a local super reduct and none of its proper subset is a local super reduct.*

*A set of all local reducts for the object $u$ is denoted by $REDLOC_{\mathcal{DT}}(u, d)$.*

For objects from the positive region, the definition of a local reduct becomes simpler.

**Lemma 2.2 ([71]).** *For $u \in POS(\mathcal{C}, d)$, we have:*
*$B \in REDLOC_{\mathcal{DT}}(u, d) \Longleftrightarrow B$ is a minimal set such that $B(u) \subseteq U_{d(u)}$.*

Also, local super reducts and local reducts are often referred to as *(super) reducts with respect to a decision and an object* ([74]).

Similarly to information systems, if the context clearly indicates a decision table, respective subscripts in corresponding symbols are omitted, e.g. $POS(B, d)$ instead of $POS_{\mathcal{DT}}(d, B)$.

*Example 2.2.* One may want to interpret the data in Table 2.1 as a decision table $(\mathcal{U}, \mathcal{C}, d) = (\{u_1, u_2, u_3, u_4, u_5, u_6, u_7, u_8\}, \{a_1, a_2, a_3, a_4\}, d)$, where $a_1, a_2, a_3, a_4$ are condition attributes and $d$ is a decision one.

Decision classes are determined by values of $d$, so that $U_0 = \{u_1, u_2, u_3\}$, $U_1 = \{u_4, u_5, u_6\}$, $U_2 = \{u_7, u_8\}$. Note that the objects $u_3$ and $u_4$ are indiscernible by means of all the condition attributes and belong to different classes. Consequently, the positive region for $\mathcal{C}$ excludes these objects and is equal to $POS(\mathcal{C}, d) = \{u_1, u_2, u_5, u_6, u_7, u_8\}$.

Further, we refer to the following relation:

$$\begin{aligned} IND(\mathcal{A}) \cup IND(\{d\}) = \{&(u_1, u_1), (u_2, u_2), (u_3, u_3), (u_4, u_4), (u_5, u_5), (u_6, u_6), \\ &(u_7, u_7), (u_8, u_8), (u_3, u_4), (u_4, u_3), (u_1, u_2), (u_2, u_1), \\ &(u_1, u_3), (u_3, u_1), (u_2, u_3), (u_3, u_2), (u_3, u_4), (u_4, u_3), \\ &(u_3, u_5), (u_5, u_3), (u_4, u_5), (u_5, u_4), (u_7, u_8), (u_8, u_7)\}. \end{aligned}$$

Let us consider the attribute sets $\{a_1, a_4\}$ and $\{a_1, a_2, a_4\}$. Since we have $IND(\{a_1, a_4\}) = IND(\{a_1, a_2, a_4\}) \subseteq IND(\mathcal{A}) \cup IND(\{d\})$, both attribute sets are global super reducts. The attribute $a_2$ is dispensable, because it does not discern any additional object pairs from different classes as compared to $\{a_1, a_4\}$. Consequently, $\{a_1, a_2, a_4\}$ is dependent. On the other hand, all the attributes of $\{a_1, a_4\}$ are indispensable and this attribute set is a global reduct. In fact, the only one for this decision table.

Local reducts for each object are listed in Table 2.2. Note that, even though $u_3$ and $u_4$ are indiscernible by means of $\mathcal{C}$, the respective sets of local reducts are different.

**Table 2.2.** Sets of local reducts computed for the objects of the decision table $(\mathcal{U}, \mathcal{C}, d) = (\{u_1, u_2, u_3, u_4, u_5, u_6, u_7, u_8\}, \{a_1, a_2, a_3, a_4\}, d)$ based on the data from Table 2.1

| $u$ | $REDLOC(u, d)$ |
|---|---|
| $u_1$ | $\{\{a_1, a_2\}, \{a_3\}, \{a_4\}\}$ |
| $u_2$ | $\{\{a_1, a_2, a_3\}, \{a_1, a_4\}\}$ |
| $u_3$ | $\emptyset$ |
| $u_4$ | $\emptyset$ |
| $u_5$ | $\{\{a_1, a_2\}, \{a_1, a_3\}, \{a_1, a_4\}\}$ |
| $u_6$ | $\{\{a_4\}\}$ |
| $u_7$ | $\{\{a_2\}, \{a_4\}\}$ |
| $u_8$ | $\{\{a_4\}\}$ |

## 2.2 Jumping Emerging Patterns

Emerging patterns belong to the field of data mining (see Section 3.1). Conceptually, they capture differences between two given sets of observations and, thus, find application in classification problems. This section covers definitions of transactional data and basic types of emerging patterns.

**Transaction Databases.** In data mining, one encounters several definitions of unclassified and classified datasets. For the purpose of this work, we provide a precise convention, which resembles concepts from the rough set theory. This makes the discussion on mutual relationship between both fields easier to comprehend. A survey of other approaches is given in Section 2.2.

The concept of a transaction database originates from the market basket problem ([2,1]), where each transaction refers to a single cart filled with certain items (products). We formalize this structure by specifying both a multiset of transactions and a space of feasible items. Note that the use of a multiset is necessary to capture repeating transactions. It requires a particular attention, for example, in cases where its cardinality is expressed by means of the $|\cdot|$ notation or multiset inclusion by means of $\subseteq$.

**Definition 2.15.** *A* transaction system *is a pair* $(\mathcal{D}, \mathcal{I})$*, where*

> $\mathcal{D}$ *is a non-empty finite multiset of transactions,*
> $\mathcal{I}$ *is a non-empty finite set of items, also called an* itemspace*,*
> $\forall_{T \in \mathcal{D}} T \subseteq \mathcal{I}$.

In order to represent decisions, we introduce a special set of distinguished items. Each of them corresponds to a single decision.

**Definition 2.16.** *A* decision transaction system *is a tuple* $(\mathcal{D}, \mathcal{I}, \mathcal{I}_d)$*, where*

$(\mathcal{D}, \mathcal{I} \cup \mathcal{I}_d)$ *is a transaction system,*
$\forall_{T \in \mathcal{D}} |T \cap \mathcal{I}_d| = 1.$

*Elements of* $\mathcal{I}$ *and* $\mathcal{I}_d$ *are called condition and decision items, respectively.*

For a transaction system $(\mathcal{D}, \mathcal{I})$ or decision transaction system $(\mathcal{D}, \mathcal{I}, \mathcal{I}_d)$, we use the traditional terms *database* and *itemset* to refer to any submultiset of $\mathcal{D}$ and any subset of $\mathcal{I}$, respectively. Also, since itemsets are the only type of patterns considered in this work, the terms *itemset* and *pattern* are used interchangeably.

Similarly to decision tables, we introduce a notation for transactions corresponding to the same decision. Note that the following databases are multisets.

**Definition 2.17.** *For* $k \in \mathcal{I}_d$*, a* decision class $k$ *is defined as*
$D_k = \{T \in \mathcal{D} : T \cap \mathcal{I}_d = \{k\}\}$.
*We use the symbols* $D_k$ *and* $D_{\{k\}}$ *interchangeably.*

To express the interestingness of an itemset in a given database, we use the following traditional measures ([1]).

**Definition 2.18.** *The* count *of* $X \subseteq \mathcal{I}$ *in a database* $D \subseteq \mathcal{D}$ *is defined as* $count_D(X) = |\{T \in D : X \subseteq T\}|$.

**Definition 2.19.** *The* support *of* $X \subseteq \mathcal{I}$ *in a database* $D \subseteq \mathcal{D}$ *is defined as* $supp_D(X) = \frac{count_D(X)}{|\mathcal{D}|}$.

**Types of Emerging Patterns.** Let us consider a transaction system $(\mathcal{D}, \mathcal{I})$ and two databases $D_1, D_2 \subseteq \mathcal{D}$ called a *background* (or *negative*) and *target* (or *positive*) database, respectively.

The following three measures express how strongly an itemset discriminates the two considered databases. These statistics rely solely on the itemset's supports in each of the databases.

**Definition 2.20.** *For an itemset* $X \subseteq \mathcal{I}$*, we define*

*the* growth rate *of* $X$ *from* $D_1$ *to* $D_2$ *as*

$$gr_{D_1 \to D_2}(X) = \begin{cases} 0, & supp_{D_1}(X) = supp_{D_2}(X) = 0 \\ \infty, & supp_{D_1}(X) = 0 \text{ and } supp_{D_2}(X) \neq 0 \\ \frac{supp_{D_2}(X)}{supp_{D_1}(X)}, & otherwise \end{cases},$$

*the* support ratio *of* $X$ *as*

$$sr(X) = max(gr_{D_1 \to D_2}(X), gr_{D_2 \to D_1}(X)),$$

*the* strength *of* $X$ *as*

$$st_{D_1 \to D_2}(X) = \frac{gr_{D_1 \to D_2}(X)}{gr_{D_1 \to D_2}(X) + 1} * supp_{D_2}(X).$$

We consider the following types of contrast patterns.

**Definition 2.21.** *For $\rho \in \mathbb{R}_+ \cup \{0, +\infty\}$, $\xi \in <0, 1>$, an itemset $X \subseteq \mathcal{I}$ is*

*a* $\rho$-emerging pattern (EP) from $D_1$ to $D_2$ *iff* $gr_{D_1 \to D_2}(X) \geq \rho$,
*a* jumping emerging pattern (JEP) from $D_1$ to $D_2$ *iff* $gr_{D_1 \to D_2}(X) = +\infty$,
*a* $\xi$-strong jumping emerging pattern (SJEP) from $D_1$ to $D_2$ *iff* $supp_{D_1}(X) = 0$ *and* $supp_{D_2}(X) > \xi$ *and this fact does not hold for any proper subset of* $X$.

*The thresholds $\rho$ and $\xi$ are referred to as* minimum growth rate *and* minimum support, *respectively. The set of all JEPs ($\xi$-SJEPs) from $D_1$ to $D_2$ is called a* JEP space *(*$\xi$-SJEP space*) and denoted by $JEP(D_1, D_2)$ ($\xi - SJEP(D_1, D_2)$). Contrary to the original definition of SJEP ([75,54]), we require the support in a target class to be* strictly *greater than $\xi$. Thanks to this approach, $\xi$-SJEPs are JEPs for any feasible value of $\xi$, in particular, for $\xi = 0\%$.*

In other words, EPs are itemsets whose support varies between the two databases as strong as it is expressed by the parameter $\rho$. Their subclass, JEPs, refers to itemsets present in the target database and absent from the background one. This property makes them highly discriminative and, thus, attractive in classification. Since every superset of a JEP is also a JEP, one usually focuses on minimal ones. Furthermore, SJEPs are minimal JEPs with an additional restriction on the support in the target class. The latter allows us to choose only the most frequent patterns in the target class and avoid noise ([75,54]).

Commonly, one considers patterns to a target database $D \subseteq \mathcal{D}$ and assumes its complement to $\mathcal{D}$ to be a background database, i.e. $D' = \mathcal{D} - D$. In this case, we only indicate the target database and talk about patterns *in* $D$ instead of: *from* $D'$ *to* $D$. In particular, the respective JEP space is denoted by $JEP(D)$.

All the above concepts remain valid for decision transaction systems. Usually, a target database is indicated by means of decision items. For example, for $(\mathcal{D}, \mathcal{I}, \mathcal{I}_d)$ and a class $k \in \mathcal{I}_d$, we have $JEP(D_k) = JEP(D'_k, D_k)$.

**Table 2.3.** A sample decision transaction system $(\mathcal{D}, \mathcal{I}, \mathcal{I}_d)$

| | | |
|---|---|---|
| $T_1$ | $abe$ | 0 |
| $T_2$ | $bd$ | 0 |
| $T_3$ | $cde$ | 1 |
| $T_4$ | $de$ | 1 |
| $T_5$ | $ac$ | 2 |
| $T_6$ | $bce$ | 2 |
| $T_7$ | $ac$ | 2 |

*Example 2.3.* The material of this section is presented for the decision transaction system $(\mathcal{D}, \mathcal{I}, \mathcal{I}_d) = \{\{T_1, T_2, T_3, T_4, T_5, T_6, T_7\}, \{a, b, c, d, e\}, \{0, 1, 2\}\}$ given in Table 2.3. We have the following databases related to the decision classes: $D_0 = \{abe, bd\}$, $D_1 = \{cde, de\}$, $D_2 = \{ac, bce, ac\}$.

Let us consider the itemset $a$. It has the following supports: $supp_{D_0}(a) = \frac{1}{2}$, $supp_{D_1}(a) = 0$, $supp_{D_2}(a) = \frac{2}{3}$. The complementary database to $D_2$ is equal to $D'_2 = D_0 \cup D_1$. One may compute the following interestingness measures for patterns in $D_2$:

$$gr_{D'_2 \to D_2}(a) = \frac{supp_{D_2}(a)}{supp_{D'_2}(a)} = \frac{8}{3},$$

$$sr(a) = max(gr_{D_2 \to D'_2}(a), gr_{D'_2 \to D_2}(a)) = max(\frac{8}{3}, \frac{3}{8}) = \frac{8}{3},$$

$$st_{D'_2 \to D_2}(a) = \frac{gr_{D'_2 \to D_2}(a)}{gr_{D'_2 \to D_2}(X) + 1} * supp_{D_2}(a) = \frac{16}{33}.$$

Consequently, $a$ is a $\rho$-EP, for any $\rho$ satisfying $0 \leq \rho \leq \frac{8}{3}$. Since $supp_{D_2}(ac) = \frac{2}{3}$ and $supp_{D'_2}(ac) = 0$, $ac$ is a JEP in $D_2$ (from $D'_2$ to $D_2$). Moreover, $ac$ is a $\xi$-SJEP, for any $\xi$ satisfying $0 \leq \rho < \frac{2}{3}$.

**Alternative Conventions.** Originally, a transaction database was considered for frequent pattern mining and defined as a sequence of transactions ([2]). This approach elegantly handles repeating transactions and is also suitable for sequential pattern mining. We used this notation in [76,77] and a conceptually similar approach, where transaction is a pair of a transaction identifier and a set of items, in [78]. Since sequences are sometimes not easily readable, whenever an order of transactions is not substantial for a considered problem, some authors prefer to use multisets ([79,80]). For the same reasons, we have chosen this convention for our work.

Furthermore, different approaches can be observed for classified transactional data. Emerging patterns are usually defined for a pair of transaction databases ([8,7,50]). In case of multiclass problems, a collection of databases is considered. Note that, no relationship between individual databases is enforced, although it is implicitly assumed they share the same itemspace. A precise proposition is given in [6]. It uses separate sets of transaction identifiers, items and class identifiers. Two functions express membership of items in transactions and transactions in classes, respectively. We employ a similar approach in [78], where each transaction is a pair of a transaction identifier and a set of items. Data are represented as a single database and a function that maps its transactions to class identifiers. Due to the complexity of this notation, we extend the traditional transaction database by adding decisions as special items. Consistency of our notation with the one used for decision tables makes the further discussion clearer.

## 2.3 Concise Representation of Convex Sets

Many problems and corresponding algorithms in rough set theory and data mining consider elements of certain power sets. For example, the power sets of a set of all attributes or all items constitute search spaces for tasks, like reduct computation (see Section 3.5) or frequent pattern mining ([2,81]). At the same time, most of the basic properties and pruning approaches employed in algorithms rely on the relation of inclusion. In order to elegantly express a set with a certain partial order, the literature of both fields adopts the formality of a lattice ([82]). Although it has not been explicitly used for emerging patterns so far, translation of respective properties is straightforward and ensures a uniform notation in our work.

Let us consider a lattice $(\mathcal{F}, \leq)$.

**Definition 2.22.** *For antichains* $\mathcal{L}, \mathcal{R} \subseteq \mathcal{F}$, *a* set interval $[\mathcal{L}, \mathcal{R}]$ *is a subset of* $\mathcal{F}$ *such that:* $[\mathcal{L}, \mathcal{R}] = \{Y \in \mathcal{F} : \exists_{X \in \mathcal{L}} \exists_{Z \in \mathcal{R}} X \leq Y \leq Z\}$.
$\mathcal{L}$ *and* $\mathcal{R}$ *are called a* left *and a* right bound, *respectively.*

With respect to $\leq$, the left and right bound of a set interval consist of minimal and maximal elements of this interval, respectively.

**Definition 2.23.** *A set* $E \subseteq \mathcal{F}$ *is* convex *(or is* interval-closed*) iff*
$\forall_{X,Z \in E} \forall_{Y \in \mathcal{F}} X \leq Y \leq Z \Rightarrow Y \in E$.

**Theorem 2.1 ([82]).**
*Every interval set is convex and each convex set can be represented by an exactly one interval set.*

The following theorem states that any power set with inclusion is a bounded lattice.

**Theorem 2.2 ([82]).**
*For a given set* $S$, $(\mathcal{P}(S), \subseteq)$ *is a lattice bounded by* $S$ *and the empty set.*

Consequently, one may consider a transaction system $(\mathcal{D}, \mathcal{I})$ and the respective lattice $(\mathcal{P}(\mathcal{I}), \subseteq)$.

**Theorem 2.3 ([50]).**
*For two databases* $D_1, D_2 \in \mathcal{D}$,
$JEP(D_1, D_2)$ *is convex in a lattice* $(\mathcal{P}(\mathcal{I}), \subseteq)$.

According to Theorem 2.1, convexity allows us to concisely represent a JEP space using a certain set interval. JEPs from its left bound are minimal.

**Lemma 2.3 ([50]).** *For* $[\mathcal{L}, \mathcal{R}] = JEP(D_1, D_2)$,
$\forall_{J \subseteq \mathcal{I}} J$ *is minimal in* $JEP(D_1, D_2) \iff J \in \mathcal{L}$.

It should be noted that, in the literature related to emerging patterns, a set interval is represented by a *border*, i.e. a pair consisting of its bounds. Also, the same concept was put forward earlier in [83,84] as a *fence* and used for reduct computation (see Section 3.5).

*Example 2.4.* Again, let us look at the decision transaction system from Table 2.3. In order to mine emerging patterns, one may want to consider the lattice $(\mathcal{P}(\mathcal{I}), \subseteq)$, where $\mathcal{I} = \{a, b, c, d, e\}$. The search space has $|\mathcal{P}(\mathcal{I})| = 2^5 = 32$ elements.

According to Theorem 2.3, the JEP space $JEP(D_k)$ for each decision class $k \in \mathcal{I}_d$ is convex and, from Theorem 2.1, can be uniquely represented by a set interval. Respective spaces are given in Table 2.4. In particular, $JEP(D_0) = JEP(D_0', D_0) = [\{bd, ae, ab\}, \{abe, bd\}] = \{ab, ae, bd\}$.

**Table 2.4.** The JEP spaces for the classes of the decision transaction system given in Table 2.3

| $k$ | $JEP(D_k)$ |
|---|---|
| 0 | $[\{bd, ae, ab\}, \{abe, bd\}]$ |
| 1 | $[\{de, cd\}, \{cde\}]$ |
| 2 | $[\{bc, ac\}, \{ac, bce\}]$ |

# 3 Literature Review

## 3.1 Contrast Patterns

Data mining is usually defined as a stage of the knowledge discovery process and is responsible for data analysis and finding interesting patterns ([1]). First and still most recognizable problems in this field referred to mining frequent itemsets and association rules ([2]). Over years, researchers have moved their interests towards other tasks that were considered earlier in disciplines, like machine learning or statistics.

Discrimination analysis is one such direction. The concept of a contrast pattern was introduced to capture differences between classes of transactions. In their pioneering works ([4,85]), the authors proposed class-association rules (CARs). The antecedent of a CAR is a frequent itemset and its consequent - a class identifier. Similarly to association rules, two characteristics, support and confidence indicate significance of a rule. CARs are used in the CBA classifier (classification based on associations), that always relies on the most confident rule to classify a new transaction.

The initial idea has been greatly improved in several ways in CMAR (classification based on multiple association rules, [5]). In terms of scalability, the training phase employs the FP-growth method ([81]). Pruning is based on correlation, confidence and coverage measures. A new structure, CR-tree, ensures economical storing of a rule set and fast lookups of rules in classification. Combined effects of matching rules, with respect to each decision class, express their popularity and correlation, and are evaluated by means of the weighted $\chi^2$ measure ([86]).

Concise representations have been another source of ideas in associative classification. In [6], a compact set of *essential rules* is discovered. These rules refer to frequent generators in respective classes of transactions. The concept of macro-items, items co-occurring in the same transactions, and MN-Tree are put forward to improve mining. On the other hand, in [87], closed itemsets are employed to generate interesting rules. Two classes, a positive and negative one, are considered. The authors look for *relevant sets of features*, which are closed itemsets in the positive class without a real subset that would also be closed in the positive class and both itemsets would have the same support in the negative class.

This work looks at probably one of the most popular types of contrast patterns - emerging patterns (EPs). This concept was introduced in [7] and, since then, it has been extended into numerous types of derivative patterns with original mining and classification techniques. EPs capture differences between a pair of transaction databases, called a *target* and *background* one. Their importance is evaluated by means of *growth-rate*, i.e. a ratio of supports in each of the databases.

Historically first EP-based classifier, CAEP, aggregates impact of matched patterns in classification (see Section 3.3, [8]). Also there, the authors put forward an EP mining method, that discovers an approximate result set using Max-miner ([88]) and border differential operations ([7]). Further works brought several enrichments of EPs with statistical concepts. Chi-EPs ([10,60]) use a chi-square test to evaluate interestingness of EPs, in addition to support, growth-rate and pattern minimality. In some cases, a complete solution can be found by traversing a search space. The IEP-Miner algorithm uses a CP-Tree ([75,54]) to take advantage of the increased pruning opportunities. Usability of Chi-EPs was demonstrated by means of a CACEP classifier analogous to CAEP.

EP-based classifiers rely on independence among pattern occurrences. This assumption is relaxed in BCEP (Bayesian Classification by Emerging Patterns), where a well-founded mathematical model of Naive Bayes is used to aggregate patterns into a single score. This adopts ideas from the Large Bayes classifier ([89]) to the field of EPs. Also, in MaxEP (Classification by Maximal Emerging Patterns, [90]), a similar aggregation strategy is employed. However, unlike in other propositions, not only patterns fully subsumed by, but also intersecting a testing transaction are taken into account.

Regarding EP discovery, in [91], the authors show that a complete set of EPs and their growth-rates can be inferred from a condensed representation based on closed patterns. Moreover, EPs are a specific type of patterns handled by MUSIC-DFS ([92]), a framework to mine patterns under constraints composable from a rich set of primitives. It uses a depth-first approach with pruning constraints expressed by set intervals.

An alternative approach is a constrained emerging pattern (CEP, [9]), where two support thresholds, minimal in the target database and maximal in the background one are used in the definition. In terms of mining scalability, the authors adopt the method from [93], that uses a structure similar to an FP-Tree ([81]) to partition the database and, then, applies a border differential algorithm ([7]) to the obtained blocks. Scoring in classification follows CAEP.

A recent proposition, delta discriminative emerging patterns ([52]) are defined similarly to CEP. The main difference is that the first threshold pertains to the entire database, not only to the target class. It is also suggested to choose a low or zero threshold in the background database. As far as discovery is concerned, patterns are represented by generators and closed frequent itemsets ([94,95]) that are mined simultaneously with a depth-first algorithm (see 3.2). Scarceness in the background classes places this type of patterns very close to jumping emerging patterns described below.

A set of all emerging patterns can be exponentially large, which limits practical analysis and usage of discovered patterns. Therefore, one important direction of the field is to investigate interesting derivatives of this general concept.

Probably the most widely known type of EPs are jumping emerging patterns (JEPs), which are supported in the target database only. It was shown ([7]) that they can be exactly represented by a set interval whose bounds are sets of minimal and maximal JEPs.

Minimal JEPs, due to high expressivity, became a basis for JEP-Classifier (see Section 3.3, [96,50]). In practice, one only uses the most supported patterns. This approach was later adopted by setting a threshold for minimal support in the target database and called a Strong Jumping Emerging Pattern (SJEP, SEP,[54]). The problem of noise in discovery of minimal patterns was further addressed by Noise-tolerant Emerging Patterns (NEPs,[54]) and Generalized Noise-tolerant Emerging Patterns (GNEPs,[54]). The former ones require patterns to satisfy minimal and maximal support in the target and background class, respectively. The latter approach, additionally modifies these conditions by applying two monotonous functions. Last but not least, minimal JEPs were employed in lazy classification in DeEPs and DeEPsNN ([97,98]).

More details on JEP discovery and related classifiers are given in Section 3.2 and 3.3. Accuracy of EP-based classifiers has been demonstrated with popular public datasets ([96,99]) and in a real-life setting, mostly in bioinformatics ([12,13,15,14]).

Evolution of emerging patterns follows general trends in data mining. This fact can be observed in discovery methods, like adopting an FP-Tree or zero-suppressed binary decision diagrams ([93,53]), as well as in addressing new tasks. In [67], Generalized Emerging Patterns are introduced to tackle hierarchical data. The patterns may consist of items from different levels of taxonomies associated with attributes. As a consequence, resulting classifiers can operate on more general information. Another interesting direction refers to the graph theory. There is a significant body of literature on finding frequent graphs in graph databases ([70]). In this case, transactions correspond to graphs and a given graph is supported by a transaction if it is a subgraph of this transaction. For a pair of such databases, the idea of emerging patterns is expressed by contrast graphs ([68,69]).

### 3.2 Jumping Emerging Pattern Discovery

In a manner similar to many problems in transaction databases, JEP discovery requires operating on exponentially-large itemset collections. Consequently, computation is often time- and memory-consuming and, for some datasets, remains infeasible. This section provides a survey of major JEP mining approaches. More space is dedicated to solutions used later for comparison purposes.

The complexity of the problem has not been derived formally. Nevertheless, it has been shown ([80]) that finding a JEP with maximal support, referred to as the *emerging pattern problem*, is MAX SNP-hard. Therefore, as long as $P \neq NP$, polynomial time approximation methods do not exist. Also, it has been proven that, for any constant $\rho < 1$, the emerging pattern problem cannot be approximated within ratio $2^{log^{\rho} n}$ in polynomial time unless $NP \subseteq DTIME[2^{polylogn}]$, where $n$ is the number of positions in a pattern.

**Algorithm 1.** JEP-Producer $([\{\emptyset\},\{A_1,..,A_k\}],[\{\emptyset\},\{B_1,..,B_m\}])$

$L_0 \Leftarrow \emptyset$, $R_0 \Leftarrow \emptyset$
**for** $(i = 1,..,k)$ **do**
  **if** $\exists_{j=1,..,m} A_i \subset B_j$ **then**
    continue
  **end if**
  $[L_{new}, R_{new}] \Leftarrow Border - Diff(<\{\emptyset\},\{A_i\}>,<\{\emptyset\},\{B_1,..,B_m\}>)$
  $L_i \Leftarrow L_{i-1} \cup L_{new}$
  $R_i \Leftarrow R_{i-1} \cup R_{new}$
**end for**
**return** $[L_k, R_k]$

**Algorithm 2.** Border-Diff $([\{\emptyset\},U],[\{\emptyset\},\{S_1,..,S_k\}])$

$L_1 \Leftarrow \{\{x\} : x \in (U - S_1)\}$
**for** $(i = 2,..,k)$ **do**
  $L_i \Leftarrow \{X \cup \{x\} : X \in L_{i-1}, x \in (U - S_i)\}$
  remove all non-minimal elements in $L_i$
**end for**
**return** $[L_k, U]$

**Border Differential.** One fundamental property of JEP spaces is their convexity (Theorem 2.3). JEP-Producer ([96]) takes advantage of this fact to find all JEPs in each of two given databases.

Let us consider a transaction system $(\mathcal{D},\mathcal{I},\mathcal{I}_d)$ and a partition of $\mathcal{D}$ into two databases $D_p, D_n$. The main idea originates from the following observation:

$JEP(D_n, D_p) = [\{\emptyset\},\{A_1,..,A_k\}] - [\{\emptyset\},\{B_1,..,B_m\}]$,

where $\{A_1,..,A_k\}$, $\{B_1,..,B_k\}$ are sets of all maximal itemsets in $D_p$ and $D_n$, respectively. Computation of these sets is straightforward and covered by a procedure called Horizon-Miner ([7]). It scans a given database and collects successive transactions. All transactions collected so far that are subsets of the currently considered one are pruned. JEP-Producer considers successive itemsets $A_i$ $(i = 1,..,k)$ pertaining to $D_p$ and performs the subtraction $[\{\emptyset\}, A_i > - < \{\emptyset\},\{B_1,..,B_m\}]$ to compute subsets of the transaction $A_i$ that are not present in $D_n$. This operation is covered by a border differential routine, Border-Diff ([7]).

The original proposition of JEP-Producer was followed up in [50], where optimizations to a minimality check in Border-Diff were put forward. In ([79]), maintenance routines for JEP spaces are covered. In particular, insertion/deletion of transactions to/from the target database, insertion of transactions to the background database and insertion/deletion of an item to/from the itemspace. Unfortunately, deletion of transactions from the background database has not found an efficient solution.

Finally, the border differential can be used to find EPs in a rectangular area of the support space, defined as: $\{X \in \mathcal{I} : \theta_{min} \leq supp_{D_p} X \leq 1 \wedge 0 \leq supp_{D_n} X \leq \delta_{min}\}$, for certain $\theta_{min}, \delta_{min} \in < 0, 1 >$ ([8,50]). This idea was employed in the training phase of CAEP ([8]) to find almost all EPs.

**CP-Tree.** Approaches based on concise representations may prevent or limit pruning opportunities. At the same time, space traversal techniques remain efficient, as long as candidates are being enumerated in the way that leverages intensive pruning.

One approach in this class is a contrast pattern tree (CP-Tree), previously also called a pattern tree (P-Tree). Considering individual itemsets gives flexibility of mining different types of patterns. The original proposition targeted strong emerging patterns (SJEPs, SEPs), noise-tolerant emerging patterns (NEPs), generalized noise-tolerant emerging patterns (GNEPs) ([75,54]) and chi-emerging patterns (chi-EPs) ([10,60]). Since SJEPs are a special type of JEPs, this method remains valid for JEP discovery as well. For this reason, we limit our description of a CP-Tree to SJEP finding.

Let us consider a decision transaction system $\mathcal{DTS} = (\mathcal{D}, \mathcal{I}, \mathcal{I}_d)$. We assume an order $\prec$ in $\mathcal{I}$. A CP-Tree is a modified prefix tree. It is multiway and ordered. Each edge from a node $N$ to one of its children is labeled with an item. The set of edge labels of $N$ is denoted by $N.items$. For $i \in N.items$, we associate the respective edge with a child (*N.child[i]*), positive count (*N.posCount[i]*) and negative count (*N.negCount[i]*). Children of each node and edges on each rooted path are ordered in the way the respective labels follow $\prec$. A set of labels on a rooted path represents a transaction prefix with counts in the positive and negative class indicated by respective counts of the last node on this path.

*Example 3.1.* Let us assume an order $\prec$ such that $b \prec d \prec c \prec a \prec e \prec f$. A sample decision transaction system and the initial structure of a CP-Tree constructed for this system are given in Figure 8.1. On the leftmost path *bd* is marked with *2,1*, since it occurs in 2 transactions in $c_0$ and 1 transaction in $c_1$.

A CP-Tree concisely represents a two-class transaction decision system and enables efficient pattern mining. In order to construct a tree for $\mathcal{DTS}$, items of each transaction from $\{T_i \cap \mathcal{I} : T_i \in \mathcal{D}\}$ are sorted according to $\prec$ and the resulting strings are inserted like to a regular prefix tree. As a part of each insertion,

| | | |
|---|---|---|
| $T_1$ | *bdca* | $c_0$ |
| $T_2$ | *bdaf* | $c_0$ |
| $T_3$ | *bce* | $c_0$ |
| $T_4$ | *ca* | $c_1$ |
| $T_5$ | *bdc* | $c_1$ |
| $T_6$ | *caef* | $c_1$ |

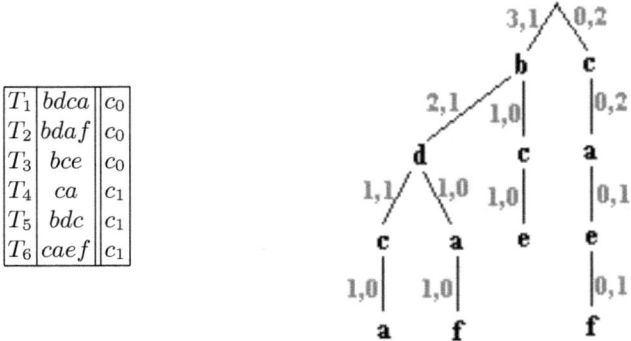

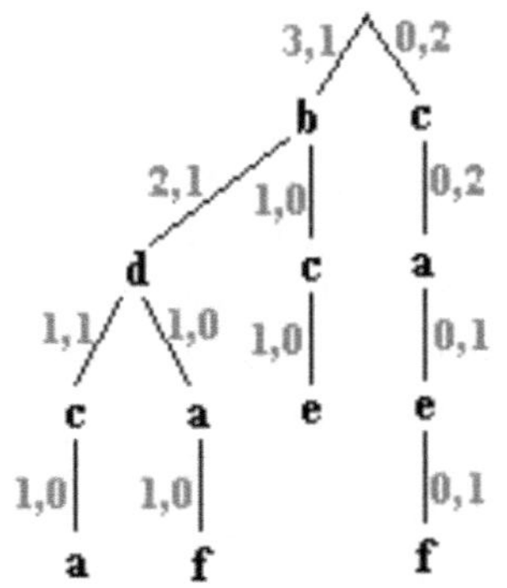


**Fig. 3.1.** A sample classified dataset with transaction items following the order $b \prec d \prec c \prec a \prec e \prec f$ and a respective CP-Tree

positive or negative counts are incremented based on the transaction's class. A tree for our sample system is given in Figure 8.1.

Patterns are mined simultaneously in both classes by traversing a tree in the pre-order fashion. Each visited node corresponds to a pattern consisting of items on the path from this node to the root. Based on counts stored in the node, one may decide if the pattern should be collected. Two result sets for positive and negative patterns are maintained.

If one considers a node and its associated pattern, children of the node correspond to supersets of this pattern. Also, the node's counts do not increase as one walks down the tree. Both facts allow us to implement pruning techniques and avoid visiting some subtrees. For example, in JEP discovery, if one of the node's counts is equal to zero, it means that it is a JEP in one of the classes. Therefore, there is no point to consider the subtree of this node, i.e. supersets of the pattern. Similarly, when SJEPs are mined, no further subtree exploration is needed if the node's counts fall below or are equal to respective thresholds of a positive and negative minimum count.

The order $\prec$ impacts effectiveness of pruning. One proposition is to sort items according to their support-ratios ([54]). Thanks to this heuristics JEPs are more likely to be located closer to the root and discovered earlier.

A simple traversal of a tree would not guarantee considering the complete search space and a region visited this way depends on the assumed order $\prec$. Therefore, the mining procedure involves a recursive operation, called tree merging, that makes the tree evolve during the process. Merging of a tree M into a tree T inserts all patterns corresponding to paths rooted in one of the children of M. It is efficient, since patterns are already represented in a prefix fashion and respective node counts can be accumulated.

*Example 3.2.* Let us start the mining process in the root of the tree given in Figure 3.2. We look at its leftmost child $b$. Its subtree is merged with the currently mined node, with the root. In other words, the subtrees of the children of $b(3,1)$, namely $d(2,1)$ and $c(1,0)$, are inserted to the tree starting in the root. New nodes are created and the counts of the respective existing nodes are appropriately incremented.

**Transversals in Hypergraphs.** The problem of finding JEPs can be expressed by means of the graph theory ([100]). For a given transaction $t \in D$ in the target class, we define a hypergraph $\mathcal{G} = (V, E)$, where vertices refer to condition items of this transaction, $V = t \cap \mathcal{I}$, and hyperedges refer to differences of this transaction and transactions in the complementary class, $E = \{t - t' : t' \in D'\}$. Each set of vertices is a pattern supported in the target class. By definition, a transversal (or a hitting set) is set of vertices that has a non-empty intersection with each hyperedge and, consequently, is a JEP. Therefore, discovery of minimal JEPs supported by a given transaction can be reduced to finding minimal transversals in $\mathcal{G}$. In order to obtain a complete JEP space, this procedure has to be repeated for all transactions in $D$ and the results summed up.

There are several algorithms for finding minimal transversals in hypergraphs. In particular, in [101], edges are considered successively. In every step, one

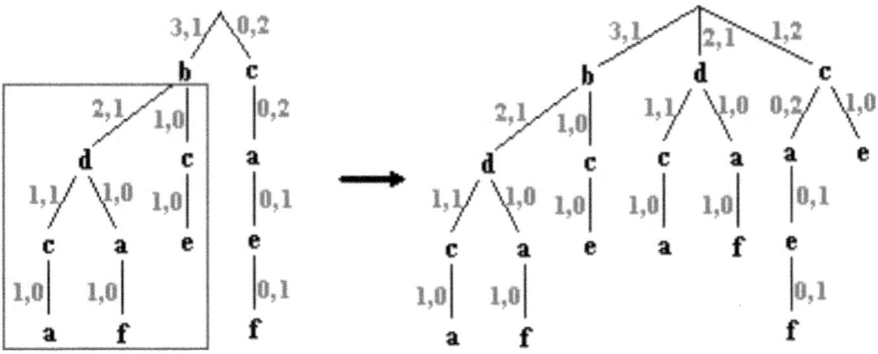


**Fig. 3.2.** Merge operation of the node $b(3, 1)$ and the root in a CP-Tree

computes a cross-product of a new edge and elements of a set of minimal transversals in the graph with edges considered so far. The result is then minimized to be a valid input for the next iteration. This approach suffers from high memory consumption and is unable to produce any results before completing computation. Both problems are addressed in [102], where a depth first search is performed. Two algorithms are proposed in [103] to solve an equivalent problem of monotone Boolean duality with pessimistic complexity of $O(n^{O(log^2(n))})$ and $O(n^{o(log(n))})$, respectively, where $n$ is the total number of implicants in the input and output function.

Taking a cross-product of edges of high cardinality leads to large intermediate results and may prevent the original method from completion for high-dimensional data. In the context of JEP mining, ideas from [101] are enriched in [51] by applying the divide and conquer strategy. The new approach recursively partitions the set of edges and stops when a resulting block is not likely to generate too many partial transversals. Namely, sets of many small edges or small sets of large edges are desirable. Each block of the resulting partition is then tackled with a border-differential approach ([7]), which is, in fact, an optimized version of the method from [101].

**Zero-suppressed Binary Decision Diagrams.** Binary decision diagrams (BDD, [104]) are a graph-based structure utilized to represent boolean functions. They are widely applied in SAT solvers ([105]) and VLSI solutions ([106]). Since a pattern collection can be expressed by a characteristic function ([107]), these diagrams allow us to concisely encode transactional datasets and speed up pattern mining in a way similar to FP-Trees or CP-Trees.

Precisely, a BDD is a canonic directed acyclic graph with one source, one true sink, one false sink and several internal nodes. Internal nodes are labeled, ordered and have two children. A node $N$ with a label $x$ and children $N_0$ and $N_1$ encodes a formula: $N = (x \wedge N_1) \vee (\overline{x} \wedge N_0)$. The edge $N - N_1$ ($N - N_0$) is called a true (false) edge. Every path from the source to one of the sink nodes gives a true or false assignment for the boolean formula. A diagram can be seen as a decision tree simplified by means of two rules. First, a node is removed from a tree if both its children are the same node. Second, isomorphic subgraphs are merged, which results in high compression.

A special type of BDD, called a zero-suppressed binary decision diagram (ZBDD, [108]) is of a special interest in data mining. The difference is the first simplification rule, namely, a node is removed when its true edge points to the sink node 0. In fact, the ZBDD obtains even higher compression rates for formulas which are almost everywhere false. Let us consider a representation of transaction database based on decision diagrams. As discussed in [109], the upper bound for the size of the ZBDD, understood as the number of its nodes, is the total number of items appearing in all transactions. The respective upper bound for the BDD is the number of transactions multiplied by the number of elements that can appear in them. Therefore, the ZBDD are especially compact in this scenario. Figure 3.3 shows a BDD and ZBDD for transactions: $\{ab, ac, c\}$, where the latter is observably smaller.

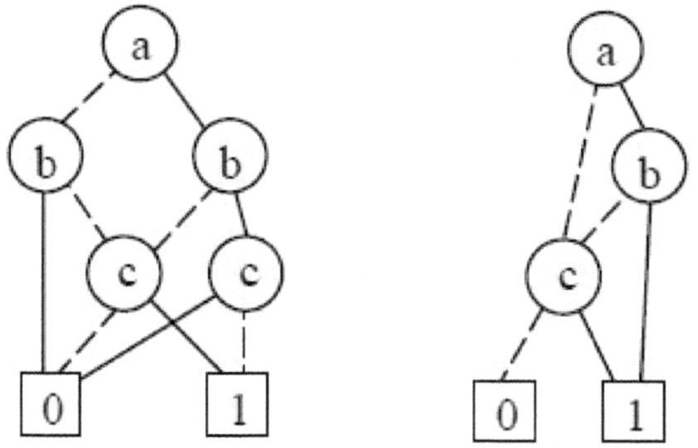


**Fig. 3.3.** A sample BDD and ZBDD for the transactions $\{ab, ac, c\}$, [109]

In [53], the authors propose a CEP mining method that traverses a search space bottom-up and depth-first, so that new candidates are created by adding new items. A ZBDD is constructed incrementally to both enumerate the space and store minimal CEPs found so far. Patterns are pruned based on support constraints and minimality. Note that, in case of JEP mining, only the maximal support constraint in a background class can be used for pruning. In addition, different heuristics to pick a good ordering of items are investigated, since the size of diagrams is very sensitive to this aspect ([104]).

**Equivalence Classes.** Commonalities between emerging patterns and frequent itemsets allow us to exploit similar mining approaches. Among others, concise representations are one of the most powerful tools of the field. As shown in [94], a set of frequent itemsets can be partitioned into equivalence classes, where each class encompasses frequent itemsets supported by the same transactions in a dataset. An equivalence class can be expressed by means of a set interval. Elements of its left bound are called generators and the only element of the right bound is called a closed frequent itemset.

One may want to apply the same methodology to datasets whose transactions are classified into two decision classes. In [52], the authors note that itemsets in a single equivalence class share the same statistical merits, such as chi-square, odds ratio, risk ratio, as long as these merits can be expressed as functions of supports in decision classes. Therefore, itemsets of given statistical properties can be mined by finding generators and closed frequent itemsets. Equivalence classes that do not meet assumed importance criteria are pruned.

While there is a large body of literature on mining generators and closed frequent itemsets ([94,95]), the authors present a single algorithm, DP-Miner ([52]), that computes the two types of patterns simultaneously and immediately forms respective equivalence classes. In order to target the most expressive patterns, it focuses on so-called non-redundant classes, whose closed patterns are minimal among closed patterns of all classes. In other words, a non-redundant class corresponds to a transaction set that is not subsumed by a transaction set corresponding to any other class.

Although DP-Miner utilizes an FP-Tree, unlike FP-Growth ([81]), it performs a depth-first search. Conditional databases are constructed for each frequent generator. In order to search for corresponding closed frequent itemsets, a tail sub-tree is maintained for these databases. Non-generators are pruned in the process, according to anti-monotonicity of generators ([95]).

On the basis of statistically important itemsets, a new type of patterns, $\delta$-discriminative patterns, was defined. As mentioned in Section 3.1, they are required to be frequent in the whole database and highly infrequent in the background database, as expressed by the threshold $\delta$. In case of many decision classes, a sum of all classes different from the target one is considered a background database. JEPs are a specific type of these patterns obtained for global minimal support and $\delta$ equal to 0. For the purpose of JEP discovery, one may use the computation framework sketched above. However, efficiency suffers, since no support-related pruning can be performed at the level of the whole database. It is worth noting, though, that only generators of non-redundant classes need to be mined if one looks for minimal JEPs, which is usually the case.

*Example 3.3.* Let us consider a dataset in Figure 8.1. Each pattern $X$ in the equivalence class $[\{ab, ad\}, \{abd\}]$ has $supp_{c_0}(X) = 2/3$ and $supp_{c_1}(X) = 0$, thus, is a JEP. The generators $ab, ad$ are minimal JEPs. Also, the only element of $[\{abcd\}, \{abcd\}]$ is a JEP. This class is redundant, though, and its generator is not a minimal JEP. On the other hand, for each $X \in [\{cd\}, \{bcd\}]$, we have $supp_{c_0}(X) = 1/3$ and $supp_{c_1}(X) = 1/3$, thus, this class does not contain any JEPs.

### 3.3 Classification Based on Emerging Patterns

Emerging patterns capture discrimination information between classes. Classifiers that employ this type of patterns have proven their accuracy in practical applications ([12,15]).

One approach is the eager classification scheme. A training decision transaction system is analyzed to construct a generalized model that can be later used to classify testing transactions. The other option is the lazy scheme, which does not have the training phase per se. The whole training system is stored and used to build ad hoc models for individual testing transactions. In both cases, the models consist of per-class sets of patterns, denoted here by $\mathcal{P}_k$, as well as of associated with these patterns interestingness measures, like growth-rate or support. The type of the patterns and, thus, an appropriate finding method depend on a specific classifier.

Since emerging patterns are naturally defined for two-class problems, multiple decision classes require special handling. In the pair-wise approach ([98]), one relates a considered class to each of the remaining classes. Classification score for this class is a sum of scores over respective pairs. Note that it is necessary to compute sets of patterns for all possible pairs of classes. This drawback is overcome in an alternative approach, where a considered class is contrasted against all transactions from other classes. This way only one set of patterns is stored per class.

The trained model can predict the classes for unseen transactions. Membership of a testing transaction in a given class, is evaluated by means of a scoring function. For this evaluation, class-specific information is employed. At the end, the resulting indicators are compared and a class with the maximal value is picked. Ties are broken by choosing the majority class and, if that fails, by a random choice.

**CAEP.** Classification by aggregating emerging patterns (CAEP) is historically the first classifier based on EPs ([8]). In the learning phase, it computes per-class collections of $\rho$-EPs and stores growth-rate and support for each of the patterns. When a testing transaction $T$ is considered, the indicator for a class $k$ is expressed by: $f_k(T) = sum_{X \subseteq T \wedge X \in \mathcal{P}_k} st(X)$.

Both sizes of classes and numbers of EPs in each class may differ significantly and introduce an undesirable bias to a model. Therefore, it is recommended to normalize indicators by dividing them by a so-called *base-score*. The baseline is defined as a fixed percentile of training transactions in the respective class.

Per-class collections of EPs are often very large. In fact, many patterns can be eliminated without a significant loss of accuracy. Therefore, it is a common practice to only use the minimal patterns with large growth-rates and supports, assuming certain thresholds.

**JEP-based Classifiers.** Simple definition and highly discriminative power make JEPs very attractive for data classification. Also, convexity of JEP spaces potentially lead to less time-consuming computation than in case of general EPs. The state of the art JEP-based classifiers employ minimal patterns as these are considered the most expressive.

JEP-Classifier ([11]) stores collections of minimal JEPs, $\mathcal{P}_k$, for each class $k$, that are computed during the training phase. Classes are scored by means of *collective impact*, defined as $f_k(T) = sum_{X \subseteq T \wedge X \in \mathcal{P}_k} supp(X)$, for a class $k$ and a testing transaction $T$. Note that this measure follows the scoring in CAEP, when we set an infinite support in the background class. The same approach can be used to subtypes of JEPs that are obtained with stricter pruning conditions but still are characterized by support only, like SJEP, NEP, GNEP ([54]).

Eager classification requires us to precompute complete sets of minimal patterns, which may be extremely expensive for some datasets. At the same time, to classify a given transaction only several patterns are eventually matched. Therefore, lazy approaches turn out to be highly effective here.

DeEPs (Decision Making by Emerging Patterns, [97,96]) is the most popular example here. It stores a whole training set. When a testing instance $T$ comes, it computes intersections of elements of this set with $T$, so that reduction is performed both in terms of dimensions and a number of transactions. Minimal JEPs are discovered in the resulting database and used to compute collective impact. Optionally, pattern collections can be enriched with interesting EPs. In this case, the remaining part of a search space can be scanned or partially scanned. Also, the CAEP scoring formula has to be applied, instead. Last but not least, DeEP can be used for continuous variables without prior discretization of a dataset. This is achieved by setting a threshold for similarity of two values.

Since original training data are always available, it is possible to combine DeEPs with the classic k-Nearest Neighbor ([110,34]). DeEPsNN ([98]) uses k-NN in the first place and, in case of an empty neighborhood, decision is left up to DeEPs. The main advantage of this approach is faster classification.

### 3.4 Rough Set Theory

Rough set theory (RST) was founded by Zdzislaw Pawlak in 1982 ([16]). Over the years, it has become a robust knowledge discovery framework that does well in handling incompleteness, vagueness and uncertainty in data ([21,22,17,18,23,24,19]). RST research has revealed numerous connections and led to hybrid concepts in important fields, such as probability theory ([25,26]), Dempster-Shafer theory of evidence ([27]), neural networks ([28]), fuzzy sets, granular computing ([29,30]) or data mining ([49,29]). Major problems include rule induction and classification ([31,32,33,34]), feature selection and data reduction ([35,36,37]), clustering ([38,39]) and discretization ([40,41]).

RST considers a universum of objects characterized by means of functions called attributes. Furthermore, knowledge is interpreted as an ability to discriminate objects and, thus, each attribute set carries certain amount of information. Assuming that only certain attributes are available, indiscernibility of objects leads to the most fundamental concept of a *rough set*. It approximates a conventional set by means of two sets, a *lower* and an *upper approximation*, which refer to objects definitely included in the set and the ones it may contain, respectively.

Among numerous consequences resulting from the rough perspective, we focus our attention to data reduction. This area is built around *reducts* - highly informative attribute sets. The most generic reducts provide the same information as all available attributes and remain minimal with respect to the inclusion relation. These requirements are relaxed in the definition of *approximate reducts*

([44,45]) that need to discriminate at least a given percentage of object pairs and, thus, are more resistant to noise. This notion, and later proposed *association reducts* ([46,47]), are directly related to association rules in transaction datasets. Another variant, *approximate entropy reducts* ([48]), employ entropy in place of indiscernibility. Finally, *dynamic reducts* ([37]) provide a different approach to handle noisy data. They refer to attribute sets that remain reducts in the universum and in many of its randomly drawn subsets, as defined by a *stability threshold*.

### 3.5 Reduct Computation

Numerous applications in feature selection and data reduction have made reduct computation one of pivotal tasks in rough set theory ([74,19]). Specifically, the reduct set and minimal reduct problem have been drawing much attention of the research world. As originally defined for the most generic reducts, both tasks have been later adopted to derivative concepts.

The reduct set problem refers to finding all reducts for a given universum. It has been proven to be NP-complete ([55]). In the literature, it is usually approached on the ground of boolean reasoning ([55,111]). In fact, one may formulate and minimize a certain boolean function and, then, interpret the result as a set of reducts. As proposed in [55], discrimination opportunities in the universum can be represented by a discernibility matrix. Each element of this matrix corresponds to a pair of objects and contains all attributes that individually discriminate these objects. If one associates each attribute with a boolean variable, an element of a matrix can be viewed as an alternative of respective variables. Also, empty elements are assigned the value *true*. As a result, two objects are discerned, as long as the corresponding alternative is satisfied. A conjunction of all alternatives is called a discernibility function ([55]) and allows us to verify if an arbitrary attribute set discerns as well as all attributes. Minimal sets with this property are reducts.

Boolean functions can be expressed by an infinite number of distinct expressions. One may want to look for the minimal ones with respect to the number of variable occurrences. In particular, discernibility functions are alternative-conjunction expressions without negation and, therefore, they are monotonous ([111]). As stated in [111], the minimum expression for a function from this class is an alternative of exactly all its prime implicants. Consequently, due to their minimality, prime implicants of a discernibility function directly correspond to reducts ([55]).

A solution to the problem based on minimization of monotonous boolean functions is described in [112,37]. First, a discernibility matrix is constructed and reduced by means of the absorption laws ([111]). The set of reducts is built iteratively by taking into account successive non-empty elements of the matrix. After each step, the set is extended, so that it contains all possible minimal attribute sets that discern pairs of objects considered so far. This method can be further optimized to avoid generation of unnecessary intermediate elements and to improve element absorption ([113]).

An alternative approach to find all prime implicants is discussed in [56,114]. It recursively applies the Shannon expansion theorem ([115]) to boolean variables corresponding to attributes. In particular, boolean quotients need to be computed in every invocation.

Apart from symbolic transformation, the problem can be solved by searching the lattice of all possible monoms. A solution based on the notions of cones and fences is proposed in ([83,84]) and further extended in ([56]). In two other algorithms, RAD and CoreRAD ([57]), all maximal monoms that do not extend any implicants need to be identified. While the latter is potentially infeasible in practice, approximate versions may be employed. Furthermore, in [56], an indirect search methodology ([116]) is applied to avoid extending implicants. Also there, an incremental strategy that adjusts a solution for a given universum to its extension with a new object, is proposed. Last but not least, the reduct set problem has been tackled with parallel methods ([117,118]).

The other interesting task, the minimal reduct problem, is to find a reduct that is minimal in terms of its cardinality. It belongs to the NP-hard class ([55]) and has mainly been addressed with approximate methods. In [112], a greedy heuristics is considered. An empty set is successively extended with new attributes, so that, in each step, one chooses the attribute that discriminates the most of still indiscernible object pairs. Other propositions employ soft computing, including genetic algorithms ([72,119,120]), simulated annealing ([121]), particle swarm optimization ([122]) and ant-colony ([123]).

## 4 Attribute Set Dependence in Reduct Computation

A reduct belongs to the most essential concepts of rough set theory. Section 3.4 briefly presents major extensions of the original idea and numerous applications, in particular, in data reduction ([35,36,37]). In addition, it reviews several computation methods.

This chapter considers the role of attribute set dependence in reduct finding. We employ this property to formulate effective pruning in our novel algorithm, RedApriori ([124]). In order to implement the pruning conditions, information on immediate subsets of a currently considered attribute set is necessary. This is naturally accomplished by following the Apriori candidate generation scheme ([2]) in search space enumeration.

Furthermore, different options for storing subsequent generations of candidate attribute sets are discussed ([125]). In particular, we describe a coverage count trie (CC-Trie), a new tree structure that provides compact representation for attribute set collections and ensures fast lookups.

It should be noted that, independently from our research, a similar space traversal scheme and pruning related to attribute set dependence were utilized for finding functional dependencies ([126,127]). Also, conceptually close pruning approach was used in EP mining based on ZBDD ([53]). Nevertheless, both works were published after our proposition and the way respective conditions are tested is substantially different.

### 4.1 Discernibility and Dependence

Let us consider an information system $(\mathcal{U}, \mathcal{A})$. In the reduct set problem one considers the power set of the attribute space $\mathcal{P}(\mathcal{A})$ and looks for all reducts $RED(\mathcal{U})$. The search space is exponentially large, therefore, many algorithms employ pruning conditions to avoid examining all possible attribute sets. Since these conditions rely on the relation of set inclusion, it is convenient to consider them in the context of the lattice of attribute sets $(\mathcal{P}(\mathcal{A}), \subseteq)$.

Basic pruning criteria originate from works related to monotonous boolean functions. In particular, the following two conditions are extensively discussed in [56], where they are combined with various methods of space traversing.

**Theorem 4.1.** *[56] Let* $B \subseteq \mathcal{A}$*, we have* $S \subset B \wedge B \notin SRED \Longrightarrow S \notin RED$.

**Theorem 4.2.** *[56] Let* $B, S \subseteq \mathcal{A}$*, we have* $B \subset S \wedge B \in SRED \Longrightarrow S \notin RED$.

The former uses the notion of discernibility and states that we do not need to examine proper subsets of a non-super reduct $B$, since they cannot differentiate more object pairs than $B$ does. The latter tells us that proper supersets of a reduct cannot be minimal, thus, they can also be excluded from examination.

Here, we propose a pruning strategy that is based solely on attribute set dependence.

**Definition 4.1.** *We define*

*an* independent set collection $ISC(\mathcal{U}) = \{B \subseteq \mathcal{A} : B \textit{ is independent}\}$,
*a* dependent set collection $DSC(\mathcal{U}) = \{B \subseteq \mathcal{A} : B \textit{ is dependent}\}$.

For brevity, we do not specify a set of objects whenever it is known from the context, e.g. $ISC$, $RED$ instead of $ISC(\mathcal{U})$, $RED(\mathcal{U})$.

The property of set dependence generates a binary partition $\{ISC, DCS\}$ in $P(\mathcal{A})$. Moreover, every subset of an independent set is independent and every superset of a dependent set is dependent. These facts are expressed formally below.

**Lemma 4.1.** *Let* $B, S \subseteq \mathcal{A}$*, we have* $S \subseteq B \wedge B \in ISC \Longrightarrow S \in ISC$.

*Proof.* Let $B, S \subseteq \mathcal{A}$, $S \subseteq B$ and $B \in ISC$.

From the definition, we observe

$$\begin{aligned} B \text{ is independent} &\Longleftrightarrow \forall_{a \in \mathcal{B}} a \text{ is indispensable in } B \\ &\Longrightarrow \forall_{a \in \mathcal{S}} a \text{ is indispensable in } B \\ &\Longleftrightarrow \forall_{a \in \mathcal{S}} IND(B - \{a\}) \neq IND(B). \end{aligned}$$

Since $IND(B) \subseteq IND(B - \{a\})$, we have $\forall_{a \in \mathcal{S}} \exists_{(u,v) \in \mathcal{U} \times \mathcal{U}} (u, v) \in IND(B - \{a\}) \wedge (u, v) \notin IND(B)$. Let us consider this $(u, v)$ for a given $a \in \mathcal{S}$. On the one hand, we have $(u, v) \in IND(B - \{a\}) \wedge (u, v) \notin IND(B) \Longrightarrow a(u) \neq a(v) \Longrightarrow (u, v) \notin IND(S)$. On the other hand, since $IND(B - \{a\}) \subseteq IND(S - \{a\})$, we have $(u, v) \in IND(B - \{a\}) \wedge (u, v) \notin IND(B) \Longrightarrow (u, v) \in IND(S - \{a\})$. Therefore, we have $(u, v) \in IND(S - \{a\}) \wedge (u, v) \notin IND(S) \Longleftrightarrow S$ is independent. □

**Lemma 4.2.** *Let $B, S \subseteq \mathcal{A}$, we have $B \subseteq S \wedge B \in DSC \Longrightarrow S \in DSC$.*

*Proof.* Analogous to the proof of Theorem 4.1. □

The following theorem states that independent set and dependent set collections can be represented by certain set intervals.

**Theorem 4.3.** *ISC and DSC are convex. There exist attribute set collections $MISC, mDSC \subseteq P(\mathcal{A})$ such that $ISC = [\{\emptyset\}, MISC]$, $DSC = [mDSC, \{\mathcal{A}\}]$, where the symbols MISC and mDSC stand for a maximal independent set collection and minimal dependent set collection, respectively.*

*Proof.* It is sufficient to show that both collections can be represented by the specified set intervals.

Let us focus on $ISC$ first. Consider $ISC \subseteq [\{\emptyset\}, MISC]$. Let $B \in ISC$. Obviously, $B \supseteq \emptyset$. Note that inclusion partially orders elements in $ISC$, so also $\exists_{S \in MISC} B \subseteq S$. Conversely, $ISC \supseteq [\{\emptyset\}, MISC]$. Let $B \in [\{\emptyset\}, MISC]$. From the definition of a set interval we have $\exists_{S \in MISC} \emptyset \subseteq B \subseteq S$. According to Lemma 4.1 $B$ is independent, so $B \in ISC$. To sum up, we have found that $ISC$ can be represented by $[\{\emptyset\}, MISC]$ and, consequently, is convex.

A proof for $DSC$ is analogous and employs Lemma 4.2. □

The Theorem 4.2 can be generalized as follows.

**Theorem 4.4.** *Let $B, S \subseteq \mathcal{A}$, we have $B \subseteq S \wedge B \in DSC \Longrightarrow S \notin RED$.*

*Proof.* Consider $B \in DSC$ and $S \subseteq \mathcal{A}$ such that $B \subseteq S$. From Lemma 4.2 we have $S \in DSC$. Thus, $S \notin ISC$ and $S$ cannot be a reduct. □

According to the definition, it is possible to test set dependence by examining all immediate subsets of a given set. In practice, it is convenient to use a coverage count for this purpose. In fact, equality of the coverage counts of attribute sets is equivalent with equality of the corresponding indiscernibility relations.

**Lemma 4.3 ([55]).** *Let $B, S \subseteq \mathcal{A}$ such that $S \subseteq B$, we have $IND(S) = IND(B) \Longleftrightarrow covcount(S) = covcount(B)$.*

As a consequence, attribute set dependence can be tested as follows.

**Theorem 4.5.** *Let $B \subseteq \mathcal{A}$, we have $\exists_{a \in B} covcount(B) = covcount(B - \{a\}) \Longleftrightarrow B \in DSC$.*

*Proof.* From the definition, an attribute $a \in B$ is dispensable in $B$ iff $IND(B) = IND(B - \{a\})$. From Lemma 4.3, where $S = B - \{a\}$, we have: $a \in B$ is dispensable iff $covcount(B) = covcount(B - \{a\})$. □

Although the latter theorem provides opportunity for pruning, every computation of *covcount* can be a costly operation. It is especially common for very large databases. Therefore, it is better to first perform pruning by means of dependent sets and reducts visited so far. We check whether all immediate subsets of a tested set are independent and are not reducts. Otherwise, the set is dependent based on Lemma 4.2 or Theorem 4.2. For brevity, we use the notation $F_k = \{B \in F : |B| = k\}$, e.g. $ISC_k$, $RED_k$, $P_k(\mathcal{A})$.

**Table 4.1.** An information system $\mathcal{IS} = (\{u_1, u_2, u_3, u_4, u_5\}, \{a, b, c, d, e\})$

| | $a$ | $b$ | $c$ | $d$ | $e$ |
|---|---|---|---|---|---|
| $u_1$ | 0 | 0 | 1 | 0 | 0 |
| $u_2$ | 1 | 1 | 1 | 1 | 0 |
| $u_3$ | 1 | 1 | 1 | 2 | 0 |
| $u_4$ | 0 | 2 | 0 | 1 | 0 |
| $u_5$ | 2 | 3 | 1 | 1 | 1 |

$MISC = \{\{a, c\}, \{a, d\}, \{b, d\}, \{c, d, e\}\}$
$mDSC = \{\{a, b\}, \{a, e\}, \{b, c\}, \{b, e\}\}$
$RED = \{\{a, d\}, \{b, d\} \{c, d, e\}\}$

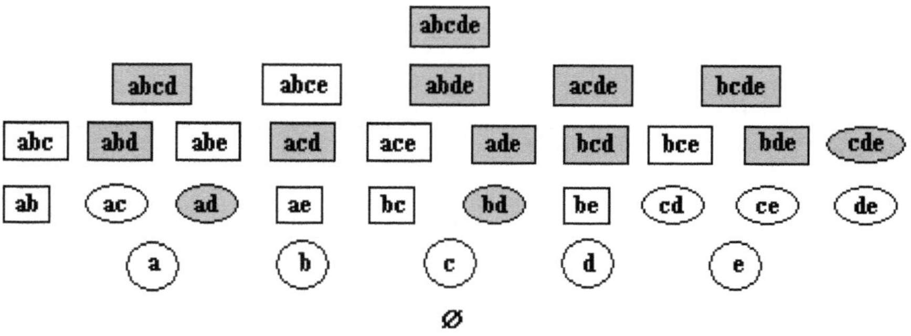


**Fig. 4.1.** The search space $P(\{a, b, c, d, e\})$ for the reduct set problem for the information system $\mathcal{IS}$ from Table 4.1. Independent sets - ovals, dependent sets - rectangles, super reducts - gray background.

**Theorem 4.6.** *Let* $B \subseteq \mathcal{A}$, *we have*
$\exists_{a \in B}(B - \{a\}) \notin (ISC_{|B|-1} - RED_{|B|-1}) \Longrightarrow B \in DSC.$

*Proof.* Let $B \subseteq \mathcal{A}$ and $a \in B$ such that $(B - \{a\}) \notin (ISC_{|B|-1} - RED_{|B|-1})$. Since $|B - \{a\}| = |B| - 1$, so $(B - \{a\}) \in P_{|B|-1}(\mathcal{A}) - (ISC_{|B|-1} - RED_{|B|-1}) = DSC_{|B|-1} \cup RED_{|B|-1}$. Therefore, $(B - \{a\}) \in DSC_{|B|-1}$ or $(B - \{a\}) \in RED_{|B|-1}$. Let us consider both cases separately.

Let $(B - \{a\}) \in DSC_{|B|-1} \subseteq DSC$. In accordance with Lemma 4.2, we have $(B - \{a\}) \subseteq B \wedge (B - \{a\}) \in DSC \Longrightarrow B \in DSC$. Let, now, $(B - \{a\}) \in RED_{|B|-1}$. It means that $IND(B - \{a\}) = IND(\mathcal{A}) = IND(B)$, so $a$ is dispensable in $B$ and, thus, $B \in DSC$. □

*Example 4.1.* We classify all attribute sets according to two binary characteristics: dependence and discernibility. The information system $\mathcal{IS}$ and the corresponding search space for the reduct set problem are depicted in Table 4.1 and Figure 4.1, respectively.

## 4.2 Reduct Computation

**Algorithm Overview.** We present a novel approach to finding all reducts of an information system. Our method is a combination of an Apriori-like attribute set

**Algorithm 3.** RedApriori($(\mathcal{U}, \mathcal{I})$)

$RED_1 = \{\text{all 1-reducts}\}$
$RED = RED_1$
$ISC_1 = \{\text{all 1-sets}\}$
$L_1 = \{\text{all 1-sets}\} - RED_1$
**for** $(k = 2; L_{k-1} \neq \emptyset; k++)$ **do**
    $C_k = \textit{apriori-gen-join}(L_{k-1})$
    $D_k = \textit{prune-with-subsets}(C_k, L_{k-1})$
    $ISC_k = D_k - \textit{find-dependent}(D_k)$
    $RED_k = \textit{find-RED}(ISC_k)$
    $L_k = ISC_k - RED_k$
    $RED = RED \cup RED_k$
**end for**
**return** $RED$

generation and an efficient pruning technique based on Theorems 4.5 and 4.6. As a consequence, the scheme of our algorithm follows the classic Apriori algorithm for finding frequent itemsets ([2]).

In every step, we generate a collection of $k$-sets and use pruning techniques to remove reducts and dependent sets. The final collection of every step, $L_k$, contains only independent sets that are not reducts. The collections $RED$ and $ISC$ are computed incrementally. In $k$-th step all their $k$-element members are identified.

The *apriori-gen-join* function is responsible for candidate set generation. A new collection of sets $C_k$ is generated according to the join step of the *apriori-gen* function described in [2]. The generation of $C_k$ is based on a collection of independent sets $L_{k-1}$ obtained in the previous iteration. For $k \geq 2$, each two elements of $L_{k-1}$ that have $k-2$ attributes in common are summed to get a single element of $C_k$. As a result, we obtain a collection of all possible $k$-element sums of two elements chosen from $L_{k-1}$. Note that a lexicographical order allows us to iterate over $L_{k-1}$ efficiently.

The *prune-with-subsets* function removes from collection $C_k$ each element $B$ that is a superset of any dependent attribute set or reduct. Direct pruning with maximal independent sets found so far would be a costly operation. However, in accordance with Theorem 4.6, it is sufficient to test whether $\{B - \{a\} \subseteq P_{k-1}(\mathcal{A}) : a \in \mathcal{A}\} \subseteq ISC_{k-1} - RED_{k-1} = L_{k-1}$. It needs at most $|B|$ membership tests in a collection $L_{k-1}$ computed in the previous step.

Even if all proper subsets of a given attribute set $B$ are independent, $B$ can be dependent. When we cannot prove dependence from Theorem 4.6, we need to arbitrate it by means of Theorem 4.5. This operation requires computing $covcount(B)$ and comparing it with $covcount(S)$, for all $S$ such that $S$ is a direct subset of $B$. Note that, each $S$ is an independent attribute set and is not a reduct, as $B$ passed through a phase of dependent superset pruning. Moreover, the value of $covcount(S)$ has already been computed to prove the independence of $S$.

Note that $RED_k \subseteq ISC_k$. Thus, in every iteration, we have to find these $B \in ISC_k$ for which $covcount(B) = covcount(\mathcal{A})$. Moreover, *covcount* is already computed for elements of $ISC_k$, thus, this step requires simply traversing $ISC_k$.

*Example 4.2.* Let us look at subsequent steps of a sample execution of the algorithm. For brevity, we use a convenient notation $ab - 14$ meaning a set $\{a, b\}$ and $covcount(\{a, b\}) = 14$.

We consider the information system $\mathcal{IS} = (\mathcal{U}, \{a, b, c, d, e\})$ from Table 4.1 and observe the following facts

$DS = \{\{a, b, e\}, \{a, b, c, e\}, \{a, b, d, e\}, \{a, b, c\}, \{a, b, d\}, \{d\}, \{b, c, d\}, \{a, b, c, d\}\}$

$covcount(\{a, b, c, d, e\}) = 8$

**Table 4.2.** Algorithm execution for $\mathcal{IS}$

| k | $L_k$ | $RED_k$ |
|---|---|---|
| 1 | $\{a-6, b-7, c-4, d-5, e-3\}$ | $\emptyset$ |
| 2 | $\{ac-7, cd-7, ce-6, de-7\}$ | $\{ad-8, bd-8\}$ |
| 3 | $\emptyset$ | $\{cde-8\}$ |

**Implementation of Pruning Conditions.** In the scheme of RedApriori (Section 4.2), we use sequences of sets to emphasize a connection between the presented theorems and particular steps of the algorithm. However, it is easy to see that additional computations can be avoided, when the steps 6, 7, 8, 9 are performed in the *apriori-gen-join* function.

Consider a $k$-th iteration and $B \in C_k$ generated from $E, F \in L_{k-1}$. First, we have to examine a collection $SUB = \{B - \{a\} : a \in B\}$ that contains direct subsets of $B$. Obviously, $E, F$ can be omitted, since they are independent and are not super reducts. Now, for each direct subset $S \in SUB - \{E, F\}$ we check $S \in L_{k-1}$. If any $S$ does not meet this condition, one rejects $B$ and repeats the whole procedure for the next candidate. Otherwise, $covcount(B)$ is calculated and the condition $covcount(B) = max_{S \in SUB}(covcount(S))$ is checked. If it holds, we reject $B$. Otherwise, $B \in ISC$. In addition to the latter, if $covcount(B) = covcount(\mathcal{A})$, $B$ is accepted as an independent set and a super reduct, i.e. a reduct. The maximum coverage count is calculated by examining elements of $SUB$. Importantly, for a given $B$, we check membership of $S \in SUB$ in the collection $L_{k-1}$ exactly once.

Another observation refers to temporary collections stored in memory. We maintain and successively update the resulting collection $RED$. Moreover, in every iteration, $L_{k-1}$ is the only needed historical collection. It is used for both: candidate generation and efficient pruning. Note that we do not have to store the collection $ISC_{k-1}$, since pruning by dependent subsets and reducts is performed in the same algorithmic step (Theorem 4.6) and only employs the collection $L_{k-1}$.

*Example 4.3.* This example demonstrates how we examine each candidate set in order to decide whether it is a reduct or member of a collection $L_k$.

Consider an information system $(\mathcal{U}, \{a,b,c,d,e,f\})$ and assume the following coverage count for all the attributes: $covcount(\{a,b,c,d,e,f\}) = 50$.

We focus on a hypothetical $k$-th iteration, where $k = 5$. From the previous iteration, we have $L_4 = \{abcd{-}26, abce{-}35, abcf{-}23, abde{-}31, acde{-}40, bcde{-}12\}$. We select two sets $E = \{a,b,c,d\}$, $F = \{a,b,c,e\}$ and generate the candidate $B = E \cup F = \{a,b,c,d,e\}$. Direct subsets of $B$ are as follows: $SUB = \{\{b,c,d,e\},\{a,c,d,e\},\{a,b,d,e\},\{a,b,c,e\},\{a,b,c,d\}\}$.

First, we have to check whether $S \in L_4$, for each $S \in SUB - \{E,F\}$. During this examination, we also compute $max_{S \in SUB}(covcount(S)) = 40$. Because $SUB \subset L_4$, we cannot take advantage of Theorem 4.6. Instead, we compute $covcount(B) = 40$. Because $covcount(B) = max_{S \in SUB}(covcount(S)) = covcount(\{a,c,d,e\})$, according to Theorem 4.5, we find that $B$ is dependent and we do not add it to $L_5$.

On the whole, our algorithm traverses $ISC$ and, additionally, examines not pruned, immediate supersets of $MISC - RED$ in order to prove their independence. We distinguish two dominant operations: *covcount* computation and membership testing.

When large databases are concerned, $|DS|$ is potentially high and makes *covcount* computations expensive. Note that, from the point of view of reduct finding, a reduced discernibility set $RDS$ contains all essential elements of $DS$. Therefore, one may want to use the reduced set in the formulation of *covcount*. Even though, this optimization requires initial preprocessing of time and space cost of $O(n^2)$, where $n = |\mathcal{U}|$, it is almost always beneficial. Cardinality difference between these two sets is usually very high. Also, the reduction is done only once, while the resulting set is employed multiple times for various candidate attribute sets.

An alternative approach to compute *covcount* operates on an information system directly. For a given $B \in \mathcal{A}$ the value of $covcount(B)$ can be obtained from the sizes of blocks of the partition $\mathcal{U}/IND_{\mathcal{IS}}(B)$. This operation involves grouping the elements of $\mathcal{U}$ taken into account only attributes $B$, e.g. by sorting, with the time cost $O(nlog(n))$, in situ. Efficiency of operations on discernibility matrices and related structures are covered in classic works of the field and we do not discuss them here ([55,56]).

Despite the optimizations, *covcount* computations remain costly and should be performed as rarely as possible. Inevitably, computation is performed for the elements of $ISC$ and for $\mathcal{A}$. Also, we compute *covcount* for each examined set only once. For efficiency reasons, following Theorems 4.5 and 4.6, we prefer to replace these computations with multiple membership tests. Such a strategy encourages us to study different data structures for storing an attribute set collection $L_{k-1}$. This discussion is given in the next section.

**Storing Attribute Set Collections.** We perform two basic operations on a collection $L_{k-1}$: enumerating its members to generate new candidates and testing whether a given attribute set belongs to the collection. As mentioned before, $C_k$ is generated from pairs of attribute sets of $L_{k-1}$ that differ with only one attribute. Thus, the collection should be traversed in the lexicographical order to avoid unnecessary examinations. On the other hand, one needs to perform efficient lookups of attribute sets in the collection to check if a given attribute set is absent or obtain its, previously computed, coverage count.

Another important issue is economical management of system memory. In fact, real life datasets usually have many attributes, which results in large search spaces. As a consequence, temporary collections become memory-consuming, especially, when each attribute set is stored separately. Thus, concise data structures are preferable.

The most natural approach is to store attribute sets in a vector. In terms of candidate generation, attribute sets can easily be traversed in the desirable order, but merging two attribute sets requires iterating over all their elements to make sure that only one is different. Definitely, the main disadvantage is a linear time of a membership test.

In a more sophisticated solution, each attribute set can be treated as a node of a balanced binary search tree. When a prefix-wise node ordering is applied, an in-order method of visiting nodes allows us to traverse attribute sets in the lexicographical order. Also, attribute set merging is performed like in a vector. Nonetheless, the dynamic nature of the tree makes candidate generation slower and consumes more memory.

Importantly, the balanced tree ensures a logarithmic time of finding a given attribute set. In fact, an access time can be worse, when the cost of attribute set comparison cannot be treated as constant. Note that, when we use prefix-wise node ordering and go down the tree to find a member set, the time needed for each comparison increases with a node level. Such a phenomenon is due to longer common prefixes of the considered set and the set associated with a current node. This undesirable property can significantly affect efficiency of building the tree and attribute set lookups.

The last approach refers to a classic trie structure that was originally designed for storing a collection of strings over a given alphabet $S$. A trie is a multi-way ordered tree. Each edge is labeled with a letter of the alphabet. There is one node for every common prefix, which can be read by scanning edge labels on the path from the tree root to this node. All leaves and some internal nodes are marked as terminating, since they are associated with prefixes that belong to $S$. The trie has two important advantages. The search time is logarithmic with a single letter comparison, not a string comparison, as a basic operation. The structure is usually more efficient than a list of strings or a balanced tree when large collections are considered. At the same time, the difference in a search time between the trie and the BST decrease with an increase of the average number of children of a trie node. Although references used to build nodes involve additional memory, the trie remains a very compact structure.

Our proposition, a coverage count trie (CC-Trie), follows a general structure of a trie. We assume certain linear order in $\mathcal{C}$, use it to sort elements of each

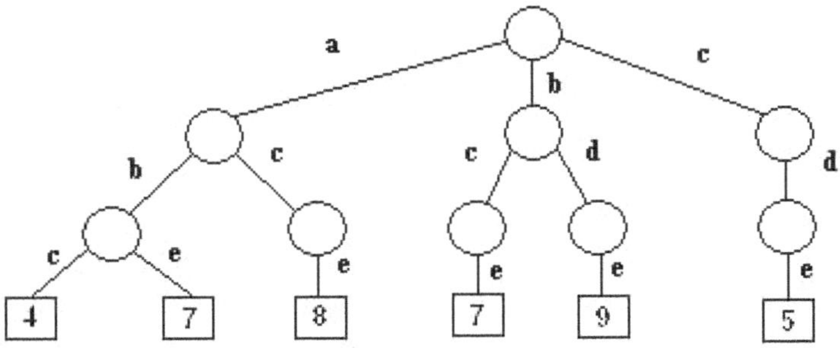


**Fig. 4.2.** The CC-Trie for a sample collection $L_{k-1}$. Each path from the root to a leaf refers to an attribute set from this collection. Ovals denote internal nodes and rectangles - terminal nodes. The values of *covcount* of respective attribute sets are given in terminal nodes.

attribute set and treat them as sequences of attributes. Now, the attribute set collection $L_{k-1}$ can be stored in a trie. Note that all terminating nodes are leaves, since all attribute sets have the same cardinality equal to $k-1$. The *covcount* of an attribute set is stored in an appropriate terminating node.

We use a balanced tree to store references from each node to its children. This allows us to conveniently traverse subsets in the candidate generation stage. In fact, each two sets that can be joined share a parent node at the last but one level. Thus, attribute set merging can be done in constant time. This approach is particularly advisable for datasets with a large number of attributes $|\mathcal{C}|$, e.g. gene expression data. Another option is to store children in a hash table. It ensures similar memory usage but outperforms dynamic structures for smaller $|\mathcal{C}|$. At the same time, it is less efficient with respect to candidate generation.

*Example 4.4.* Let us consider the set of attributes $\mathcal{C} = \{a, b, c, d, e\}$ and the step $k = 4$. We consider the collection
$L_{k-1} = L_3 = \{\{a, b, c\}, \{a, b, e\}, \{a, c, e\}, \{b, c, e\}, \{b, d, e\}, \{c, d, e\}\}$ and the values of *covcount* for these attribute sets: 4, 7, 8, 7, 9, 5, respectively. The CC-Trie for this collection is presented in Figure 4.2.

Let us consider the set $\{b, c, e\}$. Its *covcount* value can be found in the tree by going three levels down. Since all attribute sets of the collection $L_3$ have the same size, lookup costs are similar. On the other hand, when a given set does not belong to the collection, the respective computation is shorter, e.g. finding the set $\{a, d, e\}$ requires visiting only two levels.

To sum up, we employ a CC-Trie to get more economical representation of an attribute set collection, constant-time attribute set merging and a logarithmic lookup time. It is worth mentioning that the trie remains a basis for many contemporary structures proposed for transaction databases. In frequent pattern mining, a structure similar to CC-Trie was used in the Apriori algorithm to store pattern collections with respective supports ([128]). Another approach considered for the same algorithm optimizes support calculations by means of itemset trees [2]. Also, an FP-growth method [81] is completely based on another trie-like structure, a frequent pattern tree (FP-Tree). In the field of emerging patterns, a CP-Tree is utilized to mine SJEPs and chi-EPs (see Section 3.2, [54,10,60]).

### 4.3 Experimental Evaluation

Our experiments touch on two aspects of the novel algorithm: efficiency of pruning based on attribute set dependence and the overall computation time for different structures for storing temporary attribute set collections. In order to give a better picture on hardness of the considered problems, we extend information from Table A.1 with the size of the search space, number of reducts, length of a minimal reduct, size of the discernibility set and reduced discernibility set (Table 4.3). The search space directly determines computation for an exhaustive approach, whereas the length of a minimal reduct indicates when an Apriori-like algorithm starts finding reducts. In addition, the time cost of *covcount* computation for a given attribute set is determined by $|RDS|$ and the number of attributes. In the tests, all the datasets are treated as information systems.

In order to solely investigate pruning efficiency of our algorithm (RedApriori), we compare it with a similar Apriori-like method that prunes candidate collections only with reducts found so far (RedNaive). In other words, we use as a reference an algorithm based on Theorem 4.2, that is weaker than Theorem 4.4.

**Table 4.3.** Problem characteristics

| Dataset | Attrs | Search space | Min reduct | $\|RED\|$ | $\|RC_1\|$ | $\|RC_2\|$ |
|---|---|---|---|---|---|---|
| balance | 5 | 3.0e+01 | 4 | 1 | 30 | 4 |
| breast-wisc | 10 | 1.0e+03 | 10 | 1 | 997 | 10 |
| car | 7 | 1.0e+02 | 6 | 1 | 126 | 6 |
| cmc | 10 | 1.0e+03 | 10 | 1 | 1023 | 10 |
| dna | 21 | 2.0e+06 | 9 | 339 | 74633 | 50 |
| heart | 14 | 1.0e+04 | 10 | 1 | 1023 | 10 |
| irys | 5 | 3.0e+01 | 5 | 1 | 31 | 5 |
| krkopt | 7 | 1.0e+02 | 6 | 1 | 126 | 6 |
| lymn | 19 | 5.0e+05 | 9 | 75 | 8075 | 45 |
| monks-1 | 7 | 1.0e+02 | 6 | 1 | 119 | 6 |
| monks-2 | 7 | 1.0e+02 | 6 | 1 | 126 | 6 |
| monks-3 | 7 | 1.0e+02 | 7 | 1 | 127 | 7 |
| mushroom | 23 | 8.0e+06 | 15 | 1 | 19655 | 15 |
| nursery | 9 | 5.0e+02 | 8 | 1 | 510 | 8 |
| tic-tac-toe | 10 | 1.0e+03 | 8 | 9 | 1003 | 36 |
| vehicle | 19 | 5.0e+05 | 19 | 1 | 41413 | 19 |
| wine | 14 | 1.0e+04 | 12 | 1 | 3276 | 12 |
| yeast | 9 | 5.0e+02 | 8 | 1 | 227 | 8 |
| zoo | 17 | 1.0e+05 | 11 | 7 | 1118 | 12 |

**Table 4.4.** Pruning efficiency comparison between RedApriori and RedNaive

| Dataset | RedApriori | | | | | RedNaive | |
|---|---|---|---|---|---|---|---|
| | Exam. sets | Pruned Th. 4.5 | Pruned Th. 4.6 | *covcount* comput. | Member. tests | Exam. sets | *covcount* comput. |
| balance | 20 | 4 | 0 | 20 | 6 | 30 | 30 |
| breast-wisc | 1023 | 0 | 0 | 1023 | 3084 | 1023 | 1023 |
| car | 70 | 6 | 0 | 70 | 72 | 126 | 126 |
| cmc | 1023 | 0 | 0 | 1023 | 3084 | 1023 | 1023 |
| dna | 157220 | 868 | 59585 | 97635 | 526204 | 2060148 | 2057655 |
| heart | 1073 | 46 | 0 | 1073 | 3084 | 16383 | 16368 |
| irys | 31 | 0 | 0 | 31 | 23 | 31 | 31 |
| krkopt | 70 | 6 | 0 | 70 | 72 | 126 | 126 |
| lymn | 38840 | 203 | 1908 | 36932 | 175599 | 517955 | 515190 |
| monks-1 | 70 | 6 | 0 | 70 | 72 | 126 | 126 |
| monks-2 | 70 | 6 | 0 | 70 | 72 | 126 | 126 |
| monks-3 | 127 | 0 | 0 | 127 | 201 | 127 | 127 |
| mushroom | 32923 | 148 | 0 | 32923 | 180241 | 8388479 | 8388352 |
| nursery | 264 | 8 | 0 | 264 | 522 | 510 | 510 |
| tic-tac-toe | 520 | 9 | 0 | 520 | 1284 | 1012 | 1012 |
| vehicle | 524287 | 0 | 0 | 524287 | 3932181 | 524287 | 524287 |
| wine | 4122 | 25 | 0 | 4122 | 16398 | 16381 | 16380 |
| yeast | 264 | 8 | 0 | 264 | 522 | 511 | 510 |
| zoo | 7910 | 40 | 189 | 7721 | 30686 | 130991 | 130902 |

The results are in favor of our proposition (Table 4.4). First of all, RedApriori considers only a small region of a respective search space, which brings savings of 1-2 orders of magnitude. Secondly, the comparison with RedNaive shows that Theorem 4.4 has significantly better pruning capabilities than Theorem 4.2. Last but not least, RedNaive is more prone to dataset characteristics, such as the length of a minimal reduct and the number of reducts related to the size of the search space. These parameters determine how frequently the pruning is performed. Conversely, performance of our algorithm depends more on properties of a highly diversified collection $MISC$.

Time complexity is determined by two dominant operations: *covcount* computation and testing membership of an attribute set in a temporary collection. As a result of the stronger space reduction, the number of *covcount* computations performed by RedApriori is much lower in comparison to RedNaive, often by 1-2 orders of magnitude. Moreover, we do not compute *covcount* for the generated sets, which are pruned by the condition based on Theorem 4.6. For the considered datasets, this condition holds more often than the one based on Theorem 4.5.

The second part of this experiment pertains to different structures for storing a collection $L_{k-1}$. We investigate performance consequences of using a dynamic vector (java.util.ArrayList), balanced tree (java.util.TreeMap) and a CC-Trie, described in Section 4.2.

**Table 4.5.** Execution time comparison of RedApriori with different implementations of attribute set collection [ms]

| Dataset | RedApriori (array) | RedApriori (AVL) | RedApriori (trie) |
|---|---|---|---|
| balance | 141 | 109 | 78 |
| breast-wisc | 234 | 203 | 202 |
| car | 733 | 687 | 702 |
| cmc | 843 | 795 | 780 |
| dna | 904385 | 468674 | 467035 |
| heart | 141 | 109 | 109 |
| irys | 15 | 0 | 0 |
| krkopt | 174409 | 171009 | 167061 |
| lymn | 17675 | 4571 | 4181 |
| monks-1 | 78 | 94 | 78 |
| monks-2 | 109 | 94 | 93 |
| monks-3 | 78 | 93 | 78 |
| mushroom | 71277 | 61886 | 61277 |
| nursery | 48563 | 47190 | 47565 |
| tic-tac-toe | 390 | 374 | 359 |
| vehicle | 2565343 | 114598 | 150572 |
| wine | 733 | 671 | 624 |
| yeast | 562 | 530 | 546 |
| zoo | 577 | 218 | 172 |

Table 4.5 shows that the approach based on the vector tends to be slower when many membership tests, relatively to the number of generated sets, are performed (*lymn, mushroom, vehicle, zoo*). On the other hand, both solutions that employ tree structures behave similarly. However, in case of a CC-Trie, common prefixes are represented in a compact way, which makes it preferable in terms of memory.

We refrain from comparing our approach with other existing methods (see Section 3.5). In general, it is hard to conduct it reliably. The most common metric, a computation time, strongly depends on implementation details, programming language and testbed characteristics. Unfortunately, publicly available frameworks with exhaustive algorithms for reduct finding, like Rosetta ([129]) or RSES ([73]), provide limited support for taking time measurements.

## 5 JEPs in Datasets Based on Decision Tables

Emerging patterns are defined and considered in transaction databases. Meanwhile, a large portion of classified data is described by decision tables with multivariate attributes. Thus, such datasets are first transformed to transactional form, so that standard JEP discovery methods can be applied (see Section 3.2).

Decision tables are a common structure in rough set theory. We consider attribute sets in a decision table and patterns in the respective transaction database. In particular, we present several relations between JEPs and global decision reducts ([130]), and a direct analogy of minimal JEPs and local decision reducts ([131]). The latter leads to a JEP discovery method for input data given as a decision table. It performs reduct computations directly on a decision table and does not require additional transformations.

### 5.1 Transactional Dataset Construction

In our discussion, data are assumed to be complete and provided as a decision table. Below, we present a widely-known procedure to transform multivariate data into transactional form ([1]).

Consider a decision table $\mathcal{DT} = (\mathcal{U}, \mathcal{C}, d)$. We introduce the following notation to concisely represent sets of attribute-value pairs corresponding to a certain object.

**Definition 5.1.** *For $u \in \mathcal{U}$, $B \subseteq \mathcal{C}$, we define*

*a* pattern based on the object $u$ and attribute set $B$ *as* $patt(u, B) = \{(a, a(u)) : a \in B\}$,
*a* transaction based on the object $u$ *as* $T(u) = patt(u, \mathcal{C})$.

In order to represent a decision table in transactional form, each transaction corresponds to a single observation and each attribute-value pair to an item. In addition, we distinguish a set of decision items.

**Definition 5.2.** *A* decision transaction system based on $\mathcal{DT}$ *is a decision transaction system $(\mathcal{D}, \mathcal{I}, \mathcal{I}_d)$, such that*

$$\mathcal{D} = \{T(u) \cup \{(d, d(u))\} : u \in \mathcal{U}\},$$
$$\mathcal{I} = \{(a, v) : a \in \mathcal{C}, v \in V_a\}$$
$$\mathcal{I}_d = \{(d, v) : v \in V_d)\}.$$

**Lemma 5.1.** *For $u \in \mathcal{U}$, $B \subseteq \mathcal{C}$, $patt(u, B)$ is an itemset in a decision transaction system based on $\mathcal{DT}$.*

Given a special form of the itemspace, we may refer to attributes corresponding to a given pattern.

**Definition 5.3.** *For $X \subset \{(a, v) : a \in \mathcal{C} \wedge v \in V_a\}$,*
*an* attribute set based on the pattern $X$ *is a set $attr(X) = \{a \in \mathcal{C} : (a, v) \in X \wedge v \in V_a\}$.*

Obviously, for $u \in \mathcal{U}$ we have $attr(patt(u, P)) = P$ and $attr(T(u)) = \mathcal{C}$.

*Example 5.1.* Let us consider the decision table $\mathcal{DT} = (\{u_1, u_2, u_3, u_4, u_5, u_6\}, \{w, x, y, z\}, d)$ given in Table 5.1. We have:

$\mathcal{DT}$ is non-deterministic
$POS(\mathcal{C}, 0) = \mathcal{C}_*(U_0) = \{u_1, u_2\}$, $POS(\mathcal{C}, 1) = \mathcal{C}_*(U_1) = \{u_5, u_6\}$
$POS(\mathcal{C}) = \mathcal{C}_*(U_0) \cup \mathcal{C}_*(U_1) = \{u_1, u_2, u_5, u_6\}$

**Table 5.1.** A sample decision table $\mathcal{DT} = (\{u_1, u_2, u_3, u_4, u_5, u_6\}, \{w, x, y, z\}, d)$

| | $w$ | $x$ | $y$ | $z$ | d |
|---|---|---|---|---|---|
| $u_1$ | 0 | 0 | 0 | 0 | 0 |
| $u_2$ | 0 | 1 | 0 | 0 | 0 |
| $u_3$ | 0 | 1 | 1 | 1 | 0 |
| $u_4$ | 0 | 1 | 1 | 1 | 1 |
| $u_5$ | 1 | 1 | 0 | 1 | 1 |
| $u_6$ | 1 | 0 | 1 | 0 | 1 |

**Table 5.2.** The decision transaction system $\mathcal{DTS} = \{\mathcal{D}, \mathcal{I}, \mathcal{I}_d\}$ based on $\mathcal{DT}$ from Table 5.1

| $\mathcal{I}$ | $\mathcal{I}_d$ |
|---|---|
| (w,0), (x,0), (y,0), (z,0) | (d,0) |
| (w,0), (x,1), (y,0), (z,0) | (d,0) |
| (w,0), (x,1), (y,1), (z,1) | (d,0) |
| (w,0), (x,1), (y,1), (z,1) | (d,1) |
| (w,1), (x,1), (y,0), (z,1) | (d,1) |
| (w,1), (x,0), (y,1), (z,0) | (d,1) |

$POS(\{w, x\}, 0) = \{w, x\}_*(U_0) = \{u_1\}$, $POS(\{w, x\}, 1) = \{w, x\}_*(U_1) = \{u_5, u_6\}$

$SRED(\mathcal{U}, d) = \{\{w, x, y\}, \{w, x, y, z\}, \{w, x, z\}, \{w, y\}, \{w, y, z\}, \{w, z\}, \{x, y, z\}, \{y, z\}\}$

$RED(\mathcal{U}, d) = \{\{w, y\}, \{w, z\}, \{y, z\}\}$.

Let us consider the decision transaction system $\mathcal{DTS} = \{\mathcal{D}, \mathcal{I}, \mathcal{I}_d\}$ based on the decision table $\mathcal{DT}$ presented in Table 5.2. We have:

$T(u_1) = patt(u_1, \mathcal{C}) = \{(w, 0), (x, 0), (y, 0), (z, 0)\}$, $attr(T(u_1)) = \mathcal{C}$
$patt(u_1, \{w, y\}) = \{(w, 0), (y, 0)\}$, $attr(\{(w, 0), (y, 0)\}) = \{w, y\}$
$D_0 = \{T(u_1), T(u_2), T(u_3)\}$, $D_1 = \{T(u_4), T(u_5), T(u_6)\}$
$JEP(D_0) =< \mathcal{L}_0, \mathcal{R}_0 >$
$\mathcal{L}_0 = \{\{(w, 0), (y, 0)\}, \{(w, 0), (z, 0)\}, \{(y, 0), (z, 0)\}, \{(w, 0), (x, 0)\}, \{(x, 0), (y, 0)\}, \{(x, 1), (z, 0)\}\}$
$\mathcal{R}_0 = \{\{(w, 0), (x, 0), (y, 0), (z, 0)\}, \{(w, 0), (x, 1), (y, 0), (z, 0)\}\}$
$JEP(D_1) =< \mathcal{L}_1, \mathcal{R}_1 >$
$\mathcal{L}_1 = \{\{(w, 1)\}, \{(x, 0), (y, 1)\}, \{(y, 0), (z, 1)\}, \{(y, 1), (z, 0)\}\}$
$\mathcal{R}_1 = \{\{(w, 1), (x, 0), (y, 1), (z, 0)\}, \{(w, 1), (x, 1), (y, 0), (z, 1)\}\}$
$JEP(D_0, \{w, y\}) = \{\{(w, 0), (y, 0)\}\}$
$JEP(D_1, \{w, y\}) = \{\{(w, 1), (y, 1)\}, \{(w, 1), (y, 0)\}\}$

## 5.2 Relations between JEPs and Reducts

**Global Reducts.** The section looks at basic relations between global decision reducts and jumping emerging patterns. Hereinafter, we consider a decision table

$\mathcal{DT} = (\mathcal{U}, \mathcal{C}, d)$, the corresponding decision transaction system $\mathcal{DTS}$ and a target class $k \in V_d$. We assume that $JEP(D_k) = [\mathcal{L}_k, \mathcal{R}_k]$.

For convenience, we introduce the following notations to target objects or patterns of interest.

**Definition 5.4.** *For $P \subseteq \mathcal{C}$ and $D \subseteq \mathcal{D}$, we define*

$$\text{a } P,k\text{-positive region } POS(P,k) = POS(P,d) \cap U_k,$$
$$JEP(D', D, P) = JEP(D, P) = \{J \in JEP(D) : attr(J) = P\}.$$

The following theorem states that, for any attribute set $P$, an object $u$ belongs to the $B,k$-positive region, $POS(P,k)$ iff the pattern based on $u$ and $P$ is a jumping emerging pattern in $D_k$.

**Theorem 5.1.** $\forall_{P \subseteq \mathcal{C}} \forall_{u \in \mathcal{U}} u \in POS(P,k) \iff patt(u,P) \in JEP(D_k, P)$.

*Proof.* Let $P \subseteq \mathcal{C}$.

Consider $\forall_{u \in \mathcal{U}} u \in POS(P,k) \iff supp_{D'_k}(patt(u,P)) = 0 \quad (1)$.

Let $u \in POS(P,k)$. We have

$$\begin{aligned} u \in P_*(U_k) &\iff P(u) \subseteq U_k \\ &\iff \{v \in \mathcal{U} : \forall_{a \in P} a(u) = a(v)\} \subseteq U_k \\ &\iff \forall_{v \in \mathcal{U} - U_k} \exists_{a \in P} a(v) \neq a(u) \\ &\iff \forall_{v \in \mathcal{U} - U_k} \exists_{a \in P} (a, a(u)) \in patt(u,P) \wedge (a, a(u)) \notin patt(v,P) \\ &\iff \forall_{t \in D'_k} patt(u,P) \not\subseteq t \\ &\iff supp_{D'_k}(patt(u,P)) = 0. \end{aligned}$$

Examine the implication $\forall_{u \in \mathcal{U}} u \in POS(P,k) \implies patt(u,P) \in JEP(D_k, P)$.

Note that $patt(u,P) \subseteq T(u)$ and $T(u) \in D_k$, so $supp_{D_k}(patt(u,P)) \neq 0$. Based on $attr(patt(u,P)) = P$ and (1), we have $patt(u,P) \in JEP(D_k, P)$.

Now, consider the second implication $\forall_{u \in \mathcal{U}} patt(u,P) \in JEP(D_k,P) \implies u \in POS(P,k)$.

According to (1) we have:
$patt(u,P) \in JEP(D_k,P) \implies supp_{D'_k}(patt(u,P)) = 0 \iff u \in POS(P,k)$. □

*Example 5.2.* Consider the class $k = 0$, attribute set $P = \{w, x\}$ and object $u = u_1$. Both statements $u_1 \in POS(\{w,x\}, 0) = \{u_1\}$ and $patt(u_1, \{w,x\}) = \{(w,0),(x,0)\} \in JEP(D_0, \{w,x\})$ are true.

The $\mathcal{C}$-positive region contains objects that lead to consistent decision rules. The following remark shows that Theorem 5.1 can be applied to objects from this region as long as we limit our interest to super reducts.

*Remark 5.1.* $\forall_{P \in SRED(\mathcal{U},d)} \forall_{u \in \mathcal{U}} u \in POS(\mathcal{C},k) \iff patt(u,P) \in JEP(D_k,P)$

*Proof.* Let $P \in SRED(\mathcal{U},d)$. We have $POS(\mathcal{C}) = POS(P) \implies POS(\mathcal{C}) \cap U_k = POS(P) \cap U_k \iff POS(\mathcal{C},k) = POS(P,k)$. Let $u \in \mathcal{U}$. According to Theorem 5.1, we have $u \in POS(\mathcal{C},k) \iff u \in POS(P,k) \iff patt(u,P) \in JEP(D_k,P)$. □

*Example 5.3.* For the decision table $\mathcal{DT}$ from Table 5.1, we have $POS(\mathcal{C},0) = POS(\{w,x,y,z\},0) = \{u_1,u_2\}$. For the attribute set $P = \{w,y\} \in SRED(\mathcal{U},d)$, the patterns: $patt(u_1,\{w,y\}) = \{(w,0),(y,0)\}$, $patt(u_2,\{w,y\}) = \{(w,0),(y,1)\}$ belong to $JEP(D_0,\{w,y\})$. Conversely, we can choose the attribute set $P = \{w,x\} \notin SRED(\mathcal{U},d)$, for which the pattern $patt(u_1,\{w,x\}) = \{(w,0),(x,0)\}$ belongs to $JEP(D_0,\{w,x\})$. Note that the same does not hold for the pattern $patt(u_2,\{w,x\}) = \{(w,0),(x,1)\}$.

The next theorem shows that a set of transactions from the database $D_k$, corresponding to objects from any single $\mathcal{C},k$-positive region, is equal to the right bound $\mathcal{R}_k$ of the space $JEP(D_k)$.

**Theorem 5.2.** $\{T(u) \in D_k : u \in POS(\mathcal{C},k)\} = JEP(,D_k,\mathcal{C}) = \mathcal{R}_k$.

*Proof.* Firstly, consider $\{T(u) \in D_k : u \in POS(\mathcal{C},k)\} = JEP(D_k,\mathcal{C})$.

According to Theorem 5.1, we have $\{T(u) \in D_k : u \in POS(\mathcal{C},k)\} = \{T(u) \in D_k : patt(u,\mathcal{C}) \in JEP(D_k,\mathcal{C})\} = \{T(u) \in D_k : T(u) \in JEP(D_k,\mathcal{C})\} = JEP(D_k,\mathcal{C})$.

Consider $JEP(D_k,\mathcal{C}) \subseteq \mathcal{R}_k$ (1).

Let $J \in JEP(D_k,\mathcal{C})$, and assume $J \notin \mathcal{R}_k$. We have $J \in JEP(D_k,\mathcal{C}) \subseteq JEP(D_k) \Longrightarrow \exists_{t \in \mathcal{R}_k} J \subseteq t$. For this $t$, we have $attr(J) \subseteq attr(t)$. But $attr(J) = \mathcal{C}$ and $attr(t) \subseteq \mathcal{C}$, so $attr(t) = \mathcal{C}$. Finally, $attr(t) = attr(J) \wedge J \subseteq t \Longrightarrow t = J$, which contradicts the assumption $J \notin \mathcal{R}_k$.

Consider $JEP(D_k,\mathcal{C}) \supseteq \mathcal{R}_k$ (2).

Let $J \in \mathcal{R}_k$ and $J \notin JEP(D_k,\mathcal{C})$. We have $J \in \mathcal{R}_k \subseteq JEP(D_k) \Longrightarrow supp_{D_k}(J) \neq 0 \Longleftrightarrow \exists_{t \in D_k} J \subseteq t \Longleftrightarrow \exists_{u \in U_k} J \subseteq T(u)$. Also, we have $attr(J) \subseteq \mathcal{C} \wedge J \notin JEP(D_k,\mathcal{C}) \Longrightarrow \exists_{a \in \mathcal{C}} a \notin attr(J)$.

Let us now consider these $u \in U_k : J \subseteq T(u)$ and $a \in \mathcal{C} : a \notin attr(J)$. We have $J \subset J \cup \{(a,a(u))\} \subseteq T(u) \Longrightarrow supp_{D_k}(J \cup \{(a,a(u))\}) \neq 0$. Because $supp_{D'_k}(J \cup \{(a,a(u))\}) \leq supp_{D'_k}(J) = 0 \Longrightarrow supp_{D'_k}(J \cup \{(a,a(u))\}) = 0$, $J \cup \{(a,a(u))\} \in JEP(D_k)$. But this means that $J$ is not a maximal element of $JEP(D_k)$ with respect to inclusion. Therefore, $J \notin \mathcal{R}_k$, which contradicts the assumption.

According to (1) and (2) we have $JEP(D_k,\mathcal{C}) = \mathcal{R}_k$, which proves our point. □

*Example 5.4.* Consider the class $k = 0$. We have $POS(\mathcal{C},0) = \{u_1,u_2\}$ and $\{T(u) \in D_0 : u \in POS(\mathcal{C},0)\} =$
$\{\{(w,0),(x,0),(y,0),(z,0)\},\{(w,0),(x,1),(y,0),(z,0)\}\} = \mathcal{R}_0$.

The following theorem covers a relation between transactions containing JEPs from a given collection $JF$ and objects corresponding to patterns from $JF$. It makes use of the fact that there is one and only one pattern $patt(u,P)$ for a given object $u$ and an attribute set $P$.

**Theorem 5.3.** $\forall_{P \subseteq \mathcal{C}} \forall_{JF \subseteq JEP(D_k,P)} \bigcup_{J \in JF} \{u \in U_k : J \subseteq T(u)\} = \{u \in U_k : patt(u,P) \in JF\}$

*Proof.* Let $P \subseteq \mathcal{C}$ and $JF \subseteq JEP(D_k, P)$.

First, we focus on
$\forall_{u \in \mathcal{U}} \forall_{J \in \subseteq \mathcal{I}} attr(J) = P \wedge J \subseteq T(u) \iff J = patt(u, P)$ (1).

Let $u \in \mathcal{U}$ and $J \subseteq \mathcal{I}$.

Consider $attr(J) = P \wedge J \subseteq T(u) \implies J = patt(u, P)$. We have $J \subseteq T(u) = patt(u, \mathcal{C}) \implies \exists_{S \subseteq \mathcal{C}} J = patt(u, S)$. For this $S$ we have $attr(J) = attr(patt(u, S)) = S$. On the other hand, $attr(J) = P$, so $P = S$ and $J = patt(u, P)$.

Consider $J = patt(u, P) \implies attr(J) = P \wedge J \subseteq T(u)$. We have the following $attr(J) = attr(patt(u, P)) = P$ and $J = patt(u, P) \subseteq patt(u, \mathcal{C}) = T(u)$.

Now, we have $\bigcup_{J \in JF} \{u \in U_k : J \subseteq T(u)\} = \{u \in U_k : \exists_{J \in JF} J \subseteq T(u)\} = \{u \in U_k : \exists_{J \in JF} attr(J) = P \wedge J \subseteq T(u)\}$. From (1) we have $\{u \in U_k : \exists_{J \in JF} attr(J) = P \wedge J \subseteq T(u)\} = \{u \in U_k : \exists_{J \in JF} J = patt(u, P)\} = \{u \in U_k : patt(u, P) \in JF\}$. □

Consequently, we can derive a relation between the sum of the supports of all patterns in a collection $JF$ and the number of objects corresponding to these patterns.

*Remark 5.2.* $\forall_{P \subseteq \mathcal{C}} \forall_{JF \subseteq JEP(D_k, P)} \sum_{J \in JF} supp_{D_k}(J) = \frac{|\{u \in U_k : patt(u, P) \in JF\}|}{|U_k|}$

*Proof.* Let $P \subseteq \mathcal{C}$ and $JF \subseteq JEP(D_k, P)$.

Consider
$\forall_{J_1, J_2 \in JF} J_1 \neq J_2 \implies \{t \in D_k : J_1 \subseteq t\} \cap \{t \in D_k : J_2 \subseteq t\} = \emptyset$ (1).

We have $J_1, J_2 \in JEP(D_k, P) \implies \exists_{t_1, t_2 \in D_k} J_1 \subseteq t_1 \wedge J_2 \subseteq t_2 \implies \{t \in D_k : J_1 \subseteq t\} \neq \emptyset \wedge \{t \in D_k : J_2 \subseteq t\} \neq \emptyset$.

Let $t_1 \in \{t \in D_k : J_1 \subseteq t\}$ and $t_2 \in \{t \in D_k : J_2 \subseteq t\}$. We have $\exists_{u_1, u_2 \in U_k} t_1 = T(u_1) \wedge t_2 = T(u_2)$. Consider these: $u_1$, $u_2$. Because $attr(J_1) = attr(J_2) = P \wedge J_1 \neq J_2$, so $\exists_{a \in P} (a, a(u_1)) \in J_1 \wedge (a, a(u_2)) \in J_2 \wedge (a, a(u_1)) \neq (a, a(u_2))$. We have $J_1 \subseteq T(u_1) \wedge J_2 \subseteq T(u_2) \implies (a, a(u_1)) \in T(u_1) \wedge (a, a(u_2)) \in T(u_2)$. Because $(a, a(u_1)) \neq (a, a(u_2)) \wedge attr(T(u_1)) = attr(T(u_2)) = \mathcal{C}$, $T(u_1) \neq T(u_2)$. It means that $\{t \in D_k : J_1 \subseteq t\} \cap \{t \in D_k : J_2 \subseteq t\} = \emptyset$.

We have $\sum_{J \in JF} supp_{D_k}(J) = \frac{1}{|D_k|} \sum_{J \in JF} |\{t \in D_k : J \subseteq t\}|$. According to (1), each two different elements of the family $\{\{t \in D_k : J \subseteq t\} : J \in JF\}$ are exclusive. Thus, $\frac{1}{|D_k|} \sum_{J \in JF} |\{t \in D_k : J \subseteq t\}| = \frac{1}{|D_k|} |\bigcup_{J \in JF} \{t \in D_k : J \subseteq t\}|$. Because we are calculating the support of any single pattern $J$, the expression $\{t \in D_k : J \subseteq t\}$ is understood as a multiset. Hence, we have $|\{t \in D_k : J \subseteq t\}| = |\{u \in U_k : J \subseteq T(u)\}|$. From Theorem 5.3 and $|D_k| = |U_k|$ we have $\frac{1}{|D_k|} |\bigcup_{J \in JF} \{u \in U_k : J \subseteq T(u)\}| = \frac{1}{|U_k|} |\{u \in U_k : patt(u, P) \in JF\}|$. □

*Example 5.5.* Let $P = \{x, y\}$, $JEP(D_0, P) = \{(x, 0), (y, 0)\} = JF$.

We have $|\{u \in U_0 : patt(u, P) \in JF\}| = |\{u_1\}| = 1$ and $|\{U_0\}| = 3$. On the other hand, $\sum_{J \in JF} supp_{D_0}(J) = supp_{D_0}\{(x, 0), (y, 0)\} = \frac{1}{3}$, which agrees with Remark 5.2.

Finally, we obtain an expression for the sum of the supports of all patterns in a collection $JF$ and the cardinality of the positive region.

*Remark 5.3.* $\forall_{P \in SRED(\mathcal{U},d)} \forall_{JF \subseteq JEP(D_k,P)} JF = JEP(D_k, P) \iff \sum_{J \in JF} supp_{D_k}(J) = \frac{|POS(\mathcal{C},k)|}{|U_k|}$

*Proof.* Let $P \in SRED(\mathcal{U}, d)$ and $JF \subseteq JEP(D_k, P)$.

According to Remark 5.1 we have $JF = JEP(D_k, P) \iff \{u \in U_k : patt(u, P) \in JF\} = \{u \in U_k : u \in POS(\mathcal{C}, k)\} = POS(\mathcal{C}, k)$. Then, directly from Remark 5.2, we have $\{u \in U_k : patt(u, P) \in JF\} = POS(\mathcal{C}, k) \implies \sum_{J \in JF} supp_{D_k}(J) = \frac{|\{u \in U_k : patt(u,P) \in JF\}|}{|U_k|} = \frac{|POS(\mathcal{C},k)|}{|U_k|}$.

Now consider the converse implication: $|\{u \in U_k : patt(u, P) \in JF\}| = |POS(\mathcal{C}, k)| \implies \{u \in U_k : patt(u, P) \in JF\} = POS(\mathcal{C}, k)$.

On the one hand, from $JF \subseteq JEP(D_k, P)$ we obtain $\{u \in U_k : patt(u, P) \in JF\} \subseteq \{u \in U_k : patt(u, P) \in JEP(D_k, P)\} = POS(\mathcal{C}, k)$. On the other, $|\{u \in U_k : patt(u, P) \in JF\}| = |POS(\mathcal{C}, k)|$. Hence, we get an equality $\{u \in U_k : patt(u, P) \in JF\} = POS(\mathcal{C}, k)$. □

The condition in Remark 5.3 can be simplified when we deal with deterministic decision tables.

*Remark 5.4.* If $\mathcal{DT}$ is deterministic, $\forall_{P \in SRED(\mathcal{U},d)} \forall_{JF \subseteq JEP(D_k,P)} JF = JEP(D_k, P) \iff \sum_{J \in JF} supp_{D_k}(J) = 1$.

*Proof.* If $\mathcal{DT}$ is deterministic, $POS(P, k) = U_k$. The statement holds as a direct consequence of Remark 5.3. □

For further consideration, we take into account all classes of a decision table $\mathcal{DT} = (\mathcal{U}, \mathcal{C}, d)$ and the corresponding decision transaction system $\mathcal{DTS}$.

The next theorem shows how to infer whether a given attribute set $P$ is a super reduct based on the supports of patterns from the space $JEP(D_k, P)$. Note that the left-hand condition holds iff there is an equality between the sum of supports and the cardinality of the $\mathcal{C}, k$-positive region related to the cardinality of $k$-th class, for each class $k \in V_k$.

**Theorem 5.4.** $\forall_{P \subseteq \mathcal{C}} ((\forall_{k \in V_d} \sum_{J \in JEP(D_k,P)} supp_{D_k}(J) = \frac{|POS(\mathcal{C},k)|}{|U_k|}) \iff P \in SRED(\mathcal{U}, d))$

*Proof.* Let $P \subseteq \mathcal{C}$.

We start with $\forall_{P \subseteq \mathcal{C}} ((\forall_{k \in V_d} \sum_{J \in JEP(D_k,P)} supp_{D_k}(J) = \frac{|POS(\mathcal{C},k)|}{|U_k|}) \implies P \in SRED(\mathcal{U}, d))$.

Consider $\forall_{k \in V_d} (\sum_{J \in JEP(D_k,P)} supp_{D_k}(J) = \frac{|POS(\mathcal{C},k)|}{|U_k|} \implies POS(P, k) = POS(\mathcal{C}, k))$ (1).

Let $k \in V_d$ and $\sum_{J \in JEP(D_k,P)} supp_{D_k}(J) = \frac{|POS(\mathcal{C},k)|}{|U_k|}$). According to Remark 5.2, we have $|\{u \in U_k : patt(u, P) \in JEP(D_k, P)\}| = |POS(\mathcal{C}, k)|$.

Further, from Theorem 5.1 we have $\{u \in U_k : patt(u,P) \in JEP(D_k,P)\} = POS(P,k) \Longrightarrow |POS(P,k)| = |POS(\mathcal{C},k)|$. But, $POS(P,k) \subseteq POS(\mathcal{C},k)$, so $POS(P,k) = POS(\mathcal{C},k)$.

Assume that $\forall_{k\in V_d} \sum_{J\in JEP(D_k,P)} supp_{D_k}(J) = \frac{|POS(\mathcal{C},k)|}{|U_k|}$. From (1) and Lemma 2.1, we have $\forall_{k\in V_d} POS(P,k) = POS(\mathcal{C},k) \Longrightarrow \forall_{k\in V_d} POS(P) \cap U_k = POS(\mathcal{C}) \cap U_k \Longleftrightarrow POS(P) = POS(\mathcal{C}) \Longleftrightarrow P \in SRED(\mathcal{U},d)$.

The second implication $\forall_{P\subseteq C}((\forall_{k\in V_d} \sum_{J\in JEP(D_k,P)} supp_{D_k}(J) = \frac{|POS(\mathcal{C},k)|}{|U_k|}) \Longleftarrow P \in SRED(\mathcal{U},d))$ is a direct consequence of Remark 5.3. □

Remark 5.5 extends Theorem 5.4 and states precisely the conditions under which a given super reduct is minimal.

*Remark 5.5.* If $P \in SRED(\mathcal{U},d)$ and $\forall_{S\subset P} \exists_{k\in V_d} \sum_{J\in JEP(D_k,S)} supp_{D_k}(J) \neq \frac{|POS(\mathcal{C},k)|}{|U_k|}$), then $P$ is a global reduct.

*Proof.* From Theorem 5.4, we have $\forall_{S\subset P}(\exists_{k\in V_d} \sum_{J\in JEP(D_k,S)} supp_{D_k}(J) \neq \frac{|POS(\mathcal{C},k)|}{|U_k|}) \Longleftrightarrow S \notin SRED(\mathcal{U},d))$. Because $P \in SRED(\mathcal{U},d)$ and $\forall_{S\subset P} S \notin SRED(\mathcal{U},d)$, so $P \in RED$. □

*Example 5.6.* Let us see how the above relations help identify global reducts.

Consider Theorem 5.4 for $P = \{w,y\}$. We have:

$JEP(D_0,P) = \{\{(w,0),(y,0)\}\}$,
$supp_{D_0}(\{(w,0),(y,0)\}) = \frac{2}{3} = \frac{|POS(\mathcal{C},0)|}{|U_0|}$,
$JEP(D_1,P) = \{\{(w,1),(y,0)\},\{(w,1),(y,1)\}\}$,
$supp_{D_1}(\{(w,1),(y,0)\}) + supp_{D_1}(\{(w,1),(y,1)\}) = \frac{1}{3} + \frac{1}{3} = \frac{2}{3} = \frac{|POS(\mathcal{C},1)|}{|U_1|}$.

This means that $\{w,y\} \in SRED(\mathcal{U},d)$. Note that for any proper subset of $P$, we do not obtain such an equality. Remark 5.5 indicates that $P \in RED$.

On the other hand, consider $P = \{w,x\}$. We have:

$JEP(D_0,P) = \{\{(w,0),(x,0)\}\}$,
$supp_{D_0}(\{(w,0),(x,0)\}) = \frac{1}{3} \neq \frac{2}{3} = \frac{|POS(\mathcal{C},0)|}{|U_0|}$,
$JEP(D_1,P) = \{\{(w,1),(x,1)\},\{(w,1),(x,0)\}\}$,
$supp_{D_1}(\{(w,1),(x,1)\}) + supp_{D_1}(\{(w,1),(x,0)\}) = \frac{1}{3} + \frac{1}{3} = \frac{2}{3} = \frac{|POS(\mathcal{C},1)|}{|U_1|}$.

Therefore, $\{w,x\} \neq SRED(\mathcal{U},d)$.

The next theorem expresses a relation between super reducts and attribute sets associated with minimal jumping emerging patterns. It is extended in Remark 5.6 to estimate a lower bound for the cardinalities of super reducts.

**Theorem 5.5.** $POS(\mathcal{C}) \neq \emptyset \Longrightarrow \forall_{P\in SRED(\mathcal{U},d)} \forall_{k\in V_d} \exists_{J\in \mathcal{L}_k} attr(J) \subseteq P$.

*Proof.* Assume that $POS(\mathcal{C}) \neq \emptyset$ and consider $u \in POS(\mathcal{C})$.

Let $P \in SRED(\mathcal{U}, d)$ and $k \in V_d$. According to Remark 5.1, we have $patt(u, P) \in JEP(D_k, P) \subseteq JEP(D_k)$. We know that $JEP(D_k) = [\mathcal{L}_k, \mathcal{R}_k]$, so $\exists_{J \in \mathcal{L}_k} J \subseteq patt(u, P) \Longrightarrow \exists_{J \in \mathcal{L}_k} attr(J) \subseteq P$. □

*Remark 5.6.* $POS(\mathcal{C}) \neq \emptyset \Longrightarrow \forall_{P \in SRED(\mathcal{U},d)} |P| \geq max_{(k \in V_d)}(min_{(J \in \mathcal{L}_k)}|J|)$

*Proof.* Assume that $POS(\mathcal{C}) \neq \emptyset$ and consider $u \in POS(\mathcal{C})$.

Let $P \in SRED(\mathcal{U}, d)$.
According to Theorem 5.5, we have

$$\begin{aligned}\forall_{k \in V_d} \exists_{J \in \mathcal{L}_k} attr(J) \subseteq P &\Longrightarrow \forall_{k \in V_d} \exists_{J \in \mathcal{L}_k} |J| \leq |P| \\ &\Longrightarrow \forall_{k \in V_d} min_{(J \in \mathcal{L}_k)} |J| \leq |P| \\ &\Longrightarrow |P| \geq max_{(k \in V_d)}(min_{(J \in \mathcal{L}_k)}|J|).\end{aligned}$$

□

*Example 5.7.* Let us calculate a lower bound according to Remark 5.6. We have $POS(\mathcal{C}) \neq \emptyset$, $min_{(J \in \mathcal{L}_0)}|J| = 2$, $min_{(J \in \mathcal{L}_1)}|J| = 1$ and $max(2, 1) = 2$. Thus, the cardinality of any super reduct, in particular a reduct, is greater than, or equal to, 2.

**Local Reducts.** According to Theorem 5.5, every reduct $P$ generates with an object $u \in \mathcal{U}$ the pattern $patt(u, P)$ that belongs to $JEP(D_{d(u)})$. Although a reduct itself is minimal, the generated JEPs might not be. This is because global reducts are defined in the context of a whole decision table, while JEPs pertain only to supporting them transactions.

A more direct relation can be obtained by considering local reducts, i.e. minimal attribute sets that allow us to discern a certain object from objects belonging to other classes, as well as $\mathcal{C}$. Indeed, the following theorem states that every local reduct generates with an object $u \in POS(\mathcal{C}, d)$ a minimal jumping emerging pattern. In other words, this pattern belongs to $\mathcal{L}_{d(u)}$, the left bound of the set interval representing $JEP(D_{d(u)})$. Note that, in case of a consistent decision table, this statement holds for each $u \in \mathcal{U}$.

**Theorem 5.6.** $\forall_{u \in POS(\mathcal{C},d)} \forall_{P \subseteq \mathcal{C}} P \in REDLOC(u, d) \Longleftrightarrow patt(u, P) \in \mathcal{L}_{d(u)}$.

*Proof.* Let $P, B \in \mathcal{C}$, $u \in POS(\mathcal{C}, d)$ and $k = d(u)$.

Consider first $B(u) \subseteq U_k \Longleftrightarrow patt(u, B) \in JEP(D_k)$ (1). We have $B(u) \subseteq U_k \Longleftrightarrow u \in \underline{B}(U_k) \Longleftrightarrow u \in POS(B, d) \cap U_k$. But, according to Theorem 5.1, we have: $u \in POS(B, d) \cap U_k \Longleftrightarrow patt(u, B) \in \{J \in JEP(D_k) : attr(J) = B\} \Longleftrightarrow patt(u, B) \in JEP(D_k)$.

Consider $P \in REDLOC(u, d) \Longrightarrow patt(u, P) \in \mathcal{L}_k$.
Let $P \in REDLOC(u, d)$. According to Lemma 2.2, the following holds: $P \in REDLOC(u, d) \Longleftrightarrow P$ is minimal in $\{B \subseteq \mathcal{C} : B(u) \subseteq U_k\}$. Consider $R \subset P$. It means that $R(u) \not\subseteq U_k$, and, according to (1), we obtain $patt(u, R) \notin JEP(D_k)$. Summing up, according to (1) we have $patt(u, P) \in JEP(D_k)$ and for any

**Algorithm 4.** JEPRedLoc$((\mathcal{U}, \mathcal{I}, d))$

$\mathcal{L}_k = \emptyset$ for each $k \in V_d$
**for** $(i = 1; 1 <= |\mathcal{U}|; i++)$ **do**
  Compute $REDLOC(u_i, d)$
  $\mathcal{L}_k = \mathcal{L}_k \cup \{patt(u_i, R) : R \in REDLOC(u_i, d)\}$, $k = d(u_i)$
**end for**

$J \subset patt(u, P)$ we have $J \notin JEP(D_k)$. Thus, $patt(u, P)$ is minimal in $JEP(D_k)$ and, according to Lemma 2.3, we have $patt(u, P) \in \mathcal{L}_k$.

Consider $P \in REDLOC(u, d) \Longleftarrow patt(u, P) \in \mathcal{L}_k$.
Let $patt(u, P) \in \mathcal{L}_k$. According to Lemma 2.3, we have: $patt(u, P) \in \mathcal{L}_k \Longleftrightarrow patt(u, P)$ is minimal in $JEP(D_k)$. Consider $R \subset P$. It means that $patt(u, R) \subset patt(u, P) \Longrightarrow patt(u, R) \notin JEP(D_k)$, and, according to (1), we obtain $R(u) \nsubseteq U_k$. Summing up, according to (1) we have $P \in \{B \subseteq \mathcal{C} : B(u) \subseteq U_k\}$ and for any $R \subset P$ we have $R(u) \nsubseteq U_k$. Thus, $P$ is minimal in $\{B \subseteq \mathcal{C} : B(u) \subseteq U_k\}$ and, according to Lemma 2.2, we have $P \in REDLOC(u, d)$. □

*Example 5.8.* Let us consider minimal JEPs in the class 0, for which:
$JEP(D_0) = \{\{(y,0),(z,0)\},\{(x,0),(y,0)\},\{(x,1),(z,0)\},\{(w,0),(z,0)\}, \{(w,0),(y,0)\},\{(w,0),(x,0)\}\}$.
For the object $u_2$ from the positive region of $\mathcal{DT}$, we have:
$REDLOC(u_2, d) = \{\{w,y\},\{w,z\},\{x,z\},\{y,z\}\}$.
Theorem 5.6 holds for the respective patterns:
$\{\{(w,0),(y,0)\},\{(w,0),(z,0)\},\{(x,1),(z,0)\},\{(y,0),(z,0)\}\} \subseteq JEP(D_0)$.
On the other hand, the same does not hold for the object $u_3 \notin POS(\mathcal{C}, d)$. Indeed, we have $REDLOC(u_3, d) = \{\{w\},\{x,y\},\{y,z\}\}$
and $\{\{(w,0)\},\{(x,1),(y,0)\},\{(y,0),(z,0)\}\} \nsubseteq JEP(D_0)$.

### 5.3 Mining Approach

Section 3.2 presents several JEP discovery methods. Each of them assumes that input data are in transactional form. If we are provided with a decision table instead an initial transformation to a decision transaction system needs to be performed (see Section 5.1). Consequently, after the main computation is done, resulting patterns have to be transformed back to appropriate attribute-value pairs.

Theorem 5.6 talks about a close correspondence between local reducts and minimal JEPs. It allows us to directly compute desirable JEP spaces from an original decision table.

Our approach is summarized as Algorithm 4. It consists of two stages. First, for each object $u \in \mathcal{U}$, the set of local reducts $REDLOC(u, d)$ is computed. Then, for each local reduct $R \in REDLOC(u, d)$, a minimal pattern $patt(u, R)$ is added to the pattern set $\mathcal{L}_{d(u)}$. Local reduct discovery determines the total computation time. In this regard, one may use one of reduct computation methods discussed in Section 3.5 or Chapter 4.

It should be noted that our approach is related to existing rough set methods for finding minimal rules and patterns ([132,55]).

# 6 JEPs with Negation

Transactions express facts directly observed in the domain such as event occurrences or object characteristics. Consequently, itemsets from such transactions represent information that is observable. Further, we use the term *positive* to refer to this kind of knowledge. On the contrary, one may consider itemsets that complement positive transactions to a given itemspace. They cover information that is not available directly, but can be easily inferred. This approach leads to a type of knowledge referred to as *negative*.

In Chapter 5, it is demonstrated that a transformation of a decision table to a decision transaction system allows us to use reducts and local reducts to discover JEPs. A question arises if there is any relation to rough set theory, when input data are initially provided in transactional form. Interestingly, this matter is closely related to negative knowledge in transaction databases.

We generalize JEPs to *JEPs with negation (JEPNs)* by taking into account both positive and negative items. It is shown that they correspond to classic JEPs in appropriately defined transaction databases ([78]). At the same time, an information-preserving transformation of an input database to a binary decision table gives us a basis to consider relations to rough set reducts. In particular, we demonstrate that local reducts provide a way to identify minimal JEPNs.

Originally, negative relationships were introduced in [133], where a chi-square model was applied to estimate independence between two variables. As far as data mining is concerned, the vast majority of publications employ the idea of negation to formulate new interesting association rules. Consequently, many algorithms include variants of frequent itemset mining ([134]).

The extended pattern definition results in search space enlargement. Several approaches has been put forward to alleviate this effect. In [61,62], the support-confidence framework is supplemented with additional measures of interestingness, so that pruning opportunities are higher. Another option is to constrain a rule syntax. For example, in negative association rules ([61]) and confined association rules ([62]), only a complete antecedent or consequent can be negated, whereas in unexpected rules ([63]) and exception rules ([64]) negative items in antecedents are used to represent exceptions to regular associations. Some other approaches make use of domain knowledge to formulate valuable rules ([135,136]). Last but not least, the problem of mining frequent itemsets with negation can be addressed with concise data representations ([137,138]).

## 6.1 Negative Knowledge in Transaction Databases

We propose a formal apparatus to deal with positive and negative information in classified transactional datasets. In our convention new types of knowledge can be considered with the standard framework for emerging patterns, as long as the input data are appropriately transformed. Also, the new concepts provide a basis for pattern discovery approaches.

Hereinafter, we assume that our data are given by a decision transaction system $\mathcal{DTS} = (\mathcal{D}, \mathcal{I}, \mathcal{I}_d)$, where $\mathcal{D} = \{T_1, .., T_n\}$, $\mathcal{I} = \{I_1, .., I_m\}$, $\mathcal{I}_d = \{c_1, .., c_p\}$.

In order to express both positive and negative knowledge in the context of $\mathcal{DTS}$, we generalize the concepts of an itemspace, itemset and item. In this chapter, the original meaning of these terms is preserved by preceding them with the adjective *positive*, e.g. a positive item.

**Definition 6.1.** *A* positive itemspace, *a* negative itemspace *and an* extended itemspace *are defined as* $\mathcal{I}$, $\overline{\mathcal{I}} = \{\overline{i}\}_{i \in \mathcal{I}}$ *and* $\mathcal{I} \cup \overline{\mathcal{I}}$, *respectively. Their elements are called* positive, negative *and* extended items, *respectively.*

*A* positive itemset with negation, *a* negative itemset with negation *and an* itemset with negation *are any subsets of the respective itemspaces.*

Our discussion pertains mostly to itemsets with negation. Thus, for brevity, we usually use short names: *itemsets* or *patterns.*

Negative items express the absence of the corresponding positive items in transactions. Consequently, itemsets that contain at least one positive item and its corresponding negative item have a self-contradictory interpretation. Therefore, we distinguish a set of patterns that make sense in the considered setting.

**Definition 6.2.** The set of valid itemsets *is defined as* $\mathcal{P} = \{p \subseteq \mathcal{I} \cup \overline{\mathcal{I}} : \forall_{i \in \mathcal{I}} i \in p \Longrightarrow \overline{i} \notin p\}$. *Accordingly, each itemset from* $\mathcal{P}$ *is called* valid.

For brevity, we introduce the following notations.

**Definition 6.3.** *For an itemset* $X \in \mathcal{P}$, *we define*

*the* positive part $X_p = X \cap \mathcal{I}$,
*the* negative part $X_n = X \cap \overline{\mathcal{I}}$,
*the* negated pattern $\overline{X} = \{\overline{i}\}_{i \in X}$,
*the* contradictory pattern $\widehat{X} = \overline{(\mathcal{I} \cup \overline{\mathcal{I}}) - X}$,
*the* extended pattern $\widetilde{X} = X \cup \widehat{X}$.

*We assume that* $\overline{\overline{i}} = i$.

In other words, a positive (negative) part refers to the set of all positive (negative) items of a given pattern. A negated pattern is obtained by changing each positive item of a given pattern to the corresponding negative item and vice versa. A contradictory pattern is a complement of a negated pattern to the extended itemspace. Finally, an extended pattern is a sum of a given pattern and its contradictory pattern.

Now, we generalize basic measures defined for positive patterns, so that they can be used for itemsets with negation.

**Definition 6.4.** *The* extended support *of an itemset* $X \in \mathcal{P}$ *in a database* $D \subseteq \mathcal{D}$ *is defined as* $exsupp_D(X) = \frac{|\{T \in D : X_p \subseteq T \wedge X_n \subseteq \overline{\mathcal{I} - T}\}|}{|D|}$.

Having accepted this extended definition of support, we can accordingly introduce the notion of extended growth rate. The supports need to be replaced by extended supports.

**Definition 6.5.** *The* extended growth rate *of an itemset $X \subseteq \mathcal{I}$ from $D_1$ to $D_2$ is defined as follows*

$$exgr_{D_1 \to D_2}(X) = \begin{cases} 0, & exsupp_{D_1}(X) = exsupp_{D_2}(X) = 0 \\ \infty, & exsupp_{D_1}(X) = 0 \text{ and } exsupp_{D_2}(X) \neq 0 \\ \frac{exsupp_{D_2}(X)}{exsupp_{D_1}(X)}, & otherwise. \end{cases}$$

Based on the above definitions, one may consider patterns analogous to positive emerging patterns. We focus solely on JEPs.

**Definition 6.6.** *A* jumping emerging pattern with negation (JEPN) *from $D_1$ to $D_2$ is an itemset $X \in \mathcal{P}$ with infinite extended growth rate.*

For convenience, we introduce the following sets of JEPNs.

**Definition 6.7.** *For $D_1, D_2 \subseteq \mathcal{D}$, we define*

*a* JEPN space, *$JEPN_{\mathcal{DTS}}(D_1, D_2)$, as the set of all JEPNs from $D_1$ to $D_2$,*

*a* positive JEPN space *as*
$posJEPN_{\mathcal{DTS}}(D_1, D_2) = \{J \in JEPN_{\mathcal{DTS}}(D_1, D_2) : J \subseteq \mathcal{I}\}$,

*a* negative JEPNs space *as*
$negJEPN_{\mathcal{DTS}}(D_1, D_2) = \{J \in JEPN_{\mathcal{DTS}}(D_1, D_2) : J \subseteq \overline{\mathcal{I}}\}$.

Itemsets with negation are considered for a decision transaction system. We define two derivative systems to focus on just positive or negative itemsets.

**Definition 6.8.** *A* contradictory decision transaction system *based on $\mathcal{DTS}$ is a decision transaction system $\widehat{\mathcal{DTS}} = (\{T'_1, .., T'_n\}, \overline{\mathcal{I}}, \mathcal{I}_d)$, where $T'_i = \widehat{T}_i \cup (T_i \cap \mathcal{I}_d)$, for each $i = 1, .., n$.*

To construct a contradictory decision transaction system, each transaction of $\mathcal{DTS}$ is replaced by its corresponding contradictory pattern, a decision item is excluded. Items complementing transactions of $\mathcal{DTS}$ to the positive itemspace can be directly observed in transactions of $\widehat{\mathcal{DTS}}$ as negative items.

**Definition 6.9.** *An* extended decision transaction system *based on $\mathcal{DTS}$ is a decision transaction system $\widetilde{\mathcal{DTS}} = (\{T'_1, .., T'_n\}, \mathcal{I} \cup \overline{\mathcal{I}}, \mathcal{I}_d)$, where $T'_i = \widetilde{T}_i \cup (T_i \cap \mathcal{I}_d)$, for each $i = 1, .., n$.*

In other words, each transaction of $\mathcal{DTS}$ is extended to include their negated complements. As a consequence, all transactions of $\widetilde{\mathcal{DTS}}$ have the same size $|\mathcal{I}|$.

*Example 6.1.* Let us consider the decision transaction system $\mathcal{DTS}$ in Table 6.1. We have $\mathcal{I} = \{a, b, c, d, e, f\}$, $\mathcal{I}_d = \{c_0, c_1\}$, $\overline{\mathcal{I}} = \{\overline{a}, \overline{b}, \overline{c}, \overline{d}, \overline{e}, \overline{f}\}$ and $\mathcal{P} = \{p \subseteq \mathcal{I} \cup \overline{\mathcal{I}} : \forall_{i \in \mathcal{I}} i \in p \implies \overline{i} \notin p\}$. Thus, for example, we have: $ab\overline{b}c, a\overline{a}\overline{e} \notin \mathcal{P}$.

Although transactions in $\mathcal{DTS}$ cannot contain negative items, they can still support patterns belonging to $\mathcal{P}$, when the extended definition of support is being used, e.g. $exsupp_{D_0}(\overline{c}e) = 2$. The contradictory and extended decision systems based on $\mathcal{DTS}$ are also given in 6.1.

**Table 6.1.** Decision transaction systems: $\mathcal{DTS}$, $\widehat{\mathcal{DTS}}$, $\widetilde{\mathcal{DTS}}$

| | $\mathcal{I}$ | $\mathcal{I}_d$ |
|---|---|---|
| $T_1$ | cef | $c_0$ |
| $T_2$ | de | $c_0$ |
| $T_3$ | e | $c_0$ |
| $T_4$ | bcd | $c_1$ |
| $T_5$ | df | $c_1$ |
| $T_6$ | ace | $c_1$ |

| | $\overline{\mathcal{I}}$ | $\mathcal{I}_d$ |
|---|---|---|
| $T_1$ | $\overline{a}\overline{b}\overline{d}$ | $c_0$ |
| $T_2$ | $\overline{a}\overline{b}\overline{c}\overline{f}$ | $c_0$ |
| $T_3$ | $\overline{a}\overline{b}\overline{c}\overline{d}\overline{f}$ | $c_0$ |
| $T_4$ | $\overline{a}\overline{e}\overline{f}$ | $c_1$ |
| $T_5$ | $\overline{a}\overline{b}\overline{c}\overline{e}$ | $c_1$ |
| $T_6$ | $\overline{b}\overline{d}\overline{f}$ | $c_1$ |

| | $\mathcal{I} \cup \overline{\mathcal{I}}$ | $\mathcal{I}_d$ |
|---|---|---|
| $T_1$ | $cef\overline{a}\overline{b}\overline{d}$ | $c_0$ |
| $T_2$ | $de\overline{a}\overline{b}\overline{c}\overline{f}$ | $c_0$ |
| $T_3$ | $e\overline{a}\overline{b}\overline{c}\overline{d}\overline{f}$ | $c_0$ |
| $T_4$ | $bcd\overline{a}\overline{e}\overline{f}$ | $c_1$ |
| $T_5$ | $df\overline{a}\overline{b}\overline{c}\overline{e}$ | $c_1$ |
| $T_6$ | $ace\overline{b}\overline{d}\overline{f}$ | $c_1$ |

## 6.2 Transformation to Decision Table

Transactional data can be represented in the form of a decision table. We consider a transformation, in which objects are mapped to transactions and each item is assigned a binary attribute ([1]). The attribute indicates the presence of the item in the respective transaction.

**Definition 6.10.** *A* binary decision table based on $\mathcal{DTS}$ *is a decision table* $\mathcal{BDT}_{\mathcal{DTS}} = (\mathcal{U}, \mathcal{C}, d)$ *such that*

$$\mathcal{U} = \{u_1, .., u_n\},\ \mathcal{C} = \{a_1, .., a_m\},\ V_d = \{c_1, .., c_p\},$$
$$a_j(u_i) = \begin{cases} 0, & I_j \notin T_i \\ 1, & I_j \in T_i \end{cases}, \forall_{i \in 1..n, j \in 1..m},$$
$$d(u_i) = c, \text{ where } \{c\} = T_i \cap \mathcal{I}_d, \forall_{i \in 1..n}.$$

In this data representation, the fact whether an item belongs to a particular transaction or not is encoded by certain attribute values. Therefore, itemsets generated by an object and attribute set can contain negative items.

**Definition 6.11.** *For* $\mathcal{BDT}_{\mathcal{DTS}}$, $u \in \mathcal{U}$, $B = \{a_k\}_{k \in K}$ *and* $K \subseteq \{1, .., m\}$, *we define a binary pattern based on the object* $u$ *and attribute set* $B$ *as*
$binPatt(u, B) = \{I_k \in \mathcal{I} : a_k(u) = 1\} \cup \{\overline{I}_k \in \mathcal{I} : a_k(u) = 0\}$.

*Example 6.2.* In order to describe the data in the decision transaction system $\mathcal{DTS}$, we can use the binary decision table $(\mathcal{U}, \mathcal{C}, d)$ based on $\mathcal{DTS}$, where

$$\mathcal{U} = \{u_1, u_2, u_3, u_4, u_5, u_6\},$$
$$\mathcal{C} = \{a_1, a_2, a_3, a_4, a_5, a_6\},$$
$$V_d = \{c_0, c_1\}.$$

Note that we use the same symbol to denote an item and the respective attribute, even though the latter is a function. Values of the attributes are given in Table 6.2.

**Table 6.2.** The binary decision table $(\{u_1, u_2, u_3, u_4, u_5, u_6\}, \{a_1, a_2, a_3, a_4, a_5, a_6\}, d)$ based on $\mathcal{DTS}$

| | $a_1$ | $a_2$ | $a_3$ | $a_4$ | $a_5$ | $a_6$ | $d$ |
|---|---|---|---|---|---|---|---|
| $u_1$ | 0 | 0 | 1 | 0 | 1 | 1 | 0 |
| $u_2$ | 0 | 0 | 0 | 1 | 1 | 0 | 0 |
| $u_3$ | 0 | 0 | 0 | 0 | 1 | 0 | 0 |
| $u_4$ | 0 | 1 | 1 | 1 | 0 | 0 | 1 |
| $u_5$ | 0 | 0 | 0 | 1 | 0 | 1 | 1 |
| $u_6$ | 1 | 0 | 1 | 0 | 1 | 0 | 1 |

## 6.3 Properties

This section looks at basic properties of JEPNs and their relation to rough set theory. We continue to consider the decision transaction system $\mathcal{DTS} = (\mathcal{D}, \mathcal{I}, \mathcal{I}_d)$.

The following two facts demonstrate equivalence of the support (growth rate) of a given pattern with negation in the extended decision transaction system $\widetilde{\mathcal{DTS}}$ as well as the extended support (extended growth rate) of this pattern in the original decision transaction system $\mathcal{DTS}$.

**Theorem 6.1.** *Let $D \subseteq \mathcal{D}$.*

$$\forall_{X \subseteq \mathcal{I} \cup \overline{\mathcal{I}}} supp_{\widetilde{D}}(X) = exsupp_D(X).$$

*Proof.* Let $X \subseteq \mathcal{I} \cup \overline{\mathcal{I}}$.

Consider the set $\{T \in \widetilde{D} : X \subseteq T\}$. Because $X_p, T_p \subseteq \mathcal{I}$ and $X_n, T_n \subseteq \overline{\mathcal{I}}$, we have $\{T \in \widetilde{D} : X \subseteq t\} = \{(T_p \cup T_n) \in \widetilde{D} : X_p \subseteq T_p \wedge X_n \subseteq T_n\}$.

Now, from the definition of $\widetilde{D}$, we have $T_n = \overline{\mathcal{I} - T_p}$, so $\{(T_p \cup T_n) \in \widetilde{D} : X_p \subseteq T_p \wedge X_n \subseteq T_n\} = \{(T_p \cup \overline{\mathcal{I} - T_p}) \in \widetilde{D} : X_p \subseteq T_p \wedge X_n \subseteq \overline{\mathcal{I} - T_p}\}$. The database $\widetilde{D}$ is created by extending each transaction of $D$ with relevant negative items. Therefore, we have $|\{(T_p \cup \overline{\mathcal{I} - T_p}) \in \widetilde{D} : X_p \subseteq T_p \wedge X_n \subseteq \overline{\mathcal{I} - T_p}\}| = |\{T \in D : X_p \subseteq T \wedge X_n \subseteq \overline{\mathcal{I} - T}\}|$.

To sum up, the fact $|\{T \in \widetilde{D} : X \subseteq T\}| = |\{T \in D : X_p \subseteq t \wedge X_n \subseteq \overline{\mathcal{I} - T}\}|$, together with $|\widetilde{D}| = |D|$, gives $supp_{\widetilde{D}}(X) = exsupp_D(X)$. □

*Remark 6.1.* Let $D \subseteq \mathcal{D}$.

$$\forall_{X \subseteq \mathcal{I} \cup \overline{\mathcal{I}}} gr_{\widetilde{D_1} \to \widetilde{D_2}}(X) = exgr_{D_1 \to D_2}(X).$$

*Proof.* Let $X \subseteq \mathcal{I} \cup \overline{\mathcal{I}}$. According to Theorem 6.1, we have $supp_{\widetilde{D_1}}(X) = exsupp_{D_2}(X)$ and $supp_{\widetilde{D_1}}(X) = exsupp_{D_2}(X)$. Thus, we obtain $gr_{\widetilde{D_1} \to \widetilde{D_2}}(X) = exgr_{D_1 \to D_2}(X)$. □

In the light of these facts, it becomes evident that JEPNs in $\mathcal{DTS}$ are also JEPs in $\widetilde{\mathcal{DTS}}$ and vice versa.

**Theorem 6.2.** *Let $D_1, D_2 \subseteq \mathcal{D}$. $JEPN_{\mathcal{DTS}}(D_1, D_2) = JEP_{\widetilde{\mathcal{DTS}}}(D_1, D_2)$.*

*Proof.* Let $X \subseteq \mathcal{I} \cup \overline{\mathcal{I}}$.

We have $X \in JEPN(D_1, D_2) \iff exgr_{D_1 \to D_2}(X) = +\infty$. According to Theorem 6.1, we have $gr_{\widetilde{D_1} \to \widetilde{D_2}}(X) = exgr_{D_1 \to D_2}(X)$. Finally, we conclude: $exgr_{D_1 \to D_2}(X) = +\infty \iff gr_{D_1 \to D_2}(X) = +\infty \iff X \in JEP_{\widetilde{\mathcal{DTS}}}(D_1, D_2)$. □

*Example 6.3.* Let us consider the patterns $\overline{a}\overline{b}\overline{c}, \overline{d}f \in P$. In the system $\widetilde{\mathcal{DTS}}$, we have $supp_{\widetilde{D_0}}(\overline{a}\overline{b}\overline{c}) = 2$, $supp_{\widetilde{D_1}}(\overline{a}\overline{b}\overline{c}) = 1$ and $gr_{\widetilde{D_1} \to \widetilde{D_0}}(\overline{a}\overline{b}\overline{c}) = 2/1 < +\infty$, which gives $\overline{a}\overline{b}\overline{c} \notin JEP_{\widetilde{\mathcal{DTS}}}(D_1, D_0)$. In the system $\mathcal{DTS}$, we have $exsupp_{D_0}(\overline{a}\overline{b}\overline{c}) = 2$, $exsupp_{D_1}(\overline{a}\overline{b}\overline{c}) = 1$ and $exgr_{D_1 \to D_0}(\overline{a}\overline{b}\overline{c}) = 2/1 < +\infty$, therefore, $\overline{a}\overline{b}\overline{c} \notin JEPN_{\mathcal{DTS}}(D_1, D_0)$. At the same time, in $\mathcal{DTS}$, we have $supp_{\widetilde{D_0}}(\overline{d}f) = 2$, $supp_{\widetilde{D_1}}(\overline{d}f) = 0$ and $gr_{\widetilde{D_1} \to \widetilde{D_0}}(\overline{d}f) = +\infty$, thus, $\overline{d}f \in JEP_{\widetilde{\mathcal{DTS}}}(D_1, D_0)$. Now, in $\mathcal{DTS}$, we have $exsupp_{D_0}(\overline{d}f) = 2$, $exsupp_{D_1}(\overline{d}f) = 1$ and $exgr_{D_1 \to D_0}(\overline{d}f) = +\infty$, thus, $\overline{d}f \in JEPN_{\mathcal{DTS}}(D_1, D_0)$. Note that $\overline{d}f$ is also a minimal pattern.

As we can see, extended support, extended growth rate and being a JEPN in $\mathcal{DTS}$ can be equivalently concluded in the extended decision transaction system $\widetilde{\mathcal{DTS}}$, which remains consistent with Theorems 6.1, 6.2 and Remark 6.1.

As a consequence, from Theorems 2.3 and 2.1, $JEPN_{\mathcal{DTS}}(D_1, D_2)$ is a convex space and can be concisely represented by a set interval. Throughout the rest of this section, we consider spaces $JEPN_{\mathcal{DTS}}(D'_k, D_k) = [\mathcal{L}_k, \mathcal{R}_k]$ for $k \in V_d$.

Itemsets with negation are closely related to attribute sets in the respective binary decision table. The following theorem demonstrates equivalence between a JEPN in $\mathcal{DTS}$ and a pattern generated by an attribute set and object from the positive region induced by this set in $\mathcal{BDT}_{\mathcal{DTS}}$.

**Theorem 6.3.** *Let a decision table $(\mathcal{U}, \mathcal{C}, d)$ be the binary decision table based on $\mathcal{DTS}$. $\forall_{k \in V_d} \forall_{P \subseteq \mathcal{C}} \forall_{u \in \mathcal{U}} u \in POS(P, k, d) \iff binPatt(u, P) \in JEPN(D'_k, D_k)$.*

*Proof.* Let $k \in V_d$ and $P \subseteq \mathcal{C}$.

Consider $\forall_{u \in \mathcal{U}} u \in POS(P, k, d) \iff exsupp_{D'_k}(binPatt(u, P)) = 0$ (1).

Let $u \in POS(P, k, d)$. We have $u \in P_*(U_k) \iff P(u) \subseteq U_k \iff \{v \in \mathcal{U} : \forall_{a \in P} a(u) = a(v)\} \subseteq U_k \iff \forall_{v \in \mathcal{U} - U_k} \exists_{a \in P} a(v) \neq a(u)$. Because $\psi$ is a bijection, we have the following $\forall_{v \in \mathcal{U} - U_k} \exists_{a \in P} a(v) \neq a(u) \iff \forall_{v \in \mathcal{U} - U_k} \exists_{i \in \mathcal{I} \cup \overline{\mathcal{I}}} i \in binPatt(u, P) \wedge i \notin binPatt(v, P) \iff \forall_{T \in \widetilde{D'_k}} binPatt(u, P) \notin T \iff supp_{\widetilde{D'_k}} = 0$. From Theorem 6.1 we have $supp_{\widetilde{D'_k}} = 0 \iff exsupp_{D'_k} = 0$.

Consider $\forall_{u \in \mathcal{U}} u \in POS(P, k, d) \implies binPatt(u, P) \in JEPN(D'_k, D_k)$.

Note that $binPatt(u_i, P) \subseteq \widetilde{T}_i$, where $T_i \in D_k$. As a consequence, we obtain: $exsupp_{D_k}(binPatt(u_i, P)) \neq 0$. From (1), we have $exsupp_{D'_k}(binPatt(u_i, P)) = 0$; thus, $binPatt(u_i, P) \in JEPN(D'_k, D_k, P)$.

Consider the second implication:
$\forall_{u \in \mathcal{U}} binPatt(u, P) \in JEPN(D'_k, D_k, P) \implies u \in POS(P, k, d)$.

According to (1) we have the following: $binPatt(u, P) \in JEPN(D'_k, D_k, P) \implies exsupp_{D'_k}(binPatt(u, P)) = 0 \iff u \in POS(P, k, d)$. □

In addition to the above, rough set theory provides a way of finding minimal JEPNs. In fact, the left bound of a JEPN space can be generated by means of local reducts induced for each object in the positive region of $\mathcal{BDT}_{\mathcal{DTS}}$.

**Theorem 6.4.** *Let a decision table* $(\mathcal{U}, \mathcal{C}, d)$ *be the binary decision table based on* $\mathcal{DTS}$. $\forall_{P \subseteq \mathcal{C}} \forall_{u \in POS(\mathcal{C},d)} P \in REDLOC(u, d) \iff binPatt(u, P) \in \mathcal{L}_{d(u)}$.

*Proof.* Let $P, B \in \mathcal{C}$, $u \in POS(\mathcal{C}, d)$ and $k = d(u)$.

First consider $B(u) \subseteq U_k \iff binPatt(u, B) \in JEPN(D'_k, D_k)$ (1).

We have $B(u) \subseteq U_k \iff u \in B_*(U_k) \iff u \in POS(B, d) \cap U_k$. But, according to Theorem 5.1, we have $u \in POS(B, d) \cap U_k \iff binPatt(u, B) \in JEPN(D'_k, D_k)$.

Consider $P \in REDLOC(u, d) \implies binPatt(u, P) \in \mathcal{L}_k$.
Let $P \in REDLOC(u, d)$. From Lemma 2.2, we have $P \in REDLOC(u, d) \iff P$ is minimal in $\{B \subseteq \mathcal{C} : B(u) \subseteq U_k\}$. Consider $R \subset P$. It means that $R(u) \not\subseteq U_k$, and, according to (1), we obtain $binPatt(u, R) \notin JEPN(D'_k, D_k)$. To sum up, according to (1), we have $binPatt(u, P) \in JEPN(D'_k, D_k)$ and for any $J \subset binPatt(u, P)$ we have $J \notin JEPN(D'_k, D_k)$. Therefore, the pattern $binPatt(u, P)$ is minimal in $JEPN(D'_k, D_k)$ and, according to Lemma 2.3, we have $binPatt(u, P) \in \mathcal{L}_k$.

Consider $P \in REDLOC(u, d) \impliedby binPatt(u, P) \in \mathcal{L}_k$.
Let $binPatt(u, P) \in \mathcal{L}_k$. According to Lemma 2.3, we have $binPatt(u, P) \in \mathcal{L}_k \iff binPatt(u, P)$ is minimal in $JEPN(D'_k, D_k)$. Consider $R \subset P$. It means that $binPatt(u, R) \subset binPatt(u, P) \implies binPatt(u, R) \notin JEPN(D'_k, D_k)$ and, according to (1), we obtain $R(u) \not\subseteq U_k$. To sum up, according to (1) we have $P \in \{B \subseteq \mathcal{C} : B(u) \subseteq U_k\}$ and for any $R \subset P$ we have $R(u) \not\subseteq U_k$. Thus, $P$ is minimal in $\{B \subseteq \mathcal{C} : B(u) \subseteq U_k\}$ and, according to Lemma 2.2, we have $P \in REDLOC(u, d)$. □

*Example 6.4.* Let us consider the binary decision table $\mathcal{BDT}_{\mathcal{DTS}}$, the attribute set $P = \{a, b, c\}$ and the class $k = 0$. We have $POS(\{a, b, c\}, 0, d) = \{u_1, u_4, u_6\}$ and $binPatt(\{a, b, c\}, u_1) = \{\overline{a}, \overline{b}, c\}$. In addition, we have $exgr_{D_1 \to D_0}(\overline{a}\overline{b}c) = +\infty$, thus, $\overline{a}\overline{b}c \in JEPN_{\mathcal{DTS}}(D_1, D_0)$. Besides, $\{a, b, c\} \in REDLOC(u_1, d)$, $u_1 \in POS(\mathcal{C}, d)$ and $\overline{a}\overline{b}c \in \mathcal{L}_0$, where $[\mathcal{L}_0, \mathcal{R}_0]$ represents $JEPN_{\mathcal{DTS}}(D_1, D_0)$. Note that the situation is quite different for $binPatt(\{a, b, c\}, u_2) = \{\overline{a}\overline{b}\overline{c}\}$. Here, we have $u_2 \notin POS(\{a, b, c\}, 0, d)$, thus, $\overline{a}\overline{b}\overline{c}$ is not even a JEPN.

On the other hand, the right bound can be derived directly from the positive region of a binary decision table.

**Theorem 6.5.** *Let a decision table* $(\mathcal{U}, \mathcal{C}, d)$ *be the binary decision table based on* $\mathcal{DTS}$.
$\mathcal{R}_k = \{binPatt(u, \mathcal{C}) \subseteq \mathcal{P} : u \in POS(\mathcal{C}, k, d)\}$.

*Proof.* Note that, according to the definition of a binary decision table based on a decision transaction system, we have $\forall_{t \in \mathcal{D}} |\widetilde{t}| = |\mathcal{C}|$ (1).

Let k $\in V_d$.

First consider $\mathcal{R}_k \subseteq \{binPatt(u,\mathcal{C}) \subseteq \mathcal{P} : u \in POS(\mathcal{C},k,d)\}$.

Let $J \in \mathcal{R}_k$. We have $J \in \mathcal{R}_k \subseteq JEPN_{\mathcal{DTS}}(D'_k, D_k) \Longrightarrow exsupp_{D_k}(J) \neq 0 \Longleftrightarrow \exists_{T \in D_k} J \subseteq \widetilde{T}$. Because every superset of a JEP is also a JEP, we have $\widetilde{T} \in JEPN_{\mathcal{DTS}}(D'_k, D_k)$. Moreover, due to maximality of $J$ in $JEPN_{\mathcal{DTS}}(D'_k, D_k)$, we have $J = \widetilde{T}$, i.e. $J$ is a transaction in $\widetilde{D_k}$. On the other hand, from the definition of a binary decision table based on a decision transaction system, we have that $\exists_{i \in \{1,..,n\}} T = T_i$ and, then, $J = binPatt(u_i,\mathcal{C})$. Therefore, according to Theorem 6.3, we have $u_i \in POS(\mathcal{C},k,d)$.

Now, consider $\mathcal{R}_k \supseteq \{binPatt(u,\mathcal{C}) \subseteq \mathcal{P} : u \in POS(\mathcal{C},k,d)\}$.

Let $u \in POS(\mathcal{C},k,d)$. According to Theorem 6.3, we have the following: $binPatt(u,\mathcal{C}) \in JEPN(D'_k, D_k) \Longrightarrow exsupp_{D_k}(binPatt(u,\mathcal{C})) \neq \emptyset$ $\Longrightarrow \exists_{T \in D_k} binPatt(u,\mathcal{C}) \subseteq \widetilde{T}$. In addition, according to (1), we have $|\widetilde{T}| = |\mathcal{C}|$. Because $|binPatt(u,\mathcal{C})| = |\mathcal{C}|$, we have $exsupp_{D_k}(J) = 0$ for any $J \supsetneq binPatt(u,\mathcal{C})$. It means that $binPatt(u,\mathcal{C})$ is maximal in $JEPN_{\mathcal{DTS}}(D'_k, D_k)$. Finally, according to Lemma 2.3, we have $binPatt(u,\mathcal{C}) \in \mathcal{R}_k$. □

*Example 6.5.* Let us consider the binary decision table $\mathcal{BDT}_{\mathcal{DTS}}$ and class $k = 0$. We have $\{binPatt(u,\mathcal{C}) \subseteq \mathcal{P} : u \in POS(\mathcal{C},d)\} = \{binPatt(u,\mathcal{C}) \subseteq \mathcal{P} : u \in \{u_1,u_2,u_3,u_4,u_5,u_6\}\} = \{\overline{a}\overline{b}c\overline{d}ef, \overline{a}\overline{b}\overline{c}de\overline{f}, \overline{a}b\overline{c}de\overline{f}\}$. Note that this collection is equal to the right bound $\mathcal{R}_0$ .

The theorems provided so far offer two ways of finding JEPNs. First - by finding JEPs in an extended decision transaction system, and second - by finding local reducts in a binary decision table. These methods allow us to indirectly find positive or negative JEPNs by means of filtering the bounds of the set interval of a JEPN space. The following theorem demonstrates that both collections can also be obtained directly, if an original or a contradictory database is considered, respectively.

**Theorem 6.6.** *Let* $D_1, D_2 \in \mathcal{D}$. $posJEPN_{\widetilde{\mathcal{DTS}}}(D_1,D_2) = JEP_{\mathcal{DTS}}(D_1,D_2)$ *and* $negJEPN_{\widetilde{\mathcal{DTS}}}(D_1,D_2) = JEP_{\widehat{\mathcal{DTS}}}(D_1,D_2)$.

*Proof.* In the beginning, let us consider $\forall_{J \subseteq \mathcal{I}} exsupp_D(J) = supp_D(J)$ (1).

For $J \subseteq \mathcal{I}$, we have $J_p = J$ and $J_n = \emptyset$, therefore:

$exsupp_D(J) = \frac{|\{T \in D : J_p \subseteq T \wedge J_n \subseteq \overline{\mathcal{I}-t}\}|}{|D|} = \frac{|\{T \in D : J \subseteq T\}|}{|D|} = supp_D(J)$.

Now, let $J \in \mathcal{I} \cup \overline{\mathcal{I}}$. From the definition and (1), we have:

$$\begin{aligned} J \in posJEPN_{\mathcal{DTS}}(D_1,D_2) &\Longleftrightarrow exgr_{D_1 \to D_2}(J) = +\infty \wedge J \subseteq \mathcal{I} \\ &\Longleftrightarrow exsupp_{D_1}(J) = 0 \wedge exsupp_{D_2}(J) \neq 0 \wedge J \subseteq \mathcal{I} \\ &\Longleftrightarrow supp_{D_1}(J) = 0 \wedge supp_{D_2}(J) \neq 0 \\ &\Longleftrightarrow JEP_{\mathcal{DTS}}(D_1,D_2). \end{aligned}$$

The proof for negJEPNs is analogous. □

*Example 6.6.* The space $JEPN_{\mathcal{DTS}}(D_1, D_0)$ can be represented by the border $[\mathcal{L}_0, \mathcal{R}_0]$, where $\mathcal{L}_0 = \{ef, \bar{d}f, de, \bar{c}\bar{f}, \bar{c}e, \bar{c}\bar{d}, cf, \bar{b}d\bar{f}, \bar{a}e, \bar{a}\bar{d}, \bar{a}\bar{b}\bar{f}, \bar{a}\bar{b}c\}$ and $\mathcal{R}_0 = \{\bar{a}\bar{b}c\bar{d}ef, \bar{a}\bar{b}\bar{c}de\bar{f}, \bar{a}\bar{b}\bar{c}\bar{d}e\bar{f}\}$.

From the previous example, the set interval representing $JEP_{\mathcal{DTS}}(D_1, D_0)$ is equal to $[\{ef, de, cf\}, \{de, cef\}]$. This collection comprises all positive JEPNs, namely the collection $posJEPN_{\mathcal{DTS}}(D_1, D_0)$.

For the system $\widehat{\mathcal{DTS}}$, we obtain the space $JEP_{\widehat{\mathcal{DTS}}}(D_1, D_0)$ that is represented by the following set interval $[\{\bar{c}\bar{f}, \bar{c}\bar{d}, \bar{a}\bar{d}, \bar{a}\bar{b}\bar{f}\}, \{\bar{a}\bar{b}d, \bar{a}\bar{b}\bar{c}\bar{f}, \bar{a}\bar{b}\bar{c}\bar{d}\bar{f}\}]$ and equal to the collection $negJEPN_{\mathcal{DTS}}(D_1, D_0)$.

As we can observe, $JEPN_{\mathcal{DTS}}(D_1, D_0)$ contains all the itemsets from both $posJEPN_{\mathcal{DTS}}(D_1, D_0)$ and $negJEPN_{\mathcal{DTS}}(D_1, D_0)$. Besides that, it includes itemsets that contain positive and negative items at the same time.

### 6.4 Mining Approaches

Let us consider a decision transaction system $\mathcal{DTS} = (\mathcal{D}, \mathcal{I}, \mathcal{I}_d)$. We are interested in finding the space of all jumping emerging patterns with negation, $JEPN_{\mathcal{DTS}}(D'_k, D_k)$, for each decision class $k \in V_Z$. Owing to the fact that, according to Theorem 6.2, each space of this kind is convex, our task can be defined as finding the respective set intervals $[\mathcal{L}_k, \mathcal{R}_k]$.

The relations studied in the previous section provide us with two methods of finding JEPNs. The first one, given as Algorithm 5, requires building the extended decision transaction system $\widetilde{\mathcal{DTS}}$ based on $\mathcal{DTS}$. Then, for each decision class $k \in V_Z$, the databases $D_k$ and $D'_k$ are considered and the set interval for $JEP_{\widetilde{\mathcal{DTS}}}(D'_k, D_k)$ is computed. According to Theorem 6.2, the resulting set intervals are also equal to $JEPN_{\mathcal{DTS}}(D'_k, D_k)$. For a database pair, a set interval can be obtained by means of one of widely-known algorithms, like JEP-Producer (see Section 3.2) or CP-Tree mining (see Section 3.2).

The second approach, given as Algorithm 6, involves building a binary decision table $\mathcal{BDT}_{\mathcal{DTS}} = (\mathcal{U}, \mathcal{C}, d)$ and applying the rough set framework to mine minimal patterns for each class. For each $u \in \mathcal{U}$, we compute the collection of local reducts $REDLOC(u, d)$. Then, according to Theorem 6.4, the left bound of the respective set interval, for each class $k \in V_Z$, can be found by taking each object $u \in POS(\mathcal{C}, k, d)$ and generating patterns with all local reducts computed for $u$, i.e $\mathcal{L}_k = \{binPatt(u, P) : u \in POS(\mathcal{C}, k, d) \wedge P \in REDLOC(u, d)\}$. In addition, from Theorem 6.5, the respective right bounds $\mathcal{R}_k$ are trivial, i.e. $\mathcal{R}_k = \{binPatt(u, \mathcal{C}) : u \in POS(\mathcal{C}, k, d)\}$. The most important step is to efficiently mine the complete sets of local reducts for each object. One may use classic reduct computation approaches (see Section 3.5) or the method described in Chapter 4.

A JEPN space can be used to compute the corresponding JEP space. Indeed, according to Theorem 6.6, JEPs are equivalent to posJEPNs and the latter are all included in the JEPN space. From the definition, posJEPNs can be obtained by simple filtering out patterns with negative items from the bounds of the set interval representing the JEPN space. This filtering can be incorporated in the

**Algorithm 5.** JEPNBasic($\mathcal{DTS}$)

$\mathcal{L}_k = \emptyset$ for each $k \in I_d$
**for** $(k = 1; 1 <= |\mathcal{I}_d|; k++)$ **do**
  Construct the extended decision transaction system $\widetilde{\mathcal{DTS}}$
  Compute the set interval $[\widetilde{\mathcal{L}}_k, \widetilde{\mathcal{R}}_k]$ for a $JEP_{\widetilde{\mathcal{DTS}}}(D'_k, D_k)$
  $\mathcal{L}_k = \widetilde{\mathcal{L}_k}$
**end for**

**Algorithm 6.** JEPNRedLoc($\mathcal{DTS}$)

$\mathcal{L}_k = \emptyset$ for each $k \in I_d$
**for** $(i = 1; 1 <= |\mathcal{D}|; i++)$ **do**
  Construct a binary decision table $\mathcal{BDT}_{\mathcal{DTS}}$
  Compute $REDLOC(u_i, d)$ in $\mathcal{BDT}_{\mathcal{DTS}}$
  $\mathcal{L}_k = \mathcal{L}_k \cup \{binPatt_{\mathcal{DTS}}(u_i, R) : R \in REDLOC(u_i, d)\}$, $k = T_i \cup \mathcal{I}_d$
**end for**

last step of the loop in either of the algorithm. For example, one may generate a pattern $binPatt_{\mathcal{DTS}}(u, R)$ only for an object $u \in \mathcal{U}$ and a local reduct $R \subseteq \mathcal{A}$, for which $\forall_{a \in R} a(i) = 1$.

Although this approach allows us to discover JEPs, it comes with a significant overhead of additionally generated patterns and, thus, remain impractical. In Chapter 7, more efficient variants of this method are presented.

## 6.5 Experimental Evaluation

Negative knowledge is a new aspect in the field of emerging patterns. With the introduction of JEPNs, we investigate their applicability of building concise and efficient classifiers.

Our experiments employ the scheme of JEP-Classifier (see Section 3.3). Accordingly, we define posJEPN-, negJEPN- and JEPN-Classifier based on posJEPNs, negJEPNs and JEPNs, respectively. From Theorems 6.6 and 6.2, these classifiers are equivalent to JEP-Classifier in a decision transaction system $\mathcal{DTS}$, the contradictory system $\widehat{\mathcal{DTS}}$ and the extended system $\widetilde{\mathcal{DTS}}$, respectively.

We start with analyzing the structure of classifiers as expressed by the average length of a pattern and average number of patterns per class. Measurements were obtained by means of a 5-fold cross-validation scheme ([139]) and, thus, additionally averaged over the folds. Results give insights on the classifier complexity and classification capabilities of respective pattern collections.

One could expect that negative knowledge leads to a much more complex representation of a classification hypothesis. An itemspace is often orders of magnitude larger than the average transaction size, thus, the set of features that each specific instance does not possess is also large. In other words, it is easier to say what something looks like, rather than what it does not. Indeed, Table 6.3 shows that the average length for negJEPNs is higher than for the other two types of patterns. In 5 cases, *balance*, *car*, *cmc*, *nursery*, *tic-tac-toe*, the average

**Table 6.3.** The average length of a pattern and number of patterns in posJEPN-, negJEPN- and JEPN-Classifier

| Dataset | Length of a pattern | | | Number of patterns | | |
|---|---|---|---|---|---|---|
| | posJEPN | negJEPN | JEPN | posJEPN | negJEPN | JEPN |
| balance | 3.29 | 11.26 | 7.33 | 90.33 | 90.87 | 992.87 |
| breast-wisc | 3.44 | 6.77 | 6.00 | 419.70 | 5422.60 | 34132.50 |
| car | 4.99 | 11.39 | 8.41 | 84.30 | 163.40 | 4957.10 |
| cmc | 4.43 | 9.03 | 7.98 | 586.93 | 6620.87 | 87830.87 |
| heart | 4.56 | 4.56 | 5.13 | 146.70 | 146.70 | 4210.40 |
| irys | 1.97 | 3.45 | 3.06 | 4.93 | 12.53 | 43.07 |
| monks-1 | 4.12 | 7.82 | 6.72 | 31.40 | 57.80 | 1423.80 |
| monks-2 | 4.98 | 7.94 | 6.53 | 109.50 | 79.20 | 3386.40 |
| monks-3 | 3.50 | 6.72 | 6.45 | 29.50 | 40.50 | 906.40 |
| nursery | 6.34 | 15.47 | 11.57 | 413.48 | 1274.88 | 113269.60 |
| tic-tac-toe | 4.56 | 9.15 | 7.82 | 1488.40 | 7043.40 | 164861.20 |
| wine | 3.44 | 5.87 | 5.72 | 277.33 | 3192.53 | 26325.93 |
| yeast | 3.58 | 5.59 | 5.22 | 8.94 | 16.60 | 194.62 |
| zoo | 3.65 | 4.65 | 5.14 | 92.60 | 227.29 | 3503.20 |

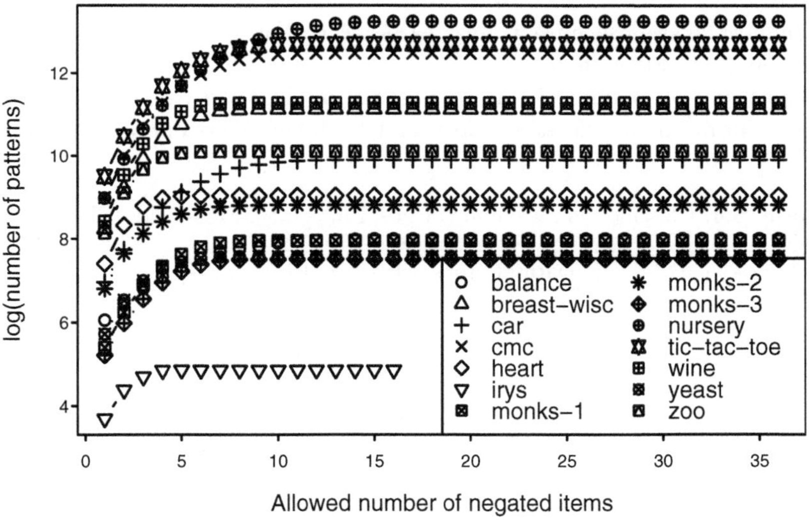


**Fig. 6.1.** Growth in number of patterns when a specified number of negated items is allowed

length increase 2-3 times towards posJEPNs. Meanwhile, the average length of a JEPN almost always places as second, which may be partially explained by the presence of both posJEPNs and negJEPNs in these JEPN spaces.

More surprising is the behavior of the number of patterns. The numbers of posJEPNs and negJEPNs (Table 6.3) suggest that a negative description does not necessarily lead to a significant growth in the number of patterns. In all the cases, the differences in the numbers of patterns do not exceed one order of magnitude and, for 5 datasets, they remain within the same one. This includes *heart*, where the numbers are equal. Moreover, for *monks-2*, the number of posJEPNs is actually higher. At the same time, the highest complexity is consistently observed for JEPN-Classifier, where in almost all the cases, the number of JEPNs is greater by 1-2 orders of magnitude in comparison to other solutions. Consequently, JEPNs are harder to manage when larger datasets are concerned.

The growth in the number of JEPNs is further analyzed by means of $k$-JEPN-Classifiers, where a class $c$ is modeled by the set $k\text{-}JEPN_{\mathcal{DTS}}(D_c) = \{X \in JEPN_{\mathcal{DTS}}(D_c) : |X_n| \leq k\}$, for natural values of $k$. In other words, any $k$-JEPN-Classifier is built out of patterns with negative parts up to the length of $k$. The results (Figure 6.1) show that the number of patterns grows very quickly with the acceptable number of negated items. It means that we can allow only a certain level of negated items and still discover almost all JEPNs. In particular, this fact could be employed in formulating efficient pruning conditions.

The second part of the experiment looks at classification capabilities of the considered methods. In Table 6.4, we present results on average classification accuracy. We observe that posJEPN-Classifier and negJEPN-Classifier achieve the highest average accuracy in 5 cases each. For *heart, irys, monks-3*, both classifiers score identically.

Depending on a dataset, either positive or negative patterns appear to be more accurate in classification. Note that both classifiers operate on patterns of

**Table 6.4.** Classification accuracy of posJEPN-, negJEPN- and JEPN-Classifier

| Dataset | posJEPN | negJEPN | JEPN |
|---|---|---|---|
| balance | 78.08 | 86.24 | 83.52 |
| breast-wisc | 95.28 | 95.14 | 94.85 |
| car | 90.39 | 96.82 | 84.09 |
| cmc | 34.28 | 36.12 | 35.03 |
| heart | 72.59 | 72.59 | 72.96 |
| irys | 85.33 | 85.33 | 85.33 |
| monks-1 | 100.00 | 99.28 | 88.13 |
| monks-2 | 77.37 | 85.19 | 79.20 |
| monks-3 | 97.83 | 97.83 | 96.39 |
| nursery | 98.64 | 99.26 | 96.84 |
| tic-tac-toe | 95.20 | 83.92 | 88.20 |
| wine | 94.38 | 87.64 | 80.34 |
| yeast | 6.87 | 6.60 | 6.60 |
| zoo | 93.07 | 92.08 | 87.13 |

significantly different generality. In fact, posJEPNs are shorter and more general, while negJEPNs are longer and more specific. Specificity means potentially higher discrimination capabilities ([90]), but also lower probability of matching. The latter effect can be balanced by a higher number of negative patterns.

In general, JEPN-Classifier seems to be inferior towards the other two approaches. It wins with posJEPN-Classifier in 3 cases (*balance, cmc, monks-2*), but only for *balance* the difference is significant. It defeats negJEPN-Classifier for *heart* and *tic-tac-toe*. At the same time, it loses evidently to both approaches for 4 datasets (*car, monks-1, wine, zoo*). Although some tests reveal accuracy gains, the classifier's complexity increases rapidly. In almost all the cases, the number of rules is greater by 1-2 orders of magnitude in comparison to other solutions. In our opinion, this excessive number of rules can be responsible for the disturbance of classification decisions.

## 7 JEP Mining by Means of Local Reducts

Originally transactional data can be transformed to the form of a binary decision table (see Section 6.2) and tackled by rough set methods. In particular, local reducts allow us to find JEPs as a side effect of JEPN discovery (Theorem 6.6). A major disadvantage of this approach is that itemspaces are usually large and result in high-dimensional decision tables. At the same time, much of computation effort is wasted on finding undesirable patterns with negative items, that need to be filtered out anyway.

*Example 7.1.* Consider the decision transaction system given in Table 7.1. A comparison of JEP and JEPN spaces given in Table 7.2 shows that an overhead of non-positive JEPNs can be overwhelming.

**Table 7.1.** A sample decision transaction system$\mathcal{DTS} = \{\{T_1, .., T_6\}, \{a, b, c, d, e, f, g, h\}, \{c_0, c_1\}\}$

| | | |
|---|---|---|
| $T_1$ | $adh$ | $c_0$ |
| $T_2$ | $afg$ | $c_0$ |
| $T_3$ | $ceg$ | $c_0$ |
| $T_4$ | $ce$ | $c_1$ |
| $T_5$ | $beh$ | $c_1$ |
| $T_6$ | $bfg$ | $c_1$ |

**Table 7.2.** JEP an JEPN spaces for $\mathcal{DTS}$

| Space | Set interval |
|---|---|
| $JEP(D_1, D_0)$ | $[\{eg, d, cg, a\}, \{adh, afg, ceg\}]$ |
| $JEP(D_0, D_1)$ | $[\{eh, b\}, \{ce, beh, bfg\}]$ |
| $JEPN(D_1, D_0)$ | $[\{\overline{f}g, \overline{e}h, \overline{e}\overline{g}, \overline{e}\overline{f}, eg, d, cg, \overline{b}h, \overline{b}g, \overline{b}f, \overline{b}\overline{e}, \overline{b}\overline{c}, a\},$ $\{a\overline{b}\overline{c}d\overline{e}\overline{f}\overline{g}h, a\overline{b}\overline{c}d\overline{e}fg\overline{h}, \overline{a}\overline{b}c\overline{d}eg\overline{h}\}]$ |
| $JEPN(D_0, D_1)$ | $[\{\overline{g}h, eh, e\overline{g}, \overline{d}h, \overline{d}\overline{g}, \overline{c}e, \overline{c}\overline{d}\overline{f}, c\overline{g}, b, \overline{a}h, \overline{a}\overline{g}, \overline{a}f, \overline{a}\overline{e}, \overline{a}\overline{c}\},$ $\{\overline{a}\overline{b}c\overline{d}e\overline{f}\overline{g}\overline{h}, \overline{a}b\overline{c}\overline{d}e\overline{f}\overline{g}h, \overline{a}b\overline{c}\overline{d}\overline{e}fg\overline{h}\}]$ |

In this chapter we discuss how to lower the dimensionality of decision tables by applying appropriate transformations. Our approaches are based on the fact that transactional data are usually sparse. Indeed, average transactions of real-life datasets contain just a few items as compared to the respective large itemspaces. Therefore, discovery methods may benefit from more concise data representations.

Hereinafter, we assume that our input data are represented by a decision transaction system $\mathcal{DTS} = (\mathcal{D}, \mathcal{I}, \mathcal{I}_d)$, where $\mathcal{D} = (T_1, .., T_n)$, $\mathcal{I} = \{I_1, .., I_m\}$, $\mathcal{I}_d = \{c_1, .., c_p\}$, $K = \{1, .., n\}$.

## 7.1 Global Condensation

In a binary decision table based on a decision transaction system, each binary attribute refers to a single item. One possible modification of this approach is to use multi-valued attributes to encode groups of items. The itemspace can be partitioned into blocks and each block is assigned a new attribute. We refer to this transformation as *global condensation.*

Local reducts in a binary decision table correspond to JEPNs. For a given target class, all the JEPNs constitute a convex space that also contains the JEPs. After performing global condensation, local reducts in the resulting decision table may no longer map to all the possible JEPNs. This fact is advantageous, since it diminishes the overhead of unnecessarily generated patterns. The method remains correct as long as the complete set of the positive JEPNs (JEPs) can be discovered.

**Condensed Decision Table.** In order to ensure that global condensation leads to a complete set of JEPs, we introduce a special type of a partition of an itemspace. Each transaction and each block are required to have at most one item in common.

**Definition 7.1.** *A partition* $\{p_1, .., p_r\}$ *of* $\mathcal{I}$ *is called* proper *iff* $\forall_{T \in \mathcal{D}} \forall_{j \in \{1,..,r\}} |T \cap p_j| <= 1$.

If a partition is proper, for each block $p_i$, we have at most $|p_i|$ different intersections of $p_i$ with transactions of $\mathcal{D}$, where $i \in \{1, .., r\}$. Each of these intersections refers to at most one item and can be mapped to a distinct value of a single multi-valued attribute. We express the transformed dataset by means of a decision table.

**Definition 7.2.** *For*

$P = \{p_1, .., p_r\}$, *a proper partition of* $\mathcal{I}$,
$F = \{f_1, .., f_r\}$, *where* $f_j : 2^{p_j} \mapsto \mathbb{N}$ *and* $f_j$ *is a bijection for each* $j \in \{1, .., r\}$

*a* condensed decision table based on $\mathcal{DTS}$, $P$ and $F$ *is a decision table* $\mathcal{CDT}_{\mathcal{DTS},P,F} = (\mathcal{U}, \mathcal{C}, d)$ *such that*

$$\mathcal{U} = \{u_1, .., u_n\},\ \mathcal{C} = \{a_1, .., a_r\},\ V_d = \{d_1, .., d_p\}$$
$$a_j(u_i) = f_j(T_i \cup p_j), \forall_{i \in 1..n, j \in 1..r}$$
$$d(u_i) = T_i \cap \mathcal{I}_d, \forall_{i \in 1..n}$$

**Table 7.3.** The binary and condensed decision table based on $\mathcal{DTS}$ from Table 7.1 and the proper partition $\{\{a,b,c\},\{d,e,f\},\{g,h\}\}$

| | $a$ | $b$ | $c$ | $d$ | $e$ | $f$ | $g$ | $h$ | $d$ |
|---|---|---|---|---|---|---|---|---|---|
| $u_1$ | 1 | 0 | 0 | 1 | 0 | 0 | 0 | 1 | 0 |
| $u_2$ | 1 | 0 | 0 | 0 | 0 | 1 | 1 | 0 | 0 |
| $u_3$ | 0 | 0 | 1 | 0 | 1 | 0 | 1 | 0 | 0 |
| $u_4$ | 0 | 0 | 1 | 0 | 1 | 0 | 0 | 0 | 1 |
| $u_5$ | 0 | 1 | 0 | 0 | 1 | 0 | 0 | 1 | 1 |
| $u_6$ | 0 | 1 | 0 | 0 | 0 | 1 | 1 | 0 | 1 |

| | $a_1$ | $a_2$ | $a_3$ | $d$ |
|---|---|---|---|---|
| $u_1$ | 0 | 0 | 0 | 0 |
| $u_2$ | 0 | 1 | 1 | 0 |
| $u_3$ | 1 | 2 | 1 | 0 |
| $u_4$ | 1 | 2 | 2 | 1 |
| $u_5$ | 2 | 2 | 0 | 1 |
| $u_6$ | 2 | 1 | 1 | 1 |

The choice of the function $F$ does not affect the structure of the decision table and is a matter of convention.

For the sake of convenience, we introduce a new notation to refer to patterns generated by an object and attribute set in a condensed decision table.

**Definition 7.3.** *For $\mathcal{CDT}_{\mathcal{DTS}}$, $u \in \mathcal{U}$, $B = \{a_k\}_{k \in K}$ and $K \subseteq \{1,..,m\}$, a* condensed pattern based on the object $u$ and attribute set $B$ *is an itemset*
$$condPatt(u,B) = \bigcup_{k \in K} f_k^{-1}(a_k(u)), \text{ where } u \in \mathcal{U},\ B = \{a_k\}_{k \in K}.$$

*Example 7.2.* In Table 7.3 we present a transformation from a sample transactional dataset, through the respective binary table, to the condensed table that is generated for the proper partition $\{\{a,b,c\},\{d,e,f\},\{g,h\}\}$. Each attribute of the condensed table refers to a block of a partition and each attribute value to an at most one item. It holds that $condPatt(u_4,a_3) = \emptyset$, $condPatt(u_5,a_3) = h$ and $condPatt(u_6,a_3) = g$. Note that the partition $\{\{a,b,c\},\{d,e,f,g,h\}\}$ is not proper, since $|T_1 \cap \{d,e,f,g,h\}| = 2 > 1$.

Let us consider a condensed decision table $\mathcal{CDT}_{\mathcal{DTS},P,F} = (\mathcal{U},\mathcal{C},d)$. The following theorem demonstrates that an object from the positive region of the condensed decision table can be used to generate a JEP, when one applies an attribute set whose each element maps to a non-empty itemset.

**Theorem 7.1.** $\forall_{R \subseteq \mathcal{C}} \forall_{u \in \mathcal{U}} (\forall_{j \in \{1,..,r\}} a_j \in R \Longrightarrow a_j(u) \neq f_j(\emptyset)) \Longrightarrow$
$(u \in POS(R,d) \cap U_{d^{-1}(u)} \Longleftrightarrow condPatt(u,R) \in JEP(C'_{d^{-1}(u)}, C_{d^{-1}(u)}))$

*Proof.* Let $R = \{a_k\}_{k \in K \subseteq \{1,..,r\}} : (\forall_{j \in \{1,..,r\}} a_j \in R \Longrightarrow a_j(u) \neq f_j(\emptyset))$.

Consider $\forall_{u \in \mathcal{U}} u \in POS(R,d) \cap U_c \Longleftrightarrow supp_{C'_c}(condPatt(u,R)) = 0$ (1).

Let $u_g \in POS(R,d) \cap U_c$ for $g \in \{1,..,n\}$ and $c = d^{-1}(u)$. We have $u_g \in R_*(U_c) \Longleftrightarrow R(u_g) \subseteq U_c \Longleftrightarrow \{v \in \mathcal{U} : \forall_{k \in K} a_k(u_g) = a_k(v)\} \subseteq U_c \Longleftrightarrow \forall_{v \in \mathcal{U}} \exists_{k \in K} a_k(u_g) \neq a_k(v) \Longleftrightarrow \forall_{u_h \in \mathcal{U} - U_c, h \in \{1,..,n\}} \exists_{k \in K} f_k(T_g \cap p_k) \neq f_k(T_h \cap p_k) \Longleftrightarrow$
$\forall_{h \in \{1,..,n\}} \exists_{k \in K} f_k(T_h \cap p_k) \in condPatt(u_g,R) \wedge f_k(T_h \cap p_k) \notin condPatt(u_g,R)$
$\Longleftrightarrow \forall_{T \in C'_c} condPatt(u_g,R) \not\subseteq T \Longleftrightarrow supp_{C'_c}(condPatt(u_g,R)) = 0.$

Examine the first implication:
$\forall_{u \in \mathcal{U}} u \in POS(R,d) \cap U_c \Longrightarrow condPatt(u,R) \in JEP(C_c', C_c)$.

Let $u_g \in POS(R,d) \cap U_c$ for $g \in \{1,..,n\}$ and $c = d^{-1}(u)$. Note that $condPatt(u_g, R) \subseteq T_g$ and $T_g \in C_c$, thus, $supp_{C_c}(condPatt(u_g, R)) \neq 0$. From (1), we have $condPatt(u_g, R) \in JEP(C_c', C_c)$.

Now, consider the second implication:
$\forall_{u \in \mathcal{U}} condPatt(u,R) \in JEP(C_c', C_c) \Longrightarrow u \in POS(R,d)$.

According to (1), we have $condPatt(u,R) \in JEP(D_c', D_c) \Longrightarrow supp_{C_c'}(condPatt(u,R)) = 0 \Longleftrightarrow u \in POS(R,d) \cap U_c$, where $c = d^{-1}(u)$. □

Furthermore, the following theorem states that if such an attribute set is a local reduct, it generates a minimal JEP.

**Theorem 7.2.** $\forall_{R \subseteq \mathcal{C}} \forall_{u \in POS(\mathcal{C},d)} (\forall_{j \in \{1,..,r\}} a_j \in R \Longrightarrow a_j(u) \neq f_j(\emptyset)) \Longrightarrow (R \in REDLOC(u,d) \Longleftrightarrow condPatt(u,R) \in \mathcal{L}_{d^{-1}(u)})$

*Proof.* Let $u \in POS(u, \mathcal{C})$, $B \subseteq \mathcal{C}$ and $k = d^{-1}(u)$, $R = \{a_k\}_{k \in K \subseteq \{1,..,r\}}$ : $(\forall_{j \in \{1,..,r\}} a_j \in R \Longrightarrow a_j(u) \neq f_j(\emptyset))$.

First consider $B(u) \subseteq U_c \Longleftrightarrow condPatt(u,B) \in JEP(C_c', C_c)$ (1).

We have $B(u) \subseteq U_c \Longleftrightarrow u \in B_*(U_c) \Longleftrightarrow u \in POS(B,d) \cap U_c$. But, according to Theorem 7.4, we have $u \in POS(B,d) \cap U_c \Longleftrightarrow condPatt(u,B) \in JEP(C_c', C_c)$.

Consider $R \in REDLOC(u,d) \Longrightarrow condPatt(u,R) \in \mathcal{L}_c$.
Let $R$ meet the left-hand side. According to Lemma 2.2, the following holds: $R \in REDLOC(u,d) \Longleftrightarrow R$ is minimal in $\{B \subseteq \mathcal{C} : B(u) \subseteq U_c\}$. Consider $S \subset R$. It means that $S(u) \not\subseteq U_c$, and, from (1), we obtain $condPatt(u,S) \notin JEP(C_c', C_c)$. To sum up, from (1), we have $condPatt(u,R) \in JEP(C_c', C_c)$ and for any $J \subset condPatt(u,R)$ we have $J \notin JEP(C_c', C_c)$. Thus, the pattern $condPatt(u,R)$ is minimal in $JEP(C_c', C_c)$ and, according to Lemma 2.3, we have $condPatt(u,R) \in \mathcal{L}_c$.

Consider $R \in REDLOC(u,d) \Longleftarrow condPatt(u,R) \in \mathcal{L}_c$.
Let $condPatt(u,R) \in \mathcal{L}_c$. According to Lemma 2.3, we have $condPatt(u,R) \in \mathcal{L}_c \Longleftrightarrow condPatt(u,R)$ is minimal in $JEP(C_c', C_c)$. Consider $S \subset R$. It means that $condPatt(u,S) \subset condPatt(u,R) \Longrightarrow condPatt(u,S) \notin JEP(C_c', C_c)$ and, according to (1), we obtain $S(u) \not\subseteq U_c$. To sum up, according to (1) we have $R \in \{B \subseteq \mathcal{C} : B(u) \subseteq U_c\}$ and for any $S \subset R$ we have $S(u) \not\subseteq U_c$. Thus, $R$ is minimal in $\{B \subseteq \mathcal{C} : B(u) \subseteq U_c\}$ and, according to Lemma 2.2, we have $R \in REDLOC(u,d)$. □

**Proper Partition Finding as Graph Coloring.** The choice of a proper partition is critical for construction of a condensed decision table. This problem can be expressed in the language of the graph theory. We construct a graph in which each vertex corresponds to an item from an itemspace. Two vertices are connected with an edge only if there is at least one transaction that contains both corresponding items. From the definition, these two items cannot belong to the same block of a proper partition, which is substantial for JEP discovery. Otherwise, one attribute value would represent both items and patterns that contain only one of these items would not be considered in the further mining.

**Definition 7.4.** *An* item-conflict graph based on $\mathcal{DTS}$ *is an undirected graph* $ICG_{\mathcal{DTS}} = (V, E)$ *such that:*
$\forall_{x,y\in\{1,..,m\}}\{v_x, v_y\} \in E \iff \exists_{T\in\mathcal{D}} i_x, i_y \in T$, *where* $V = \{v_1, .., v_m\}$.

Let us consider an item-conflict graph $ICG_{\mathcal{DTS}} = (V, E)$. In fact, every proper partition of the itemspace $\mathcal{I}$ corresponds to a coloring of this graph. For consistency, we represent colorings as partitions of the set of vertices $V$.

**Theorem 7.3.** *For a partition* $\{w_1, .., w_r\}$ *of* $V$ *and partition* $\{p_1, .., p_r\}$ *of* $\mathcal{I}$ *such that* $\forall_{j\in\{1,..,m\}}\forall_{k\in\{1,..,r\}} v_j \in w_k \iff i_j \in p_k$, *we have:*
$\{w_1, .., w_r\}$ *is a coloring of* $ICG_{\mathcal{DTS}} \iff$
$\{p_1, .., p_r\}$ *is a proper partition for* $\mathcal{DTS}$

*Proof.* Let $\{w_1, .., w_r\}$ be a partition of $V$ and $\{p_1, .., p_r\}$ be a partition of $\mathcal{I}$ such that:
$\forall_{j\in\{1,..,m\}}\forall_{k\in\{1,..,r\}} v_j \in w_k \iff i_j \in p_k \quad (1)$.

From the definition of an item-conflict graph and (1), we have:
$\{w_1, .., w_r\}$ is a coloring of $ICG_{\mathcal{DTS}} \iff \forall_{k=1,..,r}\forall_{v_x,v_y\in w_k}\{v_x, v_y\} \notin E \iff$ $\forall_{k=1,..,r}\forall_{i_x,i_y\in p_k}\neg\exists_{T\in\mathcal{D}} i_x, i_y \in T \iff \{p_1, .., p_r\}$ is a proper partition for $\mathcal{DTS}$. □

*Example 7.3.* The item-conflict graph $ICG_{\mathcal{DTS}}$ based on $\mathcal{DTS}$ from Table 7.3 is presented in Figure 7.1. Vertices connected with an edge cannot have the same color. $\{\{a, b, c\}, \{d, e, f\}, \{g, h\}\}$ is one possible coloring. Note that this coloring also determines a proper partition in which colors correspond to blocks. Each of the transactions $T_1, .., T_6$ contains at most one of the items of each block.

**Discovery Method.** We present how global condensation can be employed in identification of JEP spaces for the decision transaction system $\mathcal{DTS}$. The first stage of our method is to find a proper partition of the itemspace and use it to

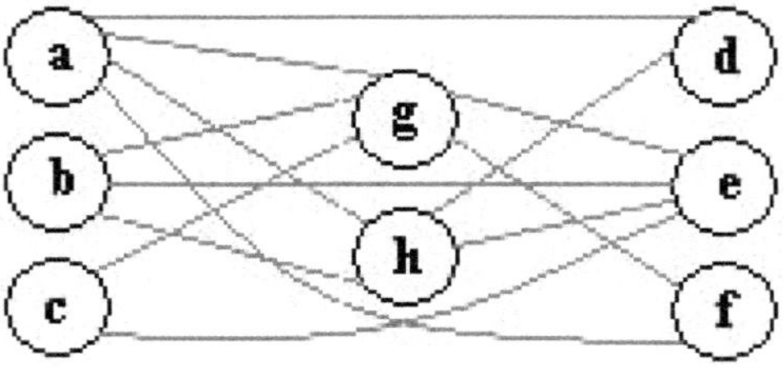


**Fig. 7.1.** The item-conflict graph based on the decision transaction system $\mathcal{DTS}$ from Table 7.3

**Algorithm 7.** JEPGlobalCond($\mathcal{DTS}$)

$\mathcal{L}_c = \emptyset$ for each $c \in I_d$
Construct an item-conflict graph $ICG_{\mathcal{DTS}}$
Find a minimal coloring $C$ in $ICG_{\mathcal{DTS}}$
Construct the condensed decision table $\mathcal{CDT}_{\mathcal{DTS},P,F} = \{\mathcal{U}, \mathcal{C}, d\}$, where $P$ corresponds to the coloring $C$ and $F$ is any fixed mapping
**for** $(k = 1; k <= |\mathcal{U}|; k++)$ **do**
  Compute $REDLOC(u_k, d)$ in $\mathcal{CDT}_{\mathcal{DTS}}$
  $\mathcal{L}_c = \mathcal{L}_c \cup \{condPatt_{\mathcal{DTS}}(u_i, R) : R \in REDLOC(u_i, d)\}$, $c = d(u_i)$
**end for**
**return** $\mathcal{L}_c$

construct the respective condensed decision table. It is not obvious which partition is optimal for a given dataset and reduct finding algorithm. Since dimensionality is usually the most significant factor, we choose a criterium stating that fewer blocks of a partition lead to better performance. Following Theorem 7.3, one may consider an item-conflict graph $ICG_{\mathcal{DTS}}$ and reduce this optimization problem to graph coloring. Furthermore, since this is a preprocessing stage and suboptimal solutions are acceptable, widely known heuristics, like LF, SLR, RLF, SR ([110]), can be applied. The resulting partition allows us to transform $\mathcal{DTS}$ to a condensed decision table $\mathcal{CDT}_{\mathcal{DTS}}$.

The most time-consuming step is discovery of minimal patterns for the condensed decision table $\mathcal{CDT}_{\mathcal{DTS}} = \{\mathcal{U}, \mathcal{C}, d\}$. A reduct finding method is used to identify the set $REDLOC(u, d)$ for each object $u \in \mathcal{U}$. Every local reduct $B \subseteq \mathcal{C}$ refers to the minimal pattern $condPatt(u, B)$. All the patterns found for the objects from a particular class constitute the left bound of a respective JEP space.

### 7.2 Local Projection

Global condensation is performed once for a whole decision transaction system and generates an insignificant additional overhead while attribute values are translated back into items. Unfortunately, this method remains very sensitive to the distribution of items across transactions and, in a general case, finding a proper partition that substantially lowers the overall dimensionality may turn out hard.

This problem can be alleviated by the following observation. Actual computation of local reducts is performed separately for each object and it involves only a subset of the universum. Therefore, instead of employing global condensation, one may want to focus only on transactions applicable to a particular reduct finding process. Conceivably, fewer transactions may lead to better partitions and overall efficiency, even though an additional pre-processing overhead is higher. Following the previous convention, we call this procedure *local condensation* ([140]).

Here, we present a different idea, *local projection*, that can potentially achieve much higher dimensionality reduction. It transforms input data with respect to each individual transaction. However, instead of grouping items into blocks, one takes into account only those from the considered transaction.

**Locally Projected Decision Table.** Let us consider a single transaction from $\mathcal{D}$. We construct a decision table that has binary attributes for all items from this transaction.

**Definition 7.5.** *For a transaction* $T_i \in \mathcal{D}$, *where* $i = 1,..,|\mathcal{D}|$, *a* locally projected decision table based on $\mathcal{DTS}$ *is a binary decision table* $\mathcal{LPDT}_{\mathcal{DTS},T_i} = \mathcal{BDT}_{\mathcal{DTS}_i}$, *where* $\mathcal{DTS}_i = (\mathcal{D}_i, T_i, \mathcal{I}_d)$ *and* $\mathcal{D}_i = (T_k \cap T_i)_{k \in K}$.

Hardness of an input decision system $\mathcal{DTS}$ can be characterized by the dimensionality of corresponding locally projected decision tables for all discernable transactions.

**Definition 7.6.** *The* average and maximum dimensionality of locally projected decision tables based on $\mathcal{DTS}$ *are defined as follows*

$$avgDim(\mathcal{DTS}) = |\{|T| : T \in \mathcal{R}_c \wedge c \in I_d\}| / \textstyle\sum_{c \in I_d} |\mathcal{R}_c|$$
$$maxDim(\mathcal{DTS}) = max_{T \in \mathcal{R}_c \wedge c \in I_d} |T|, \text{ respectively.}$$

Note that, when all transactions are discernable, these parameters refer to the average (maximum) transaction length in $\mathcal{DTS}$.

Again, we introduce a concise notation to represent patterns generated by an object and attribute set in a locally projected decision table.

**Definition 7.7.** *For*

$$T_i = \{I_k\}_{k \in M'},$$
$$u \in \mathcal{U},$$
$$B \subseteq \mathcal{C}_i = \{a_k\}_{k \in M'}$$

*a* locally projected pattern based on the object and attribute set *is an itemset* $itemPatt_{\mathcal{DTS},T_i}(u, B) = \{I_k \in T_i : a_k \in B \wedge a_k(u) = 1 \wedge k \in M'\}$, *where* $M' \subseteq \{1,..,m\}$ *and* $\mathcal{LPDT}_{\mathcal{DTS},T_i} = (\mathcal{U}, \mathcal{C}_i, d)$.

Note that $|itemPatt_{\mathcal{DTS},T_i}(u_i, B)| = |B|$. Whenever a decision transaction system is known from the context, the respective subscript is omitted.

*Example 7.4.* In Table 7.4 we present the locally projected table based on the decision transaction system $\mathcal{DTS}$ (Table 7.3) and the transactions: $T_1$,$T_2$, $T_3$, respectively. In particular, the attribute set $\{e, g\}$ generates the following patterns: $itemPatt_{T_1}(u_1, \{e, g\}) = eg$, $itemPatt_{T_1}(u_5, \{e, g\}) = e$. For this dataset, we may calculate $avgDim = 20/6 = 3.33$ and $maxDim = 5$, which gives 58.4% and 37.5% of a dimensionality gain comparing to a binary decision table.

The following theorem states that the complete JEP space for a decision transaction system and given class can be obtained by finding the locally projected table for each discernable transaction and generating patterns for the object corresponding to this transaction and all attribute sets from this table.

**Table 7.4.** The locally projected tables: $\mathcal{LPDT}_{\mathcal{DTS},T_1}$, $\mathcal{LPDT}_{\mathcal{DTS},T_2}$, $\mathcal{LPDT}_{\mathcal{DTS},T_3}$

| | $a$ | $d$ | $e$ | $g$ | $h$ | $d$ |
|---|---|---|---|---|---|---|
| $u_1$ | 1 | 1 | 1 | 1 | 1 | 0 |
| $u_2$ | 0 | 1 | 0 | 0 | 1 | 0 |
| $u_3$ | 0 | 0 | 1 | 1 | 0 | 0 |
| $u_4$ | 0 | 0 | 0 | 0 | 0 | 1 |
| $u_5$ | 0 | 1 | 1 | 0 | 0 | 1 |
| $u_6$ | 0 | 0 | 0 | 1 | 1 | 1 |

| | $b$ | $d$ | $h$ | $d$ |
|---|---|---|---|---|
| $u_1$ | 0 | 1 | 1 | 0 |
| $u_2$ | 1 | 1 | 1 | 0 |
| $u_3$ | 1 | 0 | 0 | 0 |
| $u_4$ | 1 | 0 | 0 | 1 |
| $u_5$ | 0 | 1 | 0 | 1 |
| $u_6$ | 1 | 0 | 1 | 1 |

| | $b$ | $c$ | $e$ | $g$ | $d$ |
|---|---|---|---|---|---|
| $u_1$ | 0 | 0 | 1 | 1 | 0 |
| $u_2$ | 1 | 0 | 0 | 0 | 0 |
| $u_3$ | 1 | 1 | 1 | 1 | 0 |
| $u_4$ | 1 | 0 | 0 | 0 | 1 |
| $u_5$ | 0 | 0 | 1 | 0 | 1 |
| $u_6$ | 1 | 0 | 0 | 1 | 1 |

**Theorem 7.4.** $\forall_{c\in\mathcal{I}_d}\{itemPatt_{\mathcal{DTS},T_i}(u_i,R) : i \in K \wedge \mathcal{LPDT}_{\mathcal{DTS},T_i} = (\mathcal{U},\mathcal{C}_i,d) \wedge u_i \in POS(\mathcal{C}_i,d)\cap U_c \wedge R \subseteq \mathcal{C}_i\} = JEP(C'_c,C_c)$.

*Proof.* Analogous to the proof of Theorem 7.4. □

The respective left bound of a JEP space can be found by applying local reducts for a given object rather than arbitrary attribute sets.

**Theorem 7.5.** $\forall_{c\in\mathcal{I}_d}\{itemPatt_{\mathcal{DTS},T_i}(u_i,R) : i \in K \wedge \mathcal{LPDT}_{\mathcal{DTS},T_i} = (\mathcal{U},\mathcal{C}_i,d) \wedge u_i \in POS(\mathcal{C}_i,d)\cap U_c \wedge R \in REDLOC(u_i,d)\} = \mathcal{L}_c$.

*Proof.* Let $c \in \mathcal{I}_d$.

First consider $B(u_i) \subseteq U_c \iff itemPatt_{T_i}(u_i,B) \in JEP(C'_c,C_c)$ (1), for $u_i \in \mathcal{U}$, $B \subseteq \mathcal{C}_i$, $\mathcal{LPDT}_{\mathcal{DTS},T_i} = (\mathcal{U},\mathcal{C}_i,d)$. We have $B(u_i) \subseteq U_c \iff u_i \in B_*(U_c) \iff u_i \in POS(B,d)\cap U_c$. But, according to Theorem 7.4, we have $u_i \in POS(B,d)\cap U_c \iff itemPatt_{T_i}(u_i,B) \in JEP(C'_c,C_c)$.

Consider $\{itemPatt_{\mathcal{DTS},T_i}(u_i,R) : i \in K \wedge \mathcal{LPDT}_{\mathcal{DTS},T_i} = (\mathcal{U},\mathcal{C}_i,d) \wedge u_i \in POS(\mathcal{C}_i,d)\cap U_c \wedge R \in REDLOC(u_i,d)\} \subseteq \mathcal{L}_c$

Let $i\in K$, $\mathcal{LPDT}_{\mathcal{DTS},T_i} = (\mathcal{U},\mathcal{C}_i,d)$, $u_i \in POS(\mathcal{C},d)\cap U_c$ and $R\in REDLOC(u_i,d)$. According to Lemma 2.2, we have $R \in REDLOC(u_i,d) \iff R$ is minimal in $\{B \subseteq \mathcal{C} : B(u_i) \subseteq U_c\}$. Consider $S \subset R$. It means that $S(u_i) \nsubseteq U_c$, and, according to (1), we obtain $itemPatt_{T_i}(u_i,S) \notin JEP(C'_c,C_c)$. To sum up, according to (1), we have $itemPatt_{T_i}(u_i,R) \in JEP(C'_c,C_c)$ and for any $J \subset itemPatt_{T_i}(u_i,R)$ we have $J \notin JEP(C'_c,C_c)$. Thus, the itemset $itemPatt_{T_i}(u_i,R)$ is minimal in $JEP(C'_c,C_c)$ and, according to Lemma 2.3, we have $itemPatt_{T_i}(u_i,R) \in \mathcal{L}_c$.

Let us move to the opposite inclusion. Let $J \in \mathcal{L}_c$. According to Lemma 2.3, we have $J \in \mathcal{L}_c \iff J$ is minimal in $JEP(C'_c,C_c)$. According to Theorem 7.4, $\exists_{i\in K}\exists_{R\subseteq\mathcal{C}_i} u_i \in POS(u_i,d)\cap U_c \iff J = itemPatt_{T_i}(u_i,R)$. Consider $S \subset R$. We have $itemPatt_{T_i}(u_i,S) \subset itemPatt_{T_i}(u_i,R) \implies itemPatt_{T_i}(u_i,S) \notin JEP(C'_c,C_c)$ and, according to (1), we obtain $S(u_i) \nsubseteq U_c$. To sum up, according to (1) we have $R \in \{B \subseteq \mathcal{C} : B(u_i) \subseteq U_c\}$ and for any $S \subset R$ we have $S(u_i) \nsubseteq U_c$. Thus, $R$ is minimal in $\{B \subseteq \mathcal{C} : B(u_i) \subseteq U_c\}$ and, according to Lemma 2.2, we have $R \in REDLOC(u_i,d)$. □

*Example 7.5.* In order to illustrate Theorem 7.5, let us consider the class 0. For $\mathcal{LPDT}_{\mathcal{DTS},T_1}$ we have $REDLOC(u_1,d) = \{\{a\},\{d,g\},\{d,h\},\{e,g\},\{e,h\}\}$. Each of these reducts refer to certain minimal patterns from $\mathcal{L}_0$. On the other hand, the pattern $c$ is a minimal JEP, thus, we are able to find a respective set of local reducts and a repsective locally projected table, namely: $\{c\} = REDLOC(u_3,d)$ for $\mathcal{LPDT}_{\mathcal{DTS},T_3}$.

**Discovery method.** Minimal jumping emerging patterns can be identified by local reduct computation in locally condensed tables based on consecutive transactions of a dataset. The actual procedure is straightforward and fully follows Theorem 7.2.

Let us consider a decision transaction system $\mathcal{DTS} = (\mathcal{D},\mathcal{I},\mathcal{I}_d)$. For each transaction $T_k \in \mathcal{D}$, we build the locally projected decision table $\mathcal{LPDT}_{\mathcal{DTS},T_k} = (\mathcal{U},\mathcal{C}_k,d)$, where $k = 1,..,|\mathcal{D}|$. Then, local reducts for the object $u_k$, that refers to $T_k$, are computed. Finally, each local reduct $R \in REDLOC(u_k,d)$ is mapped to the respective pattern $itemPatt_{T_k}(u,R)$ and added to the minimal JEP collection for the class $(T_k \cap \mathcal{I}_d)$. Once all transactions of $\mathcal{DTS}$ are processed, we obtain the complete resulting collections $\{\mathcal{L}_c\}_{c \in I_d}$.

Local projection can significantly reduce problem dimensionality, especially, for sparse transaction databases. Although reduct computation remains the pivotal and hardest task, additional processing of polynomial complexity may start to have noticeable negative impact on overall efficiency of the algorithm. Further, we comment on optimizations that can potentially alleviate this effect.

First of all, it is not necessary to build tables $\mathcal{LPDT}_{\mathcal{DTS},T_i}$, for each $T_i \in \mathcal{D}$, explicitly. It is sufficient to construct respective discernibility sets. Note that, when local reducts are being computed for a certain object, one takes into account only objects from other classes from the class of this object. Consequently, for a given locally projected decision table, only the elements of the discernibility set that correspond to this object are meaningful for reduct finding.

Secondly, transactions that are not maximal JEPs can be eliminated upfront, since they cannot introduce any new JEPs to the solution.

Thirdly, a significant improvement can be achieved by grouping transactions by their classes. This allows us to iterate over only these objects that are necessary for particular processing.

Fourthly, discernibility sets for objects in one class share information for common attributes. Let us consider $\mathcal{BDT}_{\mathcal{DTS}} = (\mathcal{U},\mathcal{C},d)$. Note that we have $\{X \cap \{a_i\} : DC_{\mathcal{DTS},T_1}\} = \{X \cap \{a_i\} : DC_{\mathcal{DTS},T_2}\}$, for each $c \in \mathcal{I}_d$, $T_1,T_2 \in D_c$, $T_1 \cap T_2 \neq \emptyset$ and $I_i \in T_1 \cap T_2$. This per-attribute information may be precomputed for each class and each attribute that belongs to some transaction in this class or computed lazily and stored when successive transactions are considered. As a result, one obtains a cache of the form $\{(c,I_i,\{u : u \in \mathcal{U} - U_c \wedge a(i) = 0\}) : c \in \mathcal{I}_d \wedge i \in \{1,..,m\}\}$, which is later used to build discernibility sets without examining the database every time. After construction from the cache, a set has to be reduced separately, because attribute spaces of respective locally projected tables vary.

**Algorithm 8.** JEPLocalProj($\mathcal{DTS}$)

$\mathcal{L}_c = \emptyset$ for each $c \in I_d$
**for** $(k = 1; k <= |\mathcal{D}|; k++)$ **do**
  Construct the locally projected decision table $\mathcal{LPDT}_{\mathcal{DTS},T_k}$
  Compute $REDLOC(u_k, d)$ in $\mathcal{LPDT}_{\mathcal{DTS},T_k}$
  $\mathcal{L}_c = \mathcal{L}_c \cup \{itemPatt_{\mathcal{DTS},T_k}(u_k, R) : R \in REDLOC(u_k, d)\}$, $c = T_k \cup \mathcal{I}_d$
**end for**
**return** $\mathcal{L}_c$

Last but not least, for sparse datasets with large number of transactions, elements of the discernibility collection of $LPDC_{\mathcal{DTS},T_i} = (\mathcal{U}, \mathcal{C}_i, d)$, for $i \in K$, are often close in size to $|C_i|$. Therefore, it may be more efficient to store their complements. In other words, for each object, one may generate an element consisting of attributes with values equal to 1.

### 7.3 Experimental Evaluation

In order to evaluate global condensation and local projection, we compared their efficiency on transaction databases originating from decision tables. Furthermore, local projection was tested against other widely-known methods for randomly generated sparse data.

**Table 7.5.** Efficiency comparison for originally relational datasets between table condensation and local projection with RedPrime and RedApriori, with the Array-based implementation of an attribute set

| Dataset | Global Condensation | | | | | Local Projection | | |
|---|---|---|---|---|---|---|---|---|
| | Dim | Other | Part | RedPr | RedAp | Dim | RedPr | RedAp |
| balance | 4.00 | 0 | 15 | 117 | 101 | 4.00 | 101 | 93 |
| breast-wisc | 9.00 | 0 | 8 | 453 | 398 | 9.00 | 148 | 140 |
| car | 6.00 | 0 | 8 | 741 | 787 | 6.00 | 780 | 819 |
| cmc | 9.00 | 0 | 24 | 3432 | 3861 | 9.00 | 1061 | 1185 |
| dna | 20.00 | 0 | 31 | 3384810 | 86018 | 20.00 | 3200886 | 55630 |
| heart | 13.00 | 0 | 8 | 211 | 304 | 13.00 | 70 | 124 |
| irys | 4.00 | 0 | 0 | 15 | 15 | 4.00 | 0 | 0 |
| krkopt | 6.00 | 0 | 344 | 266979 | 268554 | 6.00 | 333279 | 337381 |
| lung | 56.00 | 13860 | 32 | 6409564 | 16692 | 55.00 | 4905966 | 12480 |
| lymn | 18.00 | 0 | 7 | 3159 | 1232 | 18.00 | 2020 | 1030 |
| monks-1 | 6.00 | 0 | 15 | 163 | 156 | 6.00 | 101 | 109 |
| monks-2 | 6.00 | 0 | 0 | 202 | 234 | 6.00 | 109 | 148 |
| monks-3 | 6.00 | 0 | 0 | 140 | 140 | 6.00 | 93 | 94 |
| tic-tac-toe | 9.00 | 0 | 23 | 2254 | 2777 | 9.00 | 1896 | 2426 |
| vehicle | 20.00 | 7533 | 31 | 28969 | 23189 | 18.00 | 11567 | 10600 |
| wine | 13.00 | 0 | 8 | 405 | 280 | 13.00 | 226 | 148 |
| yeast | 8.00 | 0 | 23 | 990 | 1154 | 8.00 | 109 | 109 |
| zoo | 16.00 | 0 | 8 | 164 | 171 | 16.00 | 78 | 101 |

**Table 7.6.** A comparison of computation time for sparse datasets: local projection with RedPrime, RedApriori, with the Array-based implementation of an attribute set, Border and CP-Tree

| Dataset | Local Projection | | Border | CP-Tree |
|---|---|---|---|---|
| | RedPr | RedAp | | |
| D2.0I0.02T3P2000L1C2 | 109 | 94 | 141 | 202 |
| D2.0I0.04T5P2000L1C3 | 1123 | 826 | 3291 | 2933 |
| D2.0I0.06T6P2000L1C3 | 4493 | 1965 | 12293 | 26224 |
| D2.0I0.08T9P2000L1C2 | 46628 | 9672 | 50326 | 2184825 |
| D5.0I0.05T5P5000L1C2 | 5616 | 4009 | 24929 | 39530 |
| D10.0I0.05T4P10000L1C3 | 7129 | 6786 | 46940 | 18518 |
| D15.0I0.05T6P15000L1C2 | 148451 | 50185 | 280163 | 2356691 |
| D20.0I0.05T6P20000L1C2 | 191148 | 81058 | 468814 | 4366112 |

Table 7.5 contains results for both considered JEP discovery approaches. Each of them is coupled with two reduct computation methods: RedPrime (RedPr) and RedApriori (RedAp), for which we give the average dimensionality of a per-object subproblem (Dim) and the computation time. In addition, the number of unnecessary generated JEPNs (Other) and partial time spent on finding a good partition are presented for global condensation.

We observe that table condensation and local projection lead most often to the same subproblem dimensionality, which is equal to the number of attributes of an original decision table. As a result, both methods achieve similar efficiency.

Moreover, for this kind of input data, finding a proper partition of possibly low cardinality, by means of graph coloring heuristics (LF), is very successful. Also, the time of finding the partition is negligible. In fact, only for *vehicle*, global condensation achieves lower dimensionality reduction and produces undesirable JEPNs. Consequently, its performance is approximately two times worse than for local projection.

Global condensation strongly relies on quality of a chosen graph coloring. An overhead coming from generation and filtering of additional non-JEPs can significantly lower algorithm performance. On the other hand, subproblem dimensionality for local projection is always equal to the average transaction length. Therefore, this method is particularly applicable to sparse data. Table 7.6 shows efficiency of local projection towards Border-diff (Border, see 3.2) and CP-Tree (CP-Tree, see 3.2) for synthetically generated sparse datasets. The data preparation procedure and characteristics of databases are covered in Section A.2. Also, the rough set methods outperform the classic approaches in all the considered cases and the variant of JEPLocalProj with RedApriori is always the winner. Differences are especially visible for larger datasets and remain around one order of magnitude.

## 8 JEP Mining by Means of CP-Tree

A CP-Tree is a widely utilized structure in EP mining ([75,54]). As described in Section 3.2, one assumes a certain order in itemspace, so that transactions can be inserted into the tree in the same way strings are inserted into a suffix-tree. The tree is walked in a pre-order fashion and the rooted path to a currently visited node corresponds to a candidate pattern. Subtree merging ensures that also all suffixes of each candidate pattern are examined.

If the tree is mined exhaustively, one traverses the entire search space. Depending on the type of desirable patterns, different pruning approaches can be applied to speed up this process. One classic application is SJEP mining ([54]), where one looks for patterns frequent in a target class. The support of the pattern corresponding to an arbitrary node is not higher or equal to the supports of patterns corresponding to the node's children. Therefore, when respective supports are stored in nodes, it is possible to prune branches as soon as a given support threshold is exceeded. The resulting set needs to be post-processed to select the minimal patterns.

We extend pruning capabilities for SJEP discovery by applying concepts analogous to attribute set dependence and attribute indispensability from rough set theory ([141]). The pruning method follows the reasoning from Section 4.1, in particular Theorems 4.4 and 4.5. However, a substantially different way of search space traversal requires a new implementation of the corresponding pruning conditions.

Minimum support gives insights on the importance of desirable patterns, however, it does not restrict the size of a result set. If the threshold is too high, one may obtain few patterns, if it is too low, a mining process usually becomes time-consuming or infeasible. Therefore, searching for a certain number of best patterns may be a convenient alternative.

The problem of finding top-$k$ frequent closed patterns of length no less than a given threshold was covered in [58,59]. We introduce and discuss the problem of finding top-$k$ most supported minimal JEPs ([141]). Despite the different formulation, we exploit its resemblance to SJEP mining. The algorithm based on a CP-Tree is augmented, so that it is capable of gradually raising minimum support and, thus, intensifying pattern pruning. New values of the threshold are deduced from the minimal JEPs identified so far, which means that their minimality has to be verified at discovery time. Due to the latter, the conditions based on attribute set dependence can be easily incorporated, which pushes pruning capabilities even further. All the modifications are discussed for two and multiple classes.

## 8.1 Parameterized Mining Scheme

A CP-Tree is a structure that allows us to mine a wide variety of patterns. It can be viewed as a way of enumerating a search space in the order that enables efficient pruning. The overview of the structure, its construction and mining were covered in Section 3.2. For the sake of clarity, we follow the naming convention and example (Figure 8.1) proposed there.

Algorithm 9 is a parameterized version of the original *Mine-tree* routine used for SJEP mining ([54]). Our propositions are presented by appropriate concretizations of this scheme.

| | | |
|---|---|---|
| $T_1$ | $bdca$ | $c_0$ |
| $T_2$ | $bdaf$ | $c_0$ |
| $T_3$ | $bce$ | $c_0$ |
| $T_4$ | $ca$ | $c_1$ |
| $T_5$ | $bdc$ | $c_1$ |
| $T_6$ | $caef$ | $c_1$ |

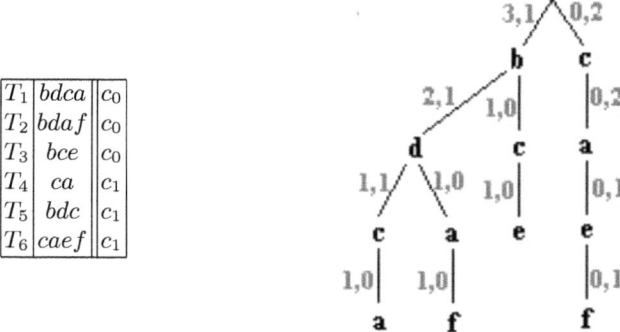

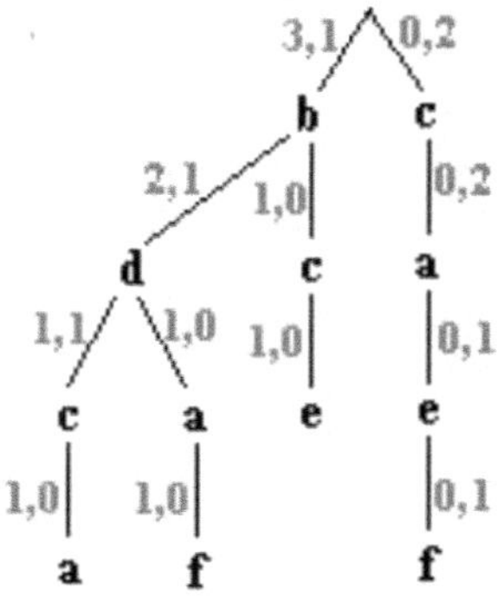


**Fig. 8.1.** A sample classified dataset and a respective CP-Tree

Throughout the mining process, two result sets, of positive and negative patterns, are maintained. The *Accept-pattern* function tests if a given pattern is a JEP, so that it can be added to the appropriate result set. Two thresholds, a positive and negative minimum count, are used to prune candidate patterns. The *Raise-count* procedure is responsible for raising their values. While a current pattern $\beta$ is considered, the *Visit-subtree* function indicates, whether the respective subtree is worth visiting. Finally, subtree merging is performed by the *Merge* procedure, as described in Section 3.2. As it is unrelated to our proposition, details are omitted ([54]).

In the original algorithm for SJEP mining, all found SJEPs are collected. Therefore, *Accept-pattern*$(\beta, T.negCount[i], T.posCount[i], minPosCount)$ basically checks the condition: $(T.negCount[i] = 0)$. Behavior for negative patterns is analogous. *Raise-count*$(minPosCount, posPatterns)$ returns $minPosCount$, because both minimum counts are constant. Also, a given subtree is traversed, if there is any chance of finding patterns that meet respective thresholds. In other words, $Visit\text{-}subtree$ tests if $(T.posCount[i] > minPosCount \vee T.negCount[i] > minNegCount)$. After the mining is finished, the result set may contain non-minimal patterns and an additional scan is required. Last but not least, the itemspace should be ordered based on support-ratio to bring JEPs closer to the top of the tree ([75]).

Frequently, decision transaction systems contain more than two classes. In this case, it is recommended ([50,75]) to induce EPs in each class separately and treat the remaining classes as a negative class. A classic approach based on a CP-Tree searches simultaneously for patterns in a positive and negative class. Since negative ones would be discarded anyway, it is economical to modify our algorithm, so that the search space and information stored in nodes are smaller.

As it was explained before, children of each node and nodes on each path are ordered by support-ratio. This strategy makes positive and negative EPs with high growth rates remain closer to the root of a tree. Since we are not interested in finding negative patterns anymore, it is better to use the order based solely on growth rate to the positive class. Also, there is no need to store an exact

---

**Algorithm 9.** Mine-tree (T,$\alpha$)

---

1: **for all** ($i \in T.items$) **do**
2:    **if** $T.child[i].items$ is not empty **then**
3:       $Merge(T.child[i], T)$
4:    **end if**
5:    $\beta = \alpha \cup T.items[i]$
6:    **if** $Accept\text{-}pattern(\beta, T.posCount[i], T.negCount[i], minNegCount)$ **then**
7:       $negPatterns := negPatterns \cup \beta$; $count(\beta) := T.negCount[i]$
8:       $minNegCount := Raise\text{-}count(minNegCount, negPatterns)$
9:    **else**
10:       **if** $Accept\text{-}pattern(\beta, T.negCount[i], T.posCount[i], minPosCount)$ **then**
11:          $posPatterns := posPatterns \cup \beta$; $count(\beta) := T.posCount[i]$
12:          $minPosCount := Raise\text{-}count(minPosCount, posPatterns)$
13:       **else**
14:          **if** $Visit\text{-}subtree(T, \beta)$ **then**
15:             $Mine\text{-}tree(T.child[i], \beta)$
16:          **end if**
17:       **end if**
18:    **end if**
19:    delete subtree $i$
20: **end for**

---

negative count in each node. A single bit that indicates if the associated pattern is a JEP or not, is sufficient. As far as mining is concerned, only one result set and one minimum count are required.

## 8.2    Pruning of Dependent Itemsets

When a CP-Tree is being mined, both minimal and non-minimal patterns are collected. In the post-processing, the latter ones have to be filtered out. In the original algorithm, the only pruning mechanism that prevents us from examining all supported patterns is based on support testing. If the negative support of a pattern is equal to zero, this pattern is a JEP and a subtree of the respective node is not traversed. In fact, it contains only non-minimal JEPs.

A consequence of this approach is that many redundant JEPs are collected, which raises mining and post-processing time. We show that it is usually possible to identify earlier if supersets of a considered pattern can be minimal JEPs.

Let us consider a database $D \subseteq \mathcal{D}$ and a pattern $X \subseteq \mathcal{I}$. The following property expresses minimality of $X$ in terms of transactions supporting $X$ in the complementary class.

**Definition 8.1.** $X$ is $D$-independent if $\forall_{Y \subset X}\{T \subseteq D' : X \subseteq T\} \subset \{T \subseteq D' : Y \subseteq T\}$.

The concept of $D$-independent is closely related to attribute set independence from rough set theory. Analogously to Theorem 4.5, it requires checking only immediate subsets of $X$. Due to the similarity of these two theorems, the proof is omitted.

**Theorem 8.1.** *$X$ is $D$-independent* $\Leftrightarrow \forall_{a \in X} supp_{D'} X < supp_{D'}(X - a)$.

If any proper subset of $X$ is supported by the same transactions in the complementary class, it means that additional items in the original pattern do not introduce any discrimination value. The following theorem states that only supersets of a $D$-independent pattern can be minimal JEPs.

**Theorem 8.2.** $\forall_{X \subseteq \mathcal{I}} X$ *is not $D$-independent* $\Longrightarrow \forall_{Y \supseteq X} Y \notin \mathcal{L}_c$.

*Proof.* Consider $X \in \mathcal{I}$ that is not $D$-independent.

From the definition of $D$-independence, we have that $\exists_{Z \subset X}\{T \subseteq D' : X \subseteq T\} \not\subset \{T \subseteq D' : Z \subseteq T\}$. Since $\forall_{Z \subset X}\{T \subseteq D' : X \subseteq T\} \subseteq \{T \subseteq D' : Z \subseteq T\}$, we have $\{T \subseteq D' : X \subseteq T\} = \{T \subseteq D' : Z \subseteq T\}$.

Now, let us assume that there exists $Y \supseteq X$, such that $Y \in \mathcal{L}_c$. We can express $Y$ with a sum $Y = X \cup B$, for a certain $B \in \mathcal{I}$, such that $X \cap B = \emptyset$. Since $Y$ is a JEP, we have $\emptyset = \{T \subseteq D' : Y \subseteq T\} = \{T \subseteq D' : (X \cup B) \subseteq T\} = \{T \subseteq D' : X \subseteq T \wedge B \subseteq T\} = \{T \subseteq D' : X \subseteq T\} \cap \{T \subseteq D' : B \subseteq T\} = \{T \subseteq D' : Z \subseteq T\} \cap \{T \subseteq D' : B \subseteq T\} = \{T \subseteq D' : (Z \cup B) \subseteq T\}$. As a consequence, $supp_{D'}(Z \cup B) = 0$. At the same time, due to $(Z \cup B) \subset Y$, we have $supp_D(Z \cup B) > 0$ and $(Z \cup B)$ is a JEP in $D$. However, this means that $Y$ is not a minimal JEP. □

For a given node, associated with a certain pattern $X$, all its descendants refer to supersets of $X$. Therefore, one may use a test for $D$-dependence in the *Visit-subtree* function to prevent unnecessary traversing and merging. According to Theorem 8.1, it requires checking counts in the negative class for all immediate subsets of $\beta$. Although this pruning approach may be very powerful, the test is performed for every pattern that is not added to result sets and, as a consequence, can impact overall efficiency. For this reason, we use a classic counting method that stores a list of transaction identifiers for each individual item and computes the count of a given pattern by intersecting the lists referring to its items ([135]). Bit operations can be employed to improve performance.

*Example 8.1.* Theorem 8.2 allows us to avoid mining subtrees of non-minimal patterns. For example, in Fig. 8.1, the patterns *bd* and *bc* are not minimal, since $supp_{D'_0} bd = supp_{D'_0} bc = supp_{D'_0} b$, and their subtrees can be pruned after merging.

## 8.3 Top-K Most Supported Minimal JEPs

Low-supported patterns are usually considered uninteresting and treated as a result of noise. The fundamental concept of frequent patterns is defined by means of minimal support that the interesting ones have to satisfy. A similar approach has been proposed for emerging patterns, where an SJEP is a frequent pattern in a considered target class ([54]). Unfortunately, the value of minimal support does not give insights on the size of a result set and, in consequence, computation time. Moreover, in case of an excessive result set, further post-processing needs to be applied to choose a subset of patterns manageable by a target application.

An alternative approach is to look for a certain number of the best patterns with respect to some importance criteria. In this case, the maximum size of a resulting collection is known upfront, whereas the lowest support over patterns of this collection depends on a given dataset. One variation, finding top-$k$ frequent closed patterns satisfying a given minimum length threshold, is considered in [58,59]. The authors modify the FP-Growth algorithm to implement appropriate pruning conditions. Along these lines, we formulate and propose a solution for finding minimal JEPs.

**Definition 8.2.** *For* $c \in \mathcal{I}_d$,
*the* top-$k$ most supported minimal JEPs problem *is to find a set* $S \subseteq JEP(D_c)$, *such that* $|S| <= k$ *and* $\forall_{X \in S, Y \in (JEP(D_c)-S)} supp_{D_c} X >= supp_{D_c} Y$.

Note that a resulting set of top-$k$ most supported minimal JEPs is not defined precisely. In other words, there may be many solutions for a given dataset, class and value $k$.

To provide an efficient algorithm, we exploit the resemblance between our problem and SJEP mining. In fact, our problem can be solved by finding a superset of $\xi$-SJEPs with a certain minimum threshold $\xi$. Higher values of this threshold limit the resulting collection and may speed up the search. On the other hand, a certain value cannot be exceeded as it would lead to elimination of patterns from the top-$k$ set. The challenge is to identify the highest correct value as early as possible during mining.

We modify the approach based on a CP-Tree, so that it is capable of gradually raising minimum support and, thus, intensifying pattern pruning. New values of the threshold are deduced from the minimal JEPs identified so far. The following theorem states that whenever one knows a pattern collection of the size at least $k$, the minimum support $\xi$ equal to the minimum over supports of patterns from this collection, a respective set of $\xi$-SJEPs contains at least $k$ elements.

**Theorem 8.3.** *For* $c \in \mathcal{I}_d$,
$\forall_{P \subseteq \mathcal{L}_c} |P| \geq k \wedge min_{X \in P}(supp_{D_c} X) > \xi \Rightarrow |\xi\text{-}SJEP(D'_c, D_c)| \geq k$.

*Proof.* Let $c \in \mathcal{I}_d$, $P \subseteq \mathcal{L}_c$, so that $|P| \geq k$.

Since $min_{X \in P}(supp_{D_c} X) > \xi$, we have $\forall_{X \in P}(supp_{D_c} X) > \xi$. Therefore, we have $P \subseteq \xi\text{-}SJEP(D'_c, D_c)$ and $|\xi\text{-}SJEP(D'_c, D_c)| \geq k$. □

Note that this condition can be utilized only if the minimality of JEPs is verified at discovery time. This test follows Theorem 8.2 and is located in the *Accept-pattern* routine. Also, if $X$ is a JEP, we have $supp_{D'_c}(X) = 0$. Thus, $X$ is minimal iff all its immediate subsets have non-zero negative counts.

Our proposition does not require significant changes to the original mining scheme. In fact, it is sufficient to store the current result in a priority queue, e.g. a heap, of at most $k$ elements ordered by non-increasing supports. For simplicity, we operate on counts instead of supports. At the beginning the minimum count is equal to 0. Whenever a new minimal JEP is identified and fewer than $k$ elements have been collected or its count is higher than the current minimum count, the pattern gets inserted to the queue. If $k$ elements are collected, one may set the minimum count to the count of the head element of the queue. This way the threshold can be raised without a significant memory overhead.

*Example 8.2.* The following JEPs are successively discovered by the original algorithm for $c_0$: $abcd, abd, bdf, abc, bce, ab, be, bf, acd, ad, df$. Additional computation is required to identify minimal patterns and pick $ab$, $ad$. If we check minimality upfront, $ab$ is the first minimal JEP, since $supp_{D'_0}(a), supp_{D'_0}(b) > 0$. For $k = 2$, one collects $ab, be$ (counts: 2, 1), and sets the minimum count to 1. Then, it prunes all patterns but $ad$, so that even the additional minimality checks are not needed.

### 8.4 Experimental Evaluation

The first part of our experimentation investigates efficiency of pruning of non-minimal patterns in exploration based on a CP-Tree. For SJEP mining and finding top-$k$ most supported minimal JEPs, we report the number of examined patterns, to demonstrate effectiveness of the additional pruning, and computation time, to show its overall influence on a given algorithm. In general, all the discussed approaches concretize Algorithm 9 from Section 8.1, however, the character of the considered datasets requires the version of the scheme that supports multiple classes. In particular, it means that the routine is invoked separately for every class of a given dataset and only positive patterns are collected.

We compare classic SJEP mining ([54]) with its variation that also checks if a visited pattern is dependent. In accordance with Theorems 8.1 and 8.2, the respective subtree can be pruned. This additional functionality is covered by the routine *Visit-subtree* and requires computing supports in a background class for all immediate subsets of a considered pattern. Apart from potential savings in the number of examined patterns, this cost can also be alleviated by the fact that all collected SJEPs are minimal and no post-processing is necessary.

The experiment was performed for different values of minimum support $\xi$. Table 8.1 gives the number of patterns examined in mining for each of the two methods. Reductions are observable in the majority of the considered problems. We highlight only these datasets, for which reductions occur for all the considered minimum supports. Table 8.2 contains corresponding computation times. Datasets without any time regressions and improvement in at least one case are given in bold.

In terms of examined patterns, efficiency of pruning of non-minimal patterns is visible for the majority of the datasets, sometimes the difference is of multiple orders of magnitude (*heart, lymn, mushroom, vehicle*). Higher minimum supports intensify original pruning and decrease the impact of the new approach. A reduction of a search space is usually followed by a lower computation time. Significant impact can be observed for *lymn*, *vehicle*, *zoo*. Nevertheless, the overhead of additional tests accompanied with relatively small reduction led to few noticeable regressions (*krkopt, nursery*). Note that, in both cases, we deal with large numbers of objects and, thus, higher computation costs.

A separate set of experiments looks at finding top-$k$ most supported minimal JEPs. The approach that gradually raises minimum count is contrasted with

**Table 8.1.** The number of patterns examined in $\xi$-SJEPs mining without and with non-minimal pattern pruning, given in two lines for each dataset, respectively

| Dataset | Examined patterns | | | | |
|---|---|---|---|---|---|
| | 0% | 5% | 10% | 20% | 30% |
| balance | 935 | 187 | 67 | 33 | 11 |
| | 895 | 187 | 67 | 33 | 11 |
| **breast-wisc** | 8024 | 1048 | 539 | 369 | 270 |
| | 3618 | 432 | 228 | 149 | 113 |
| car | 3441 | 856 | 357 | 124 | 62 |
| | 3367 | 856 | 357 | 124 | 62 |
| cmc | 53728 | 4392 | 1700 | 507 | 200 |
| | 25688 | 3763 | 1609 | 507 | 200 |
| **heart** | 125008 | 34640 | 16200 | 6968 | 3248 |
| | 6523 | 2511 | 1207 | 517 | 312 |
| **irys** | 64 | 42 | 34 | 29 | 23 |
| | 45 | 32 | 24 | 21 | 17 |
| **krkopt** | 146004 | 2790 | 1053 | 333 | 164 |
| | 134244 | 2700 | 1020 | 325 | 163 |
| **lymn** | 1670728 | 188372 | 81664 | 43780 | 26278 |
| | 36409 | 13923 | 8071 | 4483 | 1998 |
| monks-1 | 1020 | 418 | 161 | 37 | 28 |
| | 1020 | 418 | 161 | 37 | 28 |
| monks-2 | 1932 | 441 | 195 | 50 | 27 |
| | 1730 | 427 | 195 | 50 | 27 |
| monks-3 | 924 | 362 | 160 | 51 | 30 |
| | 835 | 334 | 153 | 49 | 30 |
| **mushroom** | 15279338 | 262260 | 81999 | 22872 | 5867 |
| | 64839 | 9268 | 4435 | 1702 | 631 |
| nursery | 62694 | 2282 | 855 | 315 | 206 |
| | 58123 | 2278 | 854 | 315 | 206 |
| tic-tac-toe | 45886 | 1847 | 486 | 102 | 36 |
| | 23855 | 1835 | 486 | 102 | 36 |
| **vehicle** | 19319956 | 1339050 | 484478 | 105012 | 21248 |
| | 418401 | 73801 | 31657 | 10971 | 3921 |
| **wine** | 26262 | 10574 | 7239 | 4076 | 2587 |
| | 4165 | 2050 | 1652 | 1023 | 785 |
| **yeast** | 13002 | 4886 | 3128 | 1716 | 1099 |
| | 6440 | 3283 | 2289 | 1370 | 887 |
| **zoo** | 198326 | 182267 | 166929 | 106106 | 72194 |
| | 10851 | 10452 | 9685 | 7366 | 5852 |

**Table 8.2.** Computation time of $\xi$-SJEPs mining without and with non-minimal pattern pruning, given in two lines for each dataset, respectively

| Dataset | Computation Time | | | | |
|---|---|---|---|---|---|
| | 0% | 5% | 10% | 20% | 30% |
| **balance** | 46 | 26 | 20 | 26 | 21 |
| | 36 | 26 | 20 | 26 | 21 |
| **breast-wisc** | 265 | 78 | 73 | 62 | 62 |
| | 145 | 62 | 62 | 52 | 57 |
| car | 177 | 114 | 114 | 104 | 93 |
| | 166 | 124 | 114 | 109 | 98 |
| cmc | 806 | 218 | 171 | 130 | 119 |
| | 759 | 239 | 176 | 140 | 119 |
| **heart** | 520 | 171 | 109 | 62 | 41 |
| | 234 | 119 | 78 | 52 | 36 |
| **irys** | 10 | 5 | 5 | 0 | 5 |
| | 5 | 5 | 5 | 0 | 5 |
| krkopt | 15724 | 10831 | 10239 | 9895 | 9667 |
| | 32526 | 11429 | 10727 | 10369 | 10150 |
| **lymn** | 72867 | 2917 | 1440 | 847 | 442 |
| | 1996 | 962 | 639 | 374 | 234 |
| **monks-1** | 41 | 21 | 20 | 21 | 15 |
| | 36 | 20 | 15 | 15 | 15 |
| **monks-2** | 52 | 26 | 26 | 20 | 15 |
| | 41 | 26 | 21 | 20 | 15 |
| **monks-3** | 41 | 21 | 15 | 15 | 21 |
| | 31 | 15 | 21 | 15 | 15 |
| **mushroom** | 506407 | 12761 | 6848 | 4197 | 3494 |
| | 14852 | 5553 | 4228 | 3229 | 2652 |
| nursery | 2106 | 1389 | 1263 | 1202 | 1138 |
| | 6505 | 1498 | 1342 | 1248 | 1201 |
| tic-tac-toe | 2111 | 338 | 239 | 156 | 109 |
| | 774 | 343 | 239 | 156 | 109 |
| **vehicle** | 1758198 | 34190 | 17872 | 7613 | 3562 |
| | 22677 | 6666 | 4082 | 2444 | 1492 |
| **wine** | 603 | 187 | 130 | 83 | 62 |
| | 140 | 93 | 78 | 57 | 52 |
| yeast | 280 | 223 | 213 | 202 | 192 |
| | 338 | 260 | 239 | 223 | 208 |
| **zoo** | 1690 | 1539 | 1414 | 837 | 639 |
| | 306 | 291 | 270 | 208 | 182 |

**Table 8.3.** The number of patterns examined in finding top-$k$ minimal JEPs without and with non-minimal pattern pruning, given in two lines for each dataset, respectively

| Dataset | Examined patterns | | | | |
|---|---|---|---|---|---|
| | 5 | 10 | 20 | 50 | $\infty$ |
| **balance** | 444 | 467 | 502 | 659 | 935 |
| | 416 | 435 | 470 | 623 | 895 |
| **breast-wisc** | 594 | 1095 | 1460 | 2864 | 8024 |
| | 129 | 270 | 443 | 1205 | 3618 |
| **car** | 1303 | 2187 | 2985 | 3143 | 3441 |
| | 1267 | 2140 | 2926 | 3073 | 3367 |
| **cmc** | 14965 | 18069 | 22147 | 28508 | 53728 |
| | 10298 | 12097 | 14384 | 17193 | 25688 |
| **heart** | 21193 | 33720 | 52170 | 80474 | 125008 |
| | 1306 | 1952 | 2952 | 4364 | 6523 |
| **irys** | 61 | 64 | 64 | 64 | 64 |
| | 43 | 45 | 45 | 45 | 45 |
| **krkopt** | 51017 | 58982 | 66753 | 76337 | 146004 |
| | 46984 | 54446 | 61586 | 70233 | 134244 |
| **lymn** | 15281 | 24749 | 44842 | 97792 | 1670728 |
| | 1204 | 1512 | 2328 | 4342 | 36409 |
| monks-1 | 758 | 837 | 1020 | 1020 | 1020 |
| | 758 | 837 | 1020 | 1020 | 1020 |
| **monks-2** | 872 | 1046 | 1502 | 1767 | 1932 |
| | 859 | 978 | 1333 | 1578 | 1730 |
| **monks-3** | 254 | 311 | 808 | 924 | 924 |
| | 215 | 252 | 719 | 835 | 835 |
| **mushroom** | 2274 | 5591 | 20747 | 141041 | 15279338 |
| | 162 | 306 | 474 | 2220 | 64839 |
| **nursery** | 17388 | 22641 | 25487 | 33422 | 62694 |
| | 16800 | 21700 | 24438 | 31471 | 58123 |
| **tic-tac-toe** | 2650 | 4716 | 6535 | 10251 | 45886 |
| | 2122 | 3847 | 5117 | 7335 | 23855 |
| **vehicle** | 690908 | 1277105 | 1798589 | 2878406 | 19319956 |
| | 21471 | 39374 | 56734 | 92145 | 418401 |
| **wine** | 1424 | 2336 | 3544 | 8281 | 26262 |
| | 300 | 513 | 814 | 1500 | 4165 |
| **yeast** | 10312 | 11978 | 12930 | 13002 | 13002 |
| | 5510 | 6070 | 6422 | 6440 | 6440 |
| **zoo** | 34354 | 59976 | 93980 | 146856 | 198326 |
| | 1463 | 2393 | 4062 | 7679 | 10851 |

**Table 8.4.** Computation time of finding top-$k$ minimal JEPs without and with non-minimal pattern pruning, given in two lines for each dataset, respectively

| Dataset | Computation Time | | | | |
|---|---|---|---|---|---|
| | 5 | 10 | 20 | 50 | $\infty$ |
| **balance** | 31 | 31 | 31 | 31 | 31 |
| | 26 | 31 | 31 | 31 | 31 |
| **breast-wisc** | 67 | 72 | 93 | 119 | 182 |
| | 52 | 62 | 72 | 104 | 156 |
| car | 130 | 135 | 135 | 140 | 140 |
| | 135 | 151 | 156 | 166 | 171 |
| cmc | 379 | 411 | 452 | 504 | 686 |
| | 457 | 509 | 561 | 650 | 873 |
| **heart** | 140 | 197 | 275 | 384 | 551 |
| | 93 | 114 | 151 | 197 | 260 |
| irys | 5 | 5 | 0 | 5 | 5 |
| | 5 | 5 | 0 | 5 | 5 |
| krkopt | 13834 | 14021 | 14130 | 14442 | 17464 |
| | 17171 | 18060 | 18981 | 20208 | 33348 |
| **lymn** | 442 | 577 | 868 | 1461 | 13000 |
| | 254 | 291 | 379 | 540 | 2142 |
| monks-1 | 26 | 26 | 26 | 26 | 26 |
| | 26 | 31 | 31 | 31 | 31 |
| monks-2 | 31 | 26 | 36 | 31 | 36 |
| | 31 | 31 | 41 | 36 | 46 |
| monks-3 | 20 | 15 | 26 | 20 | 26 |
| | 21 | 20 | 20 | 31 | 26 |
| **mushroom** | 2777 | 3260 | 4165 | 8393 | 148465 |
| | 1872 | 2106 | 2247 | 3354 | 14336 |
| nursery | 1934 | 1950 | 2028 | 2137 | 2450 |
| | 3432 | 3760 | 3994 | 4461 | 6911 |
| tic-tac-toe | 416 | 499 | 541 | 608 | 873 |
| | 416 | 499 | 540 | 603 | 904 |
| **vehicle** | 19224 | 28470 | 35911 | 50070 | 188895 |
| | 3234 | 4498 | 5637 | 7753 | 23265 |
| **wine** | 52 | 62 | 83 | 140 | 317 |
| | 41 | 52 | 57 | 78 | 150 |
| yeast | 265 | 275 | 281 | 275 | 281 |
| | 332 | 343 | 353 | 353 | 353 |
| **zoo** | 348 | 546 | 769 | 1123 | 1414 |
| | 88 | 124 | 161 | 260 | 332 |

its modified version that also prunes subtrees corresponding to non-minimal patterns, based on Theorems 8.1 and 8.2. As compared with SJEP mining, the additional pruning requires a potentially milder overhead. The minimum count raising already performs an independence test for all JEPs, while the modified version extends it to all examined patterns.

Different values of $k$ are considered, in particular $\infty$, to investigate the overhead of raising minimum count when all JEPs are in question. Similarly to the first part, Tables 8.3 and 8.4 give the number of patterns examined and computation times, respectively. For clarity, we use the same highlighting policy.

It is visible that the minimum count raising allows us to find top-$k$ minimal JEPs in a reasonable time for small $k$. Even without pruning of non-minimal patterns, benefits in the number of examined patterns and execution time are significant (*lymn*, *lung*, *krkopt*, *vehicle*). Additional pruning efficiently reduces the search space for *heart*, *lymn*, *mushroom*, *vehicle* and *zoo* with ratios of 1-3 orders of magnitude. Also, these datasets show the highest savings in the computation time.

## 9 Adaptive Classification Meta-scheme

The quality of training data is an important concern in classification. In the supervised approach, classified training transactions are a sole determinant of a resulting hypothesis. This strategy works well as long as the training set is representative for a considered domain. Unfortunately, in practice, only small sets of classified data may be available, e.g. due to a costly or time-consuming external expertise. This problem is potentially alleviated by semi-supervised learning that uses both classified and unclassified training data, where the latter is usually relatively cheaper to obtain.

Generative models assume that a joint distribution of features and class labels is a mixture of identifiable class distributions, e.g. normal. If this assumption holds, unlabeled data allow us to identify mixture components, which can be further assigned to classes based on labeled instances. Individual models are usually discovered by means of the Expectation-Maximization algorithm ([142]). Another classic approach, co-training ([143]), uses two classifiers learned with two, ideally independent, sets of features describing the same input data. Assuming that each set is sufficient for good classification, predictions of the classifiers should be consistent. Therefore, the classifiers can train one another with confidently classified unlabeled instances.

Our discussion is conceptually related to self-training and self-improving ([65,66]), where a classifier is iteratively learned/improved with a labeled dataset and employed to classify unlabeled instances. The most confidently predicted instances are further treated as labeled and the classifier is modified. Unlike supervised and semi-supervised approaches, we do not fix the definition of a classifier after its initial training but attempt to obtain higher accuracy by adjusting the hypothesis throughout the testing phase. In AdaAccept, our adaptive classification meta-scheme ([144]), instances are gradually classified as the underlying model evolves. The procedure is formulated using a paradigm of classification with reject option, which is a common choice for the cases with a high cost of incorrect classification, like in medical diagnosis or email filtering ([145]). It is worth mentioning that different forms of adaptation are also present in popular supervised solutions, like AdaBoost ([146]) and evolutionary methods ([71,147]).

The proposed generic scheme is independent from an internally used classifier and adaptation method. We propose a concretization based on JEP-classifier ([96,50]) with two adaptation strategies: support adjustment and border recomputation. Adaptation conditions are formulated in accordance with distance and outlier rejection approaches for probabilistic classifiers ([148,149]).

## 9.1 Classification Problem

Let us consider a decision transaction system $\mathcal{DTS} = (\mathcal{D}, \mathcal{I}, \mathcal{I}_d)$. We introduce basic definitions related to classification of transactional data. The convention partially follows [139].

**Definition 9.1.** *A* classifier *is a function* $\{T - \mathcal{I}_d : T \in \mathcal{D}\} \mapsto \mathcal{I}_d$.
*A set of possible classifiers is denoted by* $C_{\mathcal{DTS}}$.

In other words, a classifier assigns a single class to each transaction from the considered domain. In practice, there are situations, when a choice of the appropriate class for a given transaction is not unequivocal. Therefore, it is useful to consider a variant of a classifier that has a special bucket for all such confusing transactions. This approach is widely accepted in tasks, for which false positive errors are costly, like health condition classification or spam filtering ([145]).

**Definition 9.2.** *A* classifier with reject option *is a function* $\{T - \mathcal{I}_d : T \in \mathcal{D}\} \mapsto \mathcal{I}_d \cup R$, *where* $R$ *refers to a* rejection class (bucket).
*A transaction assigned to the rejection bucket is called* rejected. *The rest of transactions are called* accepted. *A set of possible classifiers with reject option is denoted by* $C^R_{\mathcal{DTS}}$.

In fact, it is often convenient to perform rejection and classification of a transaction independently, i.e. by means of separate conditions. In order to express how this choices are made, one usually uses a more general abstraction of a classifier that provides class membership indicators for all classes.

**Definition 9.3.** *A* generalized classifier *is a function* $\{T - \mathcal{I}_d : T \in \mathcal{D}\} \mapsto [0,1]^{|\mathcal{I}_d|}$.
*A set of possible generalized classifiers is denoted by* $C^G_{\mathcal{DTS}}$.

In the literature [149], generalized classifiers are also referred to as *possibilistic/fuzzy* classifiers. Further specialization, with all indicators summing up to 1, is called a *probabilistic* classifier. Finally, if exactly one indicator is equal to 1 and the rest to 0, a generalized classifier is considered *hard* and corresponds to a classifier.

A generalized classifier can be transformed to a classifier, by introducing a mapping between generalized results and decision classes.

**Definition 9.4.** *An* f,c-hardened classifier *is a classifier $c \circ f$, where*

> $f$ *is a generalized classifier, called a* labeling function,
> $c$ *is a function* $[0,1]^{|\mathcal{I}_d|} \mapsto \mathcal{I}_d$*, called a* hardening function.

**Definition 9.5.** *An* f,c,r-hardened classifier with reject option *is a classifier with reject option $g \in C^R_{\mathcal{DTS}}$, so that:*

$$g(t) = \begin{cases} (c \circ f)(t), & r(t) = 0 \\ R, & r(t) = 1. \end{cases}, \text{ for } t \in \{T - \mathcal{I}_d : T \in \mathcal{D}\},$$

*where*

> $f$ *is a generalized classifier, called a* labeling function,
> $c$ *is a function* $[0,1]^{|\mathcal{I}_d|} \mapsto \mathcal{I}_d$*, called a* hardening function,
> $r$ *is a function* $[0,1]^{|\mathcal{I}_d|} \mapsto \{0,1\}$*, called a* rejection condition.

Properties of a classifier can be evaluated by confronting its answers for a set of transactions with known classes. The correct assignment is used only for verification purposes and not available to the classifier. Note that we do not test generalized classifiers, only their hardened variants.

**Definition 9.6.** *For $D_{test} \subseteq \mathcal{D}$ and a classifier $c \in C_{\mathcal{DTS}}$ (classifier with reject option $c \in C^R_{\mathcal{DTS}}$), we define*

> classification accuracy *as:* $f_{accuracy} = \frac{|\{T \in D_{test} : \{c(T - \mathcal{I}_d)\} = T \cap \mathcal{I}_d\}|}{|D_{test}|}$,
> accept ratio *as:* $f_{accept-ratio} = \frac{|\{T \in D_{test} : \{c(T - \mathcal{I}_d)\} \neq R\}|}{|D_{test}|}$.

$D_{test}$ *is called a* testing database.

In other words, accuracy expresses a percentage of transactions that got correctly classified. Accept ratio indicates a percentage of transactions accepted in classification, i.e. assigned a class different from $R$. For classifiers without reject option, the ratio is always equal to 1.

In supervised learning classifiers are constructed based on given prior knowledge, i.e. transactions for which classes are known a priori.

**Definition 9.7.** *For $D_{train} \subseteq \mathcal{D}$,* training *of a classifier (classifier with reject option, generalized classifier) refers to finding a function $c \in C_{\mathcal{DTS}}$ ($c \in C^R_{\mathcal{DTS}}$, $c \in C^G_{\mathcal{DTS}}$) based on a $D_{train}$.*

$D_{train}$ *is called a* training database.

Normally, the structure of a classifier is fixed after the construction phase. It is no longer the case when adaptation comes into the picture. One may introduce adjustments by means of transactions with assigned classes during the testing phase. Such operation can be performed multiple times, so that the classifier can evolve and, presumably, better suit future testing transactions.

**Definition 9.8.** *For $D_{adapt} \subseteq \{T_i - \mathcal{I}_d \cup \{I_j\} : i = 1,..,|\mathcal{D}| \wedge j = 1,..,|\mathcal{I}_d|\}$ and a classifier (classifier with reject option, generalized classifier),* adaptation *refers to finding a function $c' \in C_{\mathcal{DTS}}$ ($c' \in C^R_{\mathcal{DTS}}$, $c' \in C^G_{\mathcal{DTS}}$) based on $D_{adapt}$.*

$\mathcal{D}_{adapt}$ *is called an* adaptation database.

Note that $\mathcal{D}_{adapt}$ is drawn from a set of all combinations of all transactions from $\mathcal{D}$ with all possible classes from $\mathcal{I}_d$. Also, in the above definitions, we assume that testing, training and adaptation databases are selected without replacement from respective sets of transactions.

## 9.2 Adaptive Classification

We propose an iterative self-training classification scheme that employs an internal generalized classifier (a labeling function) $f \in C^G_{\mathcal{DTS}}$. Depending on an application, this classifier can be hardened to a regular classifier or a classifier with reject option, given a certain hardening function $c$ and rejection condition $r$.

We assume that $f$ was initially trained with some training database $D_{train}$. In $i$-th iteration of our scheme, unclassified transactions are passed to an *f,c,r-hardened classifier with reject option.* Its answers for accepted transactions become answers for the whole scheme. We treat them as confident and use for adaptation. If any transactions in a particular step have been accepted, the model has possibly been enriched and can be used to classify transactions rejected so far. In this case, these transactions become an input for the next iteration. When no adaptation happens anymore, depending on the type of classification, one may reject remaining transactions or classify them with the final version of *f,c-hardened classifier.*

The procedure is shown as Algorithm 10. Given a labeling function $f$ and a testing database $D_{test}$, it produces a resulting mapping $g$ of testing transactions to decision classes. The additional flag *reject* allows us to perform classification with reject option, where $g : D_{test} \mapsto \mathcal{I}_d \cup R$, or regular classification with $g : D_{test} \mapsto \mathcal{I}_d$.

One need to define two functions in order to obtain a complete instance of the classification procedure:

*classify(set-of-transactions, resulting-mapping, labeling-function, reject)* that employs a *labeling-function*, hardened to a classifier with/without reject option, as determined by the parameter *reject*, to classify transactions from *seq-of-transactions.* It returns a new version of a resulting mapping $g$, where only answers referring to *set-of-transactions* are affected and the rest is identical to *resulting-mapping*;
*adapt(set-of-transactions, resulting-mapping, labeling-function)* that adapts a given *labeling-function* with a *set-of-transactions* and classes indicated by *resulting-mapping.*

Classification accuracy and accept ratio of our scheme are defined as they would be considered for a final mapping $g$. In fact, this mapping is a classifier for transactions from the testing database. Note that accept ratio expresses a ratio of performed adaptations related to the size of the complete testing set. Moreover, we introduce an additional evaluation criteria that refers to the degree of correct classifications within the set of all transactions accepted by a classifier with reject option and, thus, used for adaptation.

**Algorithm 10.** AdaAccept($D_{test}$, $f$, $reject$)

$g^{-1}(t) = R$, for each $t \in D_{test}$, $K^{-1}_{unclassified} \Leftarrow D_{test}\}$
$f^{-1} \Leftarrow NULL$, $f^0 \Leftarrow f$, $i \Leftarrow 0$
**while** $f^{i-1} \neq f^i$ **do**
  $K^i_{accepted} \Leftarrow \{t \in D_{test} : g^{i-1}(t) \neq R\}$
  $g^i \Leftarrow classify(K^{i-1}_{unclassified}, g^{i-1}, f^i, \mathbf{true})$
  $K^i_{unclassified} \Leftarrow \{t \in D_{test} : g^i(t) = R\}$
  $f^{i+1} \Leftarrow adapt(K^i_{adapt}, g, f^i)$,
  where $K^i_{adapt} = \{t \in D_{test} : g^i(t) \neq R\} - K^i_{accepted}$
  $i \Leftarrow i + 1$
**end while**
**if** $\neg reject$ **then**
  $g^i \Leftarrow classify(K^i_{unclassified}, D_{test}, g^{i-1}, f^i, \mathbf{false})$
**end if**
$N \Leftarrow i$
**return** $g_N$

**Definition 9.9.** Adaptation accuracy *is defined as*

$$f_{adaptation-accuracy} = \frac{|\{T \in D_{accepted} : \{c(T - \mathcal{I}_d)\} = T \cap \mathcal{I}_d\}|}{|\{T \in D_{accepted}\}|},$$

*where* $D_{accepted} = \bigcup_{i=0,..,N-1} K^i_{accepted}$.

### 9.3 Adaptive JEP-Classifier

The classification scheme proposed in this chapter is general and can be concretized in many ways. There are four choices that have to be made: a labeling function, hardening function, rejection condition and an adaptation method. Here, we present a variant inspired by emerging patterns. Our proposition focuses on JEPs and employs ideas originating from CAEP (see Section 3.3, [8]) and JEP-Classifier (see Section 3.3, [11]).

Throughout this section, we consider a decision transaction system $(\mathcal{D}, \mathcal{I}, \mathcal{I}_d)$, an internal classifier $f : \{T - \mathcal{I}_d : T \in \mathcal{D}\} \mapsto \mathcal{I}_d$ and a training set $D_{train}$.

**Labeling and Hardening Functions.** Many pattern-based classifiers follow a similar classification procedure. They employ collections of patterns characteristic for each decision class and corresponding weights that express pattern interestingness. Classification of a testing transaction has usually two stages: labeling and hardening. In the first one, membership of the transaction in each class is evaluated. Then, the resulting per-class indicators are aggregated to give a final answer.

This general procedure is leveraged in the following parametric description, which can be easily concretized to obtain classic classifiers, like JEP/CEP/SJEP-Classifier or CAEP.

Let us consider a testing transaction $X \in \mathcal{D}$. Class membership of $X$ is expressed by the value $f(X)$ of the following labeling function:

$f(X) = (f_k(X))_{k \in \mathcal{I}_d} = (\sum_{P \in \mathcal{P}_k \wedge P \subseteq X} s(m_{k,1}(P), .., m_{k,M}(P)))_{k \in \mathcal{I}_d}$, where:

class-wise pattern collections, $\{\mathcal{P}_k\}_{k \in \mathcal{I}_d}$, where $\mathcal{P} = \bigcup_{k \in \mathcal{I}_d} \mathcal{P}_k \subseteq \mathcal{D}$,
pattern interestingness measures, $m_{k,i} : \mathcal{P}_k \mapsto \mathbb{R}$, for $k \in \mathcal{I}_d$ and $i \in \{1..M\}$, where $M$ is the number of measures,
pattern scoring function, $s : \mathbb{R}^M \mapsto \mathbb{R}$.

Typically, for pattern-based classifiers resulting indicators do not have probabilistic interpretation. Hardening is performed by choosing the indicator with the highest rank, i.e. by means of the function $c(f(X)) = argmax_{k \in \mathcal{I}_d} f_k(X)$. Ties are usually broken by selecting the most frequent class and, eventually, by a random choice. We skip tie breaking in our further discussion.

*Example 9.1.* With respect to our template, JEP-Classifier can be defined by

$\{\mathcal{P}_k\}_{k \in \mathcal{I}_d} = \{\mathcal{L}_k\}_{k \in \mathcal{I}_d}$, where $[\mathcal{L}_k, \mathcal{R}_k] = JEP(D_k)$,
$m_{k,1}(X) = supp_{D_k}(X)$, for $k \in \mathcal{I}_d$, $X \in \mathcal{P}_k$, $M = 1$,
$s(m) = \sum_{k \in \mathcal{I}_d} m$, for $m \in \mathbb{R}$.

For clarity, we omit possible normalization of scores across classes ([8]).

**Adaptation method.** Adaptation allows us to modify a labeling function, so that it can be more effective in further classification. We describe two adaptation methods incorporated in JEP-Classifier: support adjustment and border recomputation.

In order to clearly present the adaptation process, let us consider a sequence of classifiers obtained by subsequent modifications. For a given step $i \in \mathbb{N} \cup \{0\}$, we have a labeling function $f^i \in C^G$ and an associated sequence $D^i_{train} = D_{train} + \sum_{j \in \{1,..,i\}} D^j_{adapt}$, where $D^j_{adapt}$, for $j \in \{1, .., i\}$, are subsequent adaptation sequences. For brevity, we denote a concatenation of sequences by addition. We also define respective decision classes $C^i_{train,k}$, for $i \in \mathbb{N} \cup \{0\}$ and $C^j_{adapt,k}$, for $j \in \{1, .., i\}$, where $k \in \mathcal{I}_d$. At the beginning, one has an initial labeling function $f^0$ trained with a sequence $D^0_{train} = D_{train}$. When a sequence of labeling functions is considered, superscripts are added only to the parameters affected by adaptation.

For convenience, we also introduce a notation analogous to growth rate of an itemset. It allows us to express the weight of a pattern by means of parameters independent from its actual supports in positive and negative databases.

**Definition 9.10.** *For $m_1, m_2 \in \mathbb{R}_+ \cup \{0\}$, we define*

$$gr(m_1, m_2) = \begin{cases} 0, & m_1 = m_2 = 0 \\ \infty, & m_1 = 0 \text{ and } m_2 \neq 0 \\ \frac{m_2}{m_1}, & \text{otherwise.} \end{cases}$$

**Lemma 9.1.** *For $X \subseteq \mathcal{I}$, $D_1, D_2 \subseteq \mathcal{D}$, we have*
$gr_{D_1 \to D_2}(X) = gr(supp_{D_1}(X), supp_{D_1}(X))$.

**Support Adjustment.** A common strategy to distinguish the importance of patterns is to assign specific weights to them. Following this idea, we put forward an approach placed among existing propositions for emerging patterns. It employs a modified JEP-Classifier. In $i$-th step, a labeling function $f^i$ can be characterized by the following parameters:

patterns: $\{\mathcal{P}_k\}_{k\in\mathcal{I}_d} = \{\mathcal{L}_k\}_{k\in\mathcal{I}_d}$ for $JEP(C'_{train,k}, C_{train,k})$,
measures: $m^i_{k,1}(P) = supp_{C^i_{train,k}}(P)$, $m^i_{k,2}(P) = supp_{C^i_{train,k}{}'}(P)$, $m^i_{k,3}(\cdot) = |C^i_{train,k}|$, $m^i_{k,4}(\cdot) = |C^i_{train,k}{}'|$, in $(D^i_{train}, \mathcal{I}, \mathcal{I}_d)$,
pattern scoring function $s(m_1, m_2) = \frac{GR(m_1,m_2)}{GR(m_1,m_2)+1} * m_1$.

Note that, once this generalized classifier is trained, we have $m_{k,2}(P) = \infty$, for $P \in \mathcal{P}_k$ and $k \in \mathcal{I}_d$. Therefore, it behaves exactly like a regular JEP-Classifier. The reason for the extension of the scoring function is that we want to modify the importance of patterns through adaptation. In fact, patterns are treated as emerging patterns with growth rate equal to $gr(m_{k,1}(P), m_{k,2}(P))$. Scoring is performed like in CAEP. Also, we do not perform normalization of scores ([8]). This modification is not present in the original JEP-Classifier.

An important advantage of our proposition is the fact that one does not need to store a training set to compute supports for each adaptation. In fact, for each transaction in an adaptation sequence, it is sufficient to increment by one respective supports of the patterns it contains.

**Theorem 9.1.** $\forall_{i\in\mathbb{N}}\forall_{k\in\mathcal{I}_d}\forall_{P\in\mathcal{P}_k} m^i_{k,1}(P) = \frac{m^{i-1}_{k,1}(P)*m^{i-1}_{k,3}(P)+|\{j:P\subseteq T_j\wedge T_j\in C^i_{adapt,k}\}|}{|m^i_{k,3}(P)|} \wedge m^i_{k,3}(P) = m^{i-1}_{k,3}(P) + |C^i_{adapt,k}|$.

*Proof.* Let $i \in \mathbb{N}$, $k \in \mathcal{I}_d$ and $P \in \mathcal{P}_k$. We have $m^i_{k,1}(P) = supp_{C^i_{train,k}}(P) = \frac{|\{j:P\subseteq T_j\wedge T_j\in C^i_{train,k}\}|}{|C^i_{train,k}|} = \frac{supp_{C^{i-1}_{train,k}}(P)*|C^{i-1}_{train,k}|+|\{j:P\subseteq T_j\wedge T_j\in C^i_{adapt,k}\}|}{|C^i_{train,k}|} = \frac{m^{i-1}_{k,1}(P)*m^{i-1}_{k,3}(P)+|\{j:P\subseteq T_j\wedge T_j\in C^i_{adapt,k}\}|}{|m^i_{k,3}(P)|}$. Also, $m^i_{k,3}(P) = |C^i_{train,k}| = |C^{i-1}_{train,k}| + |C^i_{train,k}| = m^{i-1}_{k,3}(P) + |C^i_{adapt,k}|$. □

Simple reasoning for $m_{k,2}, m_{k,4}$ is analogous. Note that the measures $m_{k,3}$ and $m_{k,4}$ are independent from a given pattern and evaluated only to make adaptation possible, they do not play any role in scoring.

*Example 9.2.* Let us consider an iteration $i$, the JEP $ab \in P_0$ and measures $m_{0,1}(ab) = 3/5$, $m_{0,2}(ab) = 0$, $m_{0,3}(\cdot) = 5$, $m_{0,4}(\cdot) = 10$ in the definition of $f^i$. Further, let the following transactions be accepted by the classifier in this iteration and become the adaptation sequence $(acd0, abe1, abg1)$. After performing the adaptation the measures for $ab$ in $f^{i+1}$ are: $m_{0,1}(ab) = (5 * 3/5)/6 = 3/6$, $m_{0,2}(ab) = (10 * 0 + 2)/12 = 1/6$, $m_{0,3}(\cdot) = 6$, $m_{0,4}(\cdot) = 12$. Since now, the pattern will be treated as an EP with growth rate equal to 3.

**Border Recomputation.** Although support adjustment enriches our knowledge on pattern importance, one is deprived of discovering new patterns and modifying the existing ones. A more invasive approach is to assume that newly accepted transactions are correctly labeled, add them to a training sequence and recompute collections of patterns and their supports.

In $i$-th step, a labeling function $f^i$ requires collections $\{\mathcal{P}_k^i\}_{k\in\mathcal{I}_d} = \{\mathcal{L}_k^i\}_{k\in\mathcal{I}_d}$ for a space $JEP(C_{train,k}^{i}{}', C_{train,k}^i)$ and supports of each pattern $P \in \mathcal{P}_k$ in each class $k \in \mathcal{I}_d$: $m_{k,1}^i(P) = supp_{C_{train,k}^i}(P)$, all computed for $(D_{train}^i, \mathcal{I}, \mathcal{I}_d)$. A matching pattern is scored with $s(m) = m$, for $m \in \mathbb{R}$.

Although conceptually straightforward, this approach requires noticeable resources. First of all, a single adaptation involves maintaining borders for each class. Efficient incremental procedures ([79,150]) provide a good solution here. Both minimal and maximal patterns have to be stored, even though only the former ones are necessary for classification. Secondly, in order to recompute positive supports a sequence $D_{train}^i$ is necessary. Instead of the whole sequence, one may use another maintainable structure that provides supports for patterns. Regardless, the definition of an adaptive classifier grows with the number of adaptations.

**Adaptation Condition.** The choice of the right adaptation condition is critical. If not strict enough, it is likely to allow modifications based on incorrectly classified transactions. On the other hand, if too hard to satisfy, it may prevent any adaptations and a classification scheme from potential improvement. Since a classifier with reject option is used internally, an adaptation condition is exactly converse to a rejection condition.

In principal, a testing transaction should be rejected when a classifier is not certain of its decision. One of possible methods is to look at a training sequence and assess how representative is the neighborhood of this transaction ([151]). In another approach, a condition can be directly based on labels predicted by a classifier ([148,149]). Interestingly, for probabilistic classifiers, whose labeling functions approximate a posteriori probabilities of classes, the problem of minimizing risk given certain costs of possible decisions has been solved ([152]). Following this path, one considers two types of rejection: outlier/distance and ambiguity. The first one expresses how certain a classifier is about each individual label, while the other one states that the choice between classes should be decisive. We formulate analogous criteria for labeling functions of EP-based classifiers. As it was pointed out, these functions do not have probabilistic interpretation, which makes a formal discussion much harder. Instead, an experimental verification is provided.

Let us perform a class-wise normalization with the largest value over training transactions in a respective class of $i$-th step, namely: $L'_k(X) = \frac{L_k(X)}{M_k^i}$ for $M_k^i = max_{T\in C_{train,k}^i} L_k(T)$, $k \in \mathcal{I}_d$ and $X \in D = \{T - \mathcal{I}_d : T \in \mathcal{D}\}$. The normalizing values $(M_k^i)_{k\in\mathcal{I}_d}$ need to be stored and maintained. Note that this normalization does not prevent labels from going beyond 1 for certain unseen transactions. If such a transaction is accepted, a respective normalizing value is updated as a part of the adaptation.

Now, let us move to the actual adaptation criteria. A given transaction $X \in D$ is accepted with a classifier's decision $c \in \mathcal{I}_d$, only if the following criteria are met: $L'_c(X) > t_{distance}$ and $\frac{L'_c(X)}{L'_k(X)} > t_{ambiguity}$, for all $k \in \mathcal{I}_d$. The values $t_{distance} \in [0,1]$, $t_{ambiguity} \in \mathbb{R}_+$ are certain thresholds. They can be estimated experimentally on a training sequence as a tipping point of accuracy.

*Example 9.3.* Let us consider a decision system $(\mathcal{D}, \mathcal{I}, \{0,1\})$ and a testing pattern $X \in D$. Let us assume that with a classifier $f^i$ we obtain the labeling $(L_0(X), L_1(X)) = (0.75, 0.27)$ and the class 0 gets chosen. After normalization with $(M_0^i, M_1^i) = (1.2, 0.9)$, we obtain $(L'_0(X), L'_1(X)) = (0.625, 0.3)$. For the thresholds $t_{distance} = 0.5$ and $t_{ambiguity} = 2.00$, we have $L'_0(X) = 0.625 > 0.5 = t_{distance}$ and $\frac{L'_c(X)}{L'_k(X)} = 2.08 > 2.00 = t_{ambiguity}$ and $X$ is accepted and used for adaptation. Note that, for $t_{ambiguity} = 2.5$, the decision would be assumed unequivocal and $X$ would be rejected.

## 9.4 Experimental Evaluation

Our tests focus on accuracy of JEP-Classifier (JEPC), Adaptive JEP-Classifier with support adjustment (AdaptJEPC-Adjust) and with border recomputation

**Table 9.1.** Accuracy, adaptation ratio and adaptation accuracy for JEP-Classifier, Adaptive JEP-Classifier with support adjustment and border recomputation by modified 5-fold cross validation

| Dataset | JEPC | AdaptJEPC-Adjust | | | AdaptJEPC-Recompute | | |
|---|---|---|---|---|---|---|---|
| | Accu | Accu | Adapt | AdAcc | Accu | Adapt | AdAcc |
| **balance** | 71.32 | 73.40 | 60.96 | 79.00 | **79.76** | 63.32 | 80.00 |
| breast-wisc | 96.17 | 96.21 | 92.45 | 98.00 | **96.92** | 94.13 | 98.00 |
| **car** | 83.17 | 87.30 | 75.27 | 93.00 | **90.35** | 84.23 | 92.00 |
| **cmc** | 42.43 | 45.03 | 22.40 | 63.00 | **47.83** | 27.10 | 64.00 |
| **dna** | 60.45 | 73.60 | **72.30** | 92.00 | 67.15 | 77.35 | 86.00 |
| **heart** | 78.70 | 79.91 | 79.26 | 87.00 | **81.57** | 82.22 | 87.00 |
| irys | **93.00** | 92.83 | 95.17 | 95.00 | **93.00** | 97.67 | 94.00 |
| **lung** | 29.69 | 31.25 | 38.28 | 40.00 | **43.75** | 68.75 | 63.00 |
| **lymn** | 71.28 | 80.57 | 78.89 | 86.00 | **81.08** | 93.75 | 86.00 |
| **monks-1** | 84.71 | 94.15 | 90.51 | 93.00 | **94.24** | 99.82 | 94.00 |
| **monks-2** | 64.89 | 67.18 | 60.11 | 71.00 | **70.84** | 57.99 | 73.00 |
| **monks-3** | 94.68 | 96.80 | 95.31 | 97.00 | **96.84** | 96.16 | 97.00 |
| mushroom | 99.93 | 99.91 | 88.18 | 99.00 | **99.95** | 98.42 | 99.00 |
| **nursery** | 94.65 | 96.30 | 73.08 | 98.00 | **97.93** | 78.66 | 99.00 |
| **tic-tac-toe** | 81.86 | **90.08** | 82.88 | 90.00 | 89.77 | 96.76 | 91.00 |
| **vehicle** | 68.29 | 70.30 | 67.49 | 75.00 | **71.72** | 87.56 | 77.00 |
| wine | 90.59 | 90.31 | 89.33 | 95.00 | **93.96** | 98.46 | 95.00 |
| **yeast** | 15.38 | 15.90 | 27.73 | 53.00 | **16.00** | 27.56 | 53.00 |
| **zoo** | 76.73 | 78.96 | 79.46 | 90.00 | **81.19** | 92.57 | 86.00 |

(AdaptJEPC-Recompute). We modify the $k$-fold cross validation scheme to investigate classification with small training sets. Our version uses one fold for training and the rest for testing. As a result, each fold is in a testing sequence $k-1$ times. Besides accuracy (*Accu*), we track the accept ratio (*Adapt*) and adaptation accuracy (*AdAcc*) for both adaptive propositions.

Table 9.1 shows results of modified 5-fold cross validation for $t_{distance} = 20\%$, $t_{ambiguity} = 2$. We observe that JEPC is outperformed by both adaptive approaches for almost all the investigated cases (in bold). Also, apart from *dna, irys* and *tic-tac-toe*, adaptation through recomputation is always the winner. Improvement of $> 8$ p.p. of any of the adaptive methods over JEPC occurs for *balance, dna, monks-1, lung, lymn.* Although each initial classifier was created with a small training sequence, AdaptJEPC-Adjust is still able to achieve a significant gain only by modifying importance of patterns. In fact, the difference in the average accuracy between the adaptation with support adjustment and JEPC, over all the datasets, is equal to 3.27 p.p. When Adaptive AdaptJEPC-Recompute is used and adaptation takes into account also pattern collections, we gain another 1.78 p.p. In addition, the high adaptation ratio and accuracy obtained for the datasets confirm the choice of the adaptation condition.

On the whole, the results show that the proposed adaptive approaches may increase classification accuracy, when a relatively small training set is available. A simple adjustment strategy used in AdaptJEPC-Adjust gives already noticeable improvement. AdaptJEPC-Recompute proves to be even more effective, however, a more expensive adaptation method is required.

## Conclusions

In this work, we have considered a specific type of contrast patterns, called Jumping Emerging Patterns (JEPs). Relations to rough set theory and JEP discovery have been of our main interest.

Rough set reducts are closely related to JEPs. We have proposed a reduct computation algorithm that follows an Apriori pattern enumeration scheme with pruning based on attribute set dependence. All information necessary for pruning is directly inferred from the previous generation. A new structure, CC-Trie, has been introduced to store compactly intermediate generations of attribute sets and optimize lookups required in pruning.

Our discussion on mutual correspondence between JEPs and reducts, as well as resulting JEP finding methods based on local reducts, have been presented in two parts. Each of them considers a different form of input data.

First, we have assumed that a decision table has been given. This representation is widely accepted in machine learning, rough set theory and other fields. Analysis on the grounds of data mining requires an information-preserving transformation to a transaction database. We have proven several relations of JEPs to global and local reducts. On this basis, we have shown that JEP finding can be solved by computing local reducts for all objects from the universum.

In the second part, we have considered a transaction database as initial input. As a matter of fact, such a database can be represented by an equivalent binary decision table. We have introduced a new type of patterns, JEPs with negation (JEPNs), and demonstrated their relations to reducts in the respective binary decision table. A direct correspondence between local reducts and JEPNs has allowed us to formulate a JEPN discovery method. In addition, classification capabilities of JEPNs have been discussed.

Theoretically, since JEPs are a special type of JEPNs, they can be discovered as a side effect of mining the latter. A significant overhead of undesirable patterns is usually prohibitive. Therefore, we have looked into more well-suited methods referred to as global condensation and local projection. The first one attempts to group attributes of a binary decision table, so that each group could be losslessly replaced by a single multi-variate attribute. Local reducts computed for the resulting decision table correspond almost directly to minimal JEPs, so that an overhead of additional patterns is minimal. Partition of binary attributes into groups is performed by reducing it to graph coloring in a so-called item-conflict graph and solving with efficient heuristics. In general, this method is very sensitive to the structure of an input transaction database. It works well with very sparse data or datasets that completely or partially originate from multi-variate attributes. The other approach, local projection, transforms a relevant part of the database from the perspective of each individual transaction. Only transactions from other classes and items from a given transaction are taken into account. Local reducts computed for such a table always correspond to minimal JEPs. As compared to global condensation, this approach does not rely on certain organization of input data. On the other hand, instead of a single pre-processing phase, it requires per-object transformations of the transaction database.

The CP-Tree provides a convenient way for mining certain types of EPs, in particular Strong JEPs (SJEPs). By analogy to our previous conditions for reduct computation that are based on attribute set dependence, we have proposed a new pruning approach for the CP-Tree. Unlike in the Apriori scheme, information on all immediate subsets of a considered pattern is not available. Therefore, attribute set dependence is tested by means of a fast counting method that intersects bit vectors of transaction identifiers containing individual items.

SJEPs are useful in practical tasks due to their potentially higher resistance to noise in the positive class. Unfortunately, setting an appropriate level of minimum support may be troublesome, since the size of a result set is unpredictable. We have proposed an alternative problem of finding top-$k$ most supported minimal JEPs. The task has been solved by modifying an SJEP mining algorithm based on the CP-Tree, so that minimal support is gradually increased as minimal JEPs are collected in a buffer of the size $k$. This approach requires checking pattern minimality at the time of collection and can be efficiently combined with our pruning technique.

Building a classifier involves a representative portion of pre-classified training instances. The latter can be hard or costly to obtain. Therefore, we have proposed an adaptive classification meta-scheme that is initialized with a potentially small training set and adjusts the underlying model in the testing phase.

The paradigm of classification with reject option has been utilized to classify new instances in rounds and perform adaptation with accepted ones. We have introduced AdaptiveJEP-Classifier, a scheme concretization based on JEPs. Two adaptation methods have been considered: adjustment of pattern supports and border recomputation. Distance and ambiguity of a classified pattern have been taken into account to formulate rejection conditions.

## Acknowledgements

The paper is a full version of my PhD thesis supervised by Krzysztof Walczak and approved in March 2009 by Warsaw University of Technology, Institute of Computer Science. It is dedicated to my parents and brother.

I wish to express my deepest gratitude to my advisor, Professor Krzysztof Walczak, for his guidance in the world of science and help in determining research objectives. Among many virtues, I value the most his expertise and timeliness throughout our work together.

Many thanks also go to Michal Nowakiewicz and Andrzej Dominik for their valuable remarks towards conceptual and implementational aspects of the presented here methods.

I am most grateful to my family and friends who have contributed to this work with their unwavering patience and heartwarming encouragement.

Last but not least, financial support for this research has been provided by the grant No 3 T11C 002 29 from the Ministry of Science and Higher Education of the Republic of Poland.

## References

1. Han, J., Kamber, M.: Data mining: Concepts and Techniques, 2nd edn. Morgan Kaufmann, San Francisco (2006)
2. Agrawal, R., Srikant, R.: Fast algorithms for mining association rules in large databases. In: Bocca, J.B., Jarke, M., Zaniolo, C. (eds.) VLDB, pp. 487–499. Morgan Kaufmann, San Francisco (1994)
3. Suzuki, E.: Autonomous discovery of reliable exception rules. In: KDD, Newport Beach, CA, USA, pp. 259–262. ACM, New York (1997)
4. Liu, B., Hsu, W., Ma, Y.: Integrating classification and association rule mining. In: KDD, pp. 80–86. AAAI Press, New York (1998)
5. Li, W., Han, J., Pei, J.: CMAR: Accurate and efficient classification based on multiple class-association rules. In: Cercone, N., Lin, T.Y., Wu, X. (eds.) ICDM, pp. 369–376. IEEE Computer Society, Los Alamitos (2001)
6. Baralis, E., Chiusano, S.: Essential classification rule sets. ACM Trans. Database Syst. 29, 635–674 (2004)
7. Dong, G., Li, J.: Efficient mining of emerging patterns: discovering trends and differences. In: KDD, San Diego, CA, United States, pp. 43–52. ACM Press, New York (1999)
8. Dong, G., Zhang, X., Wong, L., Li, J.: CAEP: Classification by aggregating emerging patterns. In: Arikawa, S., Furukawa, K. (eds.) DS 1999. LNCS (LNAI), vol. 1721, pp. 30–42. Springer, Heidelberg (1999)

9. Bailey, J., Manoukian, T., Ramamohanarao, K.: Classification using constrained emerging patterns. In: [153], pp. 226–237
10. Fan, H., Ramamohanarao, K.: Efficiently mining interesting emerging patterns. In: [153], pp. 189–201
11. Li, J., Dong, G., Ramamohanarao, K.: Making use of the most expressive jumping emerging patterns for classification. Knowl. Inf. Syst. 3, 131–145 (2001)
12. Li, J., Wong, L.: Emerging patterns and gene expression data. In: Genome Informatics Workshop, Tokyo, Japan, vol. 12, pp. 3–13. Imperial College Press, London (2001)
13. Li, J., Wong, L.: Identifying good diagnostic gene groups from gene expression profiles using the concept of emerging patterns. Bioinformatics 18, 725–734 (2002)
14. Yu, L.T.H., lai Chung, F., Chan, S.C.F., Yuen, S.M.C.: Using emerging pattern based projected clustering and gene expression data for cancer detection. In: Conference on Asia-Pacific bioinformatics, Dunedin, New Zealand, pp. 75–84. Australian Computer Society, Inc. (2004)
15. Yoon, H.S., Lee, S.H., Kim, J.H.: Application of emerging patterns for multi-source bio-data classification and analysis. In: Wang, L., Chen, K., S. Ong, Y. (eds.) ICNC 2005. LNCS, vol. 3610, pp. 965–974. Springer, Heidelberg (2005)
16. Pawlak, Z.: Rough sets. International Journal of Computer and Information Sciences 11, 341–356 (1982)
17. Demri, S.P., Orlowska, E.S.: Incomplete Information: Structure, Inference, Complexity. Springer, New York (2002)
18. Skowron, A.: Rough sets and vague concepts. Fundam. Inf. 64, 417–431 (2004)
19. Polkowski, L.: Rough Sets: Mathematical Foundations. Physica-Verlag, Heidelberg (2002)
20. Skowron, A., Suraj, Z.: Discovery of concurrent data models from experimental tables: A rough set approach. In: KDD, pp. 288–293 (1995)
21. Pawlak, Z.: Rough Sets: Theoretical Aspects of Reasoning about Data. Kluwer Academic Publishers, Dordrecht (1992)
22. Pawlak, Z.: Vagueness and uncertainty: A rough set perspective. Computational Intelligence 11, 232–277 (1995)
23. Pawlak, Z., Skowron, A.: Rudiments of rough sets. Information Sciences 177, 3–27 (2007)
24. Pawlak, Z., Skowron, A.: Rough sets: Some extensions. Information Sciences 177, 28–40 (2007)
25. Ziarko, W.: Probabilistic rough sets. In: [154], pp. 283–293
26. Yao, Y.: Probabilistic rough set approximations. Int. J. Approx. Reasoning 49, 255–271 (2008)
27. Skowron, A., Grzymala-Busse, J.: From rough set theory to evidence theory, pp. 193–236 (1994)
28. Lingras, P.: Comparison of neofuzzy and rough neural networks. Information Sciences 110, 207–215 (1998)
29. Lin, T.Y., Yao, Y.Y., Zadeh, L.A. (eds.): Data mining, rough sets and granular computing. Physica-Verlag, Heidelberg (2002)
30. Yao, Y.: Semantics of fuzzy sets in rough set theory. In: Peters, J.F., Skowron, A., Dubois, D., Grzymała-Busse, J.W., Inuiguchi, M., Polkowski, L. (eds.) Transactions on Rough Sets II. LNCS, vol. 3135, pp. 297–318. Springer, Heidelberg (2004)
31. Pawlak, Z.: Rough classification. Int. J. Hum.-Comput. Stud. 51, 369–383 (1999)
32. Bazan, J.G., Nguyen, H.S., Nguyen, S.H., Synak, P., Wroblewski, J.: Rough set algorithms in classification problem, pp. 49–88 (2000)

33. Stefanowski, J.: On combined classifiers, rule induction and rough sets. T. Rough Sets 6, 329–350 (2007)
34. Wojna, A.: Analogy-based reasoning in classifier construction. PhD thesis, University of Warsaw, Institute of Mathematics, Computer Science and Mechanics (2004)
35. Swiniarski, R.W., Skowron, A.: Rough set methods in feature selection and recognition. Pattern Recogn. Lett. 24, 833–849 (2003)
36. Bhatt, R.B., Gopal, M.: On fuzzy-rough sets approach to feature selection. Pattern Recogn. Lett. 26, 965–975 (2005)
37. Bazan, J.G., Skowron, A., Synak, P.: Dynamic reducts as a tool for extracting laws from decisions tables. In: Raś, Z.W., Zemankova, M. (eds.) ISMIS 1994. LNCS, vol. 869, pp. 346–355. Springer, Heidelberg (1994)
38. Hirano, S., Tsumoto, S.: Hierarchical clustering of non-euclidean relational data using indiscernibility-level. In: [155], pp. 332–339
39. Lingras, P., Chen, M., Miao, D.: Precision of rough set clustering. In: [156], pp. 369–378
40. Chmielewski, M.R., Grzymala-Busse, J.W.: Global discretization of continuous attributes as preprocessing for machine learning. Int. J. Approx. Reasoning 15, 319–331 (1996)
41. Nguyen, H.S.: Discretization problem for rough sets methods. In: [157], pp. 545–552
42. Skowron, A., Synak, P.: Reasoning in information maps. Fundamenta Informaticae 59, 241–259 (2004)
43. Skowron, A., Synak, P.: Hierarchical information maps. In: [154], pp. 622–631
44. Slezak, D.: Approximate reducts in decision tables. In: International Conference, Information Processing and Management of Uncertainty in Knowledge-Based Systems, Granada, Spain, vol. 3, pp. 1159–1164 (1996)
45. Nguyen, H.S., Slezak, D.: Approximate reducts and association rules - correspondence and complexity results. In: Zhong, N., Skowron, A., Ohsuga, S. (eds.) RSFDGrC 1999. LNCS (LNAI), vol. 1711, pp. 137–145. Springer, Heidelberg (1999)
46. Slezak, D.: Association reducts: A framework for mining multi-attribute dependencies. In: Hacid, M.-S., Murray, N.V., Raś, Z.W., Tsumoto, S. (eds.) ISMIS 2005. LNCS (LNAI), vol. 3488, pp. 354–363. Springer, Heidelberg (2005)
47. Slezak, D.: Association reducts: Complexity and heuristics. In: [158], pp. 157–164
48. Slezak, D.: Approximate entropy reducts. Fundam. Inf. 53, 365–390 (2002)
49. Grzymala-Busse, J.W., Ziarko, W.: Data mining based on rough sets, pp. 142–173 (2003)
50. Dong, G., Li, J.: Mining border descriptions of emerging patterns from dataset pairs. Knowledge Information Systems 8, 178–202 (2005)
51. Bailey, J., Manoukian, T., Ramamohanarao, K.: A fast algorithm for computing hypergraph transversals and its application in mining emerging patterns. In: ICDM, pp. 485–488. IEEE Computer Society, Los Alamitos (2003)
52. Li, J., Liu, G., Wong, L.: Mining statistically important equivalence classes and delta-discriminative emerging patterns. In: Berkhin, P., Caruana, R., Wu, X. (eds.) KDD, pp. 430–439. ACM, New York (2007)
53. Loekito, E., Bailey, J.: Fast mining of high dimensional expressive contrast patterns using zero-suppressed binary decision diagrams. In: Eliassi-Rad, T., Ungar, L.H., Craven, M., Gunopulos, D. (eds.) KDD, pp. 307–316. ACM, New York (2006)

54. Fan, H., Ramamohanarao, K.: Fast discovery and the generalization of strong jumping emerging patterns for building compact and accurate classifiers. IEEE Trans. on Knowl. and Data Eng. 18, 721–737 (2006)
55. Skowron, A., Rauszer, C.: The discernibility matrices and functions in information systems. In: Intelligent Decision Support. Handbook of Applications and Advances of of the Rough Sets Theory, pp. 331–362 (1992)
56. Kryszkiewicz, M.: Algorithms for knowledge reduction in information systems. PhD thesis, Warsaw University of Technology, Institute of Computer Science (1994) (in Polish)
57. Kryszkiewicz, M., Cichon, K.: Towards scalable algorithms for discovering rough set reducts. In: Peters, J.F., Skowron, A., Grzymała-Busse, J.W., Kostek, B.z., Świniarski, R.W., Szczuka, M.S. (eds.) Transactions on Rough Sets I. LNCS, vol. 3100, pp. 120–143. Springer, Heidelberg (2004)
58. Han, J., Wang, J., Lu, Y., Tzvetkov, P.: Mining top-k frequent closed patterns without minimum support. In: ICDM, pp. 211–218. IEEE Computer Society, Los Alamitos (2002)
59. Wang, J., Lu, Y., Tzvetkov, P.: Tfp: An efficient algorithm for mining top-k frequent closed itemsets. IEEE Trans. on Knowl. and Data Eng. 17, 652–664 (2005)
60. Ramamohanarao, K., Bailey, J., Fan, H.: Efficient mining of contrast patterns and their applications to classification. In: ICISIP, pp. 39–47. IEEE Computer Society, Washington (2005)
61. Wu, X., Zhang, C., Zhang, S.: Efficient mining of both positive and negative association rules. ACM Trans. Inf. Syst. 22, 381–405 (2004)
62. Antonie, M.L., Zaïane, O.R.: Mining positive and negative association rules: An approach for confined rules. In: Boulicaut, J.-F., Esposito, F., Giannotti, F., Pedreschi, D. (eds.) PKDD 2004. LNCS (LNAI), vol. 3202, pp. 27–38. Springer, Heidelberg (2004)
63. Padmanabhan, B., Tuzhilin, A.: Small is beautiful: discovering the minimal set of unexpected patterns. In: KDD, Boston, Massachusetts, United States, pp. 54–63. ACM, New York (2000)
64. Hussain, F., Liu, H., Suzuki, E., Lu, H.: Exception rule mining with a relative interestingness measure. In: Terano, T., Chen, A.L.P. (eds.) PAKDD 2000. LNCS, vol. 1805, pp. 86–97. Springer, Heidelberg (2000)
65. Li, Y., Guan, C.: An extended EM algorithm for joint feature extraction and classification in brain-computer interfaces. Neural Comput. 18, 2730–2761 (2006)
66. Jackson, Q., Landgrebe, D.: An adaptive classifier design for high-dimensional data analysis with a limited training data set. IEEE Transactions on Geoscience and Remote Sensing 39, 2664–2679 (2001)
67. Qian, X., Bailey, J., Leckie, C.: Mining generalised emerging patterns. In: Sattar, A., Kang, B.-H. (eds.) AI 2006. LNCS (LNAI), vol. 4304, pp. 295–304. Springer, Heidelberg (2006)
68. Ting, R.M.H., Bailey, J.: Mining minimal contrast subgraph patterns. In: Ghosh, J., Lambert, D., Skillicorn, D.B., Srivastava, J. (eds.) SDM. SIAM, Philadelphia (2006)
69. Dominik, A., Walczak, Z., Wojciechowski, J.: Classification of web documents using a graph-based model and structural patterns. In: Kok, J.N., Koronacki, J., Lopez de Mantaras, R., Matwin, S., Mladenič, D., Skowron, A. (eds.) PKDD 2007. LNCS (LNAI), vol. 4702, pp. 67–78. Springer, Heidelberg (2007)
70. Inokuchi, A., Washio, T., Motoda, H.: An apriori-based algorithm for mining frequent substructures from graph data. In: [159], pp. 13–23

71. Wroblewski, J.: Adaptive methods of object classification. PhD thesis, University of Warsaw, Institute of Mathematics, Computer Science and Mechanics (2001)
72. Wroblewski, J.: Finding minimal reducts using genetic algorithm. In: Joint Conference on Information Sciences, Wrightsville Beach, NC, pp. 186–189 (1995)
73. Bazan, J.G., Szczuka, M.S.: Rses and rseslib - a collection of tools for rough set computations. In: Ziarko, W.P., Yao, Y. (eds.) RSCTC 2000. LNCS (LNAI), vol. 2005, pp. 106–113. Springer, Heidelberg (2001)
74. Bazan, J.G.: Approximation inferencing methods for synthesis of decision algorithms. PhD thesis, University of Warsaw, Institute of Mathematics, Computer Science and Mechanics (1998) (in Polish)
75. Fan, H., Ramamohanarao, K.: An efficient single-scan algorithm for mining essential jumping emerging patterns for classification. In: Chen, M.-S., Yu, P.S., Liu, B. (eds.) PAKDD 2002. LNCS (LNAI), vol. 2336, pp. 456–462. Springer, Heidelberg (2002)
76. Terlecki, P., Walczak, K.: Jumping emerging pattern induction by means of graph coloring and local reducts in transaction databases. In: An, A., Stefanowski, J., Ramanna, S., Butz, C.J., Pedrycz, W., Wang, G. (eds.) RSFDGrC 2007. LNCS (LNAI), vol. 4482, pp. 363–370. Springer, Heidelberg (2007)
77. Terlecki, P., Walczak, K.: Local projection in jumping emerging patterns discovery in transaction databases. In: Washio, T., Suzuki, E., Ting, K.M., Inokuchi, A. (eds.) PAKDD 2008. LNCS (LNAI), vol. 5012, pp. 723–730. Springer, Heidelberg (2008)
78. Terlecki, P., Walczak, K.: Jumping emerging patterns with negation in transaction databases - classification and discovery. Information Sciences 177, 5675–5690 (2007)
79. Li, J., Ramamohanarao, K., Dong, G.: The space of jumping emerging patterns and its incremental maintenance algorithms. In: Langley, P. (ed.) ICML, pp. 551–558. Morgan Kaufmann, San Francisco (2000)
80. Wang, L., Zhao, H., Dong, G., Li, J.: On the complexity of finding emerging patterns. Theor. Comput. Sci. 335, 15–27 (2005)
81. Han, J., Pei, J., Yin, Y.: Mining frequent patterns without candidate generation. In: SIGMOD, Dallas, Texas, United States, pp. 1–12. ACM, New York (2000)
82. Birkhoff, G.: Lattice Theory, 3rd edn. American Mathematical Society, USA (1967)
83. Romanski, S.: Operations on families of sets for exhaustive search, given a monotonic function. In: JCDKB, Jerusalem, Israel, pp. 310–322 (1988)
84. Romanski, S.: An Algorithm Searching for the Minima of Monotonic Boolean Function and its Applications. PhD thesis, Warsaw University of Technology (1989)
85. Liu, B., Ma, Y., Wong, C.K.: Improving an association rule based classifier. In: [159], pp. 504–509
86. Li, W.: Classification based on multiple association rules (2001)
87. Garriga, G.C., Kralj, P., Lavrač, N.: Closed sets for labeled data. J. Mach. Learn. Res. 9, 559–580 (2008)
88. Bayardo Jr., R.J.: Efficiently mining long patterns from databases. In: SIGMOD, Seattle, Washington, United States, pp. 85–93. ACM, New York (1998)
89. Meretakis, D., Wüthrich, B.: Extending naïve bayes classifiers using long itemsets. In: KDD, San Diego, California, United States, pp. 165–174. ACM, New York (1999)

90. Wang, Z., Fan, H., Ramamohanarao, K.: Exploiting maximal emerging patterns for classification. In: Webb, G.I., Yu, X. (eds.) AI 2004. LNCS (LNAI), vol. 3339, pp. 1062–1068. Springer, Heidelberg (2004)
91. Soulet, A., Crémilleux, B., Rioult, F.: Condensed representation of eps and patterns quantified by frequency-based measures. In: Goethals, B., Siebes, A. (eds.) KDID 2004. LNCS, vol. 3377, pp. 173–189. Springer, Heidelberg (2005)
92. Soulet, A., Kléma, J., Crémilleux, B.: Efficient mining under rich constraints derived from various datasets. In: Džeroski, S., Struyf, J. (eds.) KDID 2006. LNCS, vol. 4747, pp. 223–239. Springer, Heidelberg (2007)
93. Bailey, J., Manoukian, T., Ramamohanarao, K.: Fast algorithms for mining emerging patterns. In: Elomaa, T., Mannila, H., Toivonen, H. (eds.) PKDD 2002. LNCS (LNAI), vol. 2431, pp. 39–50. Springer, Heidelberg (2002)
94. Bastide, Y., Taouil, R., Pasquier, N., Stumme, G., Lakhal, L.: Mining frequent patterns with counting inference. SIGKDD Explorations Newsletter 2, 66–75 (2000)
95. Pasquier, N., Bastide, Y., Taouil, R., Lakhal, L.: Discovering frequent closed itemsets for association rules. In: Beeri, C., Bruneman, P. (eds.) ICDT 1999. LNCS, vol. 1540, pp. 398–416. Springer, Heidelberg (1998)
96. Li, J.: Mining Emerging Patterns to Contruct Accurate and Efficient Classifiers. PhD thesis, University of Melbourne (2001)
97. Li, J., Dong, G., Ramamohanarao, K.: Instance-based classification by emerging patterns. In: [159], pp. 191–200
98. Li, J., Dong, G., Ramamohanarao, K., Wong, L.: DeEPs: A new instance-based lazy discovery and classification system. Mach. Learn. 54, 99–124 (2004)
99. Fan, H.: Efficient Mining of Interesting Emerging Patterns and Their Effective Use in Classification. PhD thesis, University of Melbourne (2004)
100. Merris, R.: Graph Theory. Wiley Interscience, New York (2000)
101. Berge, C.: Hypergraphs, vol. 45. Elsevier, Amsterdam (1989)
102. Kavvadias, D.J., Stavropoulos, E.C.: Evaluation of an algorithm for the transversal hypergraph problem. In: Vitter, J.S., Zaroliagis, C.D. (eds.) WAE 1999. LNCS, vol. 1668, pp. 72–84. Springer, Heidelberg (1999)
103. Elbassioni, K.M.: On the complexity of monotone dualization and generating minimal hypergraph transversals. Discrete Appl. Math. 156, 2109–2123 (2008)
104. Bryant, R.E.: Graph-based algorithms for boolean function manipulation. IEEE Transactions on Computers 35, 677–691 (1986)
105. Aloul, F.A., Mneimneh, M.N., Sakallah, K.A.: Zbdd-based backtrack search sat solver. In: IWLS, pp. 131–136 (2002)
106. ichi Minato, S.: Binary decision diagrams and applications for VLSI CAD. Kluwer Academic Publishers, Norwell (1996)
107. Cerny, E., Marin, M.A.: An approach to unified methodology of combinational switching circuits. IEEE Transactions on Computers 26, 745–756 (1977)
108. ichi Minato, S.: Zero-suppressed bdds for set manipulation in combinatorial problems. In: DAC, pp. 272–277. ACM, New York (1993)
109. Mishchenko, A.: An introduction to zero-suppressed binary decision diagrams, Tutorial (2001)
110. Cover, T.M., Hart, P.E.: Nearest neighbor pattern classification. IEEE Transactions on Information Theory 13, 21–27 (1967)
111. Wegener, I.: The complexity of Boolean functions. John Wiley & Sons, Inc., New York (1987)
112. Cykier, A.: Prime implicants of boolean functions, methods for finding and application (1997) (in polish)

113. Kryszkiewicz, M.: Fast algorithm finding reducts of monotonic boolean functions. ICS Research Report 42/93 (1993)
114. Anderson, M.: Synthesis of Information Systems. Warsaw University of Technology (1994) (in Polish)
115. Brown, F.M.: Boolean Reasoning. Kluwer Academic Publishers, Dordrecht (1990)
116. Garfinkel, R., Nemhauser, G.L.: Integer programming. John Wiley & Sons, New York (1978)
117. Susmaga, R.: Parallel computation of reducts. In: [157], pp. 450–457
118. Zhou, P.L., Mohammed, S.: A reduct solving parallel algorithm based on relational extension matrix. In: Arabnia, H.R. (ed.) PDPTA, pp. 924–931. CSREA Press (2007)
119. Bjorvand, A.T., Komorowski, J.: Practical applications of genetic algorithms for efficient reduct computation. In: IMACS
120. Walczak, Z., Dominik, A., Terlecki, P.: Space decomposition in the minimal reduct problem. In: National Conference on Evolutionary Computation and Global Optimization, Kazimierz Dolny, Poland. Warsaw University of Technology (2004)
121. Sapiecha, P.: An approximation algorithm for a certain class of np-hard problems. In: ICS Research Report 21/92 (1992)
122. Wang, X., Yang, J., Peng, N., Teng, X.: Finding minimal rough set reducts with particle swarm optimization. In: [154], pp. 451–460
123. Ke, L., Feng, Z., Ren, Z.: An efficient ant colony optimization approach to attribute reduction in rough set theory. Pattern Recogn. Lett. 29, 1351–1357 (2008)
124. Terlecki, P., Walczak, K.: Attribute set dependence in apriori-like reduct computation. In: Wang, G.-Y., Peters, J.F., Skowron, A., Yao, Y. (eds.) RSKT 2006. LNCS (LNAI), vol. 4062, pp. 268–276. Springer, Heidelberg (2006)
125. Terlecki, P., Walczak, K.: Attribute set dependence in reduct computation. Transactions on Computational Science 2, 118–132 (2008)
126. Kryszkiewicz, M., Lasek, P.: Fast discovery of minimal sets of attributes functionally determining a decision attribute. In: Kryszkiewicz, M., Peters, J.F., Rybiński, H., Skowron, A. (eds.) RSEISP 2007. LNCS (LNAI), vol. 4585, pp. 320–331. Springer, Heidelberg (2007)
127. Kryszkiewicz, M., Lasek, P.: Fun: Fast discovery of minimal sets of attributes functionally determining a decision attribute. T. Rough Sets 9, 76–95 (2008)
128. Bodon, F.: A fast apriori implementation. In: Goethals, B., Zaki, M.J. (eds.) FIMI. CEUR Workshop Proceedings, vol. 90 (2003), `CEUR-WS.org`
129. Komorowski, J., Ohrn, A., Skowron, A.: Case studies: Public domain, multiple mining tasks systems: Rosetta rough sets, pp. 554–559 (2002)
130. Terlecki, P., Walczak, K.: On the relation between rough set reducts and jumping emerging patterns. Information Sciences 177, 74–83 (2007)
131. Terlecki, P., Walczak, K.: Local reducts and jumping emerging patterns in relational databases. In: [158], pp. 358–367
132. Shan, N., Ziarko, W.: An incremental learning algorithm for constructing decision rules. In: International Workshop on Rough Sets and Knowledge Discovery, Banff, Canada, pp. 326–334. Springer, Heidelberg (1994)
133. Brin, S., Motwani, R., Silverstein, C.: Beyond market baskets: Generalizing association rules to correlations. In: Peckham, J. (ed.) SIGMOD, pp. 265–276. ACM Press, New York (1997)
134. Ruiz, I.F., Balcázar, J.L., Bueno, R.M.: Bounding negative information in frequent sets algorithms. In: Jantke, K.P., Shinohara, A. (eds.) DS 2001. LNCS (LNAI), vol. 2226, pp. 50–58. Springer, Heidelberg (2001)

135. Savasere, A., Omiecinski, E., Navathe, S.B.: Mining for strong negative associations in a large database of customer transactions. In: ICDE, pp. 494–502. IEEE Computer Society, Los Alamitos (1998)
136. Yuan, X., Buckles, B.P., Yuan, Z., Zhang, J.: Mining negative association rules. In: ISCC, pp. 623–628. IEEE Computer Society, Los Alamitos (2002)
137. Boulicaut, J.F., Bykowski, A., Jeudy, B.: Towards the tractable discovery of association rules with negations. In: FQAS, Warsaw, Poland, pp. 425–434 (2000)
138. Kryszkiewicz, M., Cichon, K.: Support oriented discovery of generalized disjunction-free representation of frequent patterns with negation. In: Ho, T.-B., Cheung, D., Liu, H. (eds.) PAKDD 2005. LNCS (LNAI), vol. 3518, pp. 672–682. Springer, Heidelberg (2005)
139. Cichosz, P.: Learning systems. WNT, Warsaw (2000) (in Polish)
140. Terlecki, P., Walczak, K.: Local table condensation in rough set approach for jumping emerging pattern induction. In: ICCS Workshop. Springer, Sheffield (2007)
141. Terlecki, P., Walczak, K.: Efficient discovery of top-k minimal jumping emerging patterns. In: [156], pp. 438–447
142. Dempster, A.P., Laird, N.M., Rubin, D.B.: Maximum likelihood from incomplete data via the em algorithm. Journal of the Royal Statistical Society, Series B 39, 1–38 (1977)
143. Blum, A., Mitchell, T.M.: Combining labeled and unlabeled sata with co-training. In: COLT, pp. 92–100 (1998)
144. Terlecki, P., Walczak, K.: Adaptive classification with jumping emerging patterns. In: [155], pp. 39–46.
145. Delany, S.J., Cunningham, P., Doyle, D., Zamolotskikh, A.: Generating estimates of classification confidence for a case-based spam filter. In: Muñoz-Ávila, H., Ricci, F. (eds.) ICCBR 2005. LNCS (LNAI), vol. 3620, pp. 177–190. Springer, Heidelberg (2005)
146. Freund, Y., Schapire, R.E.: A decision-theoretic generalization of on-line learning and an application to boosting. J. Comput. Syst. Sci. 55, 119–139 (1997)
147. Dehuri, S., Patnaik, S., Ghosh, A., Mall, R.: Application of elitist multi-objective genetic algorithm for classification rule generation. Appl. Soft Comput. 8, 477–487 (2008)
148. Vailaya, A., Jain, A.K.: Reject option for vq-based bayesian classification, pp. 2048–2051 (2000)
149. Mascarilla, L., Frélicot, C.: Reject strategies driven combination of pattern classifiers. Pattern Anal. Appl. 5, 234–243 (2002)
150. Li, J., Manoukian, T., Dong, G., Ramamohanarao, K.: Incremental maintenance on the border of the space of emerging patterns. Data Min. Knowl. Discov. 9, 89–116 (2004)
151. Fumera, G., Pillai, I., Roli, F.: Classification with reject option in text categorisation systems. In: ICIAP, pp. 582–587. IEEE Computer Society, Los Alamitos (2003)
152. Chow, C.K.: On optimum recognition error and reject tradeoff. IEEE Transactions on Information Theory 16, 41–46 (1970)
153. Asuncion, A., Newman, D.: UCI machine learning repository (2007)
154. Fayyad, U.M., Irani, K.B.: Multi-interval discretization of continuous-valued attributes for classification learning. In: IJCAI, pp. 1022–1029 (1993)
155. Kohavi, R., John, G.H., Long, R., Manley, D., Pfleger, K.: Mlc++: A machine learning library in c++. In: ICTAI, New Orleans, Louisiana, USA, pp. 740–743 (1994)

156. Karypsis, G.: Cluto. a clustering toolkit. release 2.0 (2002)
157. Dong, G., Tang, C., Wang, W.: WAIM 2003. LNCS, vol. 2762. Springer, Heidelberg (2003)
158. Zighed, D.A., Komorowski, H.J., Zytkow, J.M. (eds.): PKDD 2000. LNCS, vol. 1910. Springer, Heidelberg (2000)
159. Polkowski, L., Skowron, A. (eds.): RSCTC 1998. LNCS, vol. 1424. Springer, Heidelberg (1998)
160. Ślęzak, D., Wang, G., Szczuka, M.S., Düntsch, I., Yao, Y. (eds.): RSFDGrC 2005. LNCS (LNAI), vol. 3641. Springer, Heidelberg (2005)
161. Chan, C.-C., Grzymala-Busse, J.W., Ziarko, W.P. (eds.): RSCTC 2008. LNCS (LNAI), vol. 5306. Springer, Heidelberg (2008)
162. Wang, G., Rui Li, T., Grzymala-Busse, J.W., Miao, D., Skowron, A., Yao, Y.: RSKT 2008. LNCS (LNAI), vol. 5009. Springer, Heidelberg (2008)
163. Greco, S., Hata, Y., Hirano, S., Inuiguchi, M., Miyamoto, S., Nguyen, H.S., Słowiński, R. (eds.): RSCTC 2006. LNCS (LNAI), vol. 4259. Springer, Heidelberg (2006)

# A Experimental Conduct

A significant part of this work is dedicated to an experimental evaluation of our propositions in the areas of reduct computation, JEP mining and classification. Here, we touch on the main aspects of our methodology, like considered input data, testing procedure and environment, as well as implementation of widely known methods used as a reference.

## A.1 Real-Life Datasets

Most of the experiments have been performed on real-life datasets originating from UCI Repository ([153]). Specificity of the presented methods requires data in the form of information systems or decision tables with nominal attributes. Besides datasets immediately satisfying these requirements, we have chosen datasets with continuous attributes and performed appropriate discretization. Many papers on EPs (e.g. [79]) utilize the entropy-based approach from [154], as implemented in the MLC++ package ([155]). This has also become our method of choice. In addition, completeness and consistency of data with respect to a decision attribute were required.

The chosen datasets are summarized in Table A.1. In some cases, they have been directly used as input to the considered algorithms. The numbers of objects, attributes and classes are given. Elsewhere, when data needed to be in transactional form, the transformation from Section 5.1 has been applied. Consequently, we provide additional characteristics. The number of transactions is equal to the respective number of objects, the number of classes stays the same and the number of items results from the attributes and the sizes of their domains.

## A.2 Synthetic Sparse Datasets

Data mining usually deals with sparse data. Unfortunately, it is hard to find publicly available classified sparse datasets, whose sizes would suit various testing needs. Therefore, for the purpose of the performed experiments, a data generation pipeline has been developed ([77]). At the beginning, a transaction database is produced by means of the IBM generator ([2]). This tool was originally designed to examine algorithms for association rules mining. Its behavior can be adjusted with parameters including the number of transactions, size of an itemspace and average transaction length. Then, the CLUTO package ([156]) is applied to cluster the generated transactions and form a decision transaction system.

**Table A.1.** Characteristics of real-life datasets

| No | Dataset | Objects | Attributes | Items | Class |
|---|---|---|---|---|---|
| 1 | balance | 625 | 4 | 20 | 3 |
| 2 | breast-wisc | 699 | 9 | 29 | 2 |
| 3 | car | 1728 | 6 | 21 | 4 |
| 4 | cmc | 1473 | 9 | 29 | 3 |
| 5 | dna | 500 | 20 | 80 | 3 |
| 6 | heart | 270 | 13 | 22 | 2 |
| 7 | irys | 150 | 4 | 12 | 3 |
| 8 | kr-vs-kp | 3196 | 36 | 73 | 2 |
| 9 | krkopt | 28056 | 6 | 43 | 18 |
| 10 | lung | 32 | 55 | 220 | 4 |
| 11 | lymn | 148 | 18 | 59 | 4 |
| 12 | monks-1 | 556 | 6 | 17 | 2 |
| 13 | monks-2 | 601 | 6 | 17 | 2 |
| 14 | monks-3 | 554 | 6 | 17 | 2 |
| 15 | mushroom | 8124 | 22 | 117 | 2 |
| 16 | nursery | 12960 | 8 | 27 | 5 |
| 17 | tic-tac-toe | 958 | 9 | 27 | 2 |
| 18 | vehicle | 846 | 18 | 72 | 4 |
| 19 | wine | 178 | 13 | 35 | 3 |
| 20 | yeast | 1484 | 8 | 20 | 10 |
| 21 | zoo | 101 | 16 | 35 | 7 |

**Table A.2.** Synthetic sparse datasets

| No | Trans | Items | Classes | MaxTrans | avgDim | maxDim |
|---|---|---|---|---|---|---|
| 1 | 2000 | 20 | 2 | 326 | 5.09 | 9.00 |
| 2 | 2000 | 40 | 3 | 1075 | 6.33 | 14.00 |
| 3 | 2000 | 60 | 3 | 1551 | 6.79 | 16.00 |
| 4 | 2000 | 80 | 2 | 1858 | 9.37 | 19.00 |
| 5 | 5000 | 50 | 2 | 2918 | 6.25 | 15.00 |
| 6 | 10000 | 50 | 3 | 4119 | 5.57 | 12.00 |
| 7 | 15000 | 50 | 2 | 7920 | 7.57 | 18.00 |
| 8 | 20000 | 50 | 2 | 10300 | 7.63 | 18.00 |

Table A.2 presents the datasets used in our experiments. The density of each of them has been set up at 5-15% of the respective itemspace. The number of transactions is increasing over one order of magnitude with itemspaces of moderate cardinalities. Clustering has been performed for 2-3 classes by means of repeated bisections by $k$-way refinement. For a better picture, we provide the number of maximal transactions with respect to set inclusion, average and maximum transaction length.

## A.3 Testbed

A special framework has been developed for the purpose of the experimentation. Input data are accepted in textual form. They can be specified in a dense manner, where all attribute values of an information system or a decision table are given, or sparsely, where only non-zero values of binary attributes are provided. Main results are output as summaries ready for presentation in LaTeX. Additional, textual output allows us to track intermediate results and conduct sanity testing. Common structures and routines, like attribute/item sets and their collections or a cross-validation procedure make the implementation uniform and more reliable in experiments. The framework has been written in Java 6.

Most of the presented results have been captured by the framework itself, including a total time, characteristics of result sets or classification accuracy. In some cases, measurements refer to internal behavior of algorithms and could not be collected from outside. A number of examined patterns or individual times of specific computation phases are a good example here. We gather such information by incorporating additional variables in implementation, e.g. counters, whose maintenance do not introduce any noticeable overhead.

None of the considered algorithms has any non-deterministic elements and individual times of run repetitions are very stable. Nevertheless, every observation of a dataset that involves capturing times has been repeated 5 times and the respective values averaged. Also, reading input and producing output has not been treated as a part of execution. Since the algorithms are single-threaded and experiments have been performed one at a time, we have chosen the elapsed time as a relevant performance indicator. A dataset has been excluded from respective results, if measurements have not been possible to obtain in acceptable time.

The tests have been performed on a machine with 2 Quad CPU, 2.4GHz each and 6GB of RAM running Microsoft Windows Server 2008 in the 64-bit release. The load of the system has been kept at a possibly low level, with only system processes and the testing framework running on the virtual machine JRE 1.6.0_11-b03.

## A.4 Algorithm Implementation

All the algorithms proposed in this work have been implemented inside our testing framework. For comparison, we have also prepared several methods known from the literature. This approach is dictated by two reasons. First, it is more reliable to consider procedures coded in the same programming language and using common data structures. Especially, when the execution time is tracked. Second, in many cases, we are interested in measurements that need to be collected while a method progresses. In other words, the output has to be richer than the usual final result.

As stated before, it is important to share the same data structures across compared algorithms. As far as JEP and reduct computation methods are concerned, all known approaches process large collections of attribute/item sets. To obtain results possibly independent from what a data structure represents such a set, three implementations have been tested. The first two are characteristic vectors of an attributes/item space, one based on a regular byte array (Array) and the other one - on java.util.BitSet (BitSet). The third structure is a balanced binary tree implemented by means of java.util.TreeSet (TreeSet).

In general, an array is generous in memory allocation but it ensures the most efficient access time. Bit and dynamic structures are slower, however, they may take precedence for large attribute/item spaces when a high number of sets is created. Our tests have taken into account all these three implementations. A quantitative comparison for local projection (see Section 7.2) is given in [77]. On the whole, we have observed that the second option has been the most efficient across the considered cases and, thus, we present results of experiments conducted only for this implementation.

Finding reducts is not the main subject of this thesis, however, it plays an important role in JEP discovery. Apart from *RedApriori* (see Chapter 4), we use a method based on symbolic transformations (see Section 3.5), referred to as *RedPrime*. We follow the version of the algorithm discussed in [112]. In particular, a discernibility matrix is initially reduced by means of the absorption laws. Both methods are used for information systems and decision tables by using different definitions of an indiscernibility relation ([74]).

As far as JEP mining is concerned, two methods have been implemented. The first one is described in Section 3.2 and referred to as *Border-diff*. The code is based on [150] and takes into account all optimizations from [50]. The second method, referred to as *CP-Tree*, is given in Section 3.2 and follows [54]. Both algorithms are called individually for every decision class, while the remaining transactions of a given decision transaction system are treated as a background class. For this reason, in the latter procedure, we sort items according to their growth-rates, rather than support-ratios. In the tests, we do not compare against two new methods [53] and [52], since these ideas were published late in our research.

Border maintenance in *AdaptJEPC-Recompute* (see Section 9.3) is performed by incremental routines covered in [79].

# Author Index

GPSR Compliance

*The European Union's (EU) General Product Safety Regulation (GPSR) is a set of rules that requires consumer products to be safe and our obligations to ensure this.*

*If you have any concerns about our products, you can contact us on ProductSafety@springernature.com*

In case Publisher is established outside the EU, the EU authorized representative is:

Springer Nature Customer Service Center GmbH
Europaplatz 3
69115 Heidelberg, Germany

**Batch number: 09490878**

Printed by Printforce, the Netherlands